Deposit and Geoenvironmental Models for
Resource Exploitation and Environmental Security

NATO Science Series

A Series presenting the results of activities sponsored by the NATO Science Committee. The Series is published by IOS Press and Kluwer Academic Publishers, in conjunction with the NATO Scientific Affairs Division.

A. Life Sciences	IOS Press
B. Physics	Kluwer Academic Publishers
C. Mathematical and Physical Sciences	Kluwer Academic Publishers
D. Behavioural and Social Sciences	Kluwer Academic Publishers
E. Applied Sciences	Kluwer Academic Publishers
F. Computer and Systems Sciences	IOS Press

As a consequence of the restructuring of the NATO Science Programme in 1999, the NATO Science Series has been re-organized and new volumes will be incorporated into the following revised sub-series structure:

I. Life and Behavioural Sciences	IOS Press
II. Mathematics, Physics and Chemistry	Kluwer Academic Publishers
III. Computer and Systems Science	IOS Press
IV. Earth and Environmental Sciences	Kluwer Academic Publishers
V. Science and Technology Policy	IOS Press

NATO-PCO-DATA BASE

The NATO Science Series continues the series of books published formerly in the NATO ASI Series. An electronic index to the NATO ASI Series provides full bibliographical references (with keywords and/or abstracts) to more than 50000 contributions from international scientists published in all sections of the NATO ASI Series.
Access to the NATO-PCO-DATA BASE is possible via CD-ROM "NATO-PCO-DATA BASE" with user-friendly retrieval software in English, French and German (WTV GmbH and DATAWARE Technologies Inc. 1989).

The CD-ROM of the NATO ASI Series can be ordered from: PCO, Overijse, Belgium

Series 2. Environment Security – Vol. 80

Deposit and Geoenvironmental Models for Resource Exploitation and Environmental Security

edited by

Andrea G. Fabbri
International Institute for Geo-Information Science and Earth Observation,
ITC,
Enschede, The Netherlands

Gabor Gaál
Geological Survey of Finland,
Helsinki, Finland

and

Richard B. McCammon
U.S. Geological Survey,
Reston, VA, U.S.A.

Kluwer Academic Publishers

Dordrecht / Boston / London

Published in cooperation with NATO Scientific Affairs Division

Proceedings of the NATO Advanced Study Institute on
Deposit and Geoenvironmental Models for Resource Exploitation and Environmental
Security
Matráháza, Hungary
6–19 September 1998

A C.I.P. Catalogue record for this book is available from the Library of Congress.

ISBN 1-4020-0989-5

Published by Kluwer Academic Publishers,
P.O. Box 17, 3300 AA Dordrecht, The Netherlands.

Sold and distributed in North, Central and South America
by Kluwer Academic Publishers,
101 Philip Drive, Norwell, MA 02061, U.S.A.

In all other countries, sold and distributed
by Kluwer Academic Publishers,
P.O. Box 322, 3300 AH Dordrecht, The Netherlands.

Printed on acid-free paper

All Rights Reserved
© 2002 Kluwer Academic Publishers
No part of this work may be reproduced, stored in a retrieval system, or transmitted
in any form or by any means, electronic, mechanical, photocopying, microfilming, recording
or otherwise, without written permission from the Publisher, with the exception
of any material supplied specifically for the purpose of being entered
and executed on a computer system, for exclusive use by the purchaser of the work.

Printed in the Netherlands.

CONTENTS

PREFACE ix

ACKNOWLEDGEMENTS xi

LIST OF PARTICIPANTS xiii

PART 1. GEOENVIRONMENTAL DEPOSIT MODELS

Geoenvironmental models: An Introduction 3
    R.B. Wanty, B.R. Berger, G.S. Plumlee and T.V.V. King

PART 2. GIS/RS METHODS AND TECHNIQUES

Applications of remotely sensed data in geoenvironmental assessments 45
    G. Lee, D. Knepper Jr., A. Maccafferty, S. Miller, T. Sole, G. Swayze and K. Watson

Photo ecometrics for natural resource monitoring 65
    P. Gong

Multiple data layer modeling and analysis in assessments 81
    G.K. Lee

A strategy for sustainable development of nonrenewable resources using spatial prediction models 101
    C.F. Chung, A.G. Fabbri and K.H. Chi

PART 3. RESOURCE ASSESSMENT & MANAGEMENT

Mineral-deposit models: New Developments 121
    B.R. Berger and L.J. Drew

Geologic information for aggregate resource planning 135
    W.H. Langer, D.A. Lindsey and D.H. Knepper Jr.

Environmental impacts of mining natural aggregate 151
    W.H. Langer and B.F. Arbogast

Application of the porphyry copper/polymetallic vein kin-deposit system to mineral-resource assessment in the Mátra Mountains, northern Hungary 171
    L.J. Drew and B.R. Berger

The assessment and analysis of financial, technical and environmental risk in mineral resource exploitation 187
P.A. Dowd

PART 4. RESOURCE POLICIES AND SUSTAINABLE DEVELOPMENT

A hierarchical model of collaborative resource management 215
D.J. Shields

Environmental and resource planning: methodologies to support the decision making process 243
E. Beinat

Implementation of environmental systems in compliance with ISO 14001 standard in the context of SMEs in northwestern Italy 269
S. Olivero

Asset life-cycle in the mining industry: How to improve economic and environmental decision-making by applying ICT 277
I.L. Ritsema

Industrial ecology and bioremediation: theoretical framework and technological tools for sustainable development 291
E. Feoli

PART 5. CASE STUDIES

Land use change and vulnerability as a result of coal mining activities: application of spatial data analysis in the Upper Silesian Region of the Czech republic 305
A.G. Fabbri, T. Woldai, I.S. Babiker, M.G. Kitutu-Kimono and V. Homola

The impact of mining on the environment: a case study from the Tharsis-Lagunazo mining area, Province Huelva, SW Spain 345
T. Woldai and A.G. Fabbri

Environmental impact of mine liquidation on groundwater and surface water 365
N. Rapantova and A. Grmela

Protection of groundwater resources quality and quantity in mining areas 385
A. Grmela and N. Rapantova

PART 6. OTHER CONTRIBUTED PAPERS

Irresistible Holes in the ground: Tekst of afterdinner speech given at
Mátraháza, Hungary, on September 10, 1998   401
    L.J. Drew

Environmental security and radioactive contamination   405
    K.A. Karimov and R.D. Gainutdinova

A PC-based information system for the management and modelling of
subsurface coal fires in mining areas (COALMAN)   409
    Z. Vekerdy

Integrated modelling of acid mine drainage impact on a wetland stream using
landscape geochemistry, GIS technology and statistical methods   425
    A. Szücs, G. Jordán and U. Qvarfort

Assessment of mining induced environmental degradation using satellite data
and predictive models   449
    D. Limpitlav

Environmental impact of exploration of pyrite and stibnite in the Malé Karpaty
Mountains, Slovakia   461
    S. Tríková

Partitioning of heavy metals in sequential extraction fractions in soils developed over the historical "Sv. Jakob" silver mine, Mount Medvednica, Croatia   471
    M. Čović, G. Durn, N. Tadej and S. Miko

PART 7. WORKING GROUP REPORTS

Report of Working Group I: Geoenvironmental Models   485
    D. Limpitlaw and R.B. Wanty

Report of Working Group II: GIS/RS Methods and Techniques: A Spatial Data
Laboratory Network   501
    Z. Vekerdy and C.F. Chung

Report of Working Group III: Natural Resource Assessments and Recource
Management   507
    B.R. Berger and A. Botequilha

Report of Working Group IV: Resource East-West Relationships   511
    G. Gaál and S. Šolar

Report of Working Group V: Natural Aggregate Resources - Environmental
Issues and Resource Management 525
    W.H. Langer and S. Šolar

CD-Rom included with colour illustrations inside the backcover

PREFACE

This volume contains the edited papers prepared by lecturers and participants of the NATO Advanced Study Institute (ASI) on "Deposit and Geoenvironmental Deposit Models for Resource Exploitation and Environmental Security" held in Mátraháza, Hungary, September 6-19, 1998.

The NATO ASI provided an unique opportunity for internationally-recognized researchers from the disciplines of geology, geophysics, geochemistry, remote sensing, economics, biology, mining engineering, resource analysis, mathematics, and statistics to present recent results in geoenvironmental modeling as it relates to resource exploitation and environmental security. There were 84 participants from 32 countries represented. The participants were selected in the general area of geoenvironmental sciences based on their interests and qualifications.

The presentations centred around 4 major themes; namely, geoenvironmental models, GIS methods and techniques, assessment and resource management, and resource policies and sustainable development. Under each theme, the major topics were first introduced; these were followed by specific applications for which the lecturers had expert knowledge. Rounding out the presentations were round-table discussions that involved lecturers and participants on each of the topics under discussion. To facilitate greater interaction among the participants, the participants were divided into Working Groups organized around the 4 major themes. Each Working Group was responsible for preparing a report that was presented at the end of the ASI.

Geological processes affect the Earth and human society. Solutions to geological problems, whether they are natural or man-made, call for a high degree of global cooperation. This NATO ASI has fostered the dissemination of new approaches to current problems of environmental assessments, promoted greater interaction among those involved in addressing these problems globally, and has resulted in a publication that can be used to educate others who are involved in this important area of research.

The subject of this volume has required the introduction of many colour maps, images and diagrams. While in the text these have been printed in black and white, a CD is included inside the back cover that contains the "pdf" files of all the colour illustrations.

Andrea G. Fabbri
Gabor Gaál
Richard B. McCammon

ACKNOWLEDGEMENTS

The Scientific Affairs Division of NATO is gratefully acknowledged for having sponsored this Advanced Study Institute. Additional support was provided by the International Institute for Aerospace Survey and Earth Sciences (ITC), Enschede, The Netherlands, the United States Geological Survey (USGS), Reston VA, U.S.A., The Geological Survey of Finland, Helsinki, Finland, and the Hungarian Geological Survey, Geological Institute of Hungary (MAFI), Budapest, Hungary. Special thanks go to Janet S. Sachs, USGS, for her help in editing many of the manuscripts prepared by the authors of the papers presented and by the editors of the different Theresa van den Boogaard, Huria Almalih and Job Duim of ITC have provided valuable assistance in indexing the contributions and in retyping and graphic editing of some of the manuscripts.Working Group Reports.

The following Institutions also provided financial support:

UNESCO/DMP, Deposit Modelling Program
IUGS, International Union of Geological Sciences
IAMG, International Association for Mathematical Geology

LIST OF PARTICIPANTS

Ilghiz AITMATOV
National Academy of Sciences
Institute of Physics and Mechanics of
Rocks
98 Mederov str.
Bishkek, 720000
KYRGYZ REPUBLIC
e-mail: rjench@geol.freenet.bishkek.su

Euro BEINAT
Institute of Environmental Studies
Vrije Universiteit
de Boelelaan 1115
Amsterdam, 1081 HV
THE NETHERLANDS
e-mail: euro.beinat@ivm.vu.nl

Byron R. BERGER
U. S. Geological Survey
P.O. Box 25046
Denver, Colorado 80225
U.S.A.
e-mail: bberger@usgs.gov

André BOTEQUILHA LEITAO
Centro de Valorizacao de Recursos
Minerais, Instituto Superior Técnico -
Universidade Técnica de Lisboa
Av. Rovisco Pais
Lisboa Codex, 1096
PORTUGAL
e-mail: andre.b.leitao@alfa.ist.utl.pt

Djamila AITMATOVA
Executive Committee
Inter-State Council for Kazakhstan,
Kyrgyzstan, Uzbekistan, Tajikistan
36 Erkindik Prospect
Bishkek, 720000
KYRGYZ REPUBLIC
e-mail: john@infotel.kg

Julius BELICKAS
Information Department
Geological Survey of Lithuania
Konarskio 35
Vilnius, 2600
LITHUANIA
e-mail: julius.belickas@lgt.lt

Johann BODECHTEL
University of Munich AGF - IAAG
Luisenstr. 37
Munich, 80333
GERMANY
e-mail: joh.bod@iaag.geo.uni-muenchen.de

Fatbardha CARA
Institute of Geological Research
Blloku "Vasil Shanto"
Tirana,
ALBANIA
e-mail: ispgj@ingeol.tirana.al

Liisa CARLSON
Pietarinkatu 10, B16
Helsinki, 00140
FINLAND

Emmanuel John M. CARRANZA
Dep. of Mineral Exploration
ITC
Kanaalweg 3
Delft EB, 2626
THE NETHERLANDS
e-mail: carranza@itcdelft.itc.nl

Tariro P. CHARAKUPA
Environment and Remote Sensing Institute
ERSI -SIRDC
Box 6640
Harare,
ZIMBABWE
e-mail: ersizim@harare.iafrica.com

Chang-Jo F. CHUNG
Spatial Data Analysis Laboratory
Geological Survey of Canada
601 Booth Street
Ottawa, K1A 0E8
CANADA
e-mail: cchung@gsc.NRCan.gc.ca

Bernardas CORNELIS
SURFACES -Geomatique - Institut de Geographie
Universite de Liege
7 Place du 20-aout
Liege, 4000
BELGIUM
e-mail: cornelis@geo.ulg.ac.be

Marta COVIC
Faculty for Mining, Geology and Petroleum Engineering
University of Zagreb
Pierottieva 6
Zagreb, 10 000
CROATIA
e-mail: gdurn@rudar.rgn.hr

György CSIRIK
Geological Institute of Hungary
Stefania 14
Budapest, 1143
HUNGARY
e-mail: csirik@mafi-2.mafi.hu

Peter Alan DOWD
Department of Mining and Mineral Engineering
University of Leeds
Leeds, LS2 9JT
U.K.
e-mail: p.a.dowd@leeds.ac.uk

Lawrence J. DREW
National Center 954
U. S. Geological Survey
12201 Sunrise Valley Drive
Reston, VA 20192
U.S.A.
e-mail: ldrew@usgs.gov

Meral ERAL
Institute of Nuclear Sciences
Ege University
Bornova-Izmir, 35100
TURKEY
e-mail: eral@egeuniv.ege.edu.tr

Andrea G. FABBRI
Division of Geological Survey
ITC
Hengelosestraat 99
Enschede, 7500 AA
THE NETHERLANDS
e-mail: fabbri@itc.nl

Enrico FEOLI
Department of Biology
University of Trieste
Via Valerio 32
Trieste, 34100
ITALY
e-mail: feoli@ics.trieste.it

János FÖLDESSY
Enargit Kft.
PF. 16
Recsk, 3245
HUNGARY
e-mail: 100324.1272@compuserve.com

Razia D. GAINUTDINOVA
Institute of physics NAS
265-A Chui Prosp.
Bishkek, 720071
KYRGYZ REPUBLIC
e-mail: eco@kyrnet.kg

Nora GAL
Geological Institute of Hungary
Stefania 14
Budapest, 1142
HUNGARY
e-mail: nora@mgsz.hu

Vyda-Elena GASIUNIENÉ
Geological Survey of Lithuania
Konarskio 35
Vilnius, 2600
LITHUANIA
e-mail: Vyda.Gasiuniene@lgt.lt

Gabor GAÁL
Geological Survey of Finland
Betonimiehenkuja 14
Espoo, 02150
FINLAND
e-mail: gabor.gaal@gsf.fi

Peng GONG
Center for Assessment & Monitoring of
Forest & Env. Res.
University of California
151 Hilgard Hall
Berkeley, CA 94720-3110
U.S.A.
e-mail: gong@nature.berkeley.edu

Marek GRANICZNY
Polish Geological Institute
Rakowieczka 4
Warsaw, 00-975
POLAND
e-mail: mgra@pgi.waw.pl

Halil HALLACI
Institute of Geological Research
Blloku "Vasil Shanto"
Tirana,
ALBANIA
e-mail: ispgj@ingeol.tirana.al

Jakup HOXHA
Institute of Geological Research
Blloku "Vasil Shanto"
Tirana,
ALBANIA
e-mail: ispgj@ingeol.tirana.al

Hulya INANER
Department of Geology
Dokuz Eylül University
Bornova-Izmir, 35100
TURKEY
e-mail: inaner@izmir.eng.deu.edu.tr

Győző JORDÁN
Geological Institute of Hungary
Stefania 14
Budapest, 1143
HUNGARY
e-mail: jordan@mgsz.hu

Kestutis KADUNAS
Hydrogeological Department
Geological Survey of Lithuania
35 Konarskio
Vilnius, 2600
LITHUANIA
e-mail: Kestutis.Kadunas@lgt.lt

Roma KANOPIENE
Geological Survey of Lithuania
Konarskio 35
Vilnius, 2600
LITHUANIA
e-mail: roma.kanopiene@lgt.lt

Péter KARDEVÁN
Geological Institute of Hungary
Stefania 14
Budapest, 1143
HUNGARY
e-mail: Kardevan@euroweb.hu

Kazimir A. KARIMOV
Institute of physics NAS
265-A Chui Prosp.
Bishkek, 720071
KYRGYZ REPUBLIC
e-mail: eco@kyrnet.kg

Jacek R. KASINSKI
Polish Geological Institute
Rakowieczka 4
Warsaw, 00-975
POLAND
e-mail: jkas@pgi.waw.pl

László KORPÁS
Geological Institute of Hungary
Stefania 14
Budapest, 1143
HUNGARY
e-mail: korpasl@mafi.hu

Dénes KOVÁCS
ITC
P.O. Box 6
Enschede, 7500 AA
THE NETHERLANDS

William H. LANGER
U. S. Geological Survey
P.O. Box 25046, MS 973
Denver, Colorado 80225
U.S.A.
e-mail: blanger@usgs.gov

Gregory K. LEE
U. S. Geological Survey
P.O. Box 25046, MS 964
Denver, Colorado 80225
U.S.A.
e-mail: glee@usgs.gov

Daniel LIMPITLAW
Department of Mining Engineering
Witwatersrand University
P. Bag 3, P.O.
Witwatersrand, WITS 2050
SOUTH AFRICA
e-mail: limpitlaw@egoli.min.wits.ac.za

Árpád Ferenc LORBERER
Szentendrei út 19.
Budapest, 1035
HUNGARY
e-mail: loare@iris.geobio.elte.hu

Oleg MAKARYNSKY
Institute of Marine Sciences
Middle East Technical University
P. K. 28
Erdemli - Icel, 33731
TURKEY
e-mail: oleg@soli.ims.metu.edu.tr

Karol MARSINA
Geological Survey of Slovak Republic
Mlynská dol.
Bratislava, 817 04
SLOVAKIA
e-mail: marsina@gssr.sk

Richard B. MCCAMMON
Eastern Minerals Team
U. S. Geological Survey
12201 Sunrise Valley Drive
Reston, VA 20192
U.S.A.
e-mail: mccammon@rgborafsa.er.usgs.gov

Rumen Parvanov MIRONOV
Department of Radiocommunication
Technical University of Sofia
Sofia, 1756
BULGARIA
e-mail: rpm@vmei.acad.bg

Charlie J. MOON
Geology Department
Leicester University
Leicester, LE1 7RH
U. K.
e-mail: cjm@leicester.ac.uk

László ÓDOR
Geological Institute of Hungary, Hungarian
Geological Survey
Stefania 14
Budapest, 1143
HUNGARY
e-mail: odorl@mafi.hu

Sergio OLIVERO
Via Marco Polo 19 bis
Torino, 10129
ITALY
e-mail: soliver@tin.it

Ivars OZOLINS
Geological Survey of Latvia
5 Eksporta Iela
Riga, 1010
LATVIA
e-mail: ozolins@vgd.gov.lv

Jyrkki PARKKINEN
Geological Survey of Finland
Betonimiehenk. 4
Espoo, 021250
FINLAND
e-mail: Jyrki.Parkkinen@gsf.fi

Antonio PATERA
Via Bongiorno 48/48
Roma, 00155
ITALY
e-mail: pater@gea.geo.uniroma1.it

Petri PELTONEN
Geological Survey of Finland
Betonimiehenk. 4
Espoo, 021250
FINLAND
e-mail: petri.peltonen@gsf.fi

Nyls PONCE SEOANE
Institute of Geology and Paleontology
Vía Blanca y Carretera Central, San Miguel del Padrón
Cjudad Habana, 11000
CUBA
e-mail: cuba@matias.onat.gov.cu

Sankaran RAJENDRAN
School of Earth Sciences
Bharathidasan University
Palkalai Perur
Tiruchirapalli, Tamilnadu 620 024
INDIA
e-mail: earth@info.bdu.ernet.in

Petr RAMBOUSEK
Czech Geological Survey
Klarov 3
Prague 1, 118 21
CZECH REPUBLIC
e-mail: ramby@cgu.cz

Nadia RAPANTOVA
The University of Mining and Metallurgy, VSB
Trida 17, Listopadu
Ostrava-Poruba, 708 33
CZECH REPUBLIC
e-mail: vladimir.homola@vsb.cz

Luis RECATALÁ BOIX
Centro Investigaciones Sobre
Desertification - CIDE (CSIV, UV, GV)
C/Camí de la Marjal, s/n, Apartado Oficial,
Albal, Valencia 46470
SPAIN
e-mail: luis.recatala@uv.es

Juan REMONDO
CITIMAC - University of Cantabria
Av. de los Castros s/n
Santander, 39005
SPAIN
e-mail: remondoj@ccaix3.unican.es

Costas RIPIS
IGME
Messoghion 57
Athens, 11527
GREECE
e-mail: dkigme@compulink.gr

Ipo L. RITSEMA
Department of Geo-Information Systems
Netherlands Inst. of Applied Geosci. TNO,
National Geological Survey
Schoemakerstraat 97
Delft, 2600 JA
THE NETHERLANDS
e-mail: i.ritsema@nitg.tno.nl

Vardan SARGSYAN
Dep. of Computers and Programming
Yerevan State Institute of National
Economy
128 Nalbandyan str.
Yerevan, 375025
ARMENIA
e-mail: vardan@ysine.am

Deborah J. SHIELDS
Rocky Mountain Research Station
U. S. D. A. Forest Service
3825 E. Mulberry St.
Fort Collins, Colorado 80524
U.S.A.
e-mail: dshields@lamar.colostate.edu

Irina SHTANGEEVA
Institute of Earth Crust
St. Petersburg University
Universitetskaya nab. 7/9
St. Petersburg, 199034
RUSSIA
e-mail: Irina@ivs.usr.pu.ru

Nikolay Metodiev SIRAKOV
Instituto Superior Técnico de Lisboa
(CVRM - IST)
Av. Rovisco Pais
Lisbon, 1096
PORTUGAL
e-mail:

Slavko V. SOLAR
Institute for Geology, Geotechnics and
Geophysics
Dimiceva 14
Ljubljana, 1000
SLOVENIA
e-mail: svsolar@i-ggg.si

Delia Teresa SPONZA
Environmental Engineering Department
Dokuz Eylül University
BUCA, Izmir,
TURKEY
e-mail: dsponza@izmir.eng.deu.edu.tr

Bernhard STRIBRNY
Federal Institute for Geosciences and
Natural Resources
Stilleweg 2
Hannover, 30655
GERMANY
e-mail: Bernhard.Stribrny@bgr.de

Janos SZANYI
Hungarian Geological Survey
Sohordo u. 20
Szeged, 6721
HUNGARY
e-mail: szanyi@iif.u-szeged.hu

Géza SZEBÉNYI
Recski Ércbányák Rt.
Ércbányatelep, P. O. Box 11
Recsk, 3245
HUNGARY

Andrea SZUCS
Geological Institute of Hungary
Stefánia 14
Budapest, 1143
HUNGARY
e-mail: szucsa@mafi.hu

István SZŰCS
Geotechnique and Environmental
Protection Ltd., GEOPARD Research,
Development and Services Co.
POB 10, Pécs, 7610
HUNGARY
e-mail: geopard@mail.matav.hu

Isabel Maria TAVARES GRANADO
Centro de Valorizacao de Recursos
Minerais, Instituto Superior Técnico de
Lisboa (CVRM - IST)
Avenida Rovisco Pais
Lisboa Codex, 1096
PORTUGAL
e-mail: igranado@alfa.ist.utl.pt

Géza M. TIMCÁK
Technical University
Pk. Komeského 19,
Kosice, SK 043 84
SLOVAKIA
e-mail: gmtkg@tuke.sk

Turgay TOREN
General Directorate of Turkish Coal
Enterprises, Ministry of Energy and
Natural Resources of Turkish Republic
Eryaman 4. Etap, 17689 Ada, 11. Blok,
No. 10, Eryaman - Ankara,
TURKEY
e-mail: ttoren@superonline.com

Stanislava TRTIKOVA
Dep. of Mineralogy and Petrology
Comenius University
Mlynska Dolina G.
Bratislava, 842 25
SLOVAKIA
e-mail: trtikova@fns.uniba.sk

Ida TÖRÖK
Miskolc University
Miskolc-Egyetemváros
HUNGARY

Oleg UDOVIK
Environmental & Resources Research
Institute, National Security & Defense of
Ukraine
13 Chokolivskiy Blvd.
Kiev, 252180
UKRAINE
e-mail: udovik@topaz.kiev.ua

László VÉRTESY
ELGI
Kolumbusz u. 17-23.
Budapest, 1143
HUNGARY
e-mail: legi@elgi.hu

Tsehaie WOLDAI
ITC
P.O.Box 6
Enschede, 7500 AA
THE NETHERLANDS
e-mail: woldai@itc.nl

Hasan UCPIRTI
Department of Civil Engineering
Sakarya University
Sakarya,
TURKEY
e-mail: ucipirti@esentepe.sau.edu.tr

Zoltán VEKERDY
ITC
P.O. Box 6
Enschede, 7500 AA
THE NETHERLANDS
e-mail: vekerdy@itc.nl

Richard B. WANTY
U. S. Geological Survey
P.O. Box 25046, MS 973
Denver, CO 80225
U.S.A.
e-mail: rwanty@usgs.gov

Mehmet Ali YUKSELEN
Dep. of Environmental Engineering
Marmara University
Goztepe
Istanbul, 81040
TURKEY
e-mail: yukselen@marun.edu.tr

PART 1. GEOENVIRONMENTAL DEPOSIT MODELS

Geoenvironmental models have been recently developed that include characteristics of the ore and associated waste rocks in terms of the environmental risks, which are related to the biotic, climatic, and topographic settings. Such models serve as guides for evaluating potential environmental impacts. An important goal in the continued development of geoenvironmental models is the integration of empirical data to produce more quantitative models that can be used to predict the costs of environmental mitigation and the risks associated with mineral extraction.

In their paper, Wanty, Berger, Plumlee, and King describe the advancements made in environmental models of mineral deposits in which they stress the importance of baseline and natural backgrounds in assessing mined areas and areas where future mining may occur. The authors argue for a strategy for developing geoenvironmental models to include climatic and ecoregional effects that embody such physical environmental characteristics as precipitation, evaporation, temperature, and ground-water/surface-water interactions. The development of geoenvironmental models represents a new direction in the environmental geosciences.

GEOENVIRONMENTAL MODELS

An Introduction

R.B. WANTY, B.R. BERGER, G.S. PLUMLEE, and T.V.V. KING
United States Geological Survey
M.S. 973, Denver Federal Center, Denver, Colorado, 80225 U.S.A.
rwanty@usgs.gov

Abstract. Mineral deposits have been classified by their geologic and mineralogic characteristics for decades, but the recognition that mineral deposits could be classified by their environmental characteristics is relatively new. In the past 5 years, advancements have been made by building on the earlier work of economic geologists who classified geologic characteristics. Several approaches have been taken that range from wholesale assessments of large areas (millions of square kilometres) to detailed assessments of individual watersheds or individual mines. Current efforts in the development of geoenvironmental models include the assemblage of diverse databases and encompassing geologic, geochemical, geophysical, hydrologic, and other data. Although the geologic information layers are most important in classifying the environmental signatures of mineral deposits, other data, which include the delineation of ecological regions, or ecoregions, provide a framework within which geoenvironmental models can best be developed. For instance, although the economic classification of ore deposits is best accomplished by grouping geologic, mineralogic, and depositional characteristics, the environmental classification of ore deposits may be best accomplished by grouping according to geochemistry, mineralogy, and hydrology within the ecoregion framework. Thus, economically different deposits in one ecoregion might have greater environmental similarities than similar deposits in different ecoregions. This is because the weathering behaviour of mineral deposits is controlled by climatic, as well as geologic, properties. This presentation provides an overview of the development of mineral deposit environmental models and discussions of the most important data types to be included in the models. This paper presents the current state of the art of geoenvironmental models, suggests a new framework for the models, and sets the stage for the topical presentations that follow in this volume.

1. Introduction

This paper presents the background information necessary for geoenvironmental model development. Other papers in this volume present specific case studies from areas around the world, so it is hoped that the concepts presented here will serve as an underpinning for the more specific case studies.

Throughout most of this century and before, economic geologists have developed classification schemes for mineral deposits on the basis of their geologic and petrologic characteristics, their age, and other geologic, physical, or chemical properties. Until recently, however, no attempts have been made to classify mineral deposits by their environmental behaviour. Initial attempts at the U.S. Geological Survey (USGS) to develop geoenvironmental models of mineral deposits (duBray, 1995; Plumlee et al., 1992; Plumlee and Nash, 1995; Plumlee and Logsdon, 1999; Filipek and Plumlee, 1999) have led to useful concepts and framework development, and reveal some of the most important variables to consider; namely, deposit and host-rock geology and mineralogy, alteration styles, trace-element chemistry, topography and physiography, hydrology, and mining and milling methods. To this list, we add climate, latitude, and altitude. In fact, these latter three variables may be the primary controlling factors of the weathering of ore deposits, just as they control other weathering reactions.

To a very great extent, the development of geoenvironmental models depends on the intended application, whether for determination of baselines, mine planning, mitigation, or other purposes. Thus, the display format and regional extent of coverage may be predetermined by the intended use. As a general rule, geoenvironmental models with broader regional coverages will tend to be more qualitative, descriptive, and comparative, and models of smaller areas are generally much more quantitative in terms of surface- or ground-water quality or other environmental effects.

1.1. DEFINITIONS

The following definitions are provided as a basis for discussion of geoenvironmental models and their components. These definitions are derived from current usage of the terms by the authors and their co-workers; other authors may use other definitions. A tendency exists for scientists to conduct their research in a manner compatible with the socio-economic and regulatory environments of their country. To the extent that this is true, our definitions below also may reflect the socio-economic and regulatory environments of the United States. Modifications of these definitions may be appropriate for other countries.

1.1.1. *Mineralisation*
Mineralisation is the process or group of processes that form or accumulate minerals or elements that might be of economic interest in the present or anticipated socio-economic and technological conditions. These processes occur over geologically significant periods of time. The rocks upon which these processes have acted are said to be mineralised. The term "ore" may be synonymous with "mineralised rock", with one distinction— ore is an operationally defined term that embodies the geologic and current economic conditions and mineralised rock is a more inclusive term that depends less on the immediate economics associated with possible mining. Figures 1 and 2 show schematics of mineralised and altered systems and give a sense of the gradational contacts between mineralised and altered rock.

Mineralisation and Alteration at Creede, Colorado

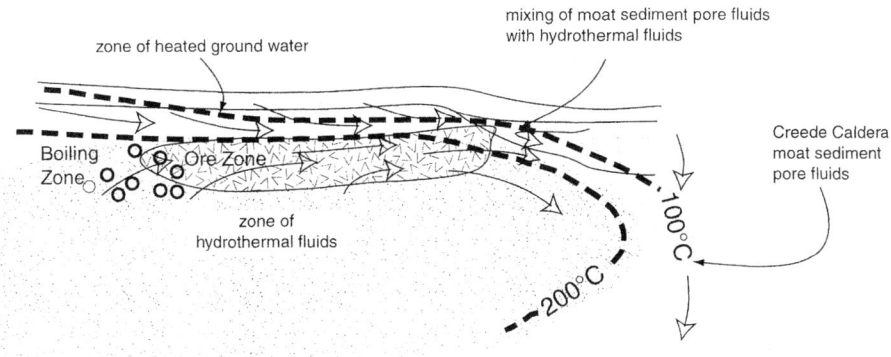

Figure 1. Cross-sectional diagram of the mineralised zone in the Creede, Colorado deposit (modified from Plumlee, 1994).

Mineralisation and Alteration in a Porphyry System

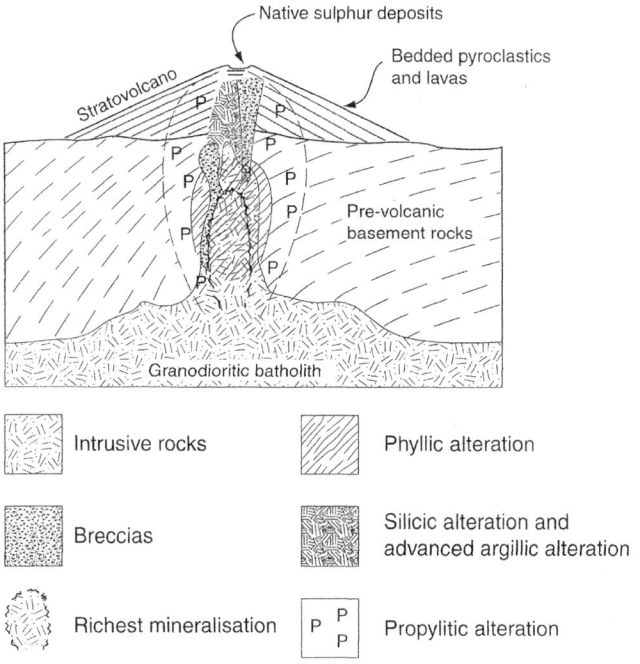

Figure 2. Schematic diagram of mineralised and altered zones in a typical porphyry deposit (modified from Sillitoe, 1973).

1.1.2. *Alteration*

Alteration is the process or group of processes that lead to chemical and physical changes in a rock by chemical reactions, such as dissolution or precipitation of minerals, or by replacement of one mineral with another. As with "mineralisation," "alteration" refers to the processes that act on a given volume of rock in the Earth's crust and occur over similar timeframes to mineralisation. In most cases, a volume of altered rocks surrounds a volume of mineralised rocks. Depending on the geologic structure and local hydrology at the time of mineralisation and alteration, the mineralised and altered zones may be roughly concentric. Contacts between these zones are gradational, and the extent of what is referred to as mineralised versus altered rock may be economically or technologically defined. "Alteration" also includes supergene weathering, which may be occurring to the present.

1.1.3. *Background*

The original condition of the environment prior to any influence of man's activities is known as background. The true preanthropogenic concentrations at the Earth's surface cannot easily be known because even some of man's earliest activities are known to have made an environmental impact. For example, lead mining and smelting activities by the ancient Greeks and Romans has led to measurable increases in the lead concentration of ice cores in Greenland (Boutron et al., 1994; Hong et al., 1994). Because samples such as those collected by the aforementioned authors are available only in a few places in polar regions, determination of reliable natural background concentrations of the elements for most areas of the Earth may not be possible.

Background concentrations are not necessarily low; areas may exist where mineralised rocks are exposed at the surface that have undergone minimal disturbance by man, yet the surface or ground waters may be unfit for aquatic life and for certain uses, such as drinking water. Documented evidence of high preanthropogenic background concentrations includes ferricrete deposits in Montana (Furniss and Hinman, 1998; Furniss, 1999) and Colorado, U.S.A. (Miller and McHugh, 1994). Carbon dating of wood fragments in the Montana ferricrete reveals ages of up to 9,000 years before present—a good indication of preanthropogenic background.

1.1.4. *Baseline*

The current environmental conditions, regardless of the nature or extent of man's activities in the area of interest, constitute the baseline. A baseline is chosen operationally for an area and then serves as a reference point for future investigations or monitoring. Anthropogenic influences are certainly present but may be minimal. In areas of minimal anthropogenic influence, a tendency exists to use the terms background and baseline interchangeably. Although it may be true that the background and baseline conditions are chemically indistinguishable (or nearly so), there may be distinct regulatory or legal connotations attributed to these terms in some countries so care should be taken with their use.

It is natural to compare the baseline with what is thought to be background conditions. In general, mining activities or other surface perturbations will lead to an increase in baseline above background because of the mineral material that is brought to the surface and exposed to air and because of the mechanical crushing of minerals that leads to increased exposed surface area. This newly exposed, relatively fine-grained material will react much more quickly in the weathering environment than if it were not brought to the surface. Thus, the natural (premining) background in a mined area is likely to be less than the post-mining baseline. In the case of sulphide minerals, exposure to air with increased surface area will lead to greater production rates of sulphuric acid and subsequently to greater dissolved trace-metal concentrations. In the case of other minerals, such as carbonates and silicates, exposure to oxygen may not affect the weathering rate appreciably, but increased surface area should. Increased dissolution rates for carbonates and some silicates may be beneficial in terms of increased acid neutralising capability, but the total dissolved solids content of the water will increase.

1.1.5. *Environmental signature*
The (perhaps) unique suite of elemental concentrations and other physical and chemical properties of waters and rocks that come in contact with specific types of mineralised or altered rock in a given region comprise the environmental signature of that rock package. The primary physical variables that determine environmental signatures include mineralogy, petrology, lithology, structure, and hydrology, as well as climate, latitude, and altitude. The latter three variables cannot be overlooked; a given deposit exposed at the surface in a warm wet climate, perhaps at low latitude or low altitude, would be expected to weather very differently from the same type of deposit in a drier or cooler climate.

1.1.6. *Mitigation*
Mitigation is the process of anticipating potential environmental impacts of planned activities and preventing those impacts by designing new processes or procedures. For example, a proposed mine may, through the exploration drilling program, determine that a great abundance of pyrite and very little acid-neutralising capacity exists in the ore rocks, which might lead to a potential acid-drainage problem. Development of a mitigation plan could include special handling or storage/stockpiling procedures for the ore rock prior to milling and special treatment and storage of the mill tailings to prevent generation of acidic drainage or to immediately neutralise and treat it. To accomplish the latter, for example, the mitigation plan might include importing limestone from another location to neutralise acid as pyrite oxidises in the waste dumps.

1.1.7. *Remediation*
Remediation is the process or act of fixing an environmental problem after it has occurred. In the context of mining operations, this usually means neutralising or preventing acid production and decreasing or preventing the migration of acid, metal-rich waters away from the mine property. Remediation also usually involves fixing downstream effects, such as loading of metals in bed sediments, revegetation of riparian habitats, re-establishing fish and insect populations, etc. In accordance with the second law of thermodynamics, remediation is a fight against entropy and is thus much more expensive and extensive than mitigation. Remediation may require extensive time periods, depending on the severity of the problem. To paraphrase an English saying, "An ounce of mitigation is worth a pound of remediation." When considering remediation issues, the concepts of background and baseline become especially important as remediation below background becomes difficult or impossible, and is certainly impractical. The question should also arise whether a mining operation should be required to remediate below background conditions.

1.1.8. *Ecosystem and Ecoregion*
The definitions of these terms have evolved over the past several decades. Bailey (1996) provided an excellent overview of the development of the art of ecosystem studies, and of the concepts of ecosystems and ecoregions. Rowe (cited in Bailey, 1996) defined ecosystems as "a topographic unit, a volume of land and air plus organic contents extended areally over a particular part of the earth's surface for a certain time." This definition captures the spatial as well as temporal nature of an ecosystem, but Bailey (1996) went one step further to demonstrate that "ecosystem" may, in fact, incorporate a scale-dependence, in which a finite system, termed an "ecosystem," is in fact part of a larger finite ecosystem. Bailey also stressed that ecosystems are not independent entities, but are interrelated to adjacent or other surrounding ecosystems and that a perturbation of one ecosystem carries forth into others. This is because ecosystems are open systems in the thermodynamic sense, with respect to fluxes of energy and mass.

For the purposes of this discussion, ecoregions may be thought of as large ecosystems. Bailey (1996) developed the progression from the smallest ecosystems ("sites"), through "landscape mosaics" (a concatenation of intimately related sites) to "ecoregions". Sites might be on the order of 10's of km^2, landscape mosaics 10^3 km^2, and ecoregions 10^5 km^2 or larger.

1.2. PURPOSE OF GEOENVIRONMENTAL MODELS

1.2.1. *Understand Environmental Behaviour of Mineralised/Altered Areas*
Geoenvironmental models, whether descriptive or quantitative, should provide the necessary data and interpretation to enable the user to anticipate the chemical and mechanical weathering behaviour of rocks within and around a given mineral deposit in a given climatic regime. In a qualitative or descriptive sense, the model should include a list of trace elements (not necessarily the elements of economic interest) likely to be found in surface and ground waters that drain the deposit. Similarly, additions of trace elements to soils and stream sediments must be considered. The list of trace elements must include those whose impact on flora and fauna would be most damaging, regardless of whether they were the elements for which the mine was developed in the first place. It may also include information on bioavailability and/or effects on flora and fauna, especially for the soils and sediments. Because climate is a primary influence on the weathering behaviour of mineral deposits, the model will include climatic dependence on weathering, perhaps with examples or case studies. Alteration styles vary, and contacts between alteration zones are usually gradational. The geoenvironmental model should, therefore, address the environmental effects of the major alteration styles associated with a given deposit type so that the user can tailor the model information to the new system under study. Quantitative geoenvironmental models would include such data as lists of trace elements expected to be found in drainage waters and expected ranges of concentrations, duration (persistence) of environmental effects, comparisons of the concentrations found in mining operations to natural backgrounds, etc.

1.2.2. *Identify and Understand Areas/Regions With High Natural Backgrounds*
Natural background concentrations of elements in surface and ground waters are determined by water-rock interactions that occur at the earth's surface. Metal-rich

deposits may be exposed naturally at the surface owing to chemical, mechanical, or biological weathering. Chemical weathering includes such processes as dissolution, ion exchange, and adsorption/desorption. Mechanical weathering includes such processes as erosion, spalling, wind, freeze/thaw destruction of rocks, and glacial processes, among others (Miller et al., 1999). Biological erosion processes may be a subset of mechanical or chemical weathering processes and include microbially aided mineral dissolution, increased rates or amounts of dissolution of minerals near root hairs of plants, and physical movement of rocks by roots of plants as the plant grows and the roots become enlarged. In any case, weathering (dissolution) of metal-bearing minerals may lead to high natural concentrations of metals. Chemical weathering may increase the rate of mechanical weathering by breaking chemical bonds in minerals and weakening the overall structural integrity of a rock (Garrels and Mackenzie, 1971) or by opening preferred flow paths for water that, in turn, may undergo freeze/thaw cycles. Conversely, in areas with increased rates of mechanical weathering, chemical weathering rates also may be increased because fresh mineral surfaces are continually exposed (Bassett et al., 1992). Mechanical weathering rates depend on rock type, slope, vegetative cover, and climate. Chemical weathering rates depend on climate and exposed reactive surface area (Paces, 1973; Paces, 1983; Anbeek, 1992; Drever and Clow, 1995; Lasaga, 1995).

Weathering rates and, therefore, natural background concentrations depend on climate in a number of ways. Mechanical weathering is increased in areas where freeze/thaw cycles are intense and in damp climates where the mechanical integrity of an exposed rock is deteriorated owing to chemical weathering along fractures, joints, or other preferred flow paths. Erosion, which is another form of mechanical weathering, depends strongly on amount and periodicity of rainfall, temperature, wind speeds and directions, and vegetation- all climatic variables. Biological weathering is a function of climate because climate is one of the principal determining factors in the assemblage of biota likely to be found in a region. Chemical weathering rates depend on climate, and especially on temperature and availability of moisture.

Figure 3 demonstrates a number of concepts discussed up to this point. First, the gradation between mineralisation and the various stages of alteration is shown. Second, the figure shows the potential for high background metal loads owing to the erosion of all the material between the reconstructed land surface and the present-day land surface. The land surface is reconstructed to be as it was during the time of mineralisation. If the mineralised and altered zones are projected to the paleosurface, then a great volume of mineralised and altered material was transported from the site since the time of mineral formation. Other mineralisation types are inferred on the basis of samples that have been collected in the area, for example the hot-spring mineralisation is documented by sinters found around the Summitville, Colorado, area. Erosion of these deposits has also contributed to natural background levels. Lastly, the figure demonstrates the gradational transition from a mineralised centre through an altered zone into unaltered country rock. In the transition between these zones, the natural water-rock interactions will lead to varying chemical characteristics of the rocks and waters with which they are in contact.

Mineralisation and Alteration at Summitville Colorado

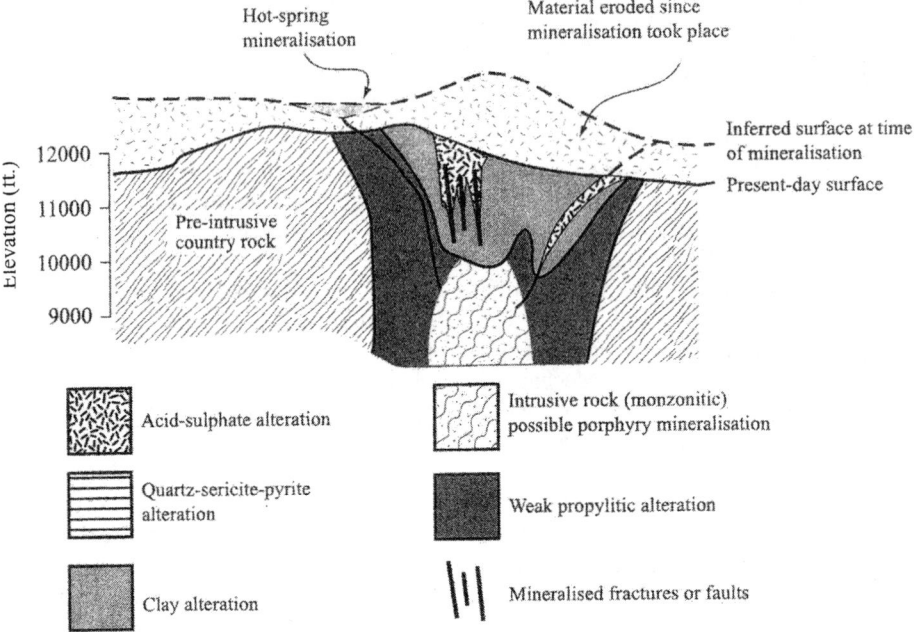

Figure 3. The Summitville, Colorado deposit has many characteristics to demonstrate concepts of background and baseline (modified from Plumlee et al., 1995).

Natural background may include metal concentrations that are not conducive to some forms of aquatic life. This is especially true for areas where sulphide-rich mineral deposits are exposed at the surface and where acid-neutralising capacity is lacking in the surrounding rock. Surface waters with high natural background may be barren of fish, insects, and many types of plants. Geoenvironmental models of mineral deposits will include data on mechanical and chemical weathering rates so that natural background concentrations, whether high or low, can be explained in a reasonable geologic and geochemical context.

1.2.3. *Determine Background/Baseline Conditions Prior to Mine Development*

A primary environmental consideration associated with a new mine development is to determine the condition of the environment prior to development, that is, to determine a premining baseline. With the recognition that natural background concentrations may be high in undisturbed areas, it becomes even more important for mining companies and government regulators to determine the condition of the environment prior to new mine development. If the natural background is high, then the mining operation should be held to the standard of higher concentrations, rather than remediating to concentrations that are lower than the natural ones. In this case, legal requirements for remediation should be examined and modified as appropriate. If legal requirements cannot be waived or modified, then the mining operation may not be profitable owing to greater remediation costs. If, however, the natural background includes low concentrations of metals, then the mining operation should be held to a standard that reflects the purity of the original water.

Given site-specific data for a proposed mine development, geoenvironmental models may be used to explain the natural weathering behaviour of the mineral deposit of interest. In this situation, geologic, geochemical, climate, and other data may be combined with case studies of similar deposit types in similar ecoregions to determine the possible range of contributions of the mineral deposit to observed element concentrations. It likely will not be possible to specify an exact background concentration, but it should be possible to provide bracketed concentration ranges with confidence limits assigned to each range. Such a procedure would allow for a more accurate determination of the possible range of natural background concentrations. At the same time, the geoenvironmental model should be able to anticipate the environmental impacts of the mine if no precautions are taken. The application of geoenvironmental models to mitigation and remediation will be discussed later.

1.2.4. *Determine the Range of Possible Premining Background and Baselines in Mined Areas*

Baseline determination is important in areas with historic mining or other perturbations where the concentrations of some elements may already have been increased above background owing to these prior operations. In this instance, the fundamental issue for the mining company or the government regulator may shift away from science and towards legal considerations. If a new mining operation is proposed in an area with historic mining, then there must be a recognition that the previous mining operations may have had an environmental impact. Depending on the applicable laws and regulations, however, the new mining company may or may not be responsible for the environmental impact of the previous operation(s). If the new mining company is responsible, then it may be useful to know the current baseline condition to satisfy an academic interest, but the mining company should be more concerned about determining what the premining natural background concentrations may have been. This is because they may be held liable for remediation to the background, rather than baseline, concentrations. Geoenvironmental models can provide a qualitative description of the probable environmental impacts of the previous mining operations. This description would include whether natural background concentrations may have been high and the likelihood of a new mining operation exacerbating any existing environmental degradation that is attributable to previous mining operations.

One way to help determine premining baselines is through an understanding of premining hydrology. In Figure 4, the premining hydrology of the Summitville area has been inferred from a number of clues. First, the water table has been drawn as a subdued expression of the land surface, which is a common observation in mountainous areas. Owing to orographic effects on weather patterns, high peaks are generally recharge areas as they gather the most moisture during storms and from snow in the winter. Second, the presence of ferricrete deposits on the mountain indicates the former presence of springs- locations where the water table intersected the surface. These ferricrete springs likely had low pH and high metals concentrations (Miller and McHugh, 1994). Thus, the Summitville area likely had relatively high natural background prior to mining (Miller and McHugh, 1999).

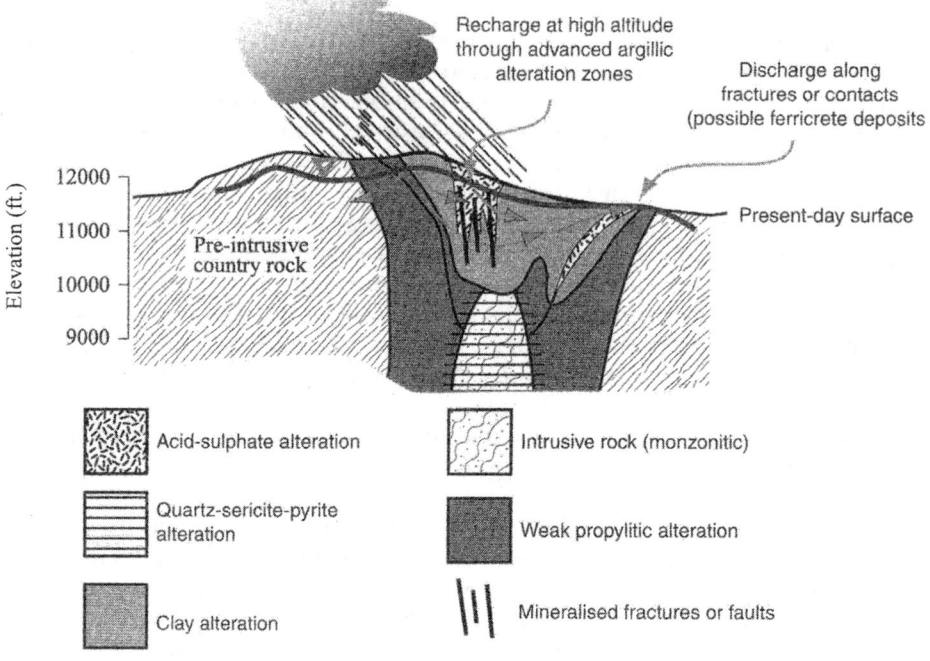

Figure 4. The inferred premining hydrology at the Summitville mine, southern Colorado (modified from Plumlee et al., 1995).

1.2.5. *Anticipate Mitigation or Remediation Requirements for Future Mine Developments*

Geoenvironmental models can be used to formulate better designs for mitigation or remediation by providing geologic and geochemical information for a specific deposit. The qualitative predictive nature of geoenvironmental models enables the user to anticipate environmental impacts of mine development and thus to develop comprehensive mitigation strategies. The holistic combination of geologic, hydrologic, and geochemical information also can be used in remediation plans.

Many remediation strategies are constructed despite an ignorance or oversimplification of fundamental geologic or hydrologic characteristics of a mining operation. Judicious use of geoenvironmental models may help overcome this problem. For example, acid-base accounting (ABA) is a method used to anticipate the acid-generating potential of a mineral or coal deposit. ABA works by conducting a quantitative inventory of minerals likely to generate acid and those likely to consume acid in the weathering environment (White et al., 1999). ABA is a method widely used in the United States, but it has many drawbacks that limit its scientific vigour (Morin, 1990; Mills, 1997; White et al., 1999). For example, the ABA method assumes that all minerals are exposed to the weathering environment so consequently the rock could, in theory, dissolve completely. Kinetics of dissolution and weathering reactions are ignored by this method. The mineralogic, petrographic, and petrologic information in a geoenvironmental model might be used to modify the ABA conclusions by providing paragenetic information that would suggest that some minerals might be shielded from the environment.

Another example of oversimplification in remediation plans is the practice of plugging adits to prevent mine drainage. In fact, this practice is seldom effective as a long-term fix because it causes the water-table level to rise to premining levels and may reactivate old springs along fracture or fault systems. Further, plugs have been known to fail under the sometimes-enormous hydraulic pressures they hold back. Abandoned mine workings may perturb the premining hydrologic system and cause new springs or seeps to emerge as the water-table level rises. The geoenvironmental model for the deposit would show the structural features of the deposit that would be expected to conduct water and that would, in turn, lead to leakage of acidic fluids at the surface. Such a situation is demonstrated at the Summitville deposit, as shown in Figure 5. The Reynolds tunnel was driven to facilitate drainage from the deposit and lowered the regional water table as shown. When the Reynolds tunnel was plugged, the water table began to rise and approach the premining water table, thus reactivating ferricrete springs and causing leakage from higher mine workings that intersected the surface. A better approach to plugging an adit is exemplified by the case of the Gyöngyösoroszi base-metal mine in northeastern Hungary, shown schematically in Figure 6, where a plug has been installed in the main haulage tunnel to divert water into a channel that leads to a treatment plant. In this situation, the plug serves as a mechanism to capture the ground water and to divert it to a treatment scheme, rather than a device to impound ground water. Thus, the mine drainage is efficiently captured, treated, and released into the natural stream drainage. Failure of the plug is less likely because the water is continuously drained through the plug, preventing the buildup of hydraulic pressure behind the plug.

14

Effect of Plugging the Reynolds Drainage Tunnel at Summitville, Colorado

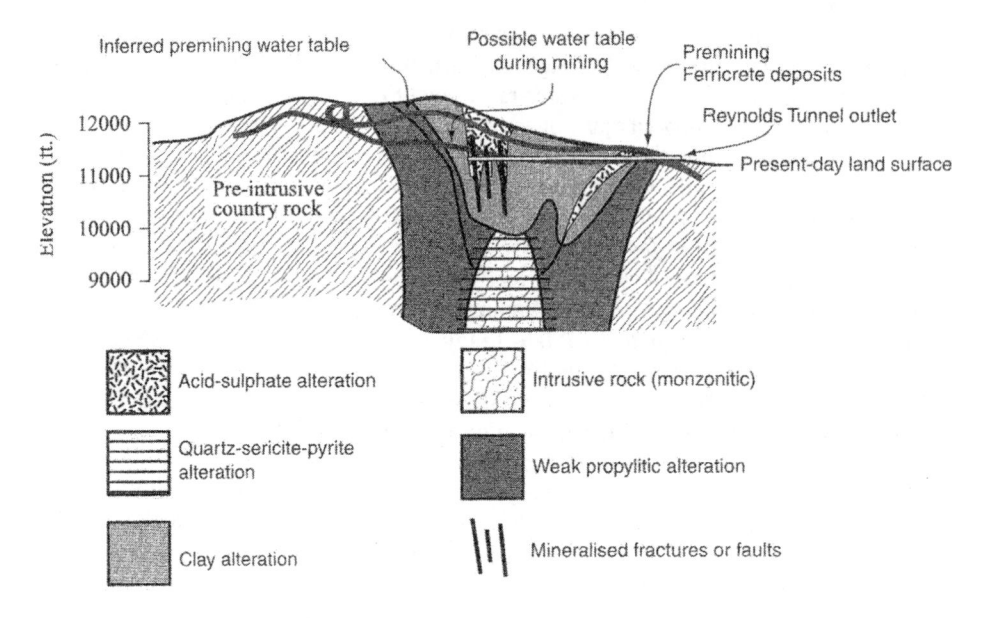

Figure 5. Plugging the Reynolds tunnel at the Summitville mine caused the water table to rise, reactivating ancient springs (base figure modified from Plumlee et al., 1995, with new information added).

The Water-Treatment Scheme at the Gyöngyösoroszi Base-Metal Mine

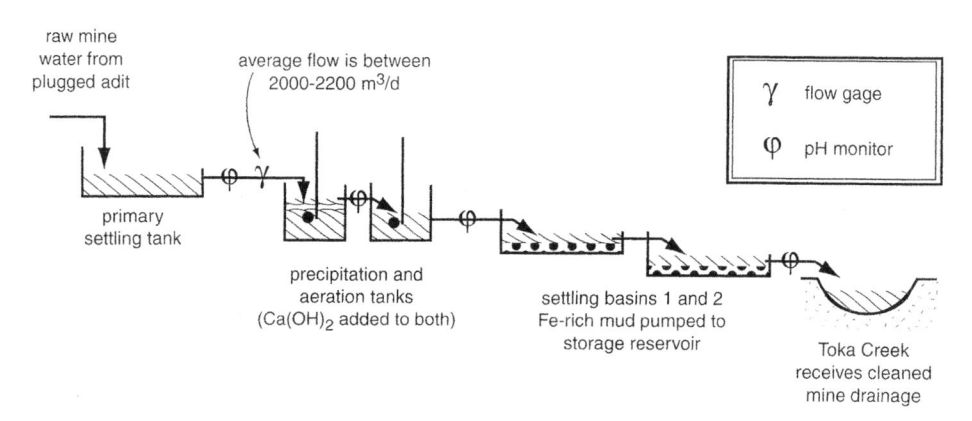

Figure 6. Schematic representation of the water capture and treatment facility at the Gyöngyösoroszi mine in northeastern Hungary (modified from Ódor et al., 1999).

2. Examples of Descriptive Geoenvironmental Models

The concepts already described can be illustrated with the descriptive geoenvironmental models as presented by duBray (1995). These models are presented for a wide variety of deposit types. The format of each chapter is built around a similar template to the deposit models presented in Cox and Singer (1986)- each presents descriptions of the geologic and mineralogic features of the deposit type, examples of the deposit type, and references to literature on specific deposits. The next section of the model describes the potential environmental effects of the mineral deposits, either naturally or when mined, in the context of geology, which includes information on mineralisation, alteration, mineralogy of ore and gangue, trace-element chemistry, and hydrology. The final section of each model describes the environmental signature of the deposit type with information (as available) on mine-drainage chemistry, metal mobility, premining baseline or background, known or expected climatic effects, and environmental effects of various mining and milling techniques. An example chapter from duBray's book is included in the Appendix. This particular chapter was selected because it is relatively complete; chapters for other deposit types may be less complete because of insufficient data. This shortage of data is one of the drawbacks to descriptive models. They are intuitively easy to use, but for many deposit types, environmental data is lacking.

3. Geoenvironmental Models- Why and How

3.1. PHILOSOPHICAL APPROACH TO GEOENVIRONMENTAL MODELS

In a published form, duBray (1995) represented the state of the art of geoenvironmental modelling, presenting a collection of chapters by various authors describing geoenvironmental models of many types of deposits. This effort was designed to be an add-on to the established methods of resource assessment at the USGS (Cox and Singer, 1986; Bliss, 1992). In particular, the first chapter (Plumlee and Nash, 1995) presents a description of the philosophical approach used in the remainder of the volume. Plumlee (1999) has added to this body of thought, expanding the scope of geologic data that may be considered in geoenvironmental models, and considering climatic effects. Perhaps the most important observation in Plumlee and Nash (1995) is that the environmental effects of ore deposits and mineralised/altered areas extend beyond the boundaries of the alteration because of transport of elements by surface or ground water. Thus, although it would be desirable to develop geoenvironmental models in the same framework as resource models, it may not be practical because the two applications have a different focus and the sphere considered in the resource model is a subset of that considered for the geoenvironmental model. Still, there must be considerable overlap between the two model types because the mineral deposit is the primary source of environmental contaminants in most cases. The altered rock surrounding the mineralised core also may be volumetrically important. However, in many cases, the concentration of sulphide minerals in the rocks decreases radially outward from the mineralised core so that the altered rocks contain a lesser mass of sulphide than the ore deposit they surround.

To summarise the first part of this paper, the main concerns which geoenvironmental models must address are as follows:

- Provide information on the geology, mineralogy, structure, geochemistry, and ecology of mineral deposits. This information will serve as a base of knowledge from which other interpretations and comparisons may be drawn.
- Provide information on the climate dependence of environmental signatures of mineralised and altered areas. Just as it will be useful to compare the environmental signatures of different deposit types within a given climatic regime, it will also be useful to compare similar deposits across a number of climatic regimes and ecoregions. This approach will allow interpolation and, perhaps, extrapolation to new deposits as they are discovered.
- Provide information on the natural background chemistry of waters, soils, and sediments expected for a mineralised and altered area. Provide explanations for the observed concentrations of various metals in the context of weathering of mineralised and altered rocks.
- Allow for the determination of the effects of mining activities to aid in the determination of what natural background may have been prior to historic mining activities.
- Allow for accurate predictions of the effects of new mining operations in mined or unmined areas so that reasonable and rigorous mitigation and remediation strategies can be developed for the new mining operation.
- Aid geochemical/geologic exploration efforts. Although this point has not been previously developed, geoenvironmental models may be useful for geochemical exploration. When environmental signatures are understood for various deposit types in various climatic regimes, they may be used to aid the interpretation of new or existing geochemical data, which is gathered as part of an exploration program.

3.2. PRACTICAL APPROACH TO GEOENVIRONMENTAL MODELS

The above list of capabilities of geoenvironmental models is at least ambitious and at worst unattainable. The challenge to the geoscientific community is to best answer the question, "How shall we gather existing data, collect new data, and assemble and present databases in a format that will be most useful for addressing the questions geoenvironmental models are supposed to address?" To some extent, the answer to this question depends on the intended use(s) for the geoenvironmental models. In turn, the intended use dictates, to some extent, the preferred format of presentation, whether it be as a map, an interactive data base, a geographic information system (GIS), a numerical model, etc.

In Section 1, the basic concepts of geoenvironmental models are presented, as well as some possible uses and applications of the models. The chapter from duBray (1995) presented in the Appendix exemplifies the current state of the art of descriptive geoenvironmental models. The descriptive models provide a qualitative overview of the environmental behaviour of mineral deposits, but much of the information is specific to a single deposit or a group of a few deposits. The general lack of environmental information from deposits around the world contributes to the lack of quantitative modelling ability. So, a logical first step in the improvement of geoenvironmental models will be to collect and assemble more environmental information from past and current mining districts and unmined, mineralised areas throughout the world.

A second approach to geoenvironmental models is a map-based approach, as exemplified by the map of the state of Colorado, by Plumlee et al. (1995a). This geoenvironmental map was the first effort undertaken by the USGS to portray a major mineralised region, the Colorado Mineral Belt, in the context of known or expected environmental conditions. A simplified version of the Colorado map is shown in Figure 7 (from the western to eastern border of Colorado is just over 600 km). Because of the complexity and detail of all the data layers shown on the original published map, the simplification is necessary for this publication venue.

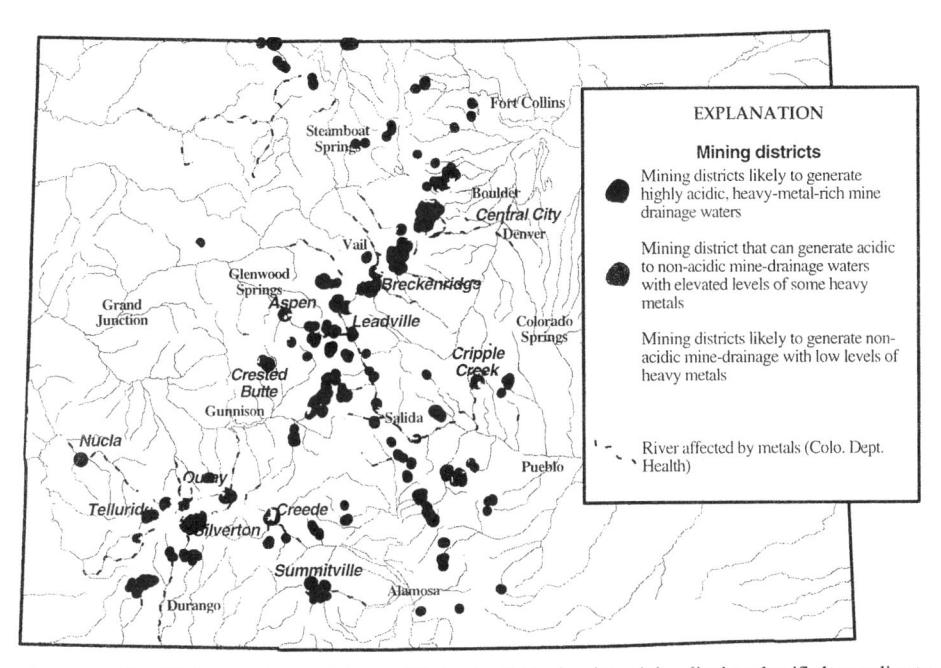

Figure 7. Simplified geoenvironmental map of Colorado, USA, showing mining districts classified according to their tendency to produce metal-rich drainage (modified from Plumlee et al., 1995a).

The published Colorado map is an highly interpretive product with objective and subjective data layers. Objective data layers include land ownership (important from the perspective of responsible management of government-controlled land), major surface-water divides and drainages, rivers and streams, average precipitation contours (for those regions with total precipitation >50 cm a^{-1}), and locations of major mining districts. The subjective data layers on the map include stream and river reaches that have degraded water quality owing to mineralisation or mining, and a colour-coded portrayal of the major mining districts based on their presumed propensity to cause adverse environmental impacts. This ranking derives from an analysis of mineralogic and geochemical investigations of representative ore-deposit types and their known or expected environmental impacts.

The various ore-deposit types were ranked on the basis of the nature and extent of alteration, mineral assemblages, metals present in the assemblage, and the presence of acid-generating minerals like pyrite versus the remnant or natural acid-consuming capacity of the host rocks. In this ranking, deposits rich in pyrite and metals but poor in acid-consuming minerals are ranked as most likely to cause environmental problems, pyrite-poor deposits being the least. Superficially, this analysis resembles that of the ABA method described above, but the actual evaluation method is more complex. For example, mineral deposits with quartz-alunite and advanced-argillic alteration were ranked as most likely to cause deleterious metal impacts because the intensely acidic mineralising/altering fluids consumed whatever natural acid-neutralising capacity the host rocks may have had prior to mineralisation. Thus, during present-day weathering of such deposits, the acid generated by sulphide oxidation is not neutralised by reaction with the host rocks. Waters flowing downgradient from such deposits, either in the ground or on the surface, are likely to have extremely low pH (usually <3) and high metals concentrations (usually in the ppm range or greater for metals such as Cu, Zn, As, Cr, Ni, Pb, Co, U and Th). Examples of such deposits in Colorado include Summitville and Red Mountain near Lake City. Deposits in this ranking within the Carpathian arc may include Lahóca, near Recsk, Hungary, and Talagiu, near Deva, Romania. To remediate the drainage waters from these deposits effectively, the acid must be neutralised and the metals removed from the drainage waters. In comparison, drainage waters in contact with carbonate-hosted deposits may be expected to have higher pH's (≥ 6), although concentrations of some metals, especially Zn with minor Pb, Cu, and As, may still be great enough to cause adverse impacts to aquatic life. For these deposits, the greatest remediation concern is to remove the metals dissolved in the water. Examples of this deposit type include parts of the Leadville district, Colorado, and Bánska Stiavnica, Slovakia.

The geoenvironmental map of Colorado was an early attempt to classify deposits in a regional and geologic framework. Climate considerations are limited to showing a threshold precipitation contour. As such, it has the advantage of showing a wide range of deposit types on a single map. This format offers the intuitive ease that comes from assimilating a relatively large amount of information into a spatial, colour-coded display. There are also some drawbacks to this approach, which offers the possibility to improve the approach for various applications and end users. For example, the data layers are displayed, but cannot be queried as they can in an interactive GIS product. Thus, interpretations can only be derived from the map upon intense inspection and reading of ancillary materials. Because the map is published on paper, it is difficult to add new layers of information that the user may need to customise the map for the intended application. Depending on the specific interests of the end user, it may be desirable to add any of a number of additional data layers to suit a specific need.

Although the Colorado map clearly shows the locations and possible environmental affects of various mining districts, many other types of information could be added to suit any particular need. It may be tempting to show as many information layers as possible, but extreme care should be taken to avoid showing too much information on one map, thereby presenting a confusing and unintelligible product. Perhaps if other information layers were to be shown and if a map format is desired, the best way would be to add a series of maps in a folio where each map is designed to show a particular effect. Alternatively, in an interactive digital product, any number of data layers could be provided and the users could choose which layers to display for their particular application. Examples of data layers that could be added include: land use / land cover data; more rigorous stream-chemistry data; ground-water chemistry; ecological information, such as habitat quality, fish-kill data, vegetation surveys, etc.; geological information, such as outcrops of formations which are carbonate-rich; climate information, such as average temperature and average number of freeze/thaw cycles per year; physiographic information, such as slope angles or slope stabilities; and data for types and thickness of soils. Certainly other types of data might also be included to suit a particular need. The sheer abundance of possibly relevant data types argues for computer-based presentation platforms wherein the data could be displayed to be customised to specific uses.

The final issue to be raised concerning the Colorado map is that of scale. The map is published at a scale of 1:750,000. This scale is appropriate for the type and specificity of information displayed and permits display of the map on a conveniently handled piece of paper- approximately 1 meter by 1.2 meter. It is inevitable, however, that the end-users of the map will direct their scrutiny to a specific part of the map (for example, "where is my house?") and attempt to extract more information than can be reasonably accomplished. The printed map format guards against this type of abuse because such detailed information is not forthcoming. At the same time, although it leaves the end user with a general regional knowledge and perspective, the lack of specific information may be construed by some as a drawback. The ideal product may be one that is available in a digital format with scale-appropriate layers of information. More discussion of scale issues will be found later in this paper.

3.3. CONSTRUCTION OF GEOENVIRONMENTAL MODELS

Ongoing efforts at the USGS are concentrated mainly on refining descriptive geoenvironmental models for specific deposits (e.g., Seal and Wandless, 1997). More work is needed to characterise deposit types in greater detail and to add additional deposit types. For example, placer gold deposits are mentioned only as a highly eroded extension of low-sulphide Au-quartz vein deposits (Goldfarb *et al.*, 1995). Little or no mention is made of certain deposit types, for example, laterites. In addition to adding to the environmental databases for specific deposit types, new research at USGS is directed at watershed-based approaches, which may include a number of ore deposit types.

The watershed approach may be more reasonable, especially because the spatial extent of environmental signatures commonly is greater than the regional extent of the mineralised and surrounding altered areas (Plumlee and Nash, 1995). Further, in watersheds that contain more than one deposit type, the suite of elements and their concentrations observed in surface or ground water may be a composite of the environmental signatures of several deposits. In this case, the local hydrology will exert a fundamental control on the observed geochemical variations, and a complete understanding of the geochemistry of the area will be unattainable without an understanding of the local and regional hydrology, as shown in Figure 8. An important characteristic of the hydrology, which is often overlooked, is the possible interactions between ground water and surface water (Winter and Woo, 1990; Winter, 1995; Winter *et al.*, 1998). Collection of a few stream samples in a watershed may not be sufficient if the individual environmental signatures are to be resolved. There may be reaches of streams where water is percolating into the ground from the stream (a "losing reach"), or there may be reaches where ground water is discharging into the streambed (a "gaining reach"). Detailed surveys of streams can be performed to determine the gaining and losing reaches by using a simple probe-manometer system (Figure 9).

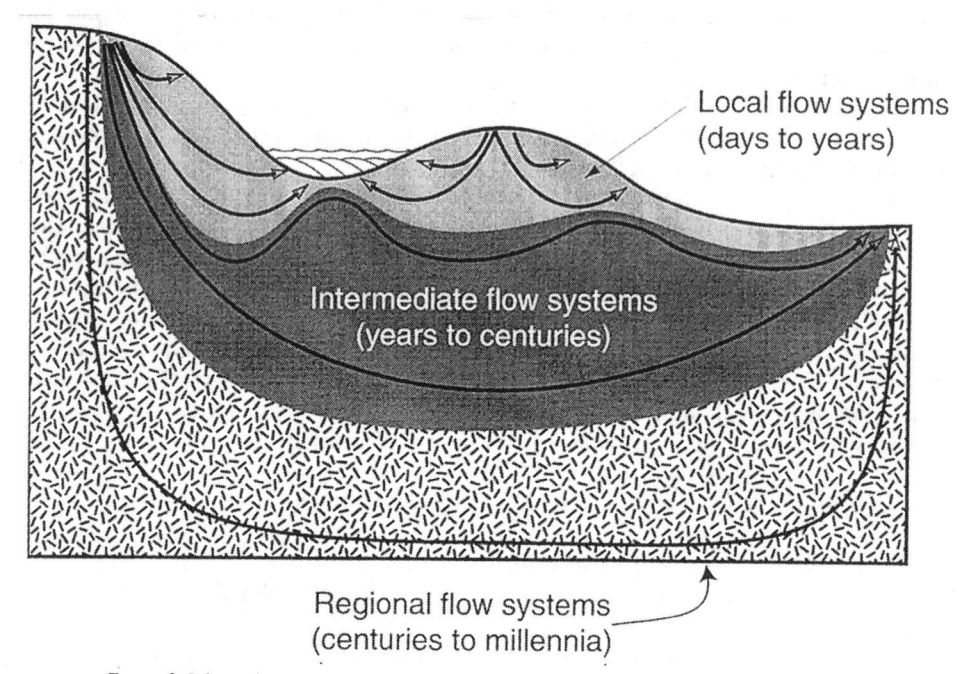

Transition From Local to Regional Flow Patterns

Figure 8. Schematic representation of local and regional flow regimes (from winter et al., 1998).

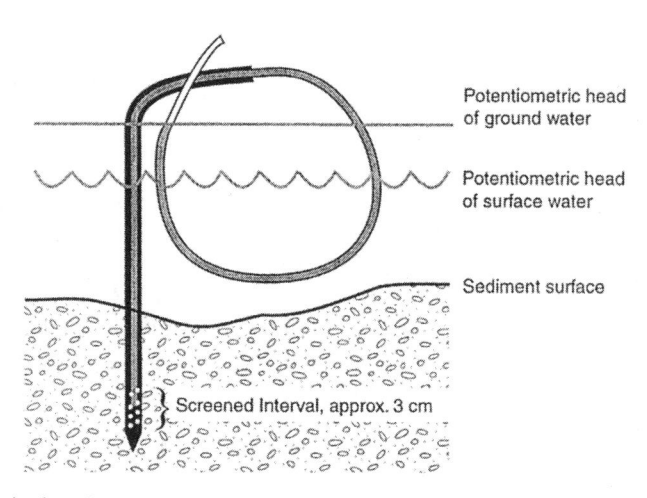

Figure 9. A simple probe manometer for measuring the hydraulic head of shallow subsurface water.

3.3.1. Ingredients and Data Layers

Various types of data may be useful components of geoenvironmental models; data layers should be chosen carefully to portray most effectively and efficiently the environmental condition of the area of study while avoiding the clutter and confusion that arises from displaying too much data. Some types of data layers that are described below provide a rather broad menu. These data may be available from numerous sources, including private companies, national, state, and local geological surveys or institutions, university researchers (especially graduate theses), and other government or private agencies.

Ecosystems and Ecoregions as Fundamental Properties of Geoenvironmental Models.
Ecosystems or ecoregions, as defined in Section 1.1.8, are a logical framework within which geoenvironmental models can be constructed. Inasmuch as the ecoregion embodies information about climate, such as temperature and nature of precipitation (amount and temporal variability), it implicitly encompasses information about the weathering environment. In turn, the nature of the weathering environment determines the environmental signature of the mineralised and altered zones. Because the spatial definition of a particular ecoregion is based on such variables as climate, latitude, altitude, and geologic framework, these four properties remain as fundamental variables in the geologic ecosystem; that is, the weathering behaviour of the rocks and their influence on water quality and on soils. Just as ecosystems have an inherent uncertainty as to scale, so do ecoregions. In general, the larger the region displayed on a map, the greater the diversity of sites (see definition for ecoregion and ecosystem in Section 1.1.8) that will be incorporated into each ecoregion. It follows, then, that larger mapped areas will be less specific with respect to weathering environments. When developing geoenvironmental models, a balance must be struck between the desire to cover large areas to broaden the utility of the model to more users while maintaining ecoregion specificity so that different weathering behaviours can be explained. The contrasting information presented in ecoregion maps of different scale is shown in Figures 10-12. Figure 10 is a map prepared only for the State of Colorado (Fleischer-Mutel and Emerick, 1984). Figure 11 is the State of Colorado excerpted from a map of the United States (Bailey, 1995). Figure 12 shows the State of Colorado (approximately) excerpted from an ecoregion map of the world (Bailey, 1996).

22

Ecoregions of Colorado

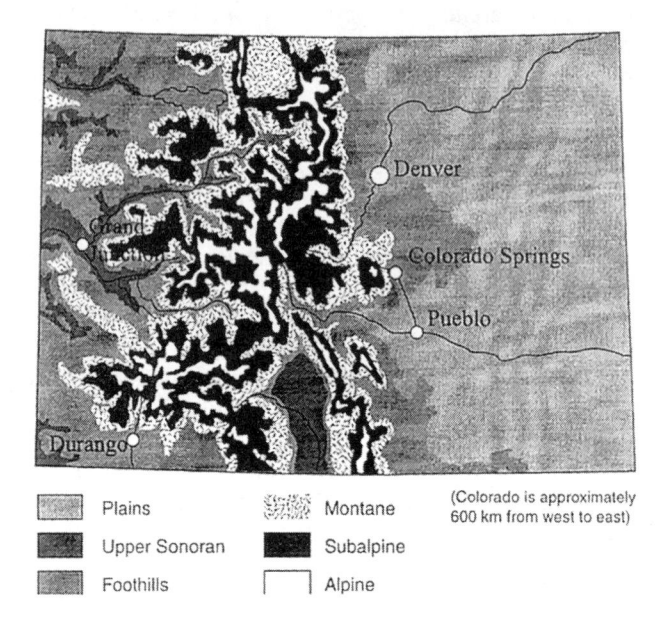

Figure 10. - Ecoregions of Colorado, USA, at the state scale (state is approx. 600 km from east to west; modified from Fleischer-Mutel and Emerick, 1984).

Ecoregions of Colorado from Map of U.S.

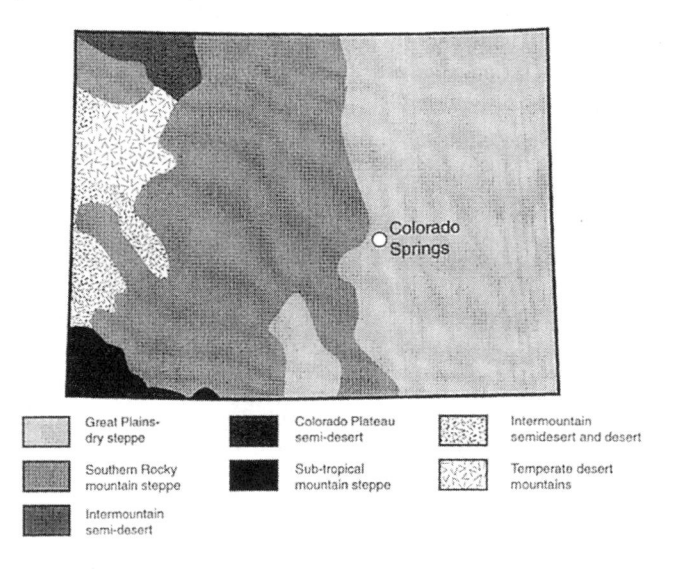

Figure 11. Ecoregions of Colorado, USA, excerpted from a map of the continental US (from Bailey, 1995).

Ecoregions of Colorado from Map of World

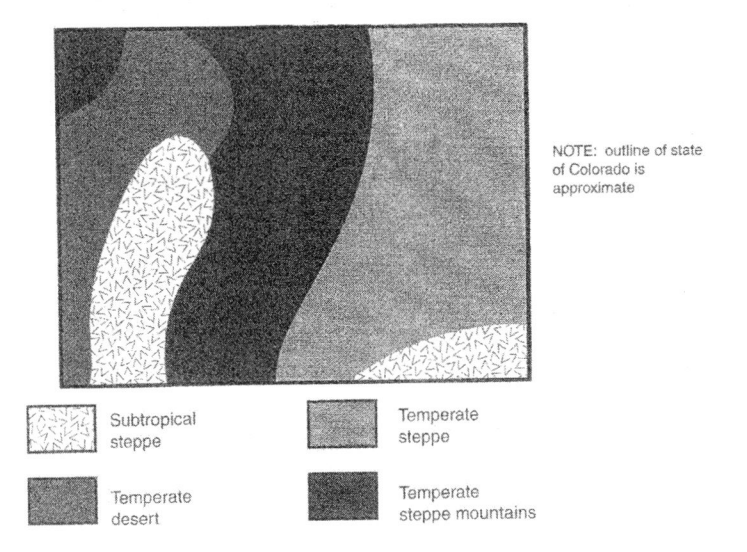

Figure 12. Ecoregions of Colorado, USA, excerpted from an ecoregion map of the world (from Bailey, 1995).

The definition of ecosystems and ecoregions is also based, in part, on the character and robustness of vegetation. In mineralised areas, vegetation may be impacted through effects on growth (stunting), density, or diversity. For example, Gough (1986) noted that in the outcroppings of mineralised or altered areas, certain species may be absent from the usual floral assemblage because they are less tolerant of high metals concentrations in soil or water. The absence of these more-fragile species may be a more commonly observed phenomenon than the presence of unique "indicator species," which are thought to signal the presence of mineralised areas. Either way, the effects of mineralised areas on vegetation patterns will be specific to a given ecoregion where a particular floral assemblage is expected.

Sources of ecoregion data range from global to local. Global perspectives on ecoregion divisions can be found in Bailey (1996, 1998). National ecoregion divisions depend on the country. For the United States, Omernik (1987) and Bailey (1995, 1998) are prime sources of mapped data. Ecoregion maps for other countries, or for smaller regions than an entire country (cf. Gallant et al., 1995, Ravichandran et al., 1996) may be available from locally published sources.

Geographic Data. Although not specifically part of geoenvironmental models, a complete base of geographic data (land and water features; anthropogenic features, such as roads and towns, topographic data; etc.) provides a political and cultural framework for the presentation of geoenvironmental data in display formats. Map- or GIS-based displays of models would be nearly useless without base geographic data, which helps the user locate the area shown, and to recognise important geographic and physiographic features in the display area. Addition of geographic data also provides an implicit sense of scale, so the user should be able to recognise the scale-dependent utility of the modelled area. In computerised interactive displays, the geographic data allows the user to recognise the need to zoom in or out to display the most appropriate area for the intended application.

Geologic Elements of Geoenvironmental Models. As considered here, geologic elements represent the broadest and most fundamental contribution to the geoenvironmental models. Table 1 lists some of the possible geologic data layers that may be used. This list is not intended to be exhaustively complete, but rather to highlight some ideas. Some of these data layers may be readily available, some may be difficult of impossible to acquire, but constitute a "wish list" of ingredients. Some of the data layers mentioned are known to be useful to geoenvironmental model development, others may be less useful, but further research is required to be sure.

Table 1. Relevant geologic components of a geoenvironmental model.

Data type	Description and Use
Regional and local geochemical data	Includes geochemical data for surface and ground waters, springs and seeps, soils, stream sediments, rocks, and plants and other biota. The chemistry of these media constitutes much of the reason for conducting environmental investigations, owing to a known, perceived, or anticipated risk of environmental degradation. Care should be taken to use and display data consistently. No data should be used without a detailed understanding of its source, accuracy, completeness, methods of collection and analysis, etc.
Regional and local lithologic variations	Focuses on lithology and chemical composition rather than stratigraphy and includes groupings of geologic formations according to lithologic and mineralogic properties. The "lithologic map" probably will be the most useful tool to explain regional geochemical variations in rocks and waters.
Geologic mapping vs. alteration / mineralisation mapping	Traditional geologic mapping groups units according to a time/stratigraphic sequence. In contrast, alteration mapping, like lithologic mapping just described, distinguishes units on the basis of their alteration mineralogy and textures. For mineralised/altered areas, alteration mapping will likely be the best tool for explaining local geochemical variations in rocks, soils, and waters.
Local geologic structure (near-field stress regime)	Fluid flow is coupled with deformation and is channelled along fractures and faults than through rock matrices. Important to modelling are all fracture sets, intensity of fracturing, brecciation (gouge) and any indicators of sense-of-fracture motion that allow the delineation of the areal fault systematics and potential hydraulic conductivity. Special note should be made of fractures that contain veins and/or sulphide minerals.
Regional geologic structure (far-field stress regime)	The regional structural setting- ancient and present-day tectonic setting, history of rock deformations, systematics of faults- provides the framework within which ground water flow may be modelled and the interaction of near- and far-field stress regimes that control fracture hydraulic conductivity understood.
Regional characteristics, such as heat flow and production	Fluid flow is coupled with heat transport. Fluid production and the rate of flow will be higher in regions with elevated geothermal gradients. For example, hot-spring areas, high-heat flow granites, etc. In high-heat flow areas, in the long run even low-level fluxes of fluids and volatiles may accumulate potentially toxic metals in surficial geologic materials.

Table 1. Continued

Data type	Description and Use
Hydrology of surface and ground water	Surface-water hydrology is reasonably well described by map information, such as channel locations, drainage divides, and zones of perennial versus ephemeral flow. Other physical data can be added, such as average discharge or stream velocities. For ground water, important properties include flow directions and quantities, hydraulic conductivity of various aquifers, average depth-of-water table, etc. As mentioned above, the interaction between ground and surface waters is an important parameter, which should be given greater consideration than it has.
Pore vs. fracture-dominated aquifers	This is an important consideration for regional, as well as local flow. In fracture-dominated systems, flow is concentrated along dilated fracture or joint planes, which means that the aquifer will be heterogeneous and anisotropic. In this context, the only important fractures are those that are sufficiently dilated to conduct water. In porous systems, flow will be distributed throughout the entire rock matrix, the homogeneity and isotropy of which may still vary.
Hydrology at time of mineralisation versus present-day hydrology	The hydrology at the time of mineralisation is governed by the usual physical rules of hydrology for the concurrent conditions. These conditions include the local and regional tectonic regime, as well as development of local and regional structures, which may be hydraulically conducting. To the extent that these conditions have changed since the time of mineralisation, the present-day hydrology may be different from that at the time of mineralisation. This is an important consideration if the shapes of alteration zones are to be understood in terms of their impact on present-day hydrology and water quality.
Spatial and temporal paragenesis of alteration and mineralisation	Paragenetic data (the order in which minerals formed) is acquired through detailed petrographic investigations of samples collected from throughout a mineralised and altered system. It is important in that the paragenesis of minerals may well determine which are exposed to pore openings and which are shielded. Shielding of minerals may prevent them from contacting ground or surface waters; thus shielded, minerals may have little or no impact on water quality.
Microscopic lithologic and geochemical variations	Information in this category includes mineral zonation, episodic incorporation of trace elements into other minerals (e.g., As in pyrite), metasomatism, overgrowths, etc. These properties may be important to the extent that they affect the reactivity of minerals (e.g., trace-element incorporation into a mineral may render it less stable) and, in turn, affect quality of ground or surface water. These data may not be readily available and can only be acquired through time-consuming investigations.
Microscopic structure and water-rock contact	This is an important property, but one that is difficult to quantify. In terms of water-rock interactions, the most important minerals are those that are reactive and in contact with the water. Unless detailed microscopic investigations are made, this property cannot be determined. As with the previous data layer, it is extremely tedious and time-consuming to conduct such investigations.
Postore geologic modifications at all scales	Postore geologic modifications include more-recent alteration events, weathering and the formation of gossans, effects of recent seismicity, regional uplift, erosional stripping, etc. In the context of geoenvironmental models, any of these postore processes that affect the environmental signature of a mineralised/altered area are germane.
Geologic engineering data	Engineering data includes information on slope stability, soil strengths or permeabilities, rock strength, etc., and may be useful in certain areas where one of these properties is an issue. For example, unstable slopes might contribute to increased rates of chemical weathering, which follows such mechanical processes as slumps or landslides.

Geophysical Elements of Geoenvironmental Models. Geophysical data may be extremely useful to the development of geoenvironmental models. Some advantages of various geophysical methods are that they may cover broad areas with relative ease and speed, that they may noninvasively determine subsurface structures or locations where chemical processes are active, and that some methods can be used to determine hydraulic pathways or to determine the shapes of subsurface contaminant plumes. If properly and rigorously tested and calibrated with ground-truth data, then geophysical methods may gain importance for the collection of geoenvironmental data. Application of any of the methods by either noninvasive (for example, airborne or satellite) or invasive (for example borehole or rigid penetrators) approaches depends on the complexity of the problem to be addressed. The costs of data acquisition will be related to the complexity and scale of the data acquisition and extent of interpretation. For a more complete discussion see Watson *et al.* (1998, in press, and included references). Table 2 outlines types of geophysical data and a brief description of their utility to address environmental problems.

Table 2. Relevant geophysical data types that may be useful components of a geoenvironmental model.

Data type	Description and Use
Electrical methods	Electrical and electromagnetic methods can be used to map subsurface variability in electrical properties caused by changes in lithology, temperature, and fluid content. Silicate minerals are insulators under near-surface temperature and conditions. Consequently, electric current flow in geologic materials is confined to fluids or clays in the pore space. A variety of techniques can be used for mapping and are characterised by: 1) the type of source and sensor (direct current or resistivity, electromagnetic induction, plane wave methods such as magnetotellurics and very-low-frequency, or VLF, resistivity, and radar) and 2) the measurement platform (ground, airborne, or borehole).
Seismic methods	Seismic methods are commonly subdivided into reflection, refraction, and direct wave categories on the basis of the path that the seismic waves take from the source to the receiver. Seismic methods are sensitive to variations in the elastic properties and density of the subsurface. On the basis of the elastic properties, seismic methods can be used to map geologic, hydrologic, and physical interfaces that can denote weathering layers, the water table, and stratigraphic or structural discontinuities in rock units. Shallow seismic methods can help define ground-water pathways, which includes faults, fractures, and stratigraphic boundaries. The geometry of the seismic experiment (ground or borehole) can be designed to provide the structural information required.
Remote sensing methods	Remote sensing methods can be used to characterise and map the Earth's surface through reflected or emitted electromagnetic energy. The data can be obtained from ground, airborne, or satellite platforms. The spatial and spectral resolution of the data is a function of the sensor and platform. Low spectral and spatial resolution data can be used to group minerals in broad classes and to depict regional distribution patterns that are based on ratios of a few spectral bands. High spectral and spatial data can be used to identify and map specific mineral species on the basis of multiple, continuous spectral bands at scales of a few tens of meters. These data can be used to map surface mineralogy, lithology, vegetation, temperature, and structural features.
Potential field methods	Gravity and magnetic methods are based on a common mathematical foundation but are sensitive to different physical rock properties. The gravity method provides information on the density of Earth's materials. Magnetic methods are sensitive to the presence of magnetic minerals that are commonly trace constituents in rocks and sediments. Variations in the distribution of density or magnetisation can provide three-dimensional information on the presence, distribution, and structure of geologic environments and man-made objects. These data can be used to construct two- or three-dimensional models. Thus, for example, providing an understanding of pathways for subsurface fluid movement and mapping of the sediment/basement interface.
Thermal methods	Thermal methods detect the production and movement of heat in the Earth. Accurate measurements of conductive heat flux require precise measurements of subsurface temperatures and thermal conductivity. Temperature determinations commonly are obtained by borehole or fixed-probe measurements. Aircraft or satellite measurements can determine surface temperature. Temperature is an important variable in a variety of geochemical processes. Subsurface temperature is a function of the type of available heat source and the mode and paths of heat flow.

3.3.2. Construction Approach (Map Format, GIS, etc.)

Specifics of construction approaches vary and, as has already been stated, may be determined according to the intended use. Table 3 summarises some of the possible approaches and gives some advantages and disadvantages of each.

Table 3. Advantages and disadvantages of various presentation formats for geoenvironmental models.

Format	Advantages	Disadvantages
Printed map	Portable, intuitively easy to understand, easy to use, requires no special hardware or software to use.	Inflexible, applies only to the area covered. The data cannot be queried as can a GIS database. Could be confusing or difficult to read. Information cannot be easily added, except through registered overlays or sketches on the map itself.
Descriptive (narrative) model	Easily understood, easiest to construct, can incorporate examples from all over the world, in different climatic zones, etc.	May not be sufficiently quantitative. Would be extremely time-consuming to construct descriptive models for all deposit types through all climatic regions.
Computerised GIS	The best solution for presenting a variety of scales- data layers can be displayed or not, depending on the current scale of display. Most flexible in terms of adding new data or modifying existing data.	For large databases, may be difficult to run on some computer systems, may be platform-specific (i.e., requires a particular brand of computer), may be easiest to produce inappropriate results through the misapplication of scale-dependent data. Cost of acquiring equipment and software to run GIS applications may be prohibitive.

3.3.3. Data Quality

As with any other exercise in which data is being assembled from numerous sources, care should be taken to ascertain that the data is of good quality. Data sources should be reliable, and data should meet the required thresholds for accuracy, precision, and completeness. The old adage, "garbage in equals garbage out" should be borne in mind. Problems may arise especially when combining data from different sources. As a test of the reliability of the data, overlapping areas of coverage should be examined closely to see that they agree. If not, then one or both data sources may be suspect and should be discarded from the model.

3.3.4. Scale Issues

Up to this point, the issue of scale has been raised several times in the course of discussion. The sequence of ecoregion maps shown in Section 3.3.1 dramatically demonstrates the variable quality of data, depending on scale. Geoenvironmental models may be for a region, for a specific deposit type, or for a specific element- all depending on the intended application. The scale of the model dictates, to some extent, the scale of the data that must be collected or assembled from existing sources. The scale of a data set is a fundamental property, one that is commonly overlooked. When presented with a map, regardless of the size of the area shown, the user of the map has an irresistible urge to find a particular point on the map, perhaps the site of his or her house. In the case of a computerised display where multiple zoom-in levels are possible, the user will attempt to zoom into that point on the map. In so doing, the scale being displayed may be inappropriate for the scale at which the data was gathered;

consequently, inappropriate or incorrect conclusions may be drawn. It may be useful in some applications to conduct statistical tests of the variance of a set of data points. Analysis-of-variance methods can be used to assign variation among sampling error, analytical error, local variations, or regional variations in the data to reduce the significance of a one-point anomaly.

Whether existing data is being assembled or new data is being collected, the scale of intended application should match the scale of the data as closely as possible. Gillman and Hails (1997, p. 6) noted that the scale of observation contributes directly to the uncertainty of the model, "so an unpredictable (and effectively random) event at one time scale may be predictable (and therefore deterministic) at another time scale. The same effect can be expected of analyses at different spatial scales." As an example we can look at the problem of forecasting the weather. Predictions may be made as to whether it will rain on a particular day next week, and the uncertainty of that prediction might be quite large. If we make a prediction on the time scale of a month or a year, however, we can use averaged data, which has been gathered over many years, to say with greater certainty that it might rain a certain amount this month. So, at the cost of specificity, we gain a measure of certainty.

The final *caveat* regarding scale pertains to the combination of data gathered at multiple scales. Care should be taken in such an exercise that the data are reasonably representative of the area displayed and that not too much importance is placed on one or two data points that happen to lie within the area displayed. To help prevent this problem, other data sources may be consulted, or other data types may be used. For example, a geologic map layer may help determine whether a particular geochemical sample point might be representative of a larger area.

3.4. TESTING GEOENVIRONMENTAL MODELS

Once a geoenvironmental model has been constructed for a chosen region, it would be useful to know some aspects of the reliability of the model. In addition to issues already mentioned (spatial scale, data types included, data reliability, etc.), certain tests may be devised to evaluate the overall reliability of the model. Perhaps the easiest way is to test the ability of the model to "predict" a known case. This can be done by arbitrarily deleting a particular mineralised area from the database for the region of interest and then reinserting it with no data. The model should be able to predict the effects of that deposit, and the prediction can be measured against the actual data. To the extent that these two sets differ, the model is deemed accurate or not.

A second test of the geoenvironmental model involves the temporal scale. Mineralised areas may have a regional environmental effect that can be measured, as well as a temporal effect. The temporal extent of the environmental effect is determined by the weathering rate, the rate of mechanical versus chemical weathering, erosion rates, hydrology, etc. Perhaps the best way to test the temporal effect is to examine the environmental behaviour of deposits whose environmental effects can be dated. For example, the dating of ferricrete deposits by Furniss and Hinman (1998) was already mentioned as an indicator of high natural background metals concentrations (sufficiently high to cause precipitation of the ferricrete itself). The same data also can be used to indicate the longevity of a measurable environmental effect- the 9,000-year history of ferricrete deposition also indicates that the environmental effect of that mineralised zone lasts on the order of 10^4 years or so. Nordstrom and Alpers (U.S

Geological Survey, 1999, pers. comm.) have dated a ferricrete deposit in California by finding a magnetic reversal that can be attributed to a polar magnetic reversal. This data set indicates that ferricrete deposition with a longevity approaching the order of 10^6 years has occurred at that site. It certainly will not be possible to test the longevity of anthropogenic disturbances on that time scale. Indeed, in the United States, the earliest mines date from the 1600's AD. In Europe and Asia, much older mining operations are known, dating back several thousand years. With these old (>1000 years) mining operations, it may be possible to test geoenvironmental models with detailed examinations of the longevity of geochemical perturbations attributable to mining.

A final test that may be applied to geoenvironmental models entails detailed environmental examinations of an area prior to and following the development of mines in an area. Few areas exist in the world today where relatively pristine conditions exist and new mines are being developed. If such areas are studied, then the geoenvironmental model might be used prior to the beginning of mine development to predict the environmental effects of mining and follow-up studies could demonstrate the veracity of the predictions.

4. Conclusions

This paper presents an overview of geoenvironmental models, their construction, and their use. As can probably be determined from reading this paper, the definition and form of geoenvironmental models is, at this writing, somewhat nebulous. However, the practical need for geoenvironmental models is quite strong. Regulatory agencies, mining companies, and citizen groups could all use reliable geoenvironmental models to ensure that the beneficial value of global mineral resources is realised while minimising local, regional, and global environmental impacts. These user groups may be at odds with one another politically, economically, or socially. Nevertheless, development of geoenvironmental models should be driven by the demonstrated need for reliable objective scientific information.

Geoenvironmental models may come in a variety of shapes and sizes, but all share a few properties. The models are based on a synthesis of a variety of data types. Thus, the individual components of the model cannot be relied on too heavily to make interpretations. As of this writing, insufficient data are available to construct complete geoenvironmental models for all deposit types in all environments. This presents two challenges to the geoscience community: first, we must use the available data most cleverly to gain maximum value from it; and second, we must work to gather new data to help complete our modelling efforts.

The papers presented at this NATO Advanced Study Institute contribute to the building blocks of the art of geoenvironmental model development. As this is the first worldwide conference devoted to the subject, many of these studies are likely to shape the future of the development and application of geoenvironmental models. To some extent, technological developments, such as new computer technologies or new geochemical or geophysical methods, might shape the future of geoenvironmental models, but for the most part, we are challenged to use existing data and methods as ingeniously as possible.

Cited References and Recommended Reading:

1. Anbeek, C. (1992) Surface roughness of minerals and implications for dissolution studies, *Geochimica et Cosmochimica Acta*, **56**, 1461-1469.
2. Bailey, R.G. (1995) *Descriptions of Ecoregions of the United States: U.S. Forest Service Miscellaneous Publication 1391*, U.S. Government Printing Office, Washington, D.C.
3. Bailey, R.G. (1996) *Ecosystem Geography*, Springer, New York.
4. Bailey, R.G. (1998) *Ecoregions map of North America: U. S. Department of Agriculture Miscellaneous Publication 1548*, U.S. Government Printing Office, Washington, D.C.
5. Bassett, R.L., Miller, W.R., McHugh, J.B., and Catts, J.G. (1992) Simulation of natural acid sulfate weathering in an alpine watershed, *Water Resources Research*, **28**, 2197-2209.
6. Berry, B.J.L., Conkling, E.C., and Ray, D.M. (1987) *Economic Geography*, Prentice-Hall, Inc., Englewood Cliffs, New Jersey.
7. Bliss, J.D. (1992) *Developments in Mineral Deposit Modelling: U.S. Geological Survey Bulletin 2004*, U.S. Government Printing Office, Washington, D.C.
8. Boutron, C.F., Candelone, J.-P., and Hong, S. (1994) Past and recent changes in the large-scale tropospheric cycles of lead and other heavy metals as documented in Antarctic and Greenland snow and ice; a review, *Geochimica et Cosmochimica Acta*, **58**, 3217-3225.
9. Cox, D.P., and Singer, D.A. (1986) *Mineral deposit models: U.S. Geological Survey Bulletin 1693*, U.S. Government Printing Office, Washington, D.C.
10. Drever, J.I. (1982) *The Geochemistry of Natural Waters*, Prentice-Hall, Inc., Englewood Cliffs, New Jersey.
11. Drever, J.I., and Clow, D.W. (1995) Weathering rates in catchments, in A.F. White and S.L. Brantley (eds.), *Chemical Weathering Rates of Silicate Minerals, Mineralogical Society of America Short Course Notes, Volume 31*, Mineralogical Society of America, Washington, DC, pp. 463-483.
12. duBray, E.A. (1995) *Preliminary compilation of descriptive geoenvironmental mineral deposit models: U.S. Geological Survey Open-File Report 95-831*, U.S. Government Printing Office, Washington, D.C.
13. Filipek, L.H., and Plumlee, G.S., eds. (1999B) The Environmental Geochemistry of Mineral Deposits, *Reviews in Economic Geology, Volume 6B, Case Studies and Research Topics.* Economic Geology Publishing Company, Littleton, Colorado, USA, 210 pp.
14. Fleischer-Mutel, C., and Emerick, J.C. (1984) *From Grassland to Glacier: The Natural History of Colorado*, Johnson Books, Boulder, Colorado.
15. Furniss, G., and Hinman, N.W. (1998) Ferricrete provides record of natural acid drainage, New World district, Montana, in G.B. Arehart and J.R. Hulston (eds.), *Water-Rock Interaction*, A.A. Balkema, Rotterdam, pp. 973-976.
16. Furniss, G. (1999) Distinguishing acid mine drainage from acid rock drainage using trace elements in ferricrete, *Geological Society of America Abstracts with Programs*, **31(7)**, A-342.
17. Gallant, A.L., Binnian, E.F., Omernik, J.M., and Shasby, M.B. (1995) *Ecoregions of Alaska: U. S. Geological Survey Professional Paper 1567*, U.S. Government Printing Office, Washington, D.C.
18. Garrels, R.M., and Mackenzie, F.T. (1971) *Evolution of Sedimentary Rocks*, W.W. Norton and Company, New York.
19. Gillman, M., and Hail, R. (1997) *An Introduction to Ecological Modelling: Putting Practice Into Theory*, Blackwell Sciences, Ltd., Oxford.
20. Goldfarb, R.J., Berger, B.R., Klein, T.L., Pickthorn, W.J., and Klein, D.P. (1995) Low sulphide Au quartz veins, in E.A. duBray (ed.), *Preliminary compilation of descriptive geoenvironmental mineral deposit models: U.S. Geological Survey Open-File Report 95-831*, U.S. Government Printing Office, Washington, DC.
21. Gough, L.P. (1986) Plant factors influencing the use of plant analysis as a tool for biogeochemical exploration, in D. Carlisle, W.L. Berry, I.R. Kaplan, and J.R. Watterson, (eds.), *Mineral Exploration: Biological Systems and Organic Matter*, Prentice-Hall, Englewood Cliffs, New Jersey, pp. 33-46.
22. Hong, S., Candelone, J.-P., Patterson, C.C., and Boutron, C.F. (1994) Greenland ice evidence of hemispheric lead pollution two millennia ago by Greek and Roman civilizations, *Science*, **265**, 1841-1843.
23. Huggett, R.J. (1995) *Geoecology: An Evolutionary Approach*, Routledge, New York.
24. Keller, E.A. (1996) *Environmental Geology, Seventh Edition*, Prentice Hall, Upper Saddle River, New Jersey.
25. Langmuir, D. (1997) *Aqueous Environmental Geochemistry*, Prentice Hall, Englewood Cliffs, New Jersey.
26. Lasaga, A. C. (1995) Fundamental approaches in describing mineral dissolution and precipitation rates, in A.F. White and S.L. Brantley (eds.), *Chemical Weathering Rates of Silicate Minerals, Mineralogical Society of America Short Course Notes, Volume 31*, Mineralogical Society of America, Washington, DC.

27. Manahan, S.E. (1991) *Environmental Chemistry, Fifth Edition*, Lewis Publishers, Chelsea, Michigan.
28. Miller, W.R. and McHugh, J.B. (1994) *Natural acid drainage from altered areas within and adjacent to the Upper Alamosa River basin, Colorado: U.S. Geological Survey Open-File Report 94-144*, U.S. Government Printing Office, Washington, DC.
29. Miller, W.R., Bassett, R.L., McHugh, J.B., and Ficklin, W.H., (1999) The behavior of trace metals in water during natural acid sulfate weathering in an alpine watershed, in Filipek, L.H., and Plumlee, G.S., (eds.) *Reviews in Economic Geology, Volume 6B*. Economic Geology Publishing Company, Littleton, Colorado, USA, pp. 493-503.
30. Miller, W.R., and McHugh, J.B., (1999b) Calculations of geochemical baselines of stream waters in the vicinity of Summitville, Colorado, before historic underground mining and prior to open-pit mining, in Filipek, L.H., and Plumlee, G.S., (eds.) *Reviews in Economic Geology, Volume 6B*. Economic Geology Publishing Company, Littleton, Colorado, USA, pp. 505-514.
31. Mills, C. (1997) *Acid-Base Accounting*, Internet website address: http://www.enviromine.com/ard/acidbase/abadiscussion.htm.
32. Morin, K.A. (1990) Problems and proposed solutions in predicting acid drainage with acid-base accounting, *Geological Association of Canada, Mineralogical Association of Canada, Joint Meeting*, Geological Association of Canada, Vancouver, B.C, p. 91.
33. Nordstrom, D.K., Jenne, E.A., and Ball, J.W. (1979) Redox equilibria of iron in acid mine waters, in E.A. Jenne (ed.), *Chemical Modelling in Aqueous Systems, American Chemical Society Symposium Series v. 93*, American Chemical Society, Washington, D.C., 51-79.
34. Nordstrom, D.K. (1982) Aqueous pyrite oxidation and the consequent formation of secondary iron minerals, in J.A. Kittrick (ed.), *Acid Sulfate Weathering*, Soil Science Society of America Special Publication Number 10, pp. 37-56.
35. Nordstrom, D.K., Alpers, C.N., and Ball, J.W. (1991) Measurement of negative pH values and high metal concentrations in extremely acidic mine waters from Iron Mountain, California: *Geological Society of America Abstract with Programs*, **23**, A383.
36. Nordstrom, D.K., and Munoz, J.L. (1994) *Geochemical Thermodynamics, 2nd edition*, Blackwell Science, Cambridge, Massachusetts.
37. Ódor, L., Wanty, R.B., Horváth, I., Fügedi, U. (1998) Mobilization and attenuation of metals downstream from a base-metal mining site in the Mátra Mountains, northeastern Hungary: *Journal of Geochemical Exploration*, **65**, 47-60.
38. Omernik, J.M. (1987) Ecoregions of the conterminous United States, *Annals of the Association of American Geographers*, **77**, 118-125.
39. Paces, T. (1973) Steady-state kinetics and equilibrium between ground water and granitic rock, *Geochimica et Cosmochimica Acta*, **37**, 2641-2663.
40. Paces, T. (1983) Rate constants of dissolution derived from the measurements of mass balance in hydrological catchments, *Geochimica et Cosmochimica Acta*, **47**, 1855-1863.
41. Plumlee, G.S., Smith, K.S., Ficklin, W.H., and Briggs, P.H. (1992) Geological and geochemical controls on the composition of mine drainages and natural drainages in mineralized areas, in Y.K. Kharaka and A.S. Maest (eds.), *Water-Rock Interaction VII*, A.A. Balkema, Rotterdam, pp. 419-422.
42. Plumlee, G.S. (1994) Fluid chemistry evolution and mineral deposition in the main-stage Creede epithermal system, *Economic Geology*, **89**, 1860-1882.
43. Plumlee, G.S., and Whitehouse-Veaux, P.H. (1994) Mineralogy, paragenesis, and mineral zoning of the Bulldog Mountain vein system, Creede District, Colorado, *Economic Geology*, **89**, 1883-1905.
44. Plumlee, G.S., Nash, J.T. (1995) Geoenvironmental models of mineral deposits— Fundamentals and applications, in E.A. duBray (ed.), *Preliminary compilation of descriptive geoenvironmental mineral deposit models: U.S. Geological Survey Open-File Report 95-831*, U.S. Government Printing Office, Washington, DC, pp. 1-9.
45. Plumlee, G.S., Streufert, R.K., Smith, K.S., Smith, S.M., Wallace, A.R., Toth, M.I., Nash, J.T., Robinson, R., Ficklin, W.H., and Lee, G.K. (1995a) *Map showing potential metal-mine drainage hazards in Colorado, based on mineral-deposit geology: U.S. Geological Survey Open-File Report 95-26*, U.S. Government Printing Office, Washington, DC.
46. Plumlee, G.S., Smith, K.S., Gray, J.E., Hoover, D.B. (1995b) Epithermal quartz-alunite deposits, in E.A. duBray (ed.), *Preliminary compilation of descriptive geoenvironmental mineral deposit models: U.S. Geological Survey Open-File Report 95-831*, U.S. Government Printing Office, Washington, DC, pp. 162-169.
47. Plumlee, G.S., and Logsdon, M., eds. (1999) The Environmental Geochemistry of Mineral Deposits, *Reviews in Economic Geology, Volume 6A, Processes, Techniques, and Health Issues*. Economic Geology Publishing Company, Littleton, Colorado, USA, 371 pp.
48. Plumlee, G.S. (1999) The environmental geology of mineral deposits, in Plumlee, G.S., and Logsdon, M., (eds.), *The Environmental Geochemistry of Mineral Deposits, Reviews in Economic Geology, Volume 6A,*

Processes, Techniques, and Health Issues. Economic Geology Publishing Company, Littleton, Colorado, USA, pp. 71-116.
49. Ravichandran, S., Ramanibai, R., Pundarikanthan, N.V. (1996) Ecoregions for describing water quality patterns in Tamiraparani basin, South India, *Journal of Hydrology*, **178**, 257-276.
50. Seal, R.R., and Wandless, G.A. (1997) Stable isotope characteristics of waters draining massive sulfide deposits in the eastern United States, in R.B. Wanty, S.P. Marsh, and L.P. Gough (eds.), *4th International Symposium on Environmental Geochemistry Proceedings: U.S. Geological Survey Open-File Report 97-496*, U.S. Government Printing Office, Washington, D.C., p. 82.
51. Sillitoe, R.H. (1973) The tops and bottoms of porphyry copper deposits, *Economic Geology*, **68**, 799-815.
52. Wanty, R.B., Marsh, S.P., and Gough, L.P. (1997) *4th International Symposium on Environmental Geochemistry Proceedings: U.S. Geological Survey Open-File Report 97-496*, U.S. Government Printing Office, Washington, D.C.
53. Watson, K., Fitterman, D., Saltus, R.W., McCafferty, A., Swayze, G., Church, S., Smith, K.S., Goldhaber, M.B., Robson, S., McMahon, P. (1998) Application of geophysical techniques to minerals-related environmental problems: *U.S. Geological Survey Open-File Report*, U.S. Government Printing Office, Washington, D.C.
54. Winter, T.C., and Woo, M.K. (1990) Hydrology of lakes and wetlands, in M.G. Wolman and H.C. Riggs (eds.), *Surface Water Hydrology: The Geology of North America, v. O-1*, Geological Society of America, Boulder, Colorado, pp. 159-187.
55. Winter, T.C. (1995) A landscape approach to identifying environments where ground water and surface water are closely interrelated, in *Groundwater Management: Proceedings of the International Symposium of the Water Resources Engineering Division/ASCE*, American Society of Civil Engineers, San Antonio, Texas, pp. 139-144.
56. Winter, T.C., Harvey, J.W., Franke, O.L., and Alley, W.M. (1998) *Ground water and surface water: A single resource, U.S. Geological Survey Circular 1139*, U.S. Government Printing Office, Washington, DC.

Appendix 1

The following is a chapter reproduced from duBray (1995). This chapter is one of many descriptive geoenvironmental models.

EPITHERMAL QUARTZ-ALUNITE AU DEPOSITS
(MODEL 25e; Berger, 1986)

by Geoffrey S. Plumlee, Kathleen S. Smith, John E. Gray, and Donald B. Hoover

SUMMARY OF RELEVANT GEOLOGIC, GEOENVIRONMENTAL, AND GEOPHYSICAL INFORMATION
Deposit geology
Central advanced argillic zone has iron-, copper-, and arsenic-rich sulphide and sulphosalt minerals with high acid-generating capacity; deposits consist of vuggy veins and breccias in highly acid-altered volcanic rocks with very low acid-consuming capacity. Central advanced-argillic zone is flanked by argillic and distal propylitic zones with some acid-buffering capacity, decreased copper and arsenic abundances, and increased zinc and lead abundances.

Examples
Summitville and Red Mountain Pass, Colo.; Goldfield, Nev.; Paradise Peak, Nev.; Julcani, Peru.

Spatially and (or) genetically related deposit types

Associated deposit types (Cox and Singer, 1986) include porphyry copper (Model 17), porphyry gold-copper (Model 20c), and porphyry copper-molybdenum (Model 21a).

Potential environmental considerations
(1) Dominant mining activity is in sulphide-mineral-rich, strongly altered volcanic rocks with negligible acid-buffering capacity.
(2) Very high potential for the generation of acid-mine drainage (pH 1.5 to 3) that contains thousands of mg/l iron and aluminum; hundreds of mg/l copper and zinc (copper>zinc); hundreds of g/l to tens of mg/l As, Co, Ni, Cr, U, Th, rare earth elements; tens to hundreds of g/l beryllium; and anomalous abundances of bismuth, antimony, thallium, selenium, and (or) tellurium (Plumlee and others 1995a).
(3) Unoxidised sulphide minerals can persist in clay-rich alteration zones to within 10 m of ground surface. Exposure of these sulphide minerals during mining can further enhance potential for acid-drainage generation.
(4) In temperate or seasonally wet climates, soluble secondary iron, aluminum, and copper sulphate minerals dissolve during storm events and snowmelt, and lead to short term pulses of highly acidic, metal-bearing water from mine sites. The sulphate salts form by evaporation of acid mine water above the water table in open pits and underground mine workings during dry periods, even in wet climates.
(5) Potential downstream environmental effects of acid drainage can be significant in magnitude and spatial extent, especially if surrounding terrane is composed primarily of volcanic rocks with low acid-buffering capacity. Dominant downstream signatures include water having low pH, and high iron, aluminum, manganese, copper, and zinc abundances.
(6) Amalgamation-extraction of gold carried out during historic operations may be a residual source of mercury.
(7) Smelter emissions at historic sites have elevated abundances of arsenic, copper, and zinc, and possibly other elements such as beryllium and tellurium.
(8) Highly oxidised deposits and (or) deposits located in arid climates probably have lower potential for acid mine drainage and other environmental problems.
(9) Cyanide heap leach solutions are composed predominantly of copper-cyanide complexes and thiocyanate.
    Mitigation and remediation strategies for potential environmental concerns presented above are described in the section below entitled "Guidelines for mitigation and remediation."

Exploration geophysics
Resistivity studies can be used to help map features such as alteration zones. Potassium contained in alteration alunite may be identified by gamma ray spectrometry. Alteration mineral assemblages and stressed vegetation can also be identified using multispectral scanning remote sensing techniques such as AVIRIS.

References
Geology: Stoffregen (1987), Vikre (1989), Ashley (1990), John and others (1991), Deen and others (1994), and Gray and Coolbaugh (1994).
Environmental geology, geochemistry: Koyanagi and Panteleyev (1993), Gray and others (1994), Smith and others (1994), and Plumlee and others (1995a,b).

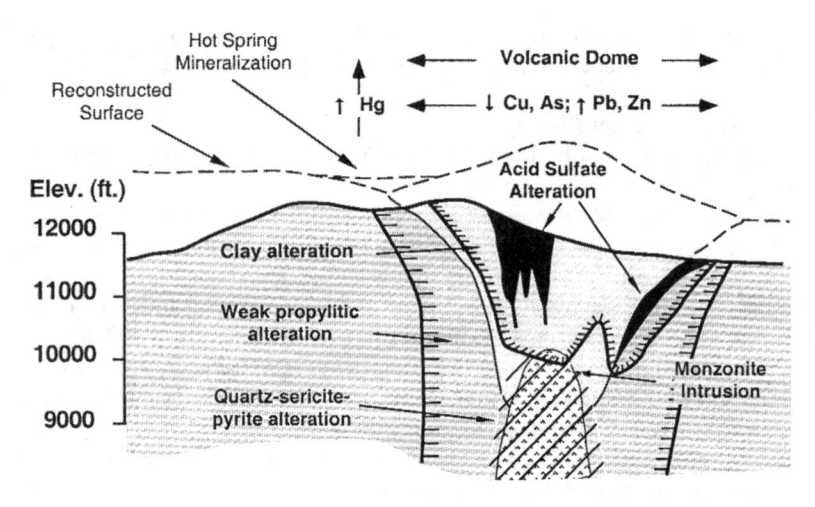

Figure A1. Simplified geologic cross section of an epithermal quartz-alunite Au deposit. Based on Summitville (Perkins and Neiman, 1982; Plumlee and others, 1995a) but modified to incorporate data from Julcani, Peru (Deen and others, 1994) and Paradise Peak, Nev., (John and others, 1991).

GEOLOGIC FACTORS THAT INFLUENCE POTENTIAL ENVIRONMENTAL EFFECTS

Deposit size
Deposit size is generally small (0.1 million tonnes) to intermediate (12 million tonnes).

Host rocks
These deposits are hosted by felsic volcanic rocks, generally intrusions or lava domes (fig. A1), that have low acid-buffering capacity. Most of these volcanic rocks are part of composite stratovolcano complexes. Some ore may be hosted in older sedimentary or crystalline rocks surrounding lava domes.

Surrounding geologic terrane
Surrounding geologic terrane is primarily volcanic but includes underlying sedimentary or crystalline rocks.

Wall-rock alteration
Wall-rock alteration reflects progressive wall-rock neutralisation of highly acidic magmatic gas condensates; alteration predates ore formation. Alteration zoning shown on Figures 1 and 2.
Intermediate-level advanced-argillic zone: Innermost vuggy silica, grading outward into quartz-alunite (± pyrophyllite), quartz-kaolinite, and montmorillonite-illite-smectite

alteration zones. Pyrite present in all zones. Phosphate minerals present primarily (?) in quartz-alunite zone.
Intermediate-level argillic zone: Montmorillonite-smectite-illite-clay minerals with pyrite.
Peripheral propylitic zone: Alteration of volcanic rocks to chlorite ± epidote ± pyrite ± calcite.
Deep phyllic zone: Quartz-sericite-pyrite.

Nature of ore
In some deposits (e.g. Summitville) disseminated sulphide minerals are focused primarily in vuggy silica, quartz-alunite, and quartz-kaolinite zones; however, significant pyrite and other sulphide minerals also are present in clay altered zones, as disseminations within breccia, and in altered wall-rock veinlets. See Figure A2 for typical sulphide mineral-sulphur content ranges for Summitville alteration zones. Although sulphide mineral-sulphur content is relatively low (<5 percent), the alteration process effectively removes nearly all of the rock's capacity to buffer acid.

Deposit trace element geochemistry
Deep: copper, ± arsenic, ± tungsten.
Intermediate inner: copper, arsenic, gold, ± tellurium.
Intermediate peripheral: copper, lead, zinc.
Shallow, near surface: Mercury, arsenic, antimony, gold, thallium.
Ore at Paradise Peak contains elevated abundances of Au, Ag, Bi, Sb, Pb, Tl, Hg, S and Ba ± Sn, Mo, Te, and Se. In addition to iron sulphide (mostly marcasite), sulphate (barite) and native sulphur were abundant throughout the deposit. Metal abundances in peripheral, argillically altered rock are essentially unchanged except that oxidised parts are enriched in iron sulphide and unoxidised rocks contain gypsum and jarosite. Mineralised rock from the deepest part of this system tended to have elevated abundances of arsenic and copper but not as elevated as those characteristic of Goldfield, Nev. or Summitville, Colo.

Ore and gangue mineralogy and zonation
Minerals listed in decreasing order of abundance. Potentially acid-generating minerals underlined. In epithermal quartz-alunite gold deposits, ore deposition usually postdates development of argillic alteration.
Deep: Pyrite, chalcopyrite, tennantite, ± wolframite-heubnerite (Julcani)
Advanced argillic, argillic alteration zones: Pyrite, enargite, covellite, chalcocite, chalcopyrite, native sulphur, marcasite, native gold, barite. Late barite, sphalerite, galena, ± siderite (Julcani), ± botryoidal pyrite.
Peripheral propylitic zone: Sphalerite, galena, ± siderite, barite, ± botryoidal pyrite
Shallow, near surface: Silica sinter, cinnabar, native mercury?, native gold, pyrite, marcasite, realgar, orpiment.

Mineral characteristics
Textures: Sulphide minerals form fine- to medium-grained (<5 mm), euhedral crystals and masses of very fine-grained, interlocking crystals. Some coarse-grained (as much as 4-5 cm) euhedral sulphide minerals also are present. Late-stage botryoidal pyrite and siderite are present in some deposits.

Trace element contents: Arsenic and antimony may be present in main-stage pyrite and (or) marcasite; late botryoidal pyrite, where present, is typically strongly enriched in arsenic, antimony, and other trace elements. Abundant stibnite-bismuthanite is the principal mineralogic site for antimony and bismuth in the Paradise Peak deposit.
General rates of weathering: Botryoidal, high trace elements > massive, fine >> coarse euhedral, low trace elements.

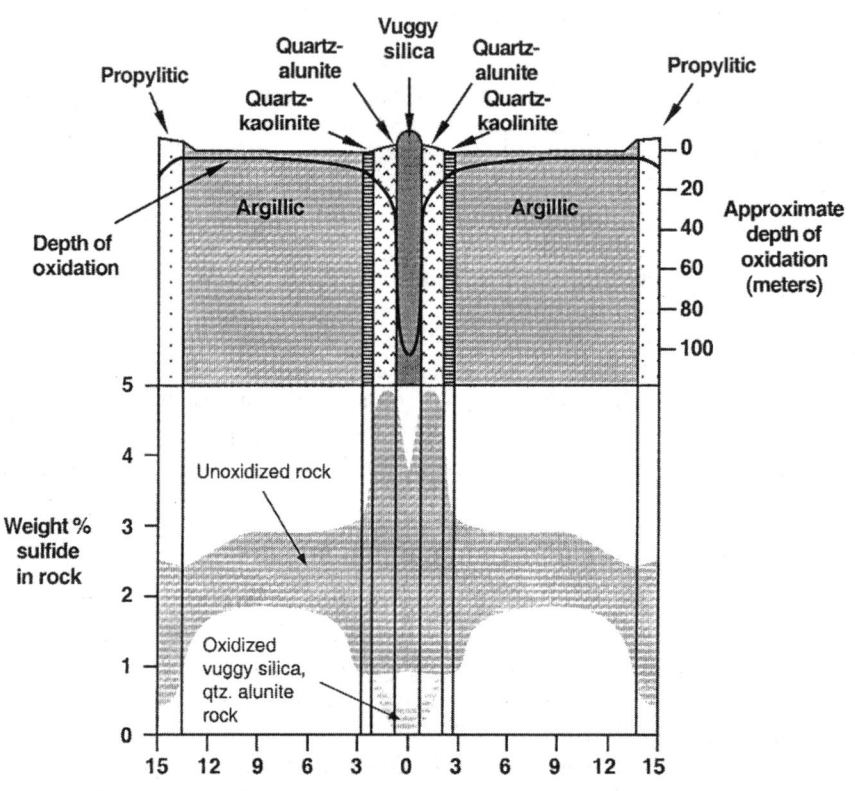

Figure A2. Schematic alteration zoning away from original fractures at Summitville, showing approximate depth of oxidation (upper plot) and range of oxidisable sulphide sulphur in sulphide minerals (lower plot). From Plumlee and others (1995a).

Secondary mineralogy
Readily soluble minerals underlined
Supergene minerals: Scorodite, goethite, limonite, K- and Na-jarosite, phosphate minerals, and plumbojarosite. Minerals formed by recent weathering: Jarosite (likely hydronium-enriched), chalcanthite, brochanthite, melanterite, alunogen, halotrichite, and phosphate minerals?. These minerals form by evaporation of acid water during dry

periods, and then redissolve during wet periods. These minerals can also form by evaporation in overbank stream sediment downstream from mine sites.

Topography, physiography
Volcanic domes generally form topographic highs. Vuggy silica zones are resistant to weathering, and form prominent knobs and pinnacles. Because vuggy silica zones are highly resistant to weathering, physical erosion immediately surrounding clay alteration zones is minimal.

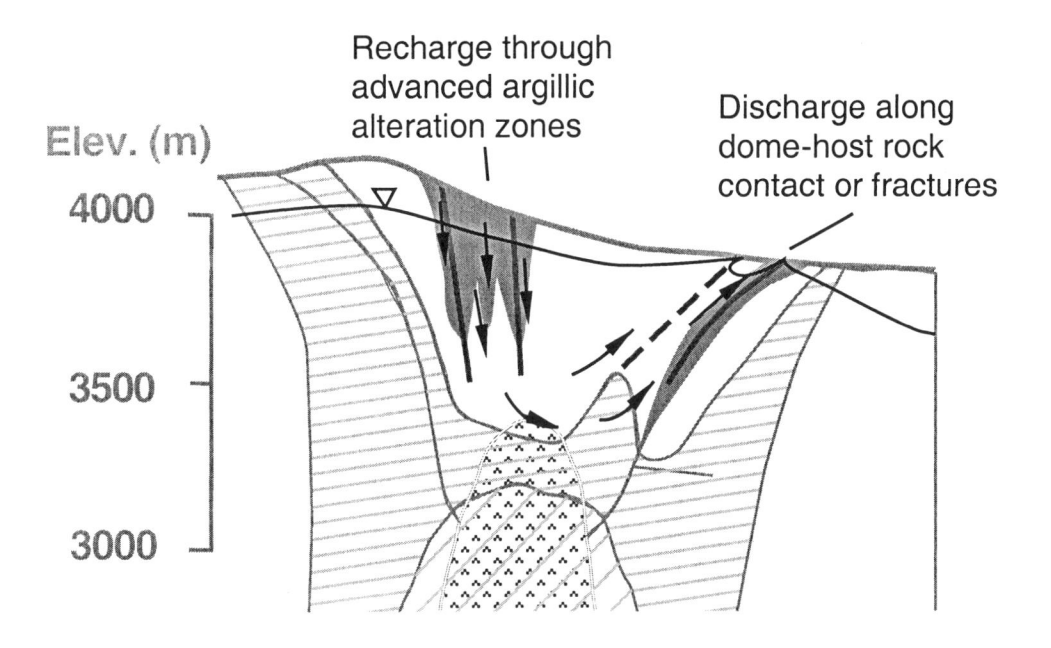

Figure A3. Inferred pre-mining hydrology at Summitville. Triangle marks position of water table.

Hydrology
Pre-mining oxidation surfaces can help identify zones of high permeability within deposits. Ferricrete deposits mark pre-mining ground water discharge points. Vuggy silica alteration zones have the highest primary permeability and therefore focus ground water flow; most are oxidised to deep levels (100 meters) by pre-mining ground water. Clay alteration zones have the lowest permeability and therefore inhibit ground water flow; most are oxidised to only shallow levels (several meters to several tens of meters) by pre-mining ground water. Post-mineralisation fractures can serve as conduits for ground water flow. Rock contacts between volcanic domes and surrounding rocks can be significant conduits for ground water flow and can also strongly influence distribution of alteration assemblages. In the vicinity of these deposits, the water table generally conforms to topography; the highest elevations are coincident with volcanic domes. At Summitville (fig. A3), ground water recharge is probably along vuggy silica zones. Ground water discharge prior to underground mining (marked by extensive ferricrete deposits) was primarily along contact between volcanic dome and host rocks,

and at scarce locations where other fractures intersected the topographic surface of the volcanic dome. In the area around the Paradise Peak deposits, the water table is locally perched, which resulted in the presence of large blocks of unoxidised rock at shallow depths.

Mining and milling methods
Historic: Underground mine workings followed vuggy silica alteration zones in most cases. Ore was processed using stamp mills and mercury amalgamation or cyanide vat leach.
Modern: Modern operations principally involve open-pit mining of vuggy silica and surrounding clay alteration zones but include some underground mining. Ore is processed primarily using cyanide heap leach techniques.

ENVIRONMENTAL SIGNATURES
Drainage signatures
Mine-drainage data (figs. A4 and A5): Summitville and Red Mountain Pass, Colo. (Plumlee and others, 1993; Plumlee and others 1995a,b). Mine water draining ore hosted by advanced argillic altered rocks is highly acidic and contains high to extreme dissolved metal abundances, including hundreds to several thousands of mg/l iron, aluminum, and manganese; hundreds of mg/l zinc and copper; and hundreds of g/l to several tens of mg/l As, Co, Ni, U, Th, Be, and REE. Water draining these deposits has elevated arsenic abundances; concentrations of uranium relative to zinc are unusual relative to those associated with many other deposit types. Preliminary data indicate that tellurium, mercury, and tungsten, though potentially enriched in advanced argillic ore, do not appear to be enriched in mine-drainage water. Limited data indicate that mine water draining argillic alteration zones has slightly higher pH and lower metal content.
Mine water draining shallow hot spring ore: No data available. The best currently available data are for water draining advanced-argillic, native sulphur-rich parts of a hot spring sulphur deposit (Leviathan, Calif.), which is acidic (pH 2-3) and has relatively low base metal contents but elevated abundances of arsenic, antimony, and thallium (Ball and Nordstrom, 1989).

Natural-drainage data: Limited data for a relatively wet climate with high dilution rates (British Columbia) suggest low acidity (pH between 2 and 3) and elevated dissolved metal abundances, including hundreds of mg/l iron and aluminum; abundances of copper and zinc, tens to hundreds of g/l, are lower than those measured in mine drainages.
Potentially economically recoverable elements: High copper abundances in drainage water could be economically extracted.

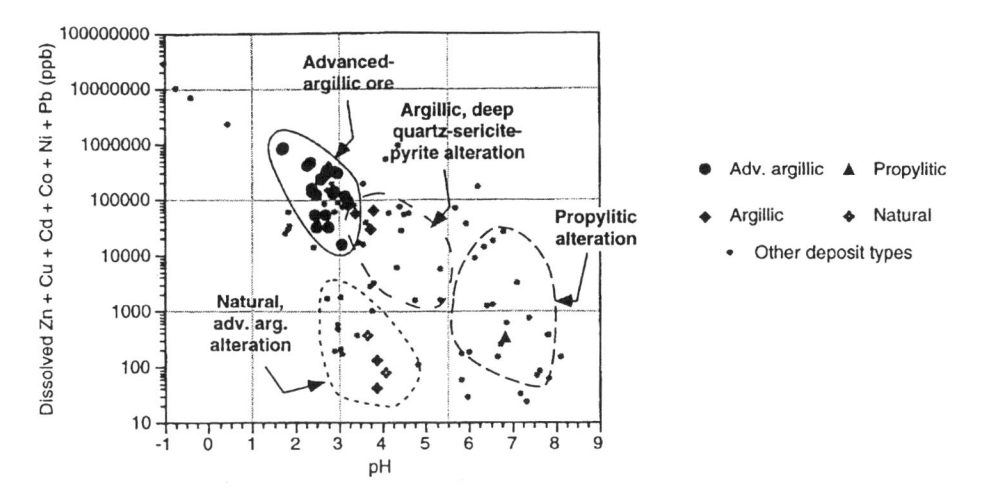

Figure A4. Plot of pH versus the sum of dissolved base metals Zn, Cu, Cd, Co, Ni, and Pb in mine and natural water draining epithermal quartz-alunite Au deposits. Lines show inferred likely ranges of metal content and pH for specific alteration zones. Samples are water draining adits and waste dumps, rain and snowmelt puddles, and seeps. Data from Koyanagi and Panteleyev (1993) and Plumlee and others (1995b).

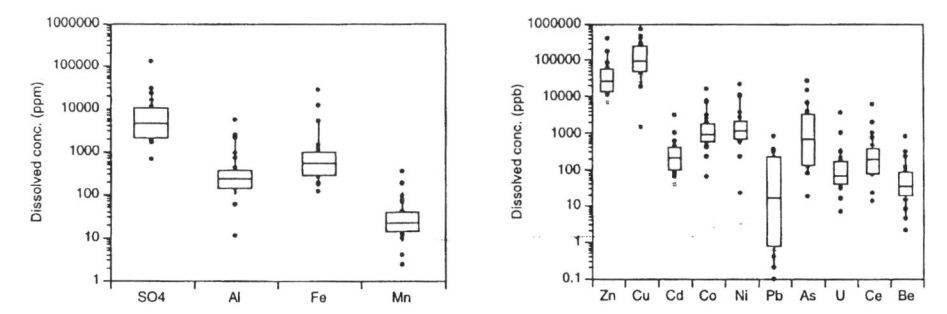

Figure A5. Box plots showing ranges of selected dissolved constituents in mine water draining ore hosted by advanced argillic alteration zones of epithermal quartz-alunite Au deposits. For each constituent, the box encloses samples falling between the 25th and 75th percentiles, the line shows the range between the 10th and 90th percentiles, and the dots show actual concentrations for samples falling outside the 25th and 75th percentiles.

Metal mobility from solid mine wastes

Metals and acid are readily liberated from sulphide-mineral-bearing mine wastes due to oxidation of sulphide minerals, mainly pyrite. During dry periods, secondary soluble salts form by evaporation. During wet periods, the soluble salts are rapidly dissolved. These salts can be present as coatings on rock material. Metal and acid are probably not liberated in significant amounts from mine wastes associated with deposits oxidised extensively prior to mining.

Storm water samples: The pH and metal contents of water in rain and snowmelt puddles, which contain dissolved soluble salts, is generally similar to that of water draining adits and waste dumps (Plumlee and others, 1995b).

Water-rock leaching: The results of a few leaching experiments with advanced argillic waste rock material from Summitville (50 g sample in 1 liter of distilled water, Plumlee and others, 1995b) show that metal concentrations and pH values of leach water rapidly (within tens of minutes) approach those of water draining adits and mine dumps.

Soil, sediment signatures prior to mining
No data currently available.

Potential environmental concerns associated with mineral processing
Mercury amalgamation of ore during historic mining may provide a source of mercury contamination not directly attributable to epithermal quartz-alunite gold deposits.
Cyanide geochemistry: Heap leach and other cyanide processing solutions are likely to include copper-cyanide complexes (containing as much as several hundred mg/l copper), with lesser zinc and silver cyanide complexes (present as weak cyanide complexes with as much as several tens of mg/l contained metals), and strong gold-cyanide complexes. Arsenic, cobalt, nickel, and iron may be present at low mg/l abundances in cyanide heap leach solutions. Thiocyanate (SCN-) abundances may be quite high in ore containing unoxidised sulphide minerals. At Summitville, degradation of cyanide accidentally released into the environment may have been enhanced by mixing with acid-mine drainage and the resulting breakdown of copper-cyanide complexes; thiocyanate likely did not degrade rapidly.

Smelter signatures
Epithermal quartz-alunite gold ore from which copper and silver were extracted during historic mining were probably smelted. No data have been identified concerning the mineralogy or chemical composition of soil affected by emissions from smelters that processed epithermal quartz-alunite gold ore. The closest analogue is Butte, Mont., where enargite-chalcocite-bornite ore from cordilleran lode deposits were smelted. There, soil proximal to smelters is very highly enriched in copper, arsenic, zinc, and lead.

Climate effects on environmental signatures
Currently available data are from moderately wet, seasonally temperate climate (Rocky Mountains). However, water draining advanced argillic ore in all climates is likely to be quite acidic and metal rich. Evaporation of acid drainage water, which leads to significant increases in metal concentrations and acidity, is important in all climates in which wet periods are interspersed with prolonged dry periods. At Summitville, evaporation has been an important process during dry Summer and Fall seasons even though total precipitation at the site exceeds 125 cm per year (Plumlee and others, 1995b). No data are available for tropical-humid, very dry, or arctic climates. Intense chemical weathering that affects exposed deposits in humid tropical climates oxidises sulphide minerals to significant depths. Unless the area is simultaneously subjected to high rates of mechanical erosion, oxidation products isolate or reduce the underlying sulphide minerals, which inhibits continued oxidation; the potential for additional acid drainage generation is thereby limited.

Potential environmental effects
Potential downstream environmental effects of acid drainage in moderately wet to moderately dry climates can be significant in magnitude and spatial extent, especially if

the surrounding geologic terrane is primarily composed of volcanic rocks with low acid buffering capacity. Predominant downstream signatures include elevated abundances of acid, iron, aluminum, manganese, and copper. Iron and aluminum form hydrous oxide precipitates as a result of dilution by downstream tributaries, and help sorb some of the dissolved metals. However, if water remains sufficiently acidic (due to limited dilution by downstream tributaries or acid generation resulting from precipitation of hydrous oxide minerals), manganese, copper, and zinc can persist (at abundances of hundreds g/l, or more) in solution well downstream from mine sites (Smith and others, 1995).

In very wet climates, dilution may significantly reduce downstream effects. In dry or seasonally wet and dry climates, off-site drainage is greatest during short-term storm events or longer-term wet periods; reactions between this water and surrounding alkaline sediment (caliche) and soil, and with alkaline water draining the sediment and soil, probably help mitigate acid drainage. Downstream storm water evaporation, however, may lead to the formation of acid- and metal-bearing salts that can themselves generate off-site acid drainage during storms (K. Stewart, unpub. data).

Guidelines for mitigation and remediation
(1) Acid drainage can be successfully remediated using lime addition and sodium-bisulphide precipitation of metals (which produce acid-generating sludge). Lime addition to iron-rich drainage water may generate sufficient suspended particulates onto which a major fraction of dissolved arsenic, lead, and copper can sorb, thereby reducing or eliminating the need for sodium-bisulphide addition; in addition, wastes are non-acid-generating.
(2) Isolation of unoxidised sulphide minerals and soluble secondary salts from oxidation and dissolution is crucial to acid drainage mitigation.
(3) Surrounding carbonate-bearing rocks, including carbonate sedimentary rocks or carbonate-bearing propylitically altered rock on the fringes of deposits, should be carefully considered for their utility in acid-mine drainage mitigation. For example, acid water could be channelled through underground fracture systems in propylitic rock to help reduce acidity.
(4) Cyanide heap-leach solutions should be treated by peroxide addition or other standard techniques. Heap-leach pads should be decommissioned by rinsing or bioremediation.
(5) Mixing cyanide heap-leach solutions with acid drainage may effectively neutralise both. Acid in drainage water breaks down copper-cyanide complexes, forming volatile free cyanide and copper-iron-cyanide particulates, which degrade photolytically. Because of the alkalinity of heap-leach solutions, iron in acidic drainage precipitates as particulates, which then effectively sorb other metals contributed from acid drainage and heap-leach solutions.

Geoenvironmental geophysics
Resistivity studies can be used to identify rocks saturated with metal-bearing ground water. Porous rocks that can focus ground water flow can be identified by microgravity studies. Heat generated by sulphide mineral oxidation may have an associated thermal anomaly measurable by borehole logging or shallow probes; measures of excess heat flux can provide an approximation of total acid generation potential. Induced

polarisation methods can provide qualitative estimates of sulphide mineral percentages and grain size. The position, volume and mineralogy of clay bodies can be identified by induced polarisation.

REFERENCES CITED

1. Ashley, R.P., 1990, The Goldfield gold district, Esmeralda and Nye Counties, Nevada, in Shawe, D.R., Ashley, R.P., and Carter, L.M.H., eds., Geology and resources of gold in the United States: U.S. Geological Survey Bulletin 1857-H, p. H1-H7.
2. Ball, J.W., and Nordstrom, D.K., 1989, Final revised analyses of major and trace elements from acid mine waters in the Leviathan Mine drainage basin, California and Nevada, October 1981 to October 1982: Water-Resources Investigations, Report No. WRI 89-4138, 46 p.
3. Berger, B.R., 1986, Descriptive model of epithermal quartz-alunite Au, in Cox, D.P., and Singer, D.A., eds., Mineral deposit models: U.S. Geological Survey Bulletin 1693, p. 158.
4. Cox, D.P., and Singer, D.A., 1986, Mineral deposit models: U.S. Geological Survey Bulletin 1693, 379 p.
5., Deen, J.A., Rye, R.O., Munoz, J.L., and Drexler, J.W., 1994, The magmatic hydrothermal system at Julcani, Peru-Evidence from fluid inclusions and hydrogen and oxygen isotopes: Economic Geology, v. 89, p. 1924-1938.
7. Gray, J.E., and Coolbaugh, M.F., 1994, Summitville, Colorado-Geologic framework and geochemistry of an epithermal acid-sulfate deposit formed in a volcanic dome: Economic Geology, v. 89, p. 1906-1923.
7. Gray, J.E., Coolbaugh, M.F., Plumlee, G.S., and Atkinson, W.W., 1994, Environmental geology of the Summitville Mine, Colorado: Economic Geology, v. 89, p. 2006-2014.
8. John, D.A., Nash, J.T., Clark, C.W., and Wulftange, W., 1991, Geology, hydrothermal alteration, and mineralization at the Paradise Peak gold-silver-mercury deposit, Nye County, Nevada, in Raines, G.L., Lisle, R.E., Schafer, R.W., and Wilkinson, W.H., eds., Geology and ore deposits of the Great Basin, Symposium proceedings: Reno, Geological Society of Nevada and U.S. Geological Survey, p. 1020-1050.
9. Koyanagi, V.M., and Panteleyev, Andre, 1993, Natural acid-drainage in the Mount McIntosh/Pemberton Hills area, northern Vancouver Island (92L/12), in Grant, B., and others, eds.: Geological fieldwork 1992; a summary of field activities and current research, Ministry of Energy, Mines and Petroleum Resources, Report No. 1993-1, p. 445-450.
10. Perkins, M., and Nieman, G.W., 1982, Epithermal gold mineralization in the South Mountain volcanic dome, Summitville, CO: Denver Region Exploration Geologists Symposium on the genesis of Rocky Mountain ore deposits: Changes with time and tectonics, Denver, Colorado, Nov. 4-5, 1982, Proceedings, p. 165-171.
11. Plumlee, G.S., Gray, J.E., Roeber, M.M., Jr., Coolbaugh, M., Flohr, M., Whitney, G., 1995a, The importance of geology in understanding and remediating environmental problems at Summitville, in Posey, H.H., Pendleton, J.A., and Van Zyl, D., eds.: Proceedings, Summitville Forum '95, Colorado Geological Survey Special Publication #38, p. 13-22.
12. Plumlee, G.S., Smith, K.S., Ficklin, W.H., Briggs, P.H., and McHugh, J.B., 1993, Empirical studies of diverse mine drainages in Colorado-Implications for the prediction of mine-drainage chemistry: Proceedings, Mined Land Reclamation Symposium, Billings, Montana, p. 176-186.
13. Plumlee, G.S., Smith, K.S., Mosier, E.L., Ficklin, W.H., Montour, M., Briggs, P.H., and Meier, A.L., 1995b, Geochemical processes controlling acid-drainage generation and cyanide degradation at Summitville, in Posey, H.H., Pendleton, J.A., and Van Zyl, D., eds.: Proceedings, Summitville Forum '95, Colorado Geological Survey Special Publication #38, p. 23-34.
14. Smith, K.S., Plumlee, G.S., and Ficklin, W.H., 1994, Predicting water contamination from metal mines and mining wastes: Notes, Workshop #2, International Land Reclamation and Mine Drainage Conference and Third International Conference on the Abatement of Acidic Drainage: U.S. Geological Survey Open-File Report 94-264, 112 p.
15. Smith, K.S., Mosier, E.L., Montour, M.R., Plumlee, G.S., Ficklin, W.H., Briggs, P.H., and Meier, A.L., 1995, Yearly and seasonal variations in acidity and metal content of irrigation waters from the Alamosa River, Colorado, in Posey, H.H., Pendleton, J.A., and Van Zyl, D., eds.: Proceedings, Summitville Forum '95, Colorado Geological Survey Special Publication #38, p. 293-298.
16. Stoffregen, R.E., 1987, Genesis of acid-sulfate alteration and Au-Cu-Ag mineralization at Summitville, Colorado: Economic Geology, v. 82, p. 1575-1591.
17. Vikre, P.G., 1989, Ledge formation at the Sandstone and Kendall gold mines, Goldfield, Nevada: Economic Geology, v. 84, p. 2115-2138.

PART 2. GIS/RS METHODS AND TECHNIQUES

Spatial analysis and image processing of remotely sensed data are indispensable tools for resource analysis and environmental risk assessment. A variety of sensors, new and old, and a diverse array of terrain data types are now available in a digital form that result in novel cartographic representations and that, in turn, lead to more effective decision-making in support of geoenvironmental planning. Image processing techniques, prediction models, and other applications are presented in the contributions that follow.

In the paper by Lee and others, an introduction to remote sensing is provided as it relates to mineral resource appraisal and geoenvironmental assessment. Multi-spectral, hyper-spectral, and thermal images are used in selected study areas in the United States to assess the impact of mining activity on the landscape. Increased resolution and the availability of integrated digital products enhances the value of remote sensing in dealing with the problems associated with rapid population growth in an area and with economic development. In his paper, Gong advocates photo-ecometrics as a new method for natural resource monitoring. This method uses digital surface models (DSM) rather than digital elevation models (DEM) in combination with 3-D surface information from stereo-images, multi-spectral, textural and contextual image information in order to achieve desired results. Detailed mapping of landslides, surface mining, land erosion/deposition, water channel incision, and coastal salt flat displacements are some of the proposed applications. In his second paper, Lee provides examples of spatial data modeling of geoenvironmental hazards and resource favourability potential. A multidisciplinary application illustrates how different probabilistic models were generated for assessing the potential for acidic, metal-rich mine drainage in the state of Montana in the United States. Chung and others propose the use of prediction-rate curves to validate numerical predictions of mineral exploration potential and the environmental impact of mining derived from integrated spatial databases. Validation of numerical predictions is a measure of the reliability of such maps and of the associated costs and benefits. The relation between economic development and its impacts is critical if sustainability is to be achieved.

APPLICATIONS OF REMOTELY SENSED DATA IN GEOENVIRONMENTAL ASSESSMENTS

G. LEE, D. KNEPPER, JR., A. MCCAFFERTY, S. MILLER, T. SOLE,
G. SWAYZE, K. WATSON
United States Geological Survey
Box 25046, Mail Stop 973
Denver Federal Center
Denver, Colorado, USA 80225

Abstract. Remote sensing data and present-day analytical methods provide excellent tools for aiding in mineral resource and geoenvironmental assessments. Modern digital imaging systems, both airborne and satellite-borne, are providing a new array of coarse to fine spatial resolution and broad to narrow spectral resolution image data that are suitable for integration into geographic information systems (GIS). GIS evaluation of these data, along with information from many other sources, must be considered when conducting these assessment studies. The scope of remote sensing applications is changing from an emphasis of interpreting photographic images to data analysis and integration with other digital geospatial information. This integration substantially leverages the value of remotely sensed data, especially in the context of rapid growth and development of the quality and availability of digital information in general, and of satellite and airborne image information in particular.

Various types of remotely sensed data have been used in mineral and environmental studies conducted by the U.S. Geological Survey (USGS). The scales of these applications have ranged from broad regional characterizations to local, site specific studies. Multispectral satellite data and airborne magnetic and radiometric data were compiled for the State of Montana, north-central USA. These layers of information were integrated to assist in identifying areas at possible risk of surface-water contamination by metals derived from past and present mining activity, from natural sources, or both. Imaging spectroscopy (hyperspectral) data have been used by the USGS to effectively map the locations of acid-producing minerals at mine sites in the Leadville mining district of the Colorado Mineral Belt, west-central USA. Thermal infrared image data have been used to identify source rocks of alluvial materials associated with environmental problems in national parks, and to produce maps of primary rock composition and alteration information.

Together, these applications of remotely sensed information have been used to (1) identify areas of potential environmental concern at regional and local scales, (2) assist in prioritizing drainage basins for mitigation efforts, (3) prioritize mine waste piles within selected drainage basins for remediation, and (4) guide remedial work by mapping geologically favorable and unfavorable repository sites.

1. Introduction

Advances in the quality and availability of remotely sensed information are promoting rapid growth in the applications of these data as integral aspects of multidisciplinary Earth science studies such as geoenvironmental assessments. Stereoscopic interpretations of large scale (1:5,000-20,000 scale) black-and-white, color, and color-infrared photographs, for example, can be used to delineate landforms, disturbed ground, vegetation distribution, and other surface features such as mine dumps and tailings, abandoned smelter sites, and active mines, that are often the focus of geoenvironmental investigations. These same photographs can also be used to map the distribution of lithologic units, geologic structures, and fracture patterns as preliminary steps toward characterizing geoenvironmental study areas, and can be invaluable for planning of remediation activities to best meet local terrain, vegetation, and habitat requirements. Modern remotely sensed image data, however, are also ideally suited for more comprehensive spectroscopy analysis, image processing techniques, and integration with other geospatial data. This integration is necessary to evaluate the myriad information sources (geologic, geochemical, hydrologic, biologic, etc.) that must be considered when assessing existing or potential geoenvironmental problems. New satellite imaging systems, scheduled for launch into Earth orbit in 1998 and 1999, will provide digital image data in broad spectral bands with 1-5 meter spatial resolution; these data promise to replace traditional aerial photography as sources of high resolution land surface information.

The following discussions provide a general overview of basic remote sensing concepts and briefly describes several of the specific categories of information which can be used in mineral and environmental investigations. Several summaries of particular USGS applications of remotely sensed data to geoenvironmental studies are included.

2. General Concepts

Remote sensing is the acquisition of data pertaining to objects or scenes by distant sensors [5]. Airborne or satellite platforms are commonly vehicles for the sensors, which usually record electromagnetic radiation. Although airborne magnetic and radiometric data applications are also considered in a later discussion, the following description of concepts is limited to electromagnetic data. Electromagnetic radiation is energy transmitted through space in the form of electric and magnetic waves [17]. Remote sensors are made up of detectors that record specific wavelengths of the electromagnetic spectrum. The electromagnetic spectrum is the range of electromagnetic radiation extending from cosmic waves to radio waves [10].

Each type of land cover absorbs a particular portion of the electromagnetic spectrum, and thereby provides a distinguishing electromagnetic profile. Knowledge of which wavelengths are absorbed by certain features and the intensity of the reflectance allows users to analyze a remotely sensed image and make fairly accurate interpretations of the scene based on the absorption-reflectance profiles.

2.1. ELECTROMAGNETIC SPECTRA

Remote sensors collect data from selected wavelength ranges (bands) of the electromagnetic spectrum. Spectral bandwidth is the width of an individual spectral channel (band) detected by a sensor. Sensors which obtain information from relatively few, wide, non-contiguous wavelength ranges are termed *broad* band, whereas sensors that record numerous, narrow, continuous bands are often called *hyperspectral*.

The selected bands vary among sensors, but are typically found in the visible (0.4-0.7 micron), infrared (0.7 micron to about 1 mm), and microwave (1 mm to beyond 1 m) regions. Together, the near-infrared (0.7-3.0 micron) and middle-infrared (3.0-30 micron) regions of the electromagnetic spectrum are sometimes referred to as the short wave infrared region (SWIR). This is to distinguish this range of wavelengths from the far infrared (30 micron to 1 mm) region, which is often referred to as the long wave infrared region (LWIR). The short wave infrared region is characterized by reflected radiation whereas the long wave infrared region is characterized by emitted radiation. The mid-infrared includes thermally emitted energy, which for the Earth starts at about 2.5 to 3 microns, peaks around 10 microns, and decreases beyond the peak [8].

2.2. SPECTROSCOPY

Spectroscopy is the study of electromagnetic radiation as a function of wavelengths that have been scattered (reflected or refracted), emitted, or absorbed by a material [2]. When radiation interacts with matter, some photons are absorbed, some pass through, and others are scattered. Photons are absorbed in materials by several processes. The variety of absorption processes and their wavelength dependence allows us to derive information about the composition of a material from its reflected or emitted light [2]. The use of imaging sensors to characterize or identify a material of interest is based on the principles of spectroscopy. Spectroscopy reveals the absorption spectra (the electromagnetic radiation wavelengths that are absorbed by specific materials) and the reflection spectra (the electromagnetic radiation wavelengths that are reflected by specific materials).

Electromagnetic radiation absorption characteristics of a material depend on the nature of the molecular bonds in the surface. These bonds, defined by the chemical composition and crystalline structure of the material, determine the wavelengths that will be absorbed. For pure compounds, these absorption bands are so specific that they provide a spectral "fingerprint" [8].

To accurately interpret the information received by a sensor, it is necessary to rigorously define the radiation that is incident to the target materials. For passive (non-transmitting) sensors, the sun is the radiation source; however, the sun does not emit the same amount of radiation at all wavelengths. Moreover, solar radiation must travel through the Earth's atmosphere before it reaches the Earth's surface. As it passes through the atmosphere, radiation is affected by four phenomena: (1) the amount of radiation absorbed by the atmosphere; (2) the amount of radiation scattered by the atmosphere away from the field of view; (3) divergent solar radiation scattered into the field of view; and (4) radiation re-emitted after absorption [7].

Atmospheric absorption is not a linear phenomenon; it is logarithmic with concentration [9]. In addition, the concentration of atmospheric gases, especially water vapor, is variable. Other major gases of importance include carbon dioxide (CO_2) and ozone (O_3), which can vary considerably around urban areas. Thus, the extent of atmospheric absorbance will vary with humidity, elevation, proximity to (or downwind of) urban smog, and other factors [8]. If the atmosphere absorbs a large percentage of the radiation, it becomes difficult or impossible to use those particular wavelengths to study the Earth.

Scattering is modeled as Rayleigh scattering with a commonly used algorithm that accounts for the scattering of short wavelength energy by the gas molecules in the atmosphere [16]. Scattering is variable with both wavelength and atmospheric aerosols [8] which, in turn, vary temporally and regionally.

Scattering source and emission source may account for only 5% of the variance. These factors are minor, but they must be considered for accurate calculation. After interaction with the target material, the reflected radiation must travel back through the atmosphere and be subjected to these phenomena a second time to arrive at the sensor. The mathematical models that attempt to quantify the total atmospheric effect on the solar illumination are called radiative transfer equations. Some of the most commonly used are LOWTRAN [11] and MODTRAN [1].

After rigorously defining the incident radiation, it is possible to study the interaction of the radiation with the target material. It is the reflected radiation, generally modeled as bidirectional reflectance [3], that is measured by the remote sensor. Remotely sensed data are made up of reflectance values which are recorded as discrete digital numbers by the sensing device. Each sensor detector is designed to record a specific part of the electromagnetic spectrum, and the bands, or channels, that are included, together with the nature of the materials of interest in the study area, determine the applicability of the sensor to an investigation.

The discrimination of target objects is based on the reflectance spectra of the materials of interest. Every material has a characteristic reflectance spectrum based on its chemical composition; however, it is the wavelengths that are *not* returned to the sensor, the absorption spectrum, that are most important in providing information about the imaged area [8].

3. System Types and Applications

Remote sensing systems are made up of a scanner and a transportation vehicle, such as a satellite, aircraft, or space shuttle. The scanner is the entire data acquisition system, and it includes sensors, which are made up of detectors. Sensors are devices that gather energy, convert it to a signal, and present it in a form suitable for obtaining information about the environment [5]. A detector is a device in a sensor system that records electromagnetic radiation.

With the exception of imaging radar, most remotely sensed data is obtained from *passive* solar imaging sensors. Passive systems measure natural radiation emitted by the target material and energy from other sources reflected from it. Passive sensors can receive, but not transmit, radiation waves, and usually operate within the visible and infrared regions, recording reflected solar energy. A difficulty with passive systems is that good quality images in the visible range of the spectrum cannot be obtained in the presence of clouds, rain, or darkness.

Active systems, on the other hand, transmit their own signal and measure the energy that is reflected or scattered back from the target material. Imaging radar sensors, for example, are active sensors that emit a burst of microwave radiation and receive the backscattered radiation. Because water droplets in fog and clouds are transparent to microwaves of the proper frequencies just as window glass is to light waves of the visible frequency, imaging radar systems can provide high quality images even in darkness or clouds. A disadvantage of active systems, however, is that they require substantial power for signal transmission and cannot operate continuously. They must spend considerable time recharging electrical supplies.

3.1. MULTISPECTRAL IMAGING

Multiple band data are referred to as *multispectral* imagery and the term is used here to represent information derived from systems that produce data from a fairly limited number of relatively broad (compared to spectrometers) spectral bands, or channels, which are often not contiguous. These broad band multispectral scanners use passive sensors which measure and record solar radiance. In contrast, *panchromatic* sensors record all wavelengths (colors) across the visible portion of the electromagnetic spectrum as a single band image [8].

Broadband multispectral satellite data, such as Landsat Multispectral Scanner (MSS) with 80 m spatial resolution and Thematic Mapper (TM) with 30 m resolution, SPOT Panchromatic, 10 m resolution, and XS, 20 m resolution, and the Indian satellites (IRS-1C, IRS-1D) with 5 m spatial resolution data, provide excellent sources of regional information suitable for analyzing the broader aspects of rock and soil type, vegetation, and landform distribution. For example, Landsat TM data, with 6 spectral bands in the visible and near infrared portions of the spectrum, can be processed to discriminate broad mineral groups, including iron oxides, clays, and other hydroxyl-bearing minerals often associated with hydrothermally altered rocks. This capability is particularly useful for geoenvironmental studies involving mine tailings, mine dumps, and hydrothermal alteration.

The airborne Thermal Infrared Mapping System (TIMS), flown by NASA, collects digital multispectral thermal infrared image data that can be broadly analyzed for general silica content of exposed rocks and minerals, as well as the soil moisture characteristics of exposed soils and unconsolidated sediments. The soil moisture mapping capability of TIMS data is especially applicable for tracing near-surface migration of fluids from potential sources of metal contamination into nearby streams.

3.2. HYPERSPECTRAL IMAGING

As remote sensing instruments and applications advance toward the detection and use of more and narrower bands, discrimination capabilities increase in detail. With current technology it has become possible to build sensors than can measure spectra as images of the Earth at high spectral and spatial resolution. *Imaging spectroscopy* has many spectral channels compared to broad band systems and is therefore also known as hyperspectral imaging. Systems such as Landsat, which measure radiance from a few broad, non-contiguous channels are not considered spectrometers, and therefore cannot distinguish narrow absorption features [2]. An imaging spectrometer, on the other hand, measures radiance from enough contiguously spaced spectral channels to resolve absorption bands in many more materials of interest.

For example, the National Aeronautics and Space Administration/Jet Propulsion Laboratory (NASA/JPL) Airborne Visible and Infrared Imaging Spectrometer (AVIRIS) system has 224 narrow (0.01 micron bandwidth) contiguous spectral channels, covering the spectral range from 0.38 to 2.5 microns (the human eye covers only about 0.4 to 0.68 microns). When the data from each detector are plotted on a graph, a spectral profile is produced for each pixel. Comparing the resulting profiles with those of known substances reveals information about the composition of the area being viewed by the instrument that allows for detailed mineralogical and vegetation mapping at about 17 m spatial resolution. These imaging spectrometer (hyperspectral) data allow the identification and mapping of many mineral species associated with hydrothermally altered rocks, and can be useful for delineating alteration zonation and characterization of mine dumps and tailings that may be associated with acid drainage or metal toxicity. The instrument has been used across the USA, plus Canada and Europe.

3.3. IMAGING RADAR

Remote sensors in the microwave region of the electromagnetic spectrum are either passive or active. The passive sensors record very low intensity microwave radiation emitted by the Earth. The extremely low intensity of the emitted energy results in images with low spatial resolution. It is the evolution of the active sensors, termed *imaging radar*, that is introducing a new generation of satellite imagery to remote sensing. Imaging radar devices produce an image by emitting a directed beam of microwave energy at the target scene and then collecting the backscattered (reflected) radiation. The microwave energy emitted by an active radar sensor is coherent and of narrow spectral bandwidth.

The spatial resolution of imagery produced by a radar sensor is a function of the antenna size. To overcome this resolution constraint, processing techniques have been developed which combine the signals received by the sensor as it travels over the target. Thus, the antenna effectively becomes as long as the sensor path during backscatter reception. This is termed a synthetic aperture, and the sensor a synthetic aperture radar (SAR). SAR images are maps in which the brightness shown is a measure of the microwave energy reflected back to the antenna.

SAR image data for geological studies can be used to make maps of geologic structural features, lithologic units, surface morphology (shape), coastal changes, volcano distributions and morphology, and surficial processes. For hydrology studies, SAR can be used to study the flux and storage of water, including snow. For ecosystem studies, SAR can be used to examine both forest and low-vegetation canopy characteristics and soil moisture. Maps of the canopy geometry, extent, and above-ground biomass can be produced using SAR data and information can be enhanced by integration with data from optical sensors.

Another imaging radar system is side-looking airborne radar (SLAR), which uses an antenna which is fixed below an aircraft and pointed to the side to transmit and receive the radar signal. SLAR data provide a means for producing up-to-date digital elevation models (DEM) at resolutions of up to about 1 meter in both vertical and horizontal directions. High resolution DEMs are a relatively new source of information for geoenvironmental studies, however a variety of potential applications are easily identified. For example, with repetitive coverage over time, these data can be processed to yield changes in elevation on the order of a few centimeters that could be applied to monitoring erosion and mass movements of mine dumps and tailings. High resolution DEMs may also be useful for detecting and identifying old dumps and tailings, as well as previously disturbed ground associated with abandoned smelters and mine processing sites.

4. Geoenvironmental Study Examples

Following are several examples of USGS mineral-environmental studies that have incorporated various types of remotely sensed information.

4.1. MULTISPECTRAL IMAGING

4.1.1. *Landsat MSS, Airborne Radiometry, and Airborne Geophysics*
Although remote sensing is generally considered to refer to measurement and recording of electromagnetic radiation, the more encompassing description offered in the previous discussion under General Concepts would include other data acquired by distant sensors. By this description, a current geoenvironmental assessment study of the State of Montana, north-central USA, has applied remotely sensed information that includes Landsat Multispectral Scanner (MSS) data, airborne magnetic data, and airborne gamma ray spectrometer data.

Landsat MSS Data. The distribution of exposed rocks and soils in Montana that contain iron oxide minerals was interpreted from 33 Landsat MSS images, using PCI image processing software. Each of the 33 scenes was georeferenced and resampled to 100 meter pixels. Iron oxide minerals derived from hydrothermal alteration and the weathering of iron-bearing minerals are predominantly red in color; other red materials at the Earth's surface are artificial and have a very restricted aerial distribution. The ratio of the green MSS band (band 4) and the red MSS band (band 5) results in very small values for materials that are red.

In a scene in the central part of the study area, band 4/band 5 ratio values below 0.093 did a good job of locating rocks and soils known to contain iron oxide minerals. Consequently, the density distribution of the iron oxide minerals in this scene was used as a standard reference for adjusting the band 4/band 5 ratio cutoff values in the adjacent scenes so that iron oxide distribution patterns have smooth transitions among scenes. Because of seasonal and atmospheric differences among scenes, the band 4/band 5 ratio cutoff values ranged between 0.093 and 0.11.

The images showing the distribution of iron oxide in each scene were digitally mosaicked to produce an iron oxide distribution map of Montana (Figure 1). The iron oxide map was filtered to enhance major concentrations and remove isolated pixels that produce a salt-and-pepper noise pattern in some areas. In areas of hydrothermal alteration, the combination of magnetic and potassium data can be a useful tool in mapping areas of magnetite destruction and potassic enrichment.

Airborne Gamma-Ray Data. Potassium concentrations for Montana were extracted from 3-km grids of regional airborne gamma ray-spectrometer data for the conterminous United States [15]. These data were acquired during the National Uranium Resource Evaluation (NURE) program of the U.S. Department of Energy and represent near-surface (to 50 cm depth) soil or rock concentrations of potassium. The potassium grid was imported by ERDAS IMAGINE image processing/GIS software where it was reprojected and resampled to 500 meter spatial resolution. Figure 2 is a map of color-encoded potassium concentration ranges draped over shaded topographic relief to provide geographic reference.

Airborne Magnetic Field Data. The entire study area is covered by aeromagnetic data which were collected as part of the NURE program data [15] and by numerous USGS surveys. These data were merged to produce a State magnetic field intensity map [13].

Inclination and declination of the Earth's magnetic field causes magnetic anomalies to be laterally skewed and shifted from their causative sources. The amount of lateral shift varies with depth of sources of anomalies -- shallow magnetic sources represented by short wavelength anomalies will tend to have less horizontal shift than anomalies caused by deeper sources. A mathematical correction procedure for the shift is called "reduction-to-pole" (RTP) and was applied to the Montana (71^0 inclination, 17^0 declination) magnetic data compilation. Data from the compilation show changes in the Earth's magnetic field as a result of variations in the magnetic-mineral content of near surface rocks.

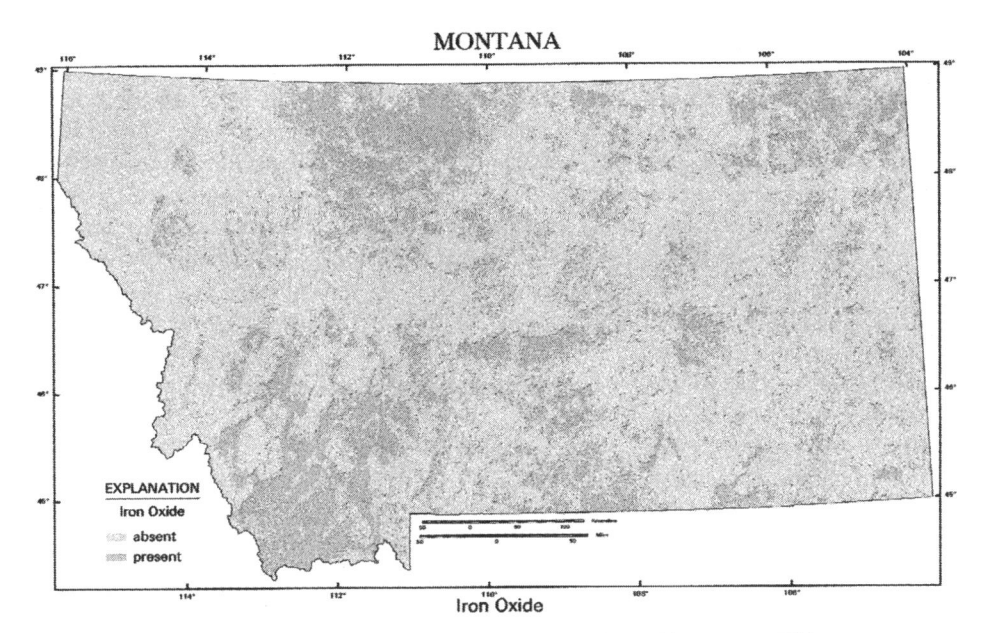

Figure 1. Distribution of Landsat MSS interpreted iron oxide, Montana, USA.

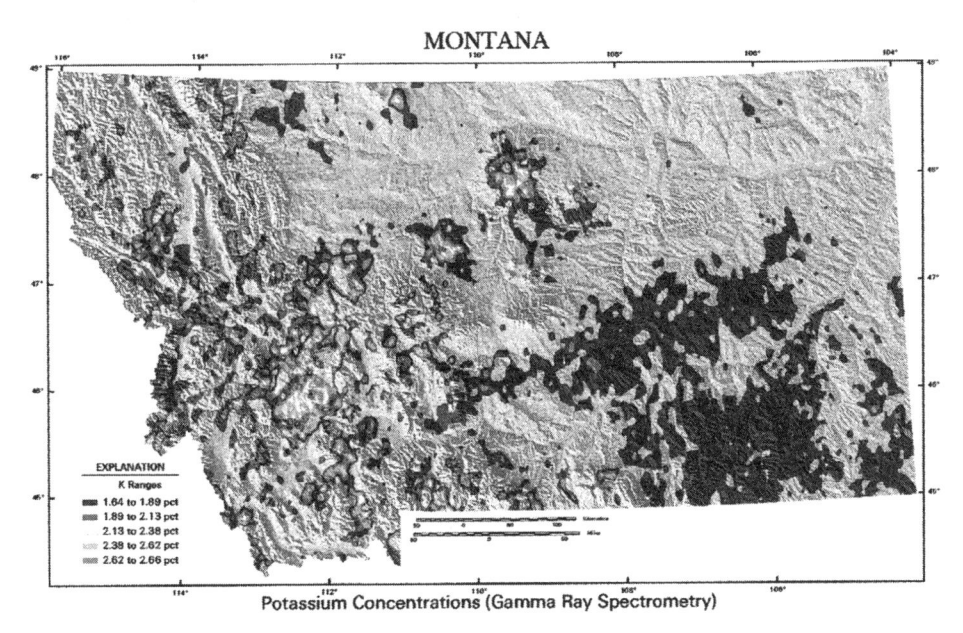

Figure 2. Distribution of potassium concentrations derived from airborne gamma-ray spectrometry, Montana, USA.

Data from the magnetic anomaly image were transformed to a map of magnetization (Figure 3) using a data processing technique called "terracing". The terrace method, described by Cordell and McCafferty [6] and Phillips [14], transforms gridded geophysical data (magnetic, gravity, or electromagnetic) into interpreted physical property (magnetization, density, or conductivity, respectively) maps that emphasize sharply-bounded domains. The domains and their edges may be used to help identify areas of lithologic and structural contrast, some of which may be related to magnetic properties of rocks that are likely to contain sufficient magnetic minerals (principally magnetite) to produce anomalies. Typically, sedimentary rocks contain few if any magnetic minerals and do not produce significant magnetic anomalies. The magnetically anomalous areas are more likely to be related to igneous rocks and may therefore be associated with sulfide-rich plutons that could produce acid drainage under appropriate environmental conditions.

Data Integration. Aeromagnetic, Landsat MSS, and airborne gamma ray spectrometer data covering the State of Montana were digitally integrated to produce an interpreted alteration map (Figure 4). This map is used to identify areas at possible risk for surface-water contamination by metals due to mining activity or natural processes, or both. The alteration map is a mathematical combination of gridded data from the magnetization map (Figure 3), iron oxide map (Figure 1), and apparent potassium concentrations (Figure 2), produced using ER-Mapper software. The primary objective in constructing the alteration map is to locate areas of possible potassic alteration and leaching or weathering of exposed and shallowly buried metal-sulfide bearing igneous rocks. Patterns of magnetization domains associated with iron oxide and relatively high potassium concentrations can be indicative of hydrothermal alteration environments that host metal sulfides. The categories that were combined to produce the image were three levels of potassium concentrations (high, moderate, low), two categories (high and low) of magnetization, and two categories (present, absent) of iron oxide. The 3 potassium classes were selected as high ≥ [mean + 2 standard deviations]; [mean + 1 standard deviation] ≤ moderate < [mean + 2 standard deviations]; and low < [mean + 1 standard deviation]. The 2 magnetization classes were selected as high > [mean + 1 standard deviation] and low ≤ [mean + 1 standard deviation]. These categories of the potassium and magnetization images were numerically coded as high magnetization = 1; low magnetization = 2; high potassium = 4; moderate potassium = 8; and low potassium = 0. Iron oxide presence was coded as 100, and absence as 0. An integrated composite image (Figure 4) was calculated from these values as follows: [coded iron oxide] x [coded magnetization] + [coded potassium].

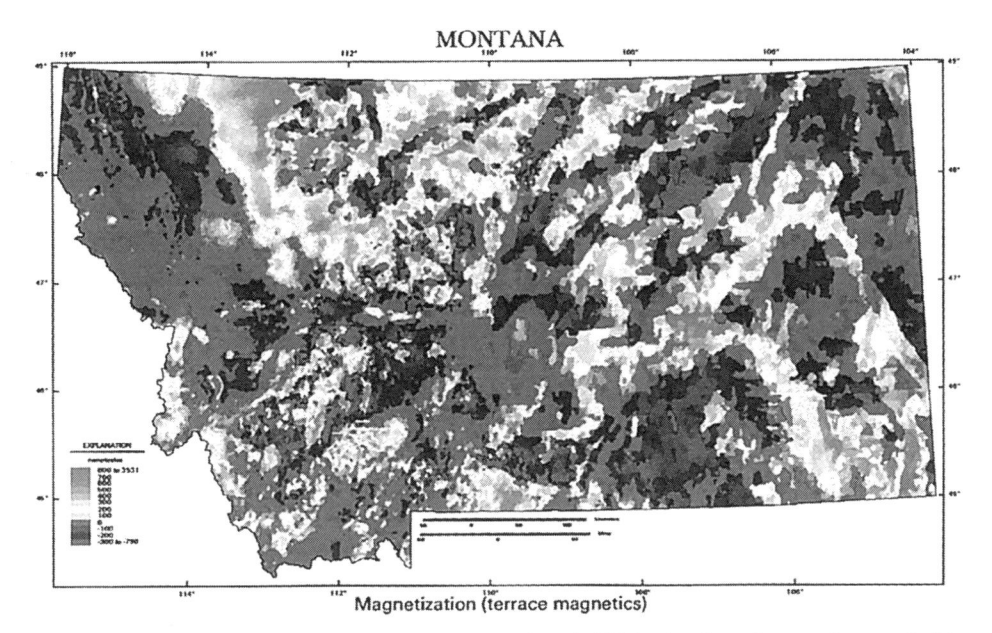

Figure 3. Magnetization map derived from terrace processing of gridded reduced-to-pole magnetic anomaly data, Montana, USA.

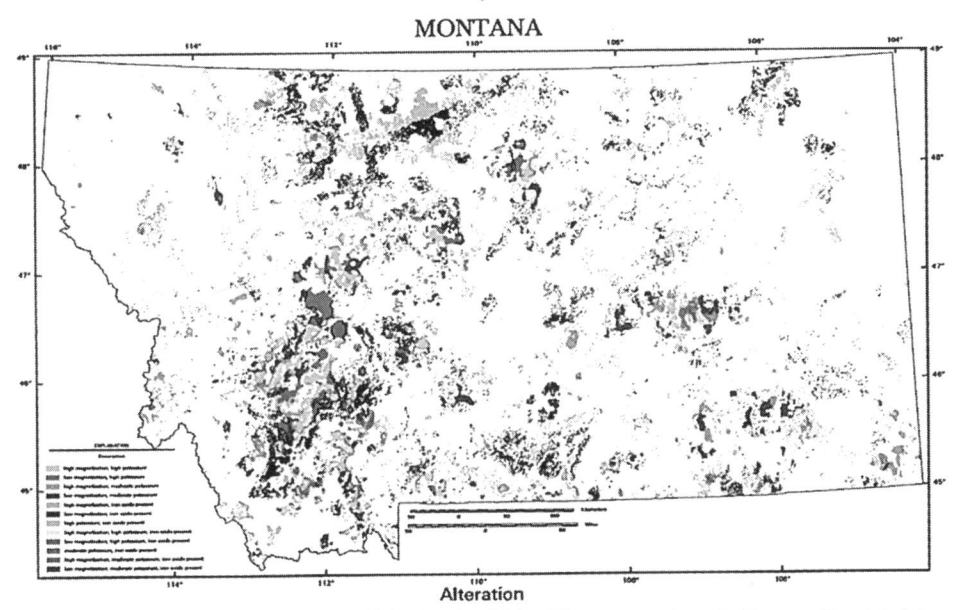

Figure 4. Alteration map derived from combining Landsat MSS, airborne magnetic, and airborne radiometric data, Montana, USA.

To incorporate the results into the Montana geoenvironmental assessment, classes of the alteration image map have been tested, using ERDAS IMAGINE software, for spatial association with known areas characterized as high in metal sulfide content and low in calcium carbonate. Such areas are thought to have relatively high potential for producing acidic, metal-rich drainage waters in the study region. These areas were found to be 41 times as likely to occur in the class of [low magnetization + high potassium + iron oxide] as elsewhere in the study area. Similarly, the likelihood ratio for the class denoting [low magnetization + high potassium] is 18; 15 for the [high potassium + iron oxide] class; 6 for the [high magnetization + moderate potassium + iron oxide] class; 5 for the [low magnetization + moderate potassium] class; 3.5 for the [low magnetization + moderate potassium] class; and 2 for the [high magnetization + moderate potassium] class.

These strong spatial associations suggest that the integration of airborne gamma ray potassium data, airborne magnetic data, and Landsat MSS data has been useful and important in the regional scale geoenvironmental assessment of potential for acidic, metal-rich drainage in Montana.

4.1.2. Landsat TM

The Landsat TM (Thematic Mapper) scanner is a multispectral scanning system that records reflected or emitted electromagnetic energy from the visible, reflective-infrared, middle-infrared, and thermal-infrared regions of the spectrum. TM detectors record electromagnetic radiation in seven bands. Bands 1, 2, and 3 are in the visible part of the spectrum and are useful in detecting cultural features, soil and geological boundaries, vegetation discrimination, water, and forest types [12]. Bands 4, 5, and 7 are in the reflective-infrared portion of the spectrum and can be used for crop identification, discrimination of soil and rock types, plant and soil moisture, plant health, and for soil-crop, land-water, and clouds-snow-ice contrasts. Band 6 is in the thermal portion of the spectrum and is used for vegetation and crop stress detection, heat intensity, insecticide applications, and for locating thermal pollution, and geothermal activity [8]. The spatial resolution of TM data is approximately 30 meters for all bands except the thermal (band 6), which has a resolution of about 120 meters. The larger pixel size of this band is necessary for adequate signal strength. However, the thermal band is resampled to 30 m to match the other bands. The radiometric resolution is 8-bit, meaning that each pixel has a possible range of 2^8 integer data values from 0 to 255 [8].

The USGS used Landsat Thematic Mapper data for a mineral resource assessment of the Caballo Resource Area in southern New Mexico, southwest USA. The TM data were used to produce color-ratio composite image maps which were used to focus limited geologic mapping support to effectively guide field efforts to areas most likely to be important to the study. Large scale (1:24,000) color-ratio composite image maps were prepared for each of eight 1:24,000-scale quadrangles where field mapping was to be concentrated. The color-ratio maps, while effective for discriminating among mappable lithologic units, were of greatest value in highlighting areas of potential hydrothermal alteration and, consequently, potential mineralization that needed to be evaluated in the field. Although the Landsat TM color-ratio composite image maps can be used in the field, they are most effectively used for planning purposes in advance of field investigations. In this way, mapping priorities were established to maximize effectiveness and efficiency of time-constrained field efforts. Figure 5 shows a color composite image map which identifies important geologic mapping features in the Caballo Resource Area.

4.1.3. *Thermal Infrared Imaging System (TIMS)*

Spectral signatures of common rock forming minerals are found in the thermal infrared (TIR) 8-14 micron region of the electromagnetic spectrum. Emitted radiation in the TIR region is a function of temporal variations in the heat fluxes that affect the temperatures of surface materials. Subtle changes in surface temperatures can be measured by remote TIR sensors. Applications of various image processing algorithms to satellite and aircraft acquired data from this spectral region provide cost effective means to characterize regional tracts of land and these characterizations can be useful for mineral and environmental assessments.

Primary compositional information can be derived about igneous rocks from analysis of thermal data. Quartz and feldspar, for example, are common constituents of igneous rocks; both have recognizable spectral signatures in the TIR but are almost featureless in the visible and near-infrared region. Also, mafic compositions can be distinguished from those that are felsic. Other important information, such as silicification, can also be mapped by applying decorrelation stretches to selected TIR bands. Presently, spectral emissivity algorithms are being developed that can detect more subtle lithologic differences, suggesting that identification of individual minerals may be possible.

The USGS has been using the TIMS (thermal infrared multispectral scanner) system for TIR studies. This aircraft-based sensor system acquires data as part of the NASA Airborne Science Program. TIMS measures six discrete channels in the TIR, and spatial resolution, which is altitude-dependent, is typically 5-30 meters. TIMS data have been acquired in key areas within several U.S. National Parks to provide background and baseline information for studying a variety of geoenvironmental problems.

Joshua Tree National Park in California was the focus of a study of the desert tortoise habitat in order to understand factors that contribute to its mortality. Two thermal modeling techniques were employed in this data analysis: the emissivity ratio method described by Watson [20]; [21] and Watson and Rowan [22], and a new procedure called the inverse wavelength method by Watson [21]. Unsupervised classification was used to identify spectral classes based on all 6 channels and the data were registered to a geographic base for comparison with other data sets. Preliminary analysis of the Joshua Tree data identified 9 classes, including several alluvial and unconsolidated units within the Pinto Basin. Desert tortoise occurrences correlated with one of these classes. The classification image (Figure 6) shows that red and green classes in the upper part of the image differ within a metamorphic unit which is predominately quartzite. The orange class corresponds to basin fill material within Pinto Basin. Black crosses indicate 1995 tortoise counts (white crosses are 1996 counts). The reduced 1996 count corresponds to a very dry springtime in that year.

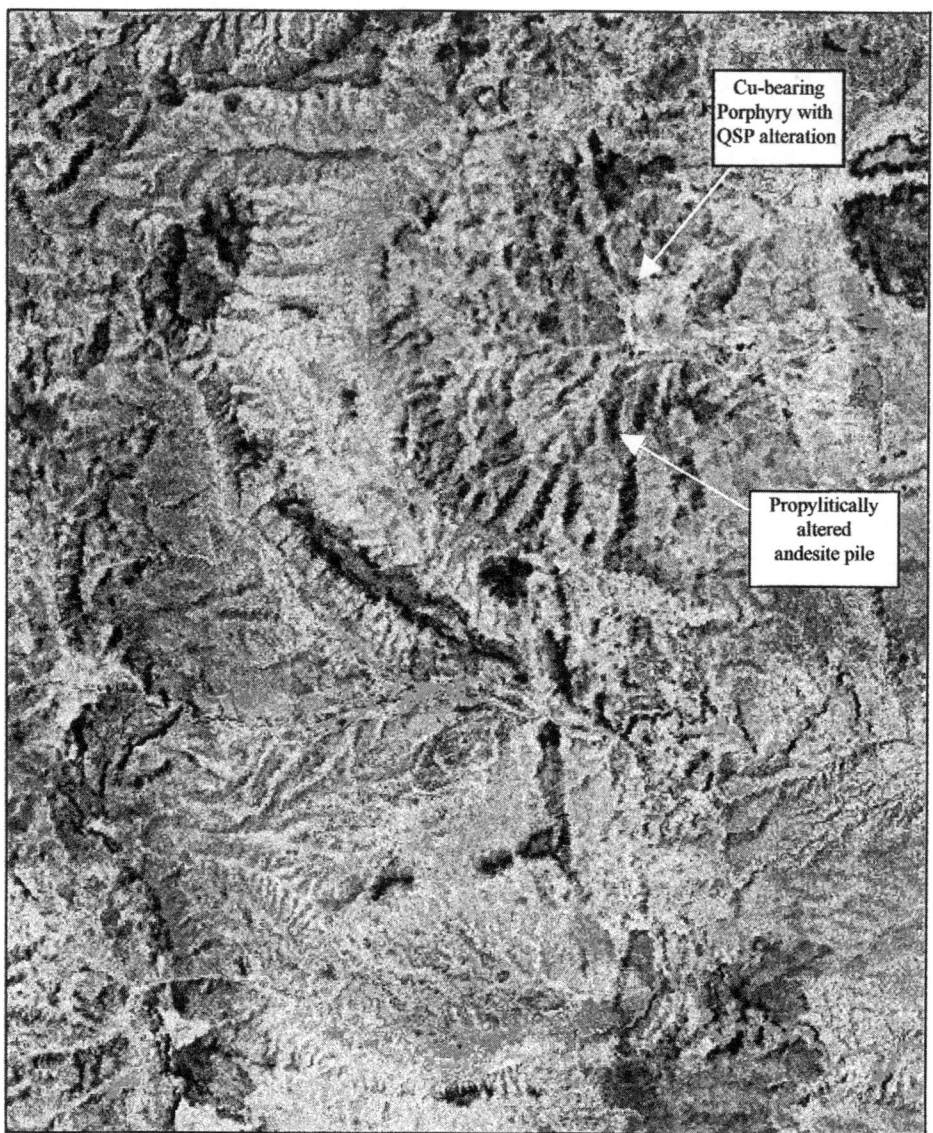

Explanation of Colors: RED=vegetation (stream bottoms, etc.); MAGENTA (red+blue)= clay minerals and other OH-bearing minerals (gypsum, alunite, etc.) and carbonates; GREEN=iron oxides; YELLOW and WHITE (green+red and green+red+blue, respectively)=presence of both clay/carbonate and iron oxides; BLUE=Non-clay and non-iron oxide bearing rocks and soils.

Figure 5. Landsat TM Color Ratio Composite Image, Hillsboro District, New Mexico, USA. [This image was prepared from ratios of band 5/band 7 (color coded red), band 3/band 1 (green), and band 5/band 4 (blue)]

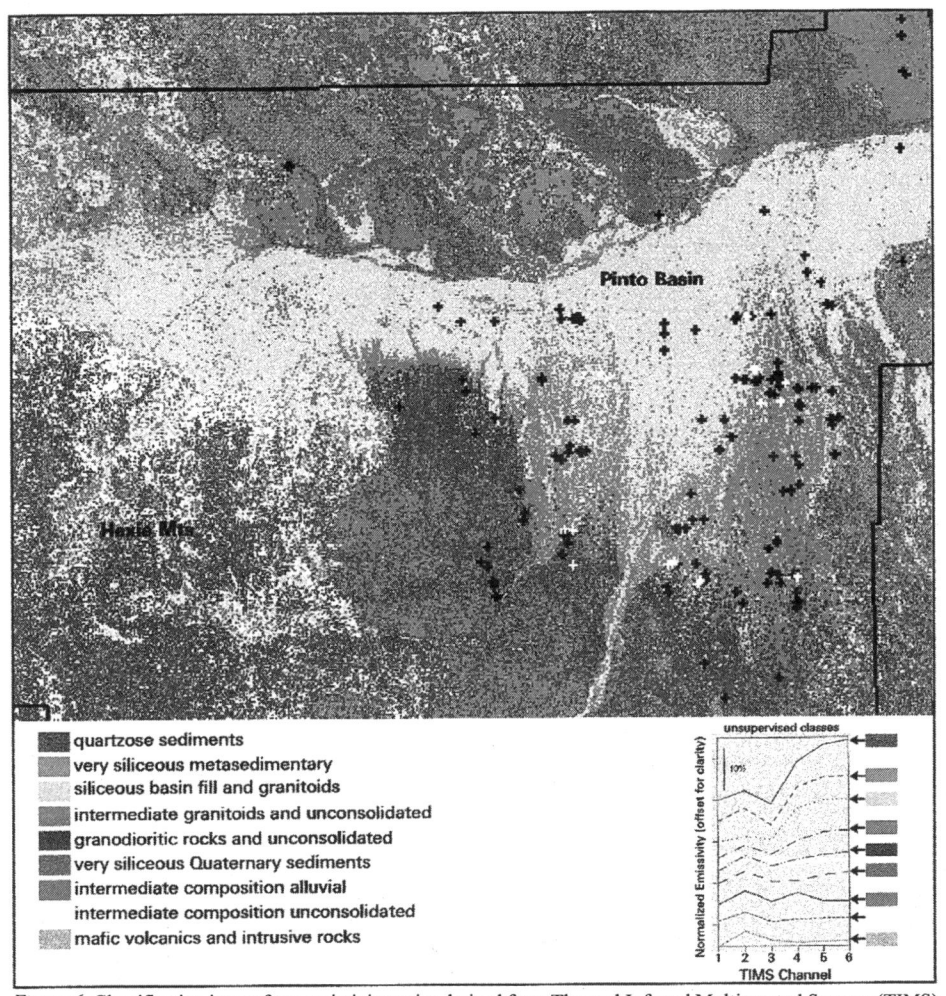

Figure 6. Classification image from emissivity ratios derived from Thermal Infrared Multispectral Scanner (TIMS) data, Joshua Tree National Park area, California, western USA.

There is an obvious association of tortoise counts with the magenta class, which is alluvial material derived from the granodiorites (cyan class).

Another geologic and environmental study in progress is of Yellowstone National Park. This study involves the integration of remote sensing, geology, geophysics, and geochemistry to investigate crustal structure and composition, hydrothermal and seismic activity, and the effects of mine waste and metal-rich geothermal waters that may enter the food chain of mammals. Figure 7 shows a Landsat TM image covering the Park with the flight lines (red) where TIMS data have been acquired. The five north-south flight lines in the western part of the Park extend from Mammoth Hot Springs to the Old Faithful Geyser area, and cover the main geyser basins. A single west-northwest to east-southeast line covers the Lamar Valley where gray wolves were recently introduced. The two northeast-southwest flight lines cover the Soda Butte

Creek drainage and extend outside the Park over areas of historic mining activities at Cooke City, Montana, and the New World mining area.

The TIR remote sensing aspect of the study is initially focused on the area around the Old Faithful Geyser, which is in rhyolitic volcanic rocks. Hydrothermal fluids have passed through the volcanic flows and the resulting alteration has formed siliceous sinter and clays (montmorillonite and kaolinite). The insert in Figure 7 is a decorrelation image of the Old Faithful area in which siliceous rocks are shown as red and mafic rocks as blue. The histories of alteration and of surficial deposits around the hot springs are important for the environmental characterization of the area, and TIR data are valuable in determining these processes.

TIMS data have been used for lithologic studies in igneous, metamorphic and sedimentary terranes, for detection of hydrothermally altered rocks to aid in mineral exploration, and for mapping surficial geology. TIR data have therefore been shown to be valuable in providing information relevant to land use management, environmental security, and mineral resource assessments.

Figure 7. Thermal Infrared Multispectral Scanner (TIMS) classification of rock types in the Old Faithful Geyser area, Yellowstone National Park, USA.

4.2. IMAGING SPECTROSCOPY

The following discussion is largely derived from Swayze et al. [19]. The Leadville mining district, located at an elevation of 3000 m in the central Colorado Rocky Mountains, west-central USA, has been mined for gold, silver, lead and zinc for more than 100 years. This activity has resulted in the dispersal of waste rock and tailings, rich in pyrite and other sulfides, over a 30 km^2 area including the city of Leadville. Oxidation of these sulfides releases lead, arsenic, cadmium, silver, and zinc into snowmelt and thunderstorm runoff which drain into the Arkansas River, a major source of water for urban centers and agricultural communities. Mineral maps, such as Figure 8, were made by the USGS from NASA/JPL Airborne Visible and Infrared Imaging Spectrometer (AVIRIS) data which were collected over this mining district. These maps were used to focus mine waste remediation efforts by locating the point sources of acid drainage.

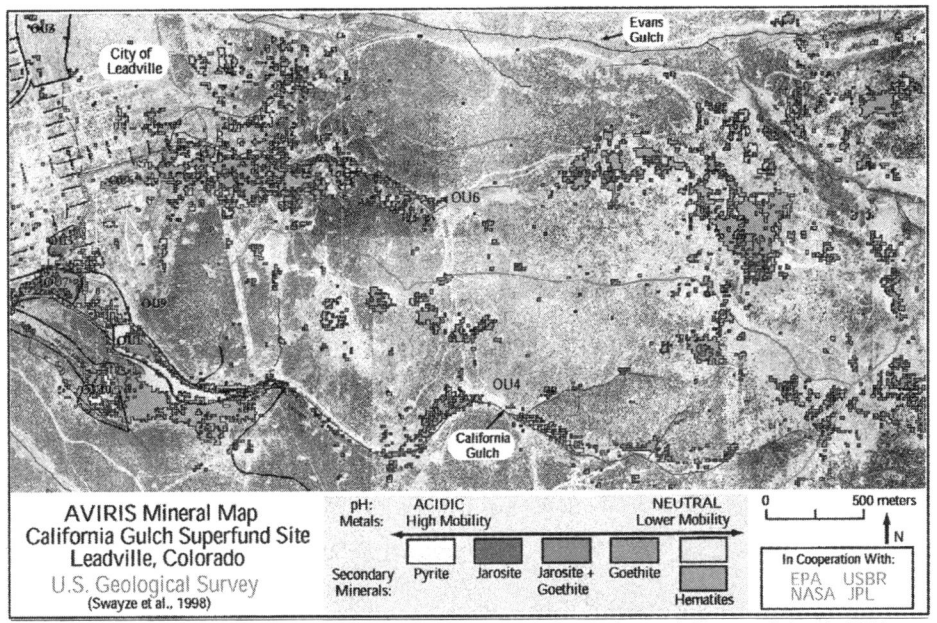

Figure 8. Mineral map derived from hyperspectral (AVIRIS) data in the Leadville area, Colorado, USA.

Release of heavy metals is facilitated by sulfide oxidation, since many of the sulfides contain lead, arsenic, cadmium, silver, and zinc. The oxidation-weathering process of many sulfide minerals produces sulfuric acid which results in low pH water in which heavy metals dissolve as aqueous phases that are then transported as runoff into nearby streams creating acid mine drainage. Secondary minerals such as copiapite ($Fe^{+2}Fe_4^{+3}(SO_4)_6(OH)_2 \cdot 20H_2O$), jarosite ($KFe_3^{+3}(OH)_6(SO_4)_2$), schwertmannite ($Fe_8O_8(OH)_6SO_4$), goethite ($\alpha$-FeO(OH)), ferrihydrite ($5Fe_2^{+3}O_3 \cdot 9H_2O$), and hematite ($\alpha$-$Fe_2O_3$) are formed by sulfide oxidation and subsequent precipitation from metal-rich water. These secondary minerals are iron rich and usually hydroxyl (OH) bearing, making it possible to identify them on the basis of their characteristic spectral signatures. Subsequent pulses of low pH water may dissolve the secondary minerals, thus remobilizing the heavy metals and transferring them downstream.

The sulfide mineral pyrite (FeS_2) is the primary source of acid drainage. Because direct spectral detection of pyrite is hampered by its low reflectance level, its broad Fe-absorption, and rapidly formed coating of secondary minerals, pyrite can only be detected when it is coarse grained or compositionally impure. However, pyrite weathers first to jarosite, and then to goethite, forming a sequence where the degree of oxidation is indicated by the type of secondary mineral exposed at the surface. Therefore, an indirect way to find oxidizing pyrite is to look for areas where the secondary minerals grade through the established oxidation sequence, that is, those areas with jarosite surrounded by goethite. Fortuitously, the presence of heavy metals and low pH, often associated with mine waste, prevent the growth of vegetation over most waste piles, leaving them exposed and suitable for detection with remote sensing.

The large size of the Leadville mining district and presence of spectrally detectable secondary minerals associated with pyrite oxidation makes imaging spectroscopy effective for locating those minerals related to the acid mine drainage sources. AVIRIS data was collected over Leadville on July 27, 1995 [18]. Calibrated AVIRIS reflectance data were spectroscopically mapped using the Tetracorder (previously called Tricorder) algorithm described by Clark and Swayze [4]. In this process, the spectra of unknown materials are compared with hundreds of reference spectra, and the best match is identified.

Mineral maps created by Tetracorder reveal characteristic mineral assemblages centered over the mine dumps and tailings which are rich in pyrite. Field checks demonstrate that highly acidic water formed in the pyrite-rich piles is gradually neutralized as it spreads away from the dumps. This process deposits progressively less acidic secondary minerals in rough concentric zones centered on the dumps, creating diagnostic patterns easily recognized on the electronic absorption mineral map. The centers of the zones are jarosite rich and have a high acid-generating capacity that produces acidic water containing heavy metals. This central zone is surrounded by a jarosite + goethite zone which is itself surrounded by an outermost goethite zone. To facilitate comparison with features on the ground, the mineral map was geometrically registered and overlaid on orthophoto information for the Leadville area (Figure 6). The map shows concentric secondary mineral zones around many of the waste piles, thereby helping to pinpoint the sources of acid drainage.

Mineral maps made from AVIRIS data provided a rapid and cost-effective method to screen entire mining districts for sources of acid and metals. It is estimated that work that took 52 days to do on the ground and in the lab for one of the particular waste piles (sample collection, preparation, and leaching) was done using AVIRIS and Tetracorder in a little over two days. This technology can be used to save time and money at other mine sites where carbonate in the waste rock is unable to neutralize the acidity created by pyrite oxidation.

5. Conclusions

Remote sensing applications have been many and varied, and, in mineral and geoenvironmental studies, have been used to successfully provide important information related to assessment of resource potential and environmental risk. The scale of the applications of remotely sensed data has ranged from broad regional characterizations to local, site specific studies. The scope of the uses of these applications is changing from photographic interpretation to data analysis and integration of multiband data with other digital geospatial information. This integration greatly enhances the value of remotely sensed data for a rapidly growing community of digital spatial data users.

The future of remote sensing technologies and applications looks bright. Many new and improved scanners are scheduled for deployment. Recent and near future developments include improving technologies for acquisition of data of higher spatial, spectral, and radiometric resolutions; increased availability; and lower user costs. These advances will continue to provide greater capabilities for discrimination of finer details about the Earth's surface. These factors, especially decreasing costs, will continue to generate growing demand for image data from a widening variety of users. Increased competition, which includes commercial civilian sectors of the World's economies, will ensure an exciting future of technological advancements and declining costs, not only for the image data, but also in the processing, analysis, and integration systems that will use them. An example:

> "Not to take away the fun of digitizing contour data, but sometime in the year 2000, we'll start making available 90 m DEMs of the world, produced from data that will be collected by the Shuttle Radar Topography Mission in September 1999. Full resolution, 30 m data will also be available for some
> areas. For more details, see our web site at: http://southport.jpl.nasa.gov click on Projects and then on SRTM."
> -- (August 6, 1998 electronic mail communication received from Tom Farr, Deputy Project Scientist, Shuttle Radar Topography Mission)

References

1. Berk, A., L.S. Bernstein, and D.C. Robertson (1989) MODTRAN: A moderate resolution model for LOWTRAN 7, Final report,GL-TR-0122, Air Force Geophysical Laboratory, Hanscom Air Force Base, Massachusetts.
2. Clark, Roger N. (in press) *Manual of Remote Sensing*, John Wiley and Sons, Inc., New York.
3. Clark, Roger N., and Roush, Ted L. (1984) Reflectance spectroscopy: Quantitative analysis techniques for remote sensing applications, *J. Geophysical Research*, **89**, no. B7, 6329-6340.
4. Clark, R.N., and Swayze, G.A. (1995) Mapping minerals, amorphous materials, environmental materials, vegetation, water, ice, and snow, and other materials: The USGS Tricorder Algorithm, *Summaries of the Fifth Annual JPL Airborne Earth Science Workshop*, January 23-26, R.O. Green (ed.), Jet Propulsion Laboratory Publication 95-1, p. 39-40.
5. Colwell, Robert N. (1983) *Manual of Remote Sensing*, American Society of Photogrammetry, Falls Church, Virginia.
6. Cordell, Lindrith, and McCafferty, A.E. (1989) A terracing operator for physical property mapping with potential field data, *Geophysics*, 54, no. 5, 621-634.
7. Elachi, Charles (1987) *Introduction to the Physics and Techniques of Remote Sensing*, John Wiley and Sons, Inc., New York.
8. ERDAS (1997) *ERDAS Field Guide* (4th ed.), ERDAS, Inc., Atlanta.
9. Flaschka, H.A. (1969) *Quantitative Analytical Chemistry: Vol. 1*, Barnes and Noble, Inc., New York.
10. Jensen, John R. (1996) *Introductory Digital Image Processing: A Remote Sensing Perspective*, Prentice-Hall, Englewood Cliffs, New Jersey.
11. Kneizys, F.X., *et al.* (1988) *Users Guide to LOWTRAN 7*, Air Force Geophysics Laboratory, Hanscom Air Force Base, Massachusetts.
12. Lillesand, Thomas M., and Kiefer, Ralph W. (1987) *Remote Sensing and Image Interpretation*, John Wiley and Sons, Inc., New York.
13. McCafferty,A.E., Bankey,Viki, and Brenner, K.C. (1998) Merged aeromagnetic and gravity data for Montana: A web site for distribution of gridded data and plot files: U.S. Geological Survey Open-File Report 98-333, 20 p. Web Address: URL: http://minerals.cr.usgs.gov/publications/ofr/98-333.
14. Phillips, J.D. (1992) TERRACE--A terracing procedure for gridded data, with FORTRAN programs, and VAX command procedure, Unix C-shell, and DOS batch file implementations: U.S. Geological Survey Open-File Report 92-5, 27 p., 1 diskette.

15. Phillips, J.D., Duval, J.S., and Ambroziak, R.A. (1993) National geophysical data grids--gamma-ray, gravity, magnetic, and topographic data for the conterminous United States: U.S. Geological Survey Digital Data Series DDS-9, 1 CD-ROM disk. [includes potential-field software version 2.1].
16. Pratt, William K. (1991) *Digital Image Processing*, John Wiley and Sons, Inc., New York.
17. Star, Jeffrey, and Estes, John (1990) *Geographic Information Systems: An Introduction*, Prentice-Hall, Englewood Cliffs, New Jersey.
18. Swayze, G.A., Clark, R.N., Pearson, R.M., and Livo, K.E. (1996) Mapping acid-generating minerals at the California Gulch Superfund Site in Leadville, Colorado using imaging spectroscopy, *Summaries of the Sixth Annual JPL Airborne Earth Science Workshop*, JPL Publication 96-4, vol.1, March 4-8, p. 231-234.
19. Swayze, Gregg A., *et al*. (in press) Using imaging spectroscopy to cost-effectively locate acid-generating minerals at mine sites: An example from the California Gulch Superfund Site in Leadville, Colorado, *Summaries of the Seventh Annual JPL Airborne Earth Science Workshop*, JPL Publication 97-21, vol. 1, AVIRIS Workshop, January 12-16, 1998.
20. Watson, Kenneth (1992) Spectral ratio method for measuring emissivity, *Remote Sensing of the Environment.*,**42**, 113-116.
21. Watson, Ken (1997) Three algorithms to extract spectral emissivity information from multispectral thermal data and their geologic applications, *First JPL Workshop on remote sensing of Land Surface Emissivity*, Pasadena, May 6-8, 1997.
22. Watson, Ken, and Rowan, L.C. (1996) Recent applications of thermal infrared data for mineral exploration and environmental studies, *Society of Exploration Geophysicists*, 66th Annual Meeting, Denver, November 10-15,1996.

PHOTO ECOMETRICS FOR NATURAL RESOURCE MONITORING

P. GONG
Center for Assessment and Monitoring of Forest and Environmental Resources (CAMFER)
Department of Environmental Science, Policy, and Management
151 Hilgard Hall, University of California, Berkeley, CA 94720-3110 USA

Abstract This chapter introduces some results obtained from new applications of digital photogrammetry for forest growth and topographic changes. They are by-products of our attempt toward the development of the field of *photo-ecometrics*. The goal is to provide low-cost yet accurate estimates of as many important biophysical parameters as can be measured and inferred with high resolution remote sensing data. Six strategies for information extraction from remotely sensed data are introduced: image classification, statistical regression, linear feature extraction, 3D surface modeling, radiative transfer modeling and inversion, and change monitoring. Accuracies of traditional multispectral image analysis algorithms of remotely sensed data are low. Traditional photo interpretation is error prone and expensive. We advocate the use of digital surface model (DSM) that contains the elevation of all surface features such as buildings and trees rather than digital elevation model (DEM) that has been traditionally used only for the terrain. Data fusion should not be considered only as integrating data acquired from different sources but also data (information) extracted from the same source of data with different strategies. This can be partly realized by new image analysis strategies that make use of the 3D surface information from stereo images and the multispectral, texture and contextual information inherent in the imagery. With digital photogrammetry, it has been proven that digital aerial images can be georeferenced and orthorectified to an accuracy of one to several meters. With the georeferenced and orthorectified digital images and many biophysical parameters accurately determined we can detect changes of species composition, height, crown closure, and diameter of forested land and topographic features. These same techniques will not only significantly improve our ability to economically assess the accuracy of vegetation and thematic maps but also provide alternatives to detailed mapping of geomorphological features such as landslides, surface mining, land erosion and material deposition. Illustrated in this contribution is the usefulness of DSM and orthophotos generated with digital photogrammetry in the monitoring of changes gully erosion, water channel incision and coastal salt flat zone displacements.

1. Introduction

The need for detailed forest parameters (species, size, canopy density and numbers of trees) and other biophysical parameters including detailed topographic models of landscape over large land holdings in the US has increased markedly in the last seven years. However, to efficiently manage forest landscapes for forest, wildlife, and other biological resources we need to have detailed information on forest composition and structure. It is usually economically infeasible to collect the requisite 5-10% field sample. On the other hand, detailed topographic data can not be obtained from regular USGS topographic maps with scales ranging from 1:24,000 to 1:100,000. Field based surveying and mapping can be done but the cost could be unbearable as the site gets bigger. Thus, it is critically important to be able to develop new advanced remote sensing technologies that allow for direct measurement of the parameters needed for landscape monitoring and substantially reduce the cost of obtaining this information compared to field sampling.

We proposed an interdisciplinary field, ecometrics, the science and technology of obtaining reliable ecological measurements over large landscapes [17]. Biologists have largely overlooked the field of ecometrics because it requires skills usually outside their expertise: remote sensing, photogrammetry, statistics, and biometrics. Nonetheless development of this field is critical to provide the tools to aquatic and wildlife biologists, ecologists, foresters, geographers and geologists to be able to measure and monitor landscape level ecosystem processes, and changes in land use and landforms. This research falls in a sub-field of ecometrics, *photo ecometrics* - the use of photogrammetry and image analysis techniques to derive ecological and landscape parameters.

In the rest of this chapter, I introduce six strategies for information extraction from remotely sensed data. They serve as the fundamental tools for the development of photo-ecometrics. The use of digital photogrammetry will then be demonstrated to the change monitoring of forest and terrain features with examples from California and Florida.

2. Information extraction strategies

Raw remotely sensed data of the Earth surface should be processed prior to their use in information extraction. Such processing is referred to as preprocessing. It includes radiometric and geometric correction, noise reduction, image enhancement, etc. Information extraction can be done through either visual interpretation or computer analysis. We briefly introduce five computer-based information extraction strategies.

2.1 IMAGE CLASSIFICATION

Image classification is one of the oldest computer-based image interpretation techniques developed for agricultural land classification in early 1970s. Earlier algorithms are primarily statistical classification methods that classify an image based on spectral information at the individual pixel level. As the spatial resolution of remotely sensed imagery improves, traditional statistical algorithms such as minimum distance classifiers (MDC) and the maximum likelihood classifier (MLC) produce classification results far less than desired.

Image classification involves two phases. The first is the design of the classification scheme. A scheme is designed as a compromise between the information expected, and the data quality and capabilities of classification techniques available. The second phase of image classification is the implementation of the classification scheme. In a typical computer-assisted classification, there are five major steps to be followed. These are:

- Step 1: Data collection and preprocessing, including radiometric and geometric correction, feature extraction and selection, data reduction, and noise elimination.
- Step 2: Training, involving either supervised or unsupervised training. Supervised training is the most commonly used approach.
- Step 3: Pixel labeling, which refers to the use of a classification algorithm to assign each image pixel to a class, according to the training statistics.
- Step 4: Post-processing for improving the visual appearance of the image. This includes filtering the classified results (e.g., majority filtering), performing geometric transformation to a specific map projection depending on the application design, and map decoration.
- Step 5: Accuracy assessment of the classified image compared with ground information. As discussed by [26], there are many different ways to assess classification results. If the results do not meet with classification requirement, some of the previous 4 steps must be adjusted.

Feature extraction in the first step usually refers to image filtering or texture image generation based on neighborhood operations [22, 16]. In the second step, supervised training involves the collection of sample pixels in the image followed by statistical signature generation for each class. Unsupervised training usually uses clustering such as c-means or the ISODATA (iterative self organizing data analysis technique A) for an initial image segmentation. Image segments are then examined and merged to form new classes. Sometimes statistical signatures are calculated based on merged image segments for a subsequent re-classification using a different classification algorithm such as the MLC. There are many pixel-labeling (classification) methods. MDC and MLCs are the traditional ones that can be found from just about any remote sensing image processing textbooks. Some of the recent ones include the artificial neural network and the fuzzy classification algorithms [27].

Over the past 30 years, computer based image classification has been evolving from conventional multi-spectral classification to the incorporation of ancillary data such as terrain and climatologic data, the use of spatial and temporal information in the classification decision with statistical, artificial neural networks and knowledge-based approaches. One of the major directions is the development of contextual classification methods. Conventional multi-spectral classification techniques perform class assignments based only on the spectral signatures of a classification unit, while contextual classification refers to the use of spatial, temporal, and other related information, in addition to the spectral information of a classification unit, in the classification of an image. When combined with other attributes in classification, multi-spectral classification is extended to multi-feature classification.

More detailed explanation on various classification techniques can be found in classical works such as [9, 15, 20, 22, 25, 28, 29, 39, 50, 52].

2.2 LINEAR FEATURE EXTRACTION METHODS

Different linear features such as roads, rivers, ridges, faults, and boundaries of vegetation communities have different extents in space. Their clarity and detectability from remote sensing imagery vary with image spatial resolution. At 10-30 m resolution, mountain ridges and faults structure may not be seen as clearly as with coarser resolution, but a road network can be seen easily as lines connected by pixels that are relatively brighter or darker than their surroundings. As the spatial resolution improves to 1-3 m, road networks become area-phenomena while even linear features at even finer scale such as vegetation boundaries, field boundaries and irrigation network can be easily observed. Therefore, the detectability of different linear features is determined by different spatial resolution of images.

Wang and Liu (1994), [48], reviewed various linear feature extraction algorithms. Most algorithms work on a single image of a particular spectral band. Edges are first enhanced from the image by applying high pass filters that are based on the calculation of gradient of the first or second order. Linear feature extraction is then conducted based on the edge-enhanced image using some advanced analysis algorithms. These include operators developed from mathematical morphology [8, 49], dynamic programming [19], gradient direction profile analysis [45, 47, 49], and knowledge-based inferring [46, 38]. As spatial resolution varies with different remote sensing imagery, linear features of certain sizes may become area features. Thus, image classification methods may be applicable to the initial detection of those linear features [18]. All these methods contain operators for the detection of edges, linking or trimming (pruning) of edges, and thinning of edges.

2.3 STATISTICAL REGRESSION METHODS

Most remote sensors measure the radiation reflected, emitted or transmitted from the atmosphere, ocean or the land surface at a short instant. For terrestrial remote sensing, in the 0.3 ~ 3 μm sensors collect primarily the reflected solar energy. The reflection capability varies with surface materials and wavelengths. Precise measurement of reflected solar energy from earth surface is important to the studies of earth energy balance and more specifically to surface evaporation, surface evapotranspiration, and atmosphere-land surface energy and matter exchange. For many years, the reflection properties by surface materials have been approximated by their *albedo* – the total solar radiation reflected in all directions to the incident solar radiation that is sometimes approximated by the average of spectral reflection in the visible and infrared region. It has been widely used in climatologic and hydrologic modeling efforts. For any given surface location, given the albedo and the solar incident energy varying with time, one can calculate the total energy reflected and absorbed. The absorbed energy determines the potential amount of water to be evaporated or evapotranspirated. Thus, a regression between absorbed or reflected energy and various climatological, ecological and hydrological parameters can be built to enable us to predict those parameters from remotely sensed data. The advent of multi-spectral remote sensing, particularly hyper-spectral remote sensing in a wide electromagnetic range, not only provides us more opportunities for *statistical regression analysis*, but also makes it possible to improve the level of accuracy in regression analysis.

Statistical regression with remotely sensed data can be done in a number of ways. The dependant variable, y, is the parameter to be estimated from remotely sensed data. The independent variable(s), $X = (x_1, x_2,..., x_k)^T$, can be brightness data from a single spectral band, a combination of bands, ratio or difference data from a number of bands or other transformed values of brightness data. When $k = 1$, we have a univariate statistical model $y = f(x)$ and when $k > 1$ we have a multivariate regression model $y = f(X)$. The function f can take any form. The widely used forms are linear, exponential, logarithmic, hyperbolic and parabolic ones. The dependant variable y ranges from physical parameters such as climatologic, hydrologic, biophysical, biogeochemical variables to socio-economic parameters such as population density, family income, environmental quality, etc. It involves five steps to establish statistical regression models:
- Step 1: Field sampling of y. This includes locating sampling sites, field measurements and/or data collection for the dependant variable.
- Step 2: Data extraction for the independent variable(s) X from remotely sensed imagery according to the field site location.
- Step 3: Searching for the most appropriate function (model) between the y and X through exploratory statistical analysis.
- Step 4: Determine the function coefficients through least squares method.
- Step 5: Statistical model validation using additional sample points.

During the search for a better function, scatterplots between y and X are helpful. At step 5, samples that have not been used in model coefficient estimation should be used to estimate the level of error of the model. If the model error is too large, steps 1-4 should be examined again for potential improvement.

[32], correlate urban settlement sizes determined from Landsat MSS images with population to estimate urban population sizes in China. [2] established statistical relationships between *vegetation index* and *photosynthetically absorbed radiation* (PAR) and *leaf area index* (LAI) of wheat in Kansas. [10] summarizes a number of methods used to estimate housing and population density and family income in Australia with Landsat MSS data. [7] examined the relationship between LAI and a number of vegetation indices derived from the red, near infrared and middle infrared for estimation of a number of agricultural variables. [40], used remote sensing derived LAI as input along with climatologic data to their Forest-Biogeochemical Cycling model to estimate forest ecosystem parameters.

2.4 INVERSION OF RADIATIVE TRANSFER MODELS

Statistical regression can be used to estimate various parameters from remotely sensed data. Statistical models are empirically built and are flexible to use. However, they need to be adjusted in time and space as they are developed based on data collected from a specific location at a given time. For a different location, a new set of model coefficients need to be calculated. Sometimes, a new model may be needed. Although generality and accuracy in modeling are usually two horses pulling in opposite directions, in practice for a given accuracy we often expect to build a model that is applicable to a larger spatial extent. To meet such an expectation, the *radiative transfer theory* based on quantum mechanics and modern physics is perhaps more helpful. Chandrasekhar (1960), [6], published the classic treatment on radiative transfer. Earlier applications of radiative transfer theory in remote sensing were made to the study of other planets in the solar system using reflectance spectroscopy and to the studies of the atmosphere and water bodies on Earth [21]. During the past 30 years, its application was expanded to the remote sensing of snow and ice [3], crops [43] and forests [30].

Radiative transfer theory is about the modeling of electromagnetic radiation propagation through various matters (media) in gaseous, liquid and solid forms. With radiative transfer theory, one can model spectral properties such as the diffuse scattering or thermal emission of matters based on (1) their internal properties such as concentration of chemical constituents, physical properties such as temperature and moisture levels, internal chemical and physical structures of each individual matter expressed in 3D geometry; (2) the abundance and geometrical structure of matters; (3) the viewing geometry; and (4) the geometry of incident energy and its intensity. At different spectral wavelength regions, different variables of the matters have different contributions. For example, chemical constituents, internal structure and moisture level may play important roles in the visible, near infrared and middle infrared, temperature plays an important role in the thermal and microwave region. Therefore the spectral properties of matter are also functions of wavelength. Because spectral properties vary with the incident energy and viewing direction, researchers use the *bi-directional reflectance distribution factor* (BRDF) to explicitly characterize scattering of matters. BRDF is the ratio of the radiance scattered by a surface into a given direction to the collimated power incident on a unit of area of a surface [21]. Modeling BRDF has attracted huge amounts of remote sensing modeler's attention.

Unfortunately, due to the complexity and lack of order of natural phenomena, modeling of their spectral properties is difficult and no exact analytic solution in closed form has been obtained. The *inversion* of such models (i.e., to derive chemical, physical and geometrical parameters of the natural phenomena from remotely sensed data), is also difficult. Sometimes it is impossible to achieve satisfactory accuracies of parameter estimation. Any radiative transfer modeling and inversion involve necessary approximations. Nevertheless, provided that certain level of accuracy can be sacrificed, mathematically simple models can be built with certain approximations and assumptions [21].

In the early days of remote sensing, inversion of canopy models using near-nadir reflectance in several spectral bands was limited in accuracy because of the fixed viewing geometry of sensors, reflectance similarity of different objects and the small number of bands. Inversion of canopy models using off-nadir data based on studies of bi-directional reflectance distribution function (BRDF) has been the primary concern for quantitative extraction of biophysical parameters and canopy architecture [11]. Directional canopy reflectance models of a canopy can be grouped into statistical models (e.g., [12]) and physically-based models. Physically-based models include radiative transfer models (e.g., [31]; [34]), geometrical-optical models (e.g., [30]), hybrid models (e.g., [44]), and computer simulation models (e.g., [35]).

Traditional inversion of canopy reflectance models employs an optimization technique [11] to estimate various model parameters through minimizing a merit function. If there is a total of n parameters to be solved for, the optimization is an n-dimensional problem. In order to find the optimal estimates of these parameters, an iterative process is necessary. At each step, the iteration direction and iteration length are important to determine. The most successful direction search algorithm was first proposed by Powell (1964) [37], and modified by [51] and [4]. Recently, neural networks were applied to invert radiative transfer models (e.g., [17]).

2.5 3D INFORMATION EXTRACTION AND GEOMETRICAL MODELING

3D geometrical information is about the morphological information of objects of interest in the image. Topographic elevation and coordinates of typical land marks are the most essential 3D geometrical information and are the primary information to be derived from aerial photogrammetry. Limited by the spatial resolution and the cross track (whisk-broom) scanning mechanism, early non-photographic remote sensing imagery were rarely used in deriving 3D information. Since the advent of SPOT HRV sensors that image the earth surface using push-broom scanners at 10 m and 20 m spatial resolution, more photogrammetric research has been made to derive digital elevation models (DEM) from satellite data.

However, DEM generation from satellite images has not been widely used. Partly because the level of accuracy is still low and partly because topographic elevation data in the forms of DEMs for most areas in the world have already been produced with aerial photogrammetric methods and topographic changes for most areas on earth are extremely slow, it is understandable that the need for similar DEM products generated from satellite imagery until recently was low. This situation is changing now. Satellite imagery at 1 m spatial resolution level are becoming available (e.g., Space Imaging's IKONOS images, and Quickbird to be launched by Earth Watch Inc.). High-resolution digital cameras are gaining a wider popularity in aerial photography.

Elevation data provided by lidars and interferometric radars have equivalent or even better accuracies when compared with those obtained with traditional aerial photogrammetric methods. In addition, height estimation method based on shading information existing in individual photographs may also be used. This technique, called "shape from shading," is developed in the field of computer vision [24]. Few applications are found in marine floor mapping from sonar imagery and mountainous topographic mapping [23]. Although it may not be as accurate as the other methods when applied to remote sensing imagery of natural scenes, the potential of this technique worth of further assessment.

Accurate surface morphological data contained in DSMs can be used as additional information for shape analysis. Unfortunately, except for the incorporation of DEM in image topographic correction, image classification and terrain structure delineation, the rich amount of spatial information that can be derived from high spatial resolution imagery using photogrammetric methods, or from other measurement methods such as radar interferometry and laser scanning, was rarely utilized [5, 14]. More research on the integrated use of spatial and spectral information in remotely sensed data is obviously possible. For example, measures about forest canopy structure such as canopy shape and size, crown closure, branch structure can be derived from DSMs. Canopy shape and size and the branch structure information may then be used to identify the tree species. Species recognition based on such spatial measures can be compared or combined with species recognition results from spectral analysis. Such kind of integration (or *fusion*) of information has not yet been reported so far.

The current situation is that stereo photography lacks spectral depth, but allows for precise photogrammetric measurements. The current generation of satellite images have more and narrower spectral bands than photographs, but are not of high spatial resolution and high geometric precision. Digital cameras bridge this gulf by providing imagery, which is of both high spatial and spectral resolution with a sufficient number of spectral bands. As high-resolution data from airborne and space-borne remote sensing become more and more available, so does DSMs derived from those data. The integrated analysis of remotely sensed data obtained by morphological analysis of DSMs and multispectral data analysis as widely used in traditional remote sensing will be a powerful tool for improving the accuracies in image classification, spatial pattern measurement and change monitoring.

2.6 CHANGE DETECTION AND HYPERTEMPORAL ANALYSIS

With the availability of daily image coverage of the globe, methods have been developed for temporal analysis of remotely sensed data. Some researchers use the *hyper-temporal remote sensing* to emphasize the huge number of images acquired at different times in image analysis [36]. In comparison with hyper-spectral image analysis that may be operated to hundreds of images in different spectral bands, hyper-temporal remote sensing involves hundreds to thousands of images in the same band or same band combination (e.g., *normalized difference vegetation index* (NDVI) between the red and near infrared band). Images acquired from different dates need to be corrected for solar angle and viewing angle differences and georeferenced at the sub-pixel accuracy. Then typical temporal analysis such as seasonal analysis and multiple-year trend can be undertaken.

2.7 SUMMARY

Although many algorithms have been developed for automated image interpretation over the past 30 years, they can be broadly grouped into one of the above six types of information extraction strategies. When checked against manual image interpretation and measurements made in the field, the accuracies of automated image interpretation methods are less satisfactory. Over the past 30 years, progress was slow on the incorporation of spatial information in remotely sensed data in computer-based image interpretation. The combined use of different information extraction strategies taking advantage of not only spectral but also spatial and temporal information in remotely sensed data is both a challenge and a new opportunity for the development of new computational algorithms for automated image processing and interpretation to extract the maximum amount of information contained in remotely sensed images. While the rest of this contribution is primarily devoted to the use of spectral information, it is hoped that the review in this section could provide us more choices of image analysis techniques in effective information extraction from hyper-spectral remotely sensed data.

3. Digital photogrammetry for change monitoring

Digital photogrammetry is a computerized technique that automates the measurement and mapping process of traditional photogrammetry. It includes all the procedures of traditional phtogrammetry such as photo orientation, stereo model construction, aerial triangulation, contour and orthophoto generation, photo mosaicing and mapping [41]. A major challenge in digital photogrammetry is image matching, a critical procedure that finds image points from the left and right photographs that correspond to the same ground points. Although many algorithms of image matching have been suggested [1], this process is error-prone in areas of trees and buildings where abrupt vertical changes are common. Some progress has been made at Berkeley to tackle the image matching problem in forest areas [42]. The two primary uses of digital photogrmmetry are digital elevation model (DEM) development and orthophoto generation. A DEM of an area is usually an array of grid points of ground elevation that exclude the heights of landscape features such as forest and buildings. For the purpose of forest measurement, a digital surface model (DSM), an array of grid points of elevation of landscape features, is necessary. An orthophoto is a photo of an area that has a constant scale, is free from point displacement caused by elevation differences. Therefore, area measurements from orthophotos are more accurate than from raw aerial photographs.

3.1 MONITORING OF FOREST CHANGES

Figure 1 shows the results from digital photogrammetry when applied to two sets of scanned aerial photographs acquired in July 1970 and August 1995. The original photos were acquired with a nominal scale of 1:12,000 at the upper Gallinas Valley, Marine County, California. They were scanned at 1000 DPI from black and white diapositives on a Vexel 3000 scanner. The digital photogrammetric software used to analyze these photos were SocetSet and VirtuoZo. Results shown here are generated automatically from digital photogrammetry. 10 ground control points were collected for photo-orientation and georeferencing. Figures 1a and 1b show the 1970 and 1995 orthophotos for the same area, respectively. Figures 1c and 1d are the 1970 and 1995 DSMs, respectively. Figures 1a and 1b were generated by projecting the raw photo onto the 1970 and 1995 DSMs, respectively.

The scanned image resolution was approximately 30 cm. The grid spacing used to extract the DSM was approximately 1 m. The DSM shown in Figures 1c and 1d were interpolated to 30 cm. The surface cover is mainly hardwood rangeland. Clusters of relatively bright areas, in Figures 1c and 1d, are coastal live oaks. Coastal live oaks can be extracted from Figures 1a and 1b through a simple image thresholding as they are much darker than their surroundings. Changes of crown closure can be obtained by comparing the two thresholded images. Were the DSMs correctly derived, changes in tree heights could be obtained by subtracting the 1970 DSM from the 1995 DSM [13]. However, this cannot be easily accomplished without modification to currently available commercial digital photogrammetry packages because they are designed to map terrain features whose elevation varies in smaller magnitudes. To overcome this problem we developed two methods, one being to modify the DSMs of trees extracted from commercial packages by introducing shadow and crown boundary information, and the other being an improvement in image matching using a model-based approach [42]. Crown closure and tree height can be extracted with better than 10% error level.

The lessons we have learnt from experiments with aerial photographs will be used to develop algorithms for analysis of images acquired from digital aerial photography. We have integrated fully digital system with a digital camera, GPS and inertial navigation systems. A preliminary test of the system was carried out with digital photography taken over the Berkeley campus in January 1997. When the system using a 28 mm camera lens is flown at 1000 m above the ground, a georeference accuracy of 2 m in the horizontal, 3.6 m in the vertical direction can be achieved [33]. This implies that the system can locate points on the ground with 2-3 m accuracy with no need of any ground control points in the area of interest.

3.2 MONITORING OF CHANGES OF GEOMORPHOLOGICAL FEATURES

Using the same sets of DSMs derived from the Marine County study site, we compared the same erosion zones and stream channels in the image. We found that the gully expansion in 1995 approximately doubled that in 1970 (Figure 2). On the other hand, the stream channel in 1995 narrowed (Figure 2) and was deeper than that in 1970 (Figure 3).

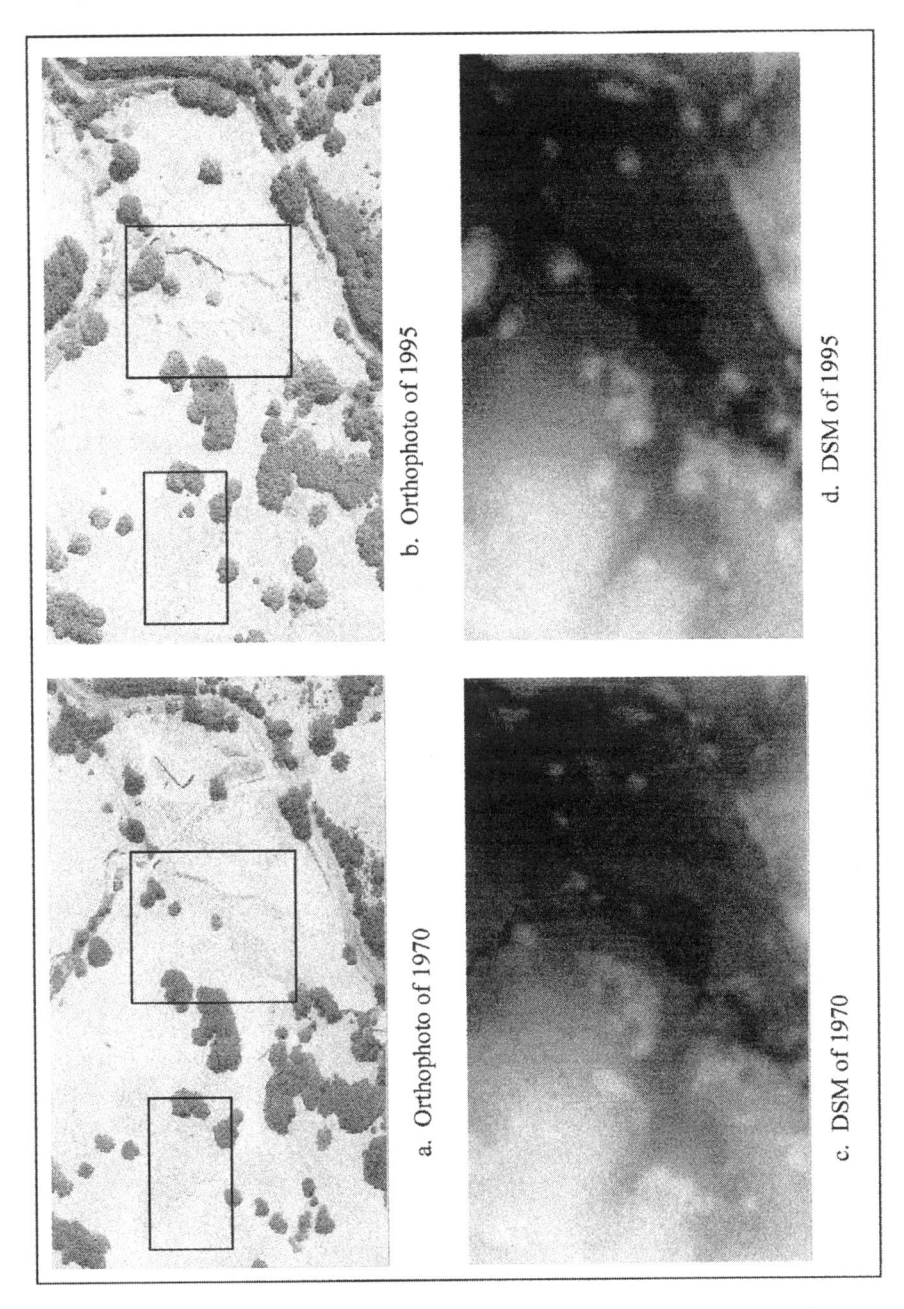

Figure 1. DSMs and orthophotos of a hardwood rangeland area. Greyscales in c and d are surface elevations. Boxed areas are shown in Figure 2.

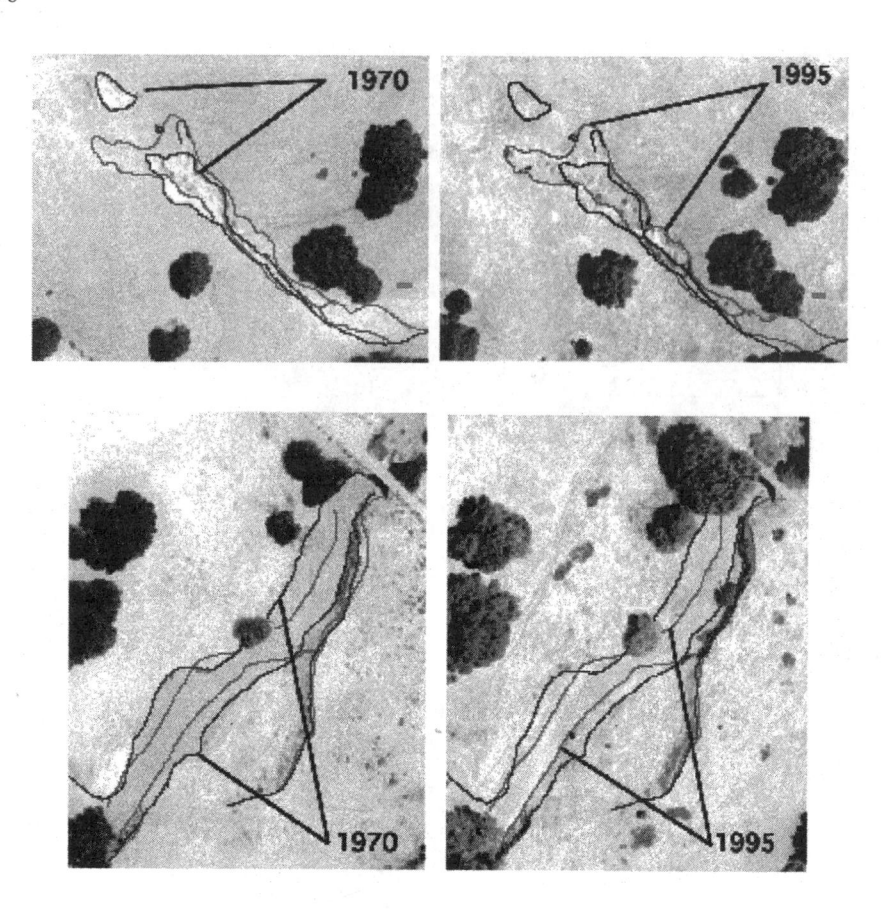

Figure 2. Slope erosion zones (upper left: 1970; upper right: 1995) and water channel incision (lower left: 1970; lower right: 1995)

Figure 2 only provides a comparison in the horizontal direction. A comparison in the vertical direction is shown by the contour lines in Figure 3. Assuming that there has been no elevation change during the past 25 years caused by geological reasons, the 70 m contour line in the study area advanced towards the upper stream for approximately 20 m during the 25-year period (Figure 3). Because the horizontal and vertical errors in the DSMs are less than 0.3 m, we are confident that both the incision of the valley and expansion of the gully are reliably measured from this data set.

Using 1:23,000 aerial stereo-pairs acquired over coastal marshland area in the St. Marks National Wildlife Preserve, Wakulla County, Florida in 1951 and 1997, we produced a digital orthophoto for each year. The marshland is transitional between in-tide and above-tide land area. At the upper edge of the marshland, salty sand flat zones can often be found without vegetation due to adequate above-tide elevation, local evaporation intensity and sand type. Through measurements, we found that the lower boundaries of the salty sand flat zones displaced toward the land for approximately 6-10 m or more from 1951 to 1997 (Figure 4). Since it was unlikely that the sand type had changed over the 46-year period, the only possible explanation is that the annual average sea level or the evaporation rate or both have undergone change in this area, leading to a conclusion that the global climate had changed over the 46-year period.

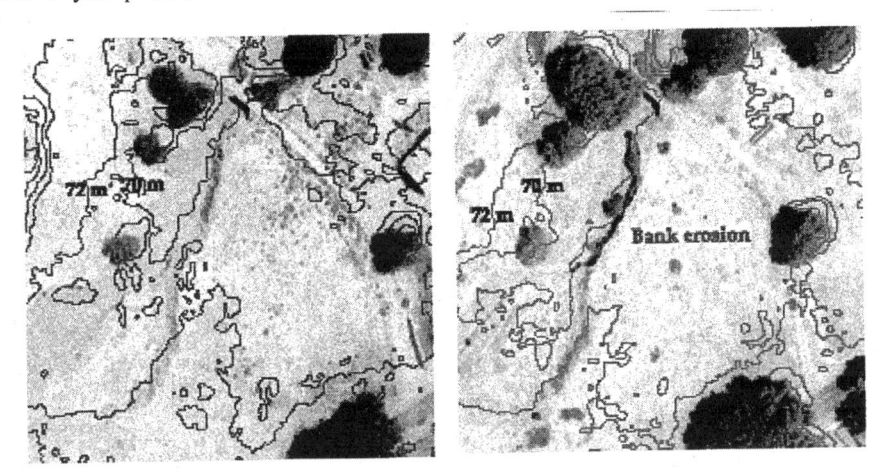

Figure 3. Channel incision mapping (Left: 1970; Right: 1995).

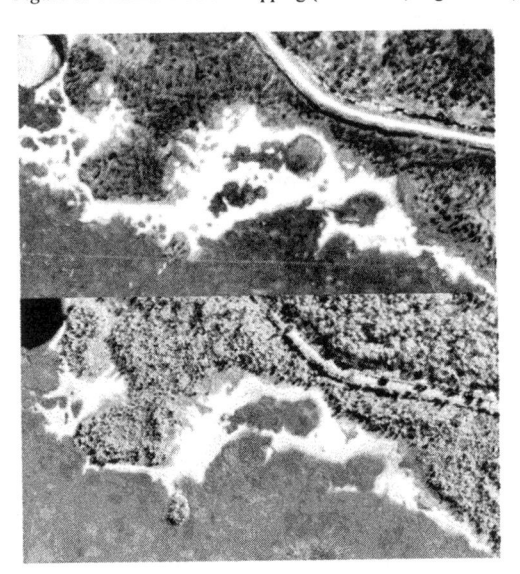

Figure 4. Marshland change in Florida as observed from orthogonal photos. Top: 1951; Bottom: 1997. The white area in the two photos are salty sand flat zones.

4. Conclusions

Undoubtedly aerial photography and digital photography are getting more popular than ever before due to the development of digital photogrammetry. Digital photogrammetry is a helpful tool for us to extract accurate DSM data from stereopairs. As high-resolution data from airborne and space-borne remote sensing become more and more available, DSMs can be easily obtained. Processing DSMs through shape analysis integrated with multispectral data analysis from traditional remote sensing will be a powerful tool for change detection. DSMs allow us to study changes not only about topography but also about canopy closure and tree heights. The potential of DSMs opens many opportunities for improving the accuracy of change detection. However, DSMs may be derived with high level of error in places where elevation changes abruptly due to poor image matching techniques used in digital photogrammetry. Further research is thus necessary to improve image-matching techniques.

Nevertheless, digital photogrammetry has the advantage of supplying 3D information from stereo aerial photographs for subsequent analysis. We demonstrated that this information is useful in estimating tree heights, removing feature displacement in the image and thus leading to more accurate crown closure estimation, measuring gully and valley developments. We will further test the potential of DSM in pattern recognition of tree species and landslides. The combined strength of high spatial resolution digital aerial photography and high spectral resolution optimal band setting will be evaluated in the context of forest species recognition and measurements.

Acknowledgements: This research is partially supported by the Integrated Hardwood Rangeland Monitoring Program of California and a USDA grant to Peng Gong and Gregory S. Biging at Berkeley. Some of the results included in this Chapter were produced by Sun Min Lee and Le Wang.

References

1. Ackermann, F. (1996) Techniques and strategies for DEM generation. Ed. Greve, Cliff, *Digital Photogrammetry: an Addendum to the Manual of Photogrammetry*, ASPRS, Bethesda, Maryland.
2. Asrar, G., M. Fuchs, E.T. Kanemasu, and J.L. Hatfield (1984) Estimating absorbed photosynthetic radiation and LAI from spectral reflectance in wheat, *Argon. J.*, 76: 300-306.
3. Bohren, C.F., and Barkstrom, B.R. (1974) Theory of the optical properties of snow, *Journal of Geophysical Research*, 55:524-533.
4. Brent, R.P., 1973. *Algorithms for Minimizing without Derivatives*, Prentice Hall, Englewood Cliff, NJ
5. Brown, D.G., and A.F. Arbogast (1999) Digital photogrammetric change analysis as applied to active coastal dunes in Michigan. *Photogrammetric Engineering and Remote Sensing*, 65(4): 467-474.
6. Chandrasekhar, S., (1960) Radiative Transfer, Dover, New York.
7. Currran, P. and H.D. Williamson (1987) GLAI estimation using measurements of red, near-infrared, and middle-infrared radiance, *Photogrammetric Engineering and Remote Sensing*, 53: 181-186.
8. Destival, I. (1986) Mathematical morphology applied to remote sensing, *Acta Astronautica*, 13(6/7): 371-385.
9. Egawa, H., and T. Kusaka (1988) Region extraction in SPOT data, *GeoCarto International*, 3: 25-30.

10. Forster, B.C. (1985) An examination of some problems and solutions in monitoring urban areas from satellite platforms. *Int. J. of Remote Sensing*, 6(1): 139-151.
11. Goel, N. (1989) Inversion of canopy reflectance models for estimation of biophysical parameters from reflectance data, in Asrar, G. Ed., *Theory and Applications of Optical Remote Sensing*, Wiley, New York.
12. Goel, N. and N. Reynolds (1989) Bidirectional canopy and its relationship to vegetation characteristics, *Int. J. Remote Sensing*, 10: 107-132.
13. Gong, P., G. Biging, R. Standiford (2000) The potential of digital surface model for hardwood rangeland monitoring, *Journal of Range Management* 53: 622-626
14. Gong, P., Greg S. Biging, S.M. Lee, X. Mei, Y. Sheng, R. Pu, B. Xu, K-P Schwarz (1999) Photoecometrics for forest inventory, *Geographic Information Sciences*, 5(1): 9-14.
15. Gong P. and P. J. Howarth (1992) Frequency-based contextual classification and gray-level vector reduction for land-use identification. *Photogrammetric Engineering and Remote Sensing*, 58(4): 423-437.
16. Gong, P., D. Marceau, and P. J. Howarth (1992) A comparison of spatial feature extraction algorithms for land-use mapping with SPOT HRV data. *Remote Sensing of Environment*, 40:137-151.
17. Gong, P., D. Wang, and S. Liang (1999) Inverting a canopy reflectance model using an artificial neural network. *International Journal of Remote Sensing*. 20(1): 111-122.
18. Gong, P., and J. Wang (1997) Road network extraction from high resolution airborne digital camera data. *Geographic Information Sciences*, 3: 51-59.
19. Gruen, A., H. Li (1995) Road extraction from aerial and satellite images by dynamic programming, *ISPRS Journal of Photogrammetry and Remote Sensing*, 50: 11-20.
20. Gurney, C.M. (1981) The use of contextual information to improve land cover classification of digital remotely sensed data, *Int. J. of Remote Sensing*, 2(4): 379-388.
21. Hapke, B. (1993) *Theory of Reflectance and Emittance Spectroscopy*, Cambridge University Press: Cambridge UK
22. Haralick, R., K. Shanmugan, I. Dinstein (1973) Texture features for image classification, *IEEE Trans. System, Man, Cybernetics*, 3(6): 610-621.
23. He, K., P. Gong, and J.A.R. Blais (1994) Shape from shading using multispectral remotely sensed data. *6th Canadian Conference on Geographic Information Systems*. Ottawa, Canada, June 4-10, 1994, vol. 1, pp.787-799.
24. Horn, B. (1986) *Robot Vision*, MIT Press, Cambridge, MA.
25. Hutchinson, C.F. (1982) Techniques for combining Landsat and ancillary data for digital classification improvement, *Photogrammetric Engineering and Remote Sensing*, 48(1): 123-130.
26. Jensen, J.R. (1996) *Digital Image Processing, a Remote Sensing Perspective*, Prentice Hall: Upper Saddle River, NJ
27. Ji, M., and J.R. Jensen (1996) Fuzzy training in supervised image classification, *Geographic Information Sciences*, 2:1-11.
28. Ketting, R. and D.A. Landgebe (1976) Classification of multispectral image data by extraction and classification of homogeneous objects, *IEEE Trans Geoc. And Electronics*, GE-14(1): 19-26.
29. Landgrebe, D. (1980) The development of spectral-spatial classifier for earth observation data, *Pattern Recognition*, 12(3): 165-175.
30. Li, X., and A. Strahler (1985) Geometric-optical modeling of a coniferous forest canopy, *IEEE Trans. Geosci. Remote Sensing*, GE-23: 705-721.
31. Liang, S., and A. Strahler (1995) An Analytic Radiative Transfer Model for a Coupled Atmosphere and Leaf Canopy, *J. Geophys. Res.* 100 (D3): 5085-5094.
32. Lo, C.P., and R. Welch (1977) Chinese urban population estimates, *Annals of Association of American Geographers*, 47:246-253
33. Mostafa, M., K-P, Schwarz, P. Gong (1997) A fully digital camera for airborne mapping. *Proceedings of International Conference on Kinematic Systems in Geodesy, Geomatics and Navigation*, Banff, Canada, June 1997, pp. 463-471.
34. Myneni, R.B., J. Ross and G. Asrar (1990) A review on the theory of photon transport in leaf canopies, *Agric. For Meteorol.*, 45:1-151.
35. Oikawa, T (1977) Light Regime in Relation to Plant Population Geometry. II. Light Penetration in a Square-Planted Population, *Bot Mag. Tokyo*, 90, 11-22.
36. Piwarwar, J., and E.F. LeDrew (2002) in press.
37. Powell, M. (1964) An Efficient Method for Finding the Minimum of a Function of Several Variables without Calculating Derivatives, *Computer J.*, 7: 155-162

38. Qian, J.Z., W.E. Roger, and J.B. Campbell (1990) DNESYS: an expert system for automatic extraction of drainage networks from digital elevation data, *IEEE Transactions on Geosciences and Remote Sensing*, 28: 29-44.
39. Richards, J., D. Landgrebe, P. Swain (1982) A means for utilizing ancillary information in multispectral classification, *Remote Sensing of Environment*, 12: 463-477.
40. Running, S. W., R. R. Nemani, D. L. Peterson, L. E.Band, D. F. Potts, L. L. Pierce, and M. A. Spanner (1989) Mapping regional forest evapotranspiration and photosynthesis by coupling satellite data with ecosystem simulation, *Ecology*, 70(4): 1090-1101.
41. Saleh, R.A., and F.L. Scarpace (1994) *Softcopy Photogrammetry, the Concepts and the Technology*, Lecture Notes, University of Wisconsin, Madison, 88p.
42. Sheng, Y., P. Gong, G.S. Biging (2001), A model-based image matching algorithm for 3D reconstruction of conifer crown, *Photogrammetric Engineering and Remote Sensing*, 67 (8): 957-965.
43. Suits, G.H. (1972) The calculation of the directional reflectance of vegetation canopy, Remote Sensing of Environment, 2: 117-125.
44. Suits, G.H. (1983) Extension of a Uniform Canopy Reflectance Model to Include Row Effects, *Remote Sens. Environ*, 13, 112-129
45. Wang D., D.C. He, L. Wang, and D. Morin (1996) Extraction of urban and road networks from SPOT HRV images, *International Journal of Remote Sensing*, 17: 827-833.
46. Wang, F., R. Newkirk (1988) Knowledge-based system for highway network extraction, *IEEE Transactions on Geoscience and Remote Sensing*, 26: 525-531.
47. Wang, J. (1989) *A New Automated Linear-Feature Network Detection and Analysis (LINDA) System and Its Applications*. PhD Dissertation, Department of Geography, University of Waterloo.
48. Wang, J.F., W. Liu (1994) Road detection from multispectral satellite imagery, *Canadian Journal of Remote Sensing*, 22: 180-191.
49. Wang, J., P. Treitz, P.J. Howarth, 1992. Road network detection from SPOT imagery for updating geographical information systems in the rural-urban fringe. *International Journal of Geographic Iinformation Science*, 6(2): 141-157.
50. Weszka, J. R. Drer, A. Rosenfeld (1976) A comparative study of texture measures for terrain classification. IEEE Trans. Systems, Man and Cybernetics, 6(4): 269-285.
51. Zhangwill, W. (1967) Minimizing a Function without Calculating Derivatives, *Computer J.*, 10:293-296.
52. Zheng X., P. Gong, and M. Strome (1995) Characterizing spatial structure of tree canopy using colour photographs and mathematical morphology. *Canadian Journal of Remote Sensing*, 21(4): 420-428.

MULTIPLE DATA LAYER MODELING AND ANALYSIS IN ASSESSMENTS

GREGORY K. LEE
United States Geological Survey
Box 25046, Mail Stop 973
Denver Federal Center
Denver, Colorado, USA 80225

Abstract. Geoscientific information is essential for assessing not only mineral resource favorability, but also geoenvironmental hazard potential. The health of any ecosystem is directly related to its underlying geology. Within every watershed, concentrations of metals and other elements in ground and surface waters, sediments, soils, plants, many animals, and humans are intimately connected to the geologic "landscape" of the area. Comprehensive, unbiased geoscientific information is therefore fundamentally and critically important in effectively assessing current, as well as past and future impacts on the environment caused by natural and/or anthropogenic changes.

A multidisciplinary team of U.S. Geological Survey (USGS) scientists was assembled to study the geoenvironmental assessment process and to develop applications for the State of Montana, north-central USA. Areas of expertise represented in the team include geology, geochemistry, remote sensing, and geophysics. A variety of "layers" of digital information related to these various disciplines was compiled for the project.

Several questions typically arise in assessment projects that involve numerous information components: (1) what categories of which data are relevant to the study objectives; (2) what relative importance (weights) should appropriately be ascribed to the data layers; and (3) how should the layers be combined to effectively achieve the objectives of the study. The Montana project has relied heavily on geographic information systems (GIS) applications to attempt to address these questions in pursuit of the objective of mapping geoenvironmental potential for acidic, metal-rich drainage. However, methodologies utilized in this project could effectively be applied to other geoenvironmental questions or for assessing potential for mineral resources.

Mining districts within the State were characterized according to their acid generating and acid buffering potentials. These properties were assigned according to considerations of abundance of sulfides and of calcium carbonate in the mining areas. Those areas that were characterized as high in acid generating potential and low to moderate in acid buffering capacity were chosen as prototypical of areas to be identified in the geoenvironmental assessment. Various classes of each information layer were considered as "candidates" for inclusion in the overall assessment model and were tested for significance of spatial association with the reference, or prototype, areas. This was accomplished using the GIS to calculate the "probability ratio" of each class, representing the relative likelihood of finding the class within the reference areas versus elsewhere in the study area. These computations not only guided determination of which data layers were relevant to the assessment, but also what weights should appropriately be assigned to them. The resulting weighted "submodels" were combined, using various methods, to produce composite models and derivative maps showing relative potential for acidic, metal-rich drainage.

1. Introduction

In the early 1990s a rising concern about the environmental effects of abandoned mines became an issue with many State and Federal land-use agencies, especially in the western United States. Because of some ongoing acid mine drainage problems in Colorado, west-central USA, a statewide assessment of environmental effects related to abandoned mines in Colorado was made by the U.S. Geological Survey [12] utilizing digital data from many sources including State and Federal agencies. The resulting 1:500,000 scale assessment maps were well received by both the public and private sectors. Building on the success of the Colorado assessment, a statewide geoenvironmental assessment of Montana, northwest-central USA, was proposed. Montana was selected because the State has had a long history of mining and includes many districts, abandoned mines, and large areas of State-managed lands and Federal lands administered by the U.S. Forest Service, the U.S. Bureau of Land Management, the U.S. Fish and Wildlife Service, and the National Park Service.

The following discussions and examples pertain to a resulting geoenvironmental assessment project for the State of Montana, and the objective has been to assess statewide potential for acidic, metal-rich drainage. However, similar procedures could as effectively be directed toward assessing potential for other types of environmental hazards or for mineral deposits. Geographic information systems applications have been extensively used in the project to assemble, evaluate, interpret, and integrate the various digital data layers; to construct assessment models and maps; and to graphically display results. Emphasis in the following descriptions will be focused less on particular assessment results of the project, and more on describing general methods that have been used to systematically evaluate and combine various geoscientific information layers to produce composite models.

2. Objectives and Hypothesis

The purpose of the Montana project has been to map statewide potential for acidic, metal-rich drainage. In other words the objective of the study has been to measure and map, for all areas in the State, the predicted success of the following, underlying hypothesis: *this area is favorable for acidic, metal-rich drainage.* This premise has been used to guide data selection and modeling procedures for the investigation.

3. Data Sources

The data used for consideration in the modeling process were selected according to 2 separate criteria: (1) from a scientific judgment perspective, what information might be *useful*, and (2) from a practical viewpoint, what information was *available*, particularly in digital form. Numerous data layers of various types were obtained both from within the USGS and from outside sources. Geochemical information was obtained from NURE (National Uranium Resource Evaluation) data archives [16] part of the National Geochemical Database; geophysics from the Mineral Resources Program geophysical databases [10]; digital geology was modified from a digital version [13] of the State Geologic Map produced by Ross *et al.* [14]; mineral occurrence data from the MAS/MILS (Minerals Availability System/Mineral Industry Location System) and MRDS (Mineral Resources Data System) databases [15]; precipitation data from Oregon State University; geography and topography from the USGS EROS (Earth Resources Observation Systems) Data Center; numerous vector coverages from the Montana State Library Natural Resource Information System (NRIS) – an excellent source of on-line spatial data; Science Applications International Corporation (SAIC) through contractual agreement with the Environmental Protection Agency (EPA); ecoregion boundaries from the U.S. Forest Service; mining areas from in-house digitizing by the USGS; ground and surface water pH from NURE archives [16] and from the USGS-Water Resources Division.

These data were obtained and processed in a variety of formats including ERDAS IMAGINE, MapInfo, Arc/Info, GS-Map, ER-Mapper, EarthVision, DLG, DEM, in-house geophysics grids, Arc GRID, database and spreadsheet files, PCI, and STATPAC. ERDAS IMAGINE was used as the GIS for data compilation, integration, processing, interpretation, modeling, and map composition in the project. The diversities of data types and formats that were assembled for the project were readily assimilated by the IMAGINE system. Most of the data layers used in the modeling exercises are raster (grid cell) based information. The exceptions, geology and mining areas, were also converted to raster form to facilitate the model formulations.

4. Procedures

The general investigative approach used in the Montana study has been, after compilation of the various digital data layers, to systematically attempt to determine which categories of what layers are predictive of the identified issue, geoenvironmental potential for acidic, metal-rich drainage. It was felt that a number of areas could be identified in the State that are known to satisfy the hypothesis, and could therefore be used as characteristic references, or prototypes, of the sorts of areas to be identified in the multidisciplinary assessment model. Because these references are intended to represent areas where the hypothesis is held to be true, levels of spatial association with the reference area are taken to be measures of association with success of the hypothesis throughout the study area. The spatial association of data layer classes with these prototype, or reference areas, has therefore been used to serve as a testing criterion to help guide determination of what categories of which layers are important in satisfying the hypothesis, and moreover, what weights should appropriately be assigned to these classifications. In addition, the reference areas can be used in testing the validity of assessment models.

An important assumption is made in using prototype(s) for weighting and validity analysis: the reference is assumed to adequately characterize the modeling objective (assessment goal). If this assumption is valid, the following working hypothesis may be derived from the original premise: *this characteristic is predictive of the prototype* (and therefore is predictive of areas with similar attributes). By using this working premise as a surrogate for the original hypothesis (*this area is favorable for acidic, metal-rich drainage*), testing of multi class data layers can be performed.

If weights will be assigned to data layers based on measures of spatial associations with the prototype, it is important to note that the validity of the model depends directly on the efficacy of the reference. Therefore, in this treatment, the prototype, whether areal, as in the present case, or punctual, as used by Elliott *et al.* [3], is the foundation upon which the model edifice is reared. As a result, this methodology is not applicable for areas in which adequate references cannot be established.

4.1. PROTOTYPE AREAS

Polygonal boundaries of 142 mining areas within the state were digitized by team members knowledgeable about the economic geology of Montana. The boundaries were delineated from interpretations of maps and mineral occurrence data (mines and prospects) found in numerous previous reports such as those by Elliott *et al.* [3]; Frishman *et al.* [4], [5]; Green and Tysdal [6]; Ryder [15]; and Tysdal *et al.* [17]. Information related to these areas has been provided in the resulting digital coverage and includes attributes describing primary commodities, ore deposit type, geology, and history. In addition, the mining areas were characterized according to their acid-generating and acid-buffering characteristics.

Acid producing potential was assigned to each area based on a three-fold ranking (high, medium, low) of estimated metal-sulfide concentrations within the district—metal sulfides, when oxidized, produce sulfuric acid in the environment. Similarly, ranked acid-consuming potential was ascribed to each of the mining areas according to estimated calcium carbonate content--calcium carbonate provides a natural acid-consuming, or buffering influence in the environment. The rankings for acid-generating and buffering potential for the mining areas were assigned by the economic geology specialists on the team.

Those areas that were characterized as high in acid generating potential and low to moderate in acid buffering capacity were selected as prototypical of areas to be identified in the geoenvironmental assessment of potential for acidic, metal-rich drainage. These areas are shown as red and orange in Figure 1. The prototype, or reference mining areas include Radersburg, Wickes, Basin, Boulder, Butte, New World, Heddleston, Jardine, Elliston, Emigrant-Mill Creek, Rimini, and Sheridan districts.

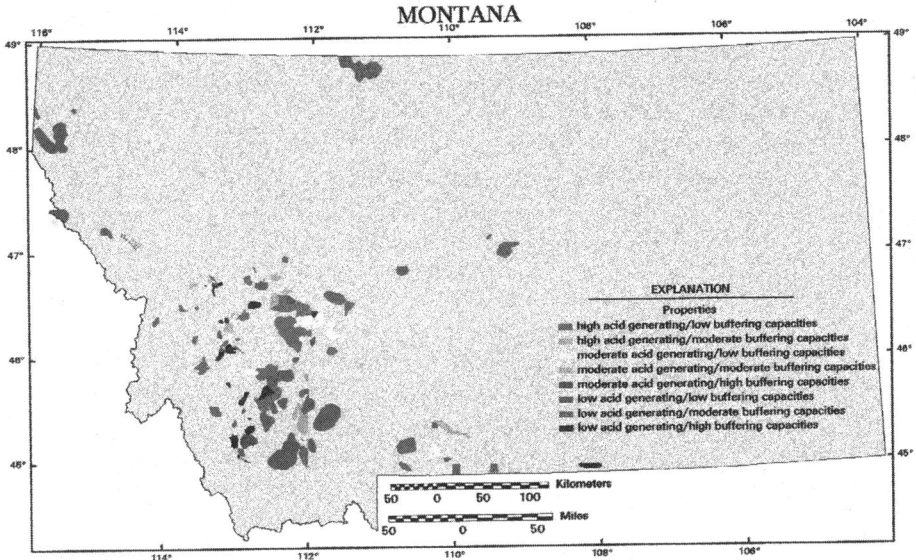

Figure 1. Acid generating and buffering characteristics of mining areas, Montana, USA.

4.2 SUBMODELS

All classes of each information layer were considered as "candidates" for inclusion in the overall assessment model and were tested for significance of spatial association with the prototype, or reference, areas. Those layers containing classes of information determined to be significantly associated with the reference areas were reclassified according to the strengths of their correlations, creating constituents, or "submodels," to eventually be combined with other submodels to produce overall composite assessment models. Testing for associative significance was performed using the GIS to calculate the ratio of the *favorability* of each class to the *prior* probability of finding a reference area within the State. This calculation is a variation of the "normalized density" computation as described by Elliott *et al.* [3] for the case where the prototype consists of points (see below).

The *prior* probability is simply the probability of finding oneself in the reference area if he were put at a random location in the study area. *Favorability* is described by Bonham-Carter [1] as the *conditional*, or *posterior*, probability which, in the present study, is the probability of finding oneself in the reference area if he were known to be at a location where a particular class of information exists. For this study the favorability:prior probability ratio, or posterior probability:prior probability ratio will be simply termed the *probability ratio*, PR.

The probability ratio is a measure of the relative likelihood of finding the reference condition within the tested class (candidate) area versus outside the class. Equivalently, the probability ratio measures the relative likelihood of finding the tested class within the reference area compared to elsewhere in the study area. Several ways of expressing the probability ratio calculation are listed below:

(a) the ratio of the percentage of the prototype area occupied by the candidate to the percentage of the study area occupied by the candidate. This calculation is shown in the following:

$$\begin{aligned}PR &= [100(A_{ref} \cap A_{test})/A_{ref}] \div [100(A_{test}/A_{sa})] \\ &= [(A_{ref} \cap A_{test})/A_{ref}] \div [(A_{test}/A_{sa})] \\ &= [(A_{ref} \cap A_{test})(A_{sa})] \div [(A_{ref})(A_{test})] \end{aligned} \quad (1)$$

where PR = probability ratio
A_{ref} = area occupied by prototype
A_{test} = area occupied by tested class
A_{sa} = study area
$A_{ref} \cap A_{test}$ = area within the prototype area occupied by the tested class
           = area occupied by *both* the prototype *and* by the tested class

(b) the ratio of the percentage of the tested class area occupied by the prototype to the percentage of the study area occupied by the prototype. This calculation is shown in the following:

$$\begin{aligned}PR &= [100(A_{test} \cap A_{ref})/A_{test}] \div [100(A_{ref}/A_{sa})] \\ &= [(A_{test} \cap A_{ref})/A_{test}] \div [(A_{ref}/A_{sa})] \\ &= [(A_{test} \cap A_{ref})(A_{sa})] \div [(A_{ref})(A_{test})] \end{aligned} \quad (2)$$

where $A_{test} \cap A_{ref}$ = area within the test class area occupied by the prototype
           = area occupied by *both* the prototype *and* by the tested class

Because $A_{test} \cap A_{ref} = A_{ref} \cap A_{test}$, it can be seen that expressions (1) and (2) are equivalent.

(c) In the case where the reference, or prototype, consists of points, rather than area(s), the expressions in (1) become the following:

$$\begin{aligned}ND &= [100(N_{ref} \cap A_{test})/N_{ref}] \div [100(A_{test}/A_{sa})] \\ &= [(N_{ref} \cap A_{test})/N_{ref}] \div [(A_{test}/A_{sa})] \\ &= [(N_{ref} \cap A_{test})(A_{sa})] \div [(N_{ref})(A_{test})] \end{aligned} \quad (3)$$

where ND = normalized density
N_{ref} = number of reference points

A_{test} = area occupied by tested class
A_{sa} = study area
$N_{ref} \cap A_{test}$ = number of reference points found within the tested class area

A probability ratio value of 1 suggests random association between the tested class and the prototype; a value less than 1 indicates negative association; and a value greater than 1 implies positive association of the tested class with respect to the reference, or prototype. For example, a probability ratio of 20 for a class implies that the class is 20 times as likely to be found within the reference area as in other (non-reference) areas in the study area. A value of 0.3 for a class, on the other hand, indicates that the class is approximately one-third as likely to be found within the reference area as elsewhere in the study area.

The GIS (ERDAS IMAGINE) has been used to perform probability ratio calculations for every class of each tested data layer. Values for $A_{test} \cap A_{ref}$, areas occupied by both tested class and prototype, were derived from summary statistics computations which provide spatial cross-correlations. Values for A_{test}, A_{ref}, and A_{sa}, were readily obtained from histogram attributes of the image layers, and the GIS functions provide for calculations of the PR formulas based on these values as above.

4.2.1. *Geology*

Geologic information used in the study was derived from digital [13] and original paper versions of the geologic map of Montana [14]. Team geologists familiar with Montana geology interpreted the State map in the context of the modeling objectives of the study. Underlying lithology influences the quality of surface and ground waters, and may naturally exacerbate or mitigate the effects of concentrations of metals and acidity. Lithologic attributes were added to the digital geologic map that describe the rock units in terms of three-fold rankings (high, medium, low) of concentrations of calcium carbonate and of pyrite (iron sulfide). The rock unit map was reclassified according to these added characteristics to create a derivative map of environmental geology (Figure 2), which depicts acid-producing (pyrite-related) versus acid-consuming (carbonate-related), capacities of the lithologic categories in the study area. This interpretation provides a view of the relative abilities of geologic map units to moderate or increase acidic, metal-rich drainage in the state.

Each of the environmentally-reclassified geologic map units was tested for significance of association with respect to the reference mining areas by calculating the probability ratio of each class (category). The weighted geological submodel was produced by reclassifying the environmental classes according to their computed probability ratios. The highest weights were assigned to rock units characterized as having relatively high potential to generate acidic, metal-rich waters, with a probability ratio of approximately 20. These units are shown as red in Figure 2.

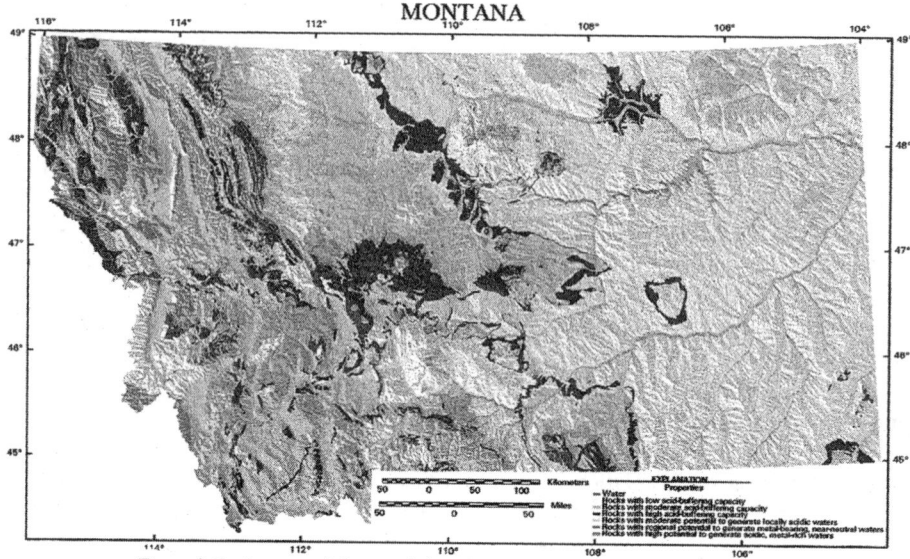

Figure 2. Environmental characteristics of geologic map units, Montana

4.2.2. Geochemistry

Stream-sediment and soil samples were obtained from over 48,000 sites in and around Montana as part of the NURE program conducted by the U.S. Department of Energy in the early 1980s. Nearly 35,000 of these sites are found within the State boundary. The analytical data for these samples were obtained from the NURE part of the National Geochemical Data Base [7] and were compiled, corrected, and reformatted by Smith [16].

For this study it was felt that stream-sediment and soil data could reasonably be combined to form a single geochemical data set. Justification derives from 3 considerations: (1) many of the samples encoded as "soil" were collected from sites lying along stream courses, implying that they are probably stream sediments from dry, intermittent streams; (2) the statistical distributions of element concentrations in the soil and stream-sediment data were found to be very similar, so that geochemical "landscapes" generated from the combined data set would not likely be distorted by the integration; and (3) the regional scale of the study serves to attenuate distortions if they do exist.

Because the mapping objective of the project is to show statewide potential for success of the stated hypothesis, it is desirable to consider classes of each data layer as areal in extent. Although the geochemical data are punctual, several methods can be used to transform the scattered point information into area representations. In this study, a sophisticated surface-generating algorithm utilized in Dynamic Graphics EarthVision software was applied to the geochemical data set to produce continuous surface model images of interpolated element values across the State. While it may be argued that stream-sediment data should be considered to represent the chemistry of the drainage basins upstream from the sample sites, at the statewide scale these basin areas are quite small ("point-like"). Furthermore, the large number (high density) of quite uniformly distributed sample locations is ideally suited for the EarthVision surface modeling application, and the regional scale of the study area serves to minimize the effects of outlier values.

The element surface models were "sliced" into standard deviation-wide concentration ranges by transforming the raster images to standard normal form. This transformation subtracts the mean of the image from each grid cell value and divides the result by the standard deviation of the image values. This was done to reduce the number of classes for each element to a more easily managed set of values for probability ratio computations. The PR calculations, as previously described, were performed for each of the class ranges of every selected element. This procedure helped guide decisions of not only which elements should be included in the geochemical submodel, but also what concentration thresholds should appropriately be assigned to indicate significant spatial association with the reference area.

Geochemical Submodel. Instead of producing a separate submodel for each element, a geochemical combination image was produced to develop a single geochemical submodel that avoids conditional dependence problems. Geochemical elements are often inter-correlated, and combining numerous single-element submodels risks double weighting of factors. The positively associated elements, Cu, Pb, Zn, Fe, Mn, Al, and V, were "bitmapped" by assigning unique values in the mathematical set $\{2^0, 2^1, 2^2, 2^3, 2^4, ..., 2^n\}$ to concentration ranges at or above thresholds of significant spatial association in the sliced element images. These values were summed to form a new image whose geochemical combination classes are uniquely identifiable. It can be seen that $[2^l + 2^m + ... + 2^n]$, where l, m, and n are integers ≥ 0 and $l \neq m \neq n$, results in a positive integer for which l, m, and n are uniquely identifiable. For example, a sum of 11 can only be produced from addends within the set by $[2^0 + 2^1 + 2^3]$; 33 can only result from $[2^0 + 2^5]$; and 47 from only $[2^0 + 2^1 + 2^2 + 2^3 + 2^5]$. A schematic diagram of the bitmapping procedure used for combining geochemical elements is shown in Figure 3.

For the summation of 7 bitmapped elements there are 2^7 (128) possible combination results (including 0 for the presence of none). Probability ratios were calculated for each of the element combination categories and weights in the geochemical submodel (Figure 4) were derived directly from the computed PR values, which range to approximately 64, indicating strong association with the hypothesis.

pH Submodel. Ground and surface water pH data were modeled similarly to the geochemical elements. It was felt that, at least at a regional scale, these data, which are thought to be related to the composition of rocks proximal to the sample sites, could be useful in modeling, but there is insufficient confidence in the quality and consistency of these data to use them as a prototype basis for assessment. The pH image was divided into discrete ranges and PR was calculated for each range. Weights were assigned to various pH ranges to produce the pH submodel. Figure 5 shows a color-encoded map of the pH submodel.

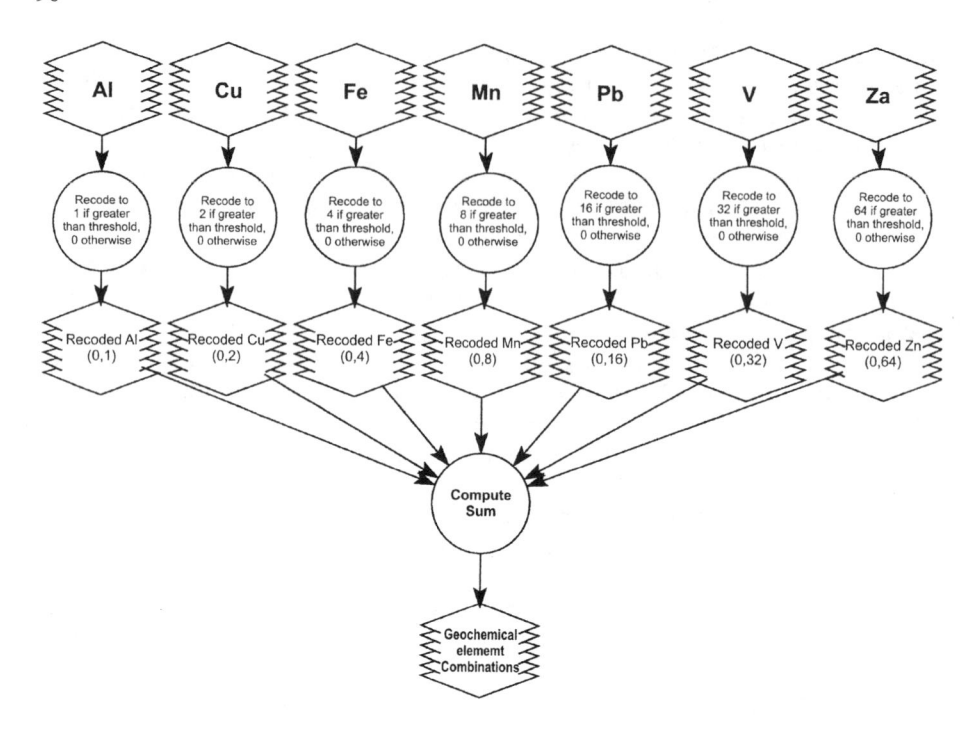

Figure 3. Schematic diagram of bitmapping procedure for combining multiple geochemical elements.

4.2.3. *Geophysics*

The mining areas that comprise the reference area are predominantly associated with deposit types that are related to the presence of plutons. Therefore, geophysical data that can be indicative of exposed or near surface intrusive rocks are potentially important in helping to identify areas that may have similarities to the prototype characteristics. The following discussion of magnetic data is provided by A.E. McCafferty, USGS; the gravity data descriptions by V. Bankey, USGS; and radiometric data information by J.A. Pitkin, USGS.

Magnetic Submodel. The entire study area is covered by aeromagnetic data collected as part of the NURE program of the U.S. Department of Energy. These data are available in digital form [10]. At this geomagnetic latitude, the inclination (71^0) and declination (17^0) of the earth's magnetic field causes magnetic anomalies to be laterally skewed and shifted from their causative sources. The amount of lateral shift will vary from anomaly to anomaly; shallow magnetic sources represented by short wavelength anomalies tend to have less horizontal shift than anomalies caused by deeper sources. A mathematical operation to correct for the shift is called "reduction-to-pole" (RTP) and was applied to the Montana magnetic data compilation.

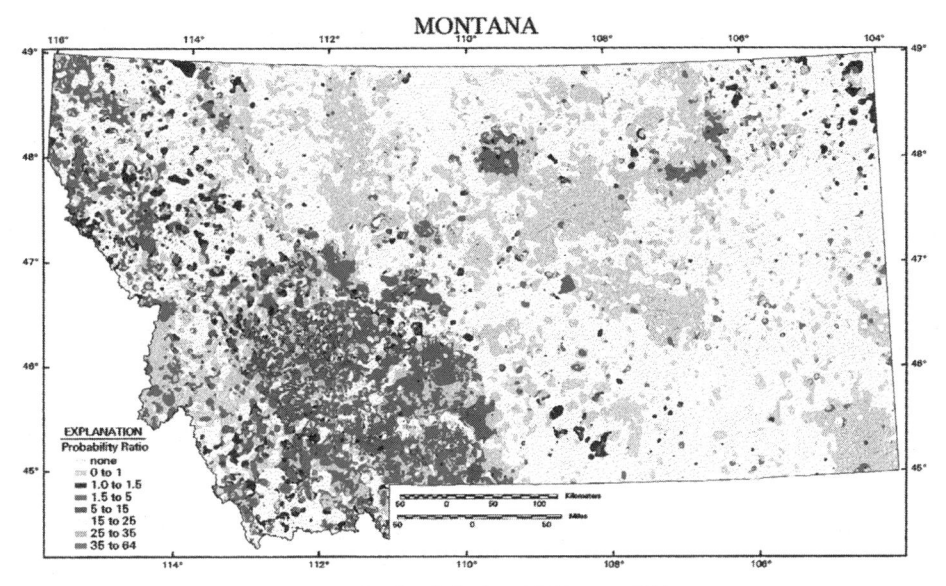

Figure 4. Geochemical submodel, Montana, USA.

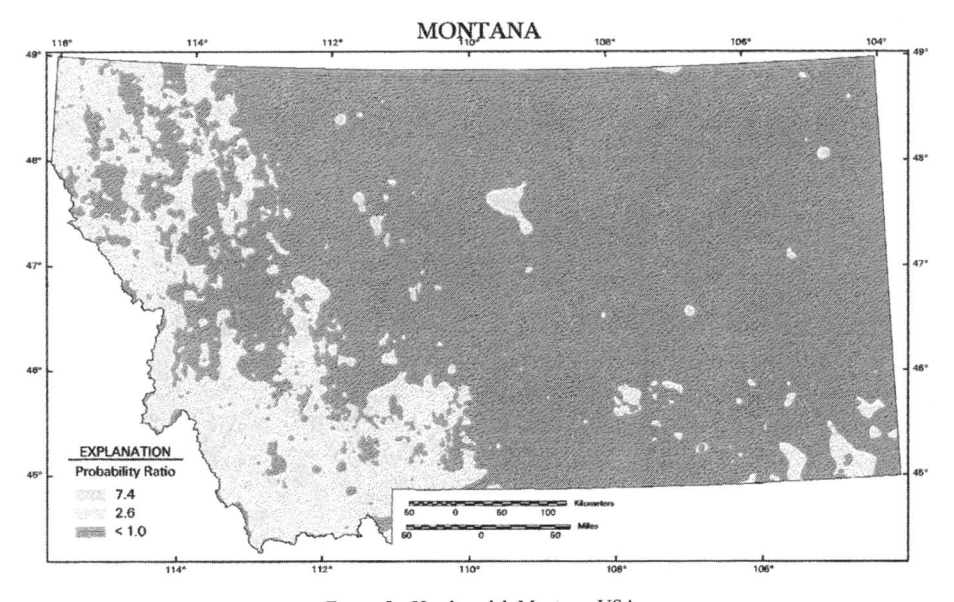

Figure 5. pH submodel, Montana, USA.

Data from the compilation show changes in the Earth's magnetic field as a result of variations in the magnetic-mineral content of near surface rocks. Typically, sedimentary rocks contain few if any magnetic minerals and do not produce significant magnetic anomalies. Therefore, the reduced-to-pole magnetic anomaly map primarily reflects lithologic and structural changes related to magnetic properties of crystalline and volcanic rocks that are likely to contain sufficient magnetic minerals (principally magnetite) to produce anomalies. Many of the aeromagnetic anomalies may correspond to metal-sulfide rich plutons that could produce acid drainage under appropriate environmental conditions.

The magnetic anomaly grid image was bandpass filtered, using ER-Mapper image processing software, to emphasize short wavelength contrasts associated with exposed and near surface features. This filtered image was imported by ERDAS IMAGINE and the result was transformed to standard normal form and tested, using probability ratio calculations, for significance of spatial association with the reference mining areas. Based on these calculations, weights were assigned to the image, producing a magnetic anomaly submodel (Figure 6), whose maximum probability ratio value is approximately 4.

Magnetic Gradient Submodel. The horizontal magnetic gradient was calculated as a derivative of the reduced-to-pole magnetic anomaly map and measures the change in magnetic field intensity as a function of change in horizontal distance. Steep magnetic gradients may be indicative of zones of contact between plutons and surrounding rocks. The values of the calculated gradient image were transformed into standard normal form, and the probability ratio of each of the resulting standard deviation ranges was calculated using IMAGINE. The correspondingly-weighted magnetic gradient submodel (Figure 7) has a maximum probability ratio of about 20, suggesting fairly strong association with the prototype area, and therefore with the underlying hypothesis. It is interesting to note that the magnetic gradient submodel has considerably higher values than does the magnetic anomaly submodel. This may suggest that contact zones around the peripheries of plutons are more supportive of the hypothesis than are the interiors.

Gravity Submodel. To explore the possibility that gravity signatures might exist in the study area that could be used to map locations of features such as intrusive bodies that could be associated with the hypothesis, a gravity anomaly image was produced from the compilation of data that includes observations from 32,152 stations. Data reduction procedures, briefly described here, are discussed in more detail in McCafferty *et al.* [9]. The Internet address for access to the report, which includes access to the gravity data, is URL:http://minerals.cr.usgs.gov/publications/ofr/98-333.

Distances between gravity stations are highly variable and range from a few hundred meters apart to tens of kilometers. Observed gravity values were calculated from field observations relative to the IGSN-71 datum [8]. Terrain corrections were computed using a program by R.H. Godson (USGS, unpublished computer program, 1978), correcting for the gravity effects of terrain from each station to a radius of 166.7 km

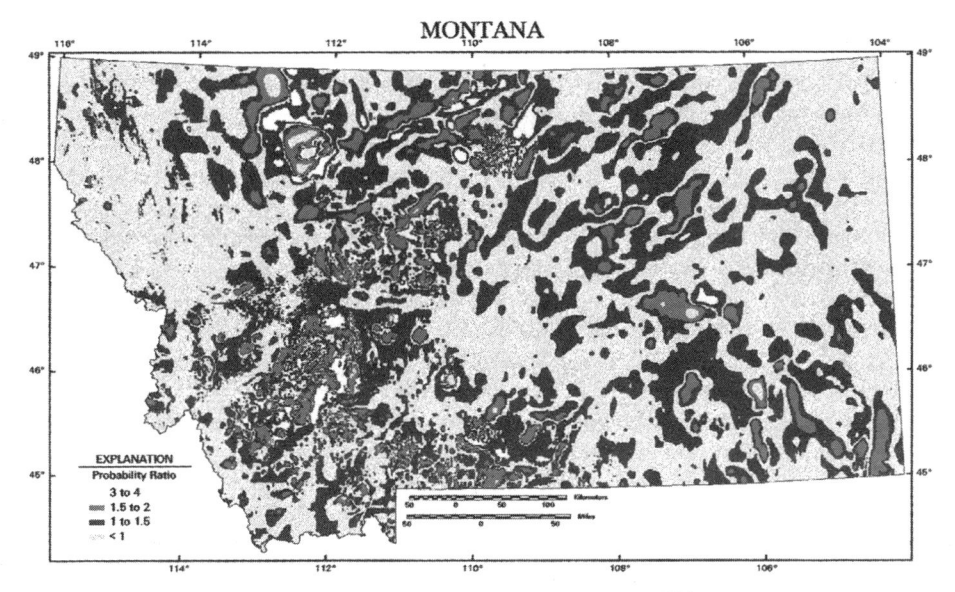

Figure 6. Magnetic anomaly submodel, Montana, USA.

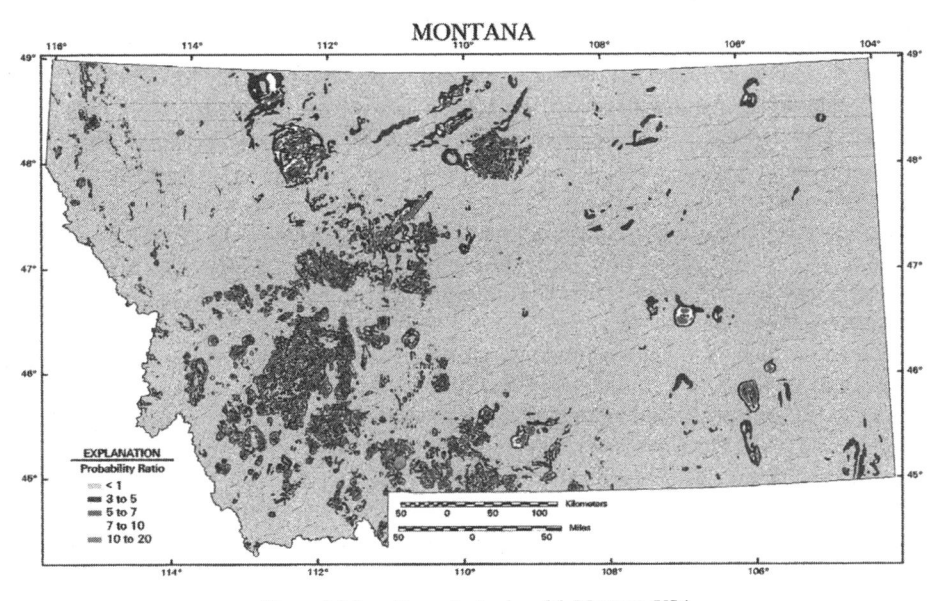

Figure 7. Magnetic gradient submodel, Montana, USA.

using the method of Plouff [11]. Bouguer gravity anomaly values were calculated using a 2.67/cm^3 reduction density employing standard USGS reduction equations. A continuous grid of 1-km interval was calculated from the irregularly spaced data using a minimum curvature interpolation technique [2].

Many of the variations in the Bouguer gravity anomalies reflect density discontinuities related to geologic features in the mid to upper crust but are masked by longer wavelength anomalies related to topography and isostacy (deficiencies in mass that support the topographic load). A regional field based on the Airy-Heiskanen isostatic compensation model was calculated and subtracted from the gridded Bouguer anomaly data to remove the long wavelength anomalies derived from deep crustal sources. Isostatic corrections were made using the following parameters: an upper crustal density of 2.67g/cm^3, a sea level crustal thickness of 30 km, and a density contrast between the lower crust and upper mantle of 0.35 g/cm^3.

Large, broad gravity anomalies caused by regional geologic features can often hide small anomalies that may be more geologically significant for locating plutons. The gravity model used in the GIS processing was calculated using wavelength filtering described by Phillips *et al.* [10]. This filtering process retains gravity anomalies that have shorter wavelengths, including any caused by shallow intrusions, whereas longer wavelengths are removed. These wavelength-filtered data, therefore, emphasize gravity anomalies produced by shallow sources and suppress the longer wavelength anomalies related to deep sources. The filtered image was imported by ERDAS IMAGINE and was transformed to standard normal form. Probability ratios were computed for each range class and weights derived from the calculations were assigned to the gravity submodel shown in Figure 8. The maximum value is approximately 24, indicating fairly strong support for the hypothesis, and is associated with areas of moderately low gravity (2.5 to 3.5 standard deviations below mean). These relatively low gravity values may be associated with low-density intrusive rocks, although anomalies are not limited to unique sources.

Radiometric Submodel. Airborne radiometric element data were investigated to discover if they would be helpful in predicting locations of successful hypothesis testing. It was felt that areas of potassic alteration or possible pluton locations might be indicated by these data.

Potassium, thorium, and uranium concentrations for Montana were extracted from 3-km grids of regional gamma ray-spectrometer data for the conterminous United States [10]. Data were acquired during the NURE program of the U.S. Department of Energy. The grid values are in ppm (parts per million) and represent near-surface (to 50 cm depth) soil or rock concentrations.

The grids were imported by IMAGINE, resampled to 500 meter resolution, and transformed to standard normal form. As in the geochemical submodel formulation, a single radiometric element combination image was produced to avoid possible interdependence problems. Probability ratios with respect to the prototype mining areas were calculated for every class of each of the three elements and concentration thresholds of significant spatial association with the reference mining areas were determined. The concentrations at and above these thresholds were bitmapped, as in the geochemical combination procedure, shown in Figure 3, and the results were summed to produce a derivative map of uniquely identifiable radiometric element combinations. Probability ratios were computed for each combination category to form the radiometric submodel, shown in Figure 9, whose weights were assigned according to the calculated results. The highest weight, approximately 30, was computed for the class denoting the presence of all 3 elements at concentrations above thresholds.

4.2.4. Remote Sensing

Iron oxide minerals derived from hydrothermal alteration and the weathering of iron-bearing minerals are predominantly red in color; other red materials at the earth's surface are artificial and have a very restricted aerial distribution. Landsat multispectral scanner (MSS) remotely sensed data were processed to map the interpreted distribution of iron oxide across the State.

The ratio of the green MSS band (band 4) and the red MSS band (band 5) results in very small values for materials that are red. A map of interpreted iron oxide was produced by mapping ratio values in the 0.093-0.11 range for 33 scenes of MSS data. The probability ratio of the resulting interpreted iron oxide presence with respect to the reference mining areas was calculated, and the result was approximately 1. This suggests *random* association, and indicates that the presence of interpreted iron oxide is not, by itself, predictive of the hypothesis. This lack of correlation may be due to vegetation cover in the western part of the State, which would interfere with MSS sensor measurements in the part of the study area where the prototype mining areas are located. A submodel was therefore not produced for this data layer.

4.3. COMPOSITE MODELS

The various submodels discussed in the preceding were combined to produce composite models and map to display relative statewide potential for acidic, metal-rich drainage. Several different combining methods have been applied, although the outcomes have generally appeared to be quite similar. Following are descriptions of various approaches used to create composite models for the assessment. The values in each of the models provide a relative measure of the degree to which the working hypothesis is successful at any location (grid cell) in the State.

4.3.1. Sum-of-Submodel-Weights Model

This composite was calculated by summing the values (weights) of the submodel images at each point (grid cell) in the State. Additive compilations reflect the notion that multiple evidence factors that are positively associated with a hypothesis reinforce each other and are therefore more supportive than any single factor taken separately.

4.3.2. Sum-of-Scaled-Submodel-Weights Model

The sum of scaled submodel weights model compilation is shown in Figure 10 as an example of a composite result. This composite was created as a variation of the sum-

96

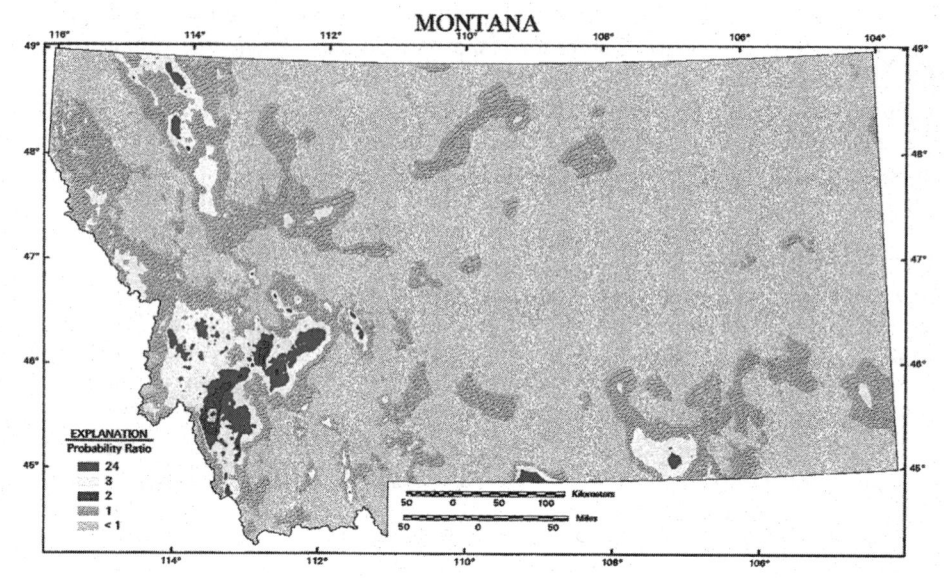

Figure 8. Gravity submodel, Montana, USA.

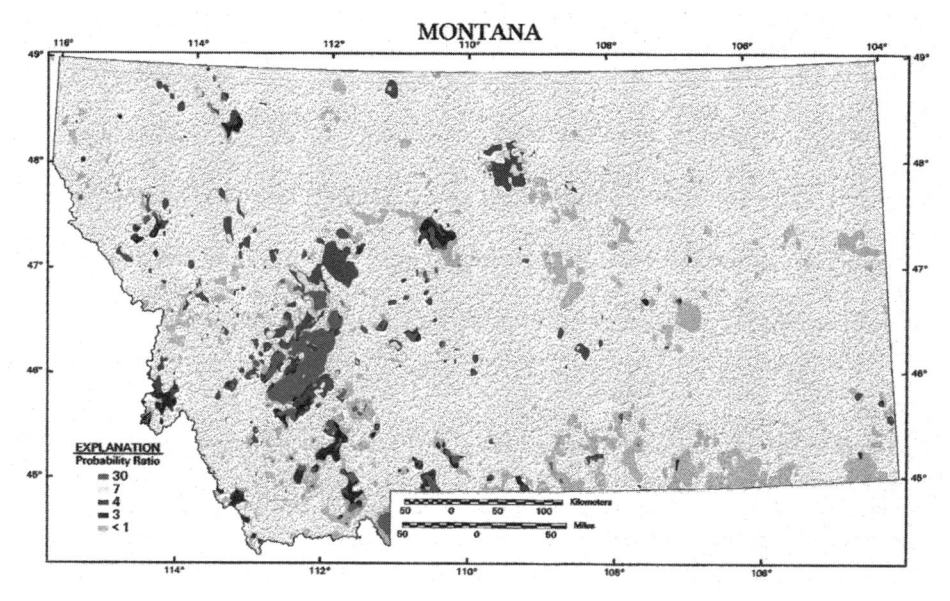

Figure 9. Radiometric submodel, Montana, USA.

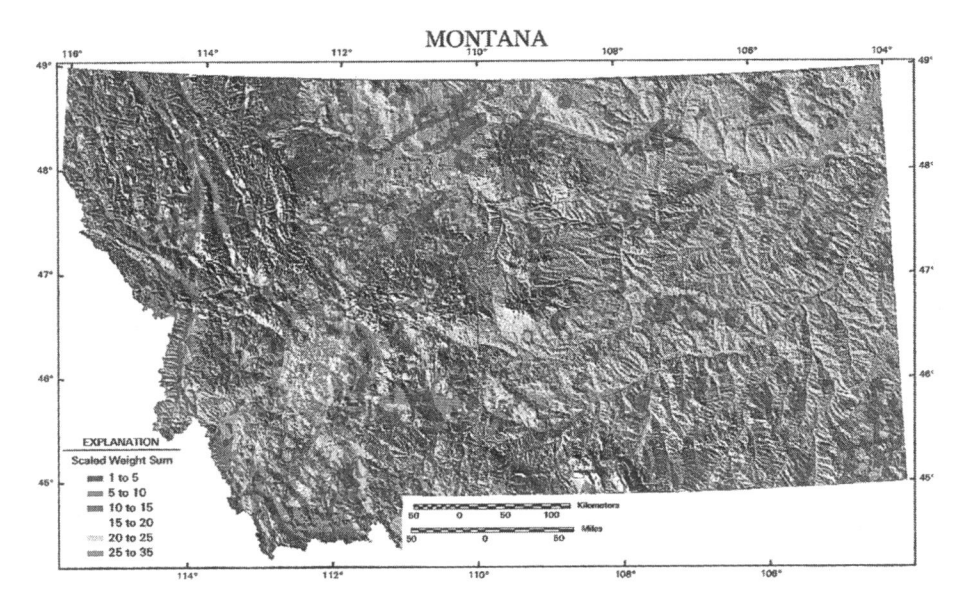

Figure 10. Sum of scaled submodel weights model of geoenvironmental potential for acidic, metal-rich drainage, Montana, USA.

of-submodel-weights compilation by assigning each submodel weights which are smaller than, but roughly proportional to the computed probability ratios of the multiple classes. This was done to not only produce a map of smaller, more easily interpreted values, but to also facilitate modification of weights based on judgment considerations of project team members. For example, it was felt that the gravity submodel might be over-emphasized if it were scaled in direct proportion to its probability ratio weights, which showed relatively strong association with the reference mining areas. This concern was related to the fact that similar gravity anomalies can derive from varied and dissimilar sources, many of which may be unrelated to potential sources of acidic, metal rich drainage. Conversely, the pH submodel was not down-scaled to the extent that direct proportion would dictate. It was felt that low interpreted pH values are directly related to the crux of the modeling effort and should therefore retain relatively high significance in the compilation.

4.3.3. *Focused Sum-of-Scaled-Weights Model*
This compilation was produced by calculating probability ratios of the multiple classes of the sum-of-scaled-submodel-weights model, described above, with respect to the reference mining areas. The sum-of-scaled-weights composite was then recoded with the results of the probability ratio computations to produce a derivative model compilation. The model is "focused" inasmuch as the effect of the second probability ratio calculation is to reduce the areal extent of the higher model values to more closely approximate the reference areas.

4.3.4. Maximum-of-Scaled-Weights Model

This model compilation was created by calculating the maximum value of scaled weights of all constituent submodels for every 500 meter grid cell in the state. As was described for the sum-of-scaled-submodel-weights compilation above, weights were assigned to each submodel, which are smaller than, but roughly proportional to the computed probability ratios of the multiple classes. In contrast to the preceding model compilations, this computation is not additive in its effects. Single submodel category weights control the output at each point. The effect of this procedure is to calculate and assign to the composite model the single most significant contribution from all constituent submodels at each point.

4.3.5. Fuzzy "OR" Model

This model was derived from fuzzy membership values [1], which were assigned to each class of the scaled submodels. Fuzzy membership is related to the degree of certainty of an observation or the confidence in a proposition. In this case, assigned membership values were primarily derived from consideration of the relative likelihood of finding the classes within the reference areas versus other areas within the state. The fuzzy OR computation produces output membership values that are controlled by the maximum values of any of the input maps for each location [1]. Thus, the fuzzy OR is given by $M_{model} = \max(M_1, M_2, M_3, \ldots, M_i)$, where M_{model} is the fuzzy membership of the composite model and M_i are membership values associated with the constituent submodels. Like the calculation of the maximum-of-scaled-weights model, the fuzzy OR calculation is not additive in its effects, and the value of the composite model never greater than the highest contributing membership value.

4.3.6. Fuzzy Algebraic Sum Model

This model compilation was also produced from computations involving fuzzy membership values assigned to the multiple classes of the submodels. The fuzzy algebraic sum is given by $1-[(1-M_1)(1-M_2)\ldots(1-M_i)]$, where M_i are fuzzy membership values of the constituent submodels [1]. Unlike the fuzzy OR model, the fuzzy Algebraic Sum is additive with regard to the effects of multiple submodel constituents. The result is always larger than or equal to the highest contributing submodel fuzzy membership value, although the maximum algebraic sum value is limited to 1.0 [1].

5. Summary

The modeling efforts presented here have incorporated systematic approaches for multiple class data analysis, weighting, and integration for the purposes of hypothesis testing and assessment modeling. Computation of probability ratios or normalized densities can be used to guide decisions of data selection, threshold evaluation, and ranking if appropriate references can be identified in the area of study. Furthermore, the calculations can be performed for multiple prototypes.

However, there are important limitations to this type of approach. It is not applicable in study areas for which no suitable reference areas or points exist to serve as testing criteria for the working hypothesis. If such references do exist, pitfalls remain inasmuch as somewhat subjective judgements may be required to characterize the reference areas or points. For example, three-fold rankings of acid generating and buffering capacities used in the Montana study are imprecise evaluation criteria, yet were fundamental to the formulation of the prototype. Scientific expertise and judgement are critical in the process of defining a suitable reference, for it provides the essential basis for the modeling efforts.

Nevertheless, the methodologies have been fruitful in this statewide regional assessment and have helped guide prioritization of drainage basins within the State for more detailed investigation and remediation efforts. The integration of scientific judgement and expertise, together with utilization of GIS-based techniques for performing investigative data analysis can be a potent set of tools in exploring relationships among multiple layers of geoscientific data. Such methods can be effectively applied to a variety of geoscientific assessment and exploration investigations.

References

1. Bonham-Carter, Graeme F. (1994) *Geographic information systems for geoscientists: modelling with GIS*, Elsevier Science Inc., Computer methods in the geosciences series, Tarrytown, New York, U.S.A.
2. Briggs, R.J. (1974) Machine contouring using minimum curvature, *Geophysics*, **39**, no. 1, 39-48.
3. Elliott, J.E., Wallace, C.A., Lee, G.K., Antweiler, J.C., Lidke, D.J., Rowan, L.C., Hanna, W.F., Trautwein, C.M., Dwyer, J.L., and Moll, S.H. (1992) Mineral resource assessment map for vein and replacement deposits of gold, silver, copper, lead, zinc, manganese, and tungsten in the Butte 1° x 2° quadrangle, Montana: U.S. Geological Survey Map I-2050-D-E, scales 1:250,000 and 1:500,000.
4. Frishman, David, Elliott, J.E., Foord, E.E., Pearson, R.C., and Raymond, W.H. (1990), Preliminary map showing the location of productive lode and placer gold mines in Montana: U.S. Geological Survey Open-File Report 90-0241, 62 p., 2 plates.
5. Frishman, David, Elliott, J.E., Foord, E.E., Pearson, R.C., and Raymond, W.H. (1992) Map showing the location of productive lode and placer gold mines in Montana: U.S. Geological Survey Mineral Investigations Resources Map MR-96, 48 p. 1 plate.
6. Green, Gregory N. and Tysdal, Russell G. (1996), Digital maps and figures on CD-ROM for mineral and energy resource assessment of the Helena National Forest, west-central Montana: U.S. Geological Survey Open-File Report 96-683-B, CD-ROM.
7. Hoffman, J.D. and Buttleman, Kim (1996) National Geochemical Data Base: 1. National Uranium Resource Evaluation (NURE) Hydrogeochemical and Stream Sediment Reconnaissance (HSSR) data for Alaska, formatted for GSSEARCH data base search software; 2. NURE HSSR data formatted as dBASE files for Alaska and the conterminous United States; 3. NURE HSSR data as originally compiled by the Department of Energy for Alaska and the conterminous United States; *with* MAPPER display software by R.A. Ambroziak and MAPPER documentation by C.A. Cook: U.S. Geological Survey Digital Data Series DDS-0018-B, CD-ROM.
8. International Association of Geodesy (1974) The International Gravity Standardization Net 1971: International Association of Geodesy Special Publication No. 4, 194 p.
9. McCafferty, A.E., Bankey, Viki, and Brenner, K.C. (1998) Merged aeromagnetic and gravity data for Montana: A web site for distribution of gridded data and plot files: U.S. Geological Survey Open-File Report 98-333, 20 p. [Web address: URL:http://minerals.cr.usgs.gov/publications/ofr/98-333].
10. Phillips, J.D., Duval, J.S., and Ambroziak, R.A. (1993) National geophysical data grids--gamma-ray, gravity, magnetic, and topographic data for the conterminous United States: U.S. Geological Survey Digital Data Series DDS-9, 1 CD-ROM disk. [includes potential-field software version 2.1].

11. Plouff, Donald (1977) Preliminary documentation for a FORTRAN program to compute gravity terrain corrections based on topography digitized on a geographic grid: U.S. Geological Survey Open-File Report 77-535, 43 p.
12. Plumlee, Geoffrey S., Streufert, Randall K., Smith, Kathleen S., Smith, Steven M., Wallace, Alan R., Toth, Margo I., Nash, J. Thomas, Robinson, Rob, Ficklin, Walter H., and Lee, Gregory K. (1995) Map showing potential metal-mine drainage hazards in Colorado, based on mineral deposit geology: U.S. Geological Survey Open-File Report 95-26, 1 sheet, scale 1:750,000.
13. Raines, Gary L. and Johnson, Bruce R., (1995) Digital representation of the Montana state geologic map: a contribution to the Interior Columbia River Basin Ecosystem Management Project: U.S. Geological Survey Open-File Report 95-691.
14. Ross, C.P., Andrews, D.A., and Witkind, I.J. (1955) Geologic map of Montana: U.S. Geological Survey, 2 plates, scale 1:500,000.
15. Ryder, Jean L. (1995) Active, inactive, and abandoned mine information for Montana with selected geochemical data: U.S. Geological Survey Open-File Report 95-229, diskette. [also available for electronic transfer at ftp://greenwood.cr.usgs.gov/pub/open-file-reports/ofr-95-0229/ (see URL http://minerals.cr.usgs.gov for links)]
16. Smith, S.M. (1997) Geochemistry of Montana, National Uranium Resource Evaluation, Hydrogeochemical and Stream-Sediment Reconnaissance Program, U.S. Geological Survey, National Geochemical Database: U.S. Geological Survey Open-File report 97-492. [to be published as a World Wide Web site]
17. Tysdal, R., Ludington, S., and McCafferty, A.E. (1996) Mineral and energy resource assessment of the Helena National Forest, west-central Montana: U.S. Geological Survey Open-File Report 96-683-A.

A STRATEGY FOR SUSTAINABLE DEVELOPMENT OF NONRENEWABLE RESOURCES USING SPATIAL PREDICTION MODELS

C. F. Chung[1], A. G. Fabbri[2] and K. H. Chi[3]
[1]*Geological Survey of Canada, Ottawa, Canada*
[2]*ITC, Enschede, The Netherlands*
[3]*Korea Institute of Geosciences and Mineral Resources, Taejon, Korea*

Abstract. This contribution provides an analytical strategy applicable in mineral exploration to not only predicting the location of undiscovered mineral resources but also estimating the probability of the next discovery at that location. In addition, the strategy is applicable to the likely environmental impacts of developing the resources as a result of the exploration. General concepts of spatial prediction, of the likelihood ratio model, and of a two-stage approach to derive the probability of the next discovery in each prediction class are introduced.

Two examples of predictions of undiscovered deposits are discussed for the respective extreme situations of rich and poor spatial databases. They are not yet fully developed to cover environmental-impact prediction; nevertheless, they provide the basic decisional elements as the estimator of the conditional probability of the next discovery applicable to either resource exploration or to environmental protection. The estimation of the probability of the next discovery through the validation technique is the most critical element in spatial prediction modeling. A review of deposit and geoenvironmental deposit models provides the foundations of predictive analysis in sustainable development terms.

1. Introduction

Exploration and development of mineral resources expose to both economic and environmental risk. Sustainability is defined as "meeting the needs of today without compromising the ability of future generations to meet their needs." For a sustainable development of nonrenewable resources, two critical issues are: (1) depletion of nonrenewable resources and hence the need of continuing exploration, and (2) direct economic-environmental effects and hence the need of managing the impact of development. Understanding those issues is particularly relevant in environmentally fragile areas. Figure 1 illustrates relationships and implications.

The balancing act between the economic benefits of development and the associated environmental damages is the essential strategy for sustainable development of nonrenewable resources. For the balancing act, what we need the most is a spatial prediction model to identify the conveniently smallest areas that are likely to contain undiscovered resources. This should be obtained with measurements of the reliability or the uncertainty associated with the prediction model. The spatial prediction model should also delineate fragile areas to future environmental impacts by the development, be it either exploration, or exploitation, or decommissioning of mineral deposits.

Spatial prediction is reviewed in this contribution as follows: (1) in its general terms, applicable to resource exploration and to environmental security, (2) by introducing the likelihood ratio prediction model, and (3) by proposing an analytical strategy termed two-stage approach.

Specific mineral deposit models have been proposed to identify typical metallogenic settings to aid in mineral exploration. Furthermore, related geoenvironmental deposit models have been developed to anticipate the environmentally related characteristics of the deposits. These models prepare the ground for the development of spatial prediction applications.

The spatial prediction methods that we have developed can identify both the potential areas for exploration and the ones that are environmentally fragile to development. Exploration of mineral resources is carried out by a stepwise approach with increasing levels of detail. At each step, a selection of the target area for the next exploration strategy is made that is based on all the data harnessed in the previous step. The uncertainty of the selected target area, containing undiscovered resources, is a critical factor for estimating the exploration risk and for subsequent decision-making. The estimation of the uncertainties associated with the predictions is the most critical part of the models.

Two case studies are discussed that represent two extreme situations of rich and poor spatial databases. Critical characteristics of the prediction patterns are extracted to assist decision makers in successive steps of increasing detail, cost and risk. Considering the multidisciplinary aspects of mining lifecycles, industrial ecology and multi-criteria analysis in environmental impact assessment can expand the application of deposit and geoenvironmental models in spatial predictions.

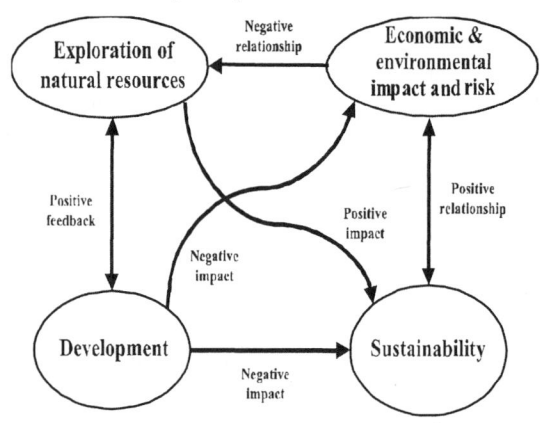

Figure 1. . Relationships among four activities: exploration, development, economic and environmental impact and sustainability, related to nonrenewable natural resources

2. Spatial prediction

2.1 GENERAL CONCEPTS

Spatial prediction models have been developed to delineate areas that are likely to contain undiscovered mineral deposits. The models have been built on mathematical foundations using multi-layered spatial databases consisting of either discrete data such as geological maps or continuous data such as geophysical maps. Traditionally in mineral exploration, a team of specialists constructs a thematic map identifying potential areas that are likely to contain mineral deposits. The thematic maps have been built on expert's knowledge and experience on previous discoveries and interpreting geophysical, geochemical and geological maps. The thematic maps are to guide further exploration by predicting where the deposits are likely to occur. Although all predictions related to undiscovered deposits are subjected to uncertainties, the thematic maps produced by experts do not expressly include such uncertainties.

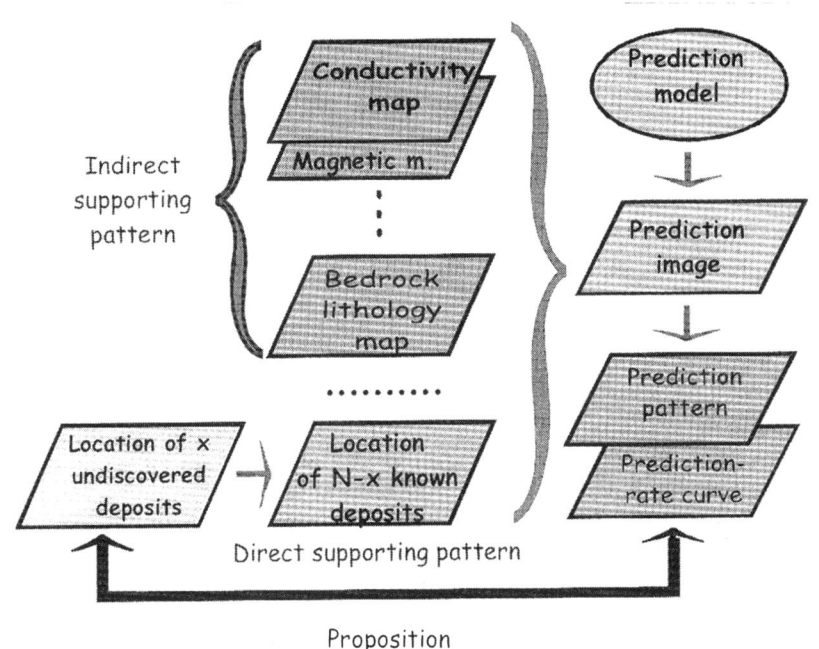

Figure 2. General flow diagram for Favourability Function modeling. The direct supporting pattern is partitioned into N-x "known deposit locations" and x deposit locations to be considered as "undiscovered" for validation purposes. The prediction pattern validated with the prediction-rate curve is a representation of the target pattern.

Our quantitative approach to spatial predictions stems from the same needs to generate thematic maps based on the same spatial data used by the experts. We have termed our approach "Favourability Function" or FF [1,2,3] that is based on a unified mathematical framework. Figure 2 illustrates the procedure to construct and interpret a prediction map of undiscovered mineral deposits. It provides an example of the proposition that a pixel (or a unit area) in the study area contains a mineral deposit of type D (i.e., a specific type of mineralization that can be within a recognizable geologic setting). It also provides the related symbolic relationships of the target pattern (the areas containing what we want to discover (e.g., the location of the undiscovered mineral deposits of type D), the supporting patterns (the components or layers of the database including the distribution of discovered mineral deposits of the same type), the prediction model (the mathematical formulation), the prediction image (the resulting set of predicted values), the prediction pattern (a way to visualize and analyze the prediction image), and the prediction-rate curve (that expresses the relationship between the prediction image and the target pattern, i.e., the undiscovered deposits).

The FF approach requires the two assumptions that: (i) the spatial target pattern (the distribution of undiscovered deposits of type D) can be characterized by the input spatial data (the supporting patterns), and (ii) the deposits to be discovered have formed under similar geologic circumstances and are located in similar geologic settings. The occurrences of undiscovered deposits are reflected in the mathematical formulation through the proposition. In the FF then, the fact that a pixel in the prediction image identifies the undiscovered deposits is expressed as a mathematical function, a favourability function, through a proposition given the information at the pixel from the multi-layered spatial database. Several mathematical functions have been proposed and explored, such as probability functions, fuzzy set membership functions, belief and plausibility functions [1,3,4].

2.2 THE LIKELIHOOD RATIO FUNCTION

To study the role played by an individual layer (of indirect supporting patterns), let us consider as an example a geophysical map of electromagnetic EM conductivity of low frequencies (i.e., deep penetration). We want to know whether it provides useful information to identify the location of undiscovered volcanogenic massive sulfide deposits (an example of type D). We also want to separate mineralized sub-areas and non-mineralized sub-areas. If the conductivity map derived from an EM survey is to provide useful information, then the pixel data in the map from mineralized sub-areas should have unique characteristics that are different from the data from non-mineralized sub-areas. This suggests that the probability frequency distribution function of the mineralized sub-areas should be distinctly different. The likelihood ratio function, which is the ratio of two frequency distribution functions, can highlight this difference [2,5,6].

Consider a number of map layers (often termed channels, images or variables) contained in the database of the study area. Based on the likelihood ratio function of the multivariate frequency distribution functions calculated from these layers, a prediction model identifying the areas likely to be mineralized, i.e., the sites of new discoveries can be generated.

To formalize the idea, let us consider a pixel p with m pixel values, $c_1, \cdots, c_m$ in the whole study area A consisting of two sub-areas, the mineralized area $\mathbf{M}$ and the remaining non-mineralized area $\overline{\mathbf{M}}$. Let $f\{c_1, \cdots, c_m \mid \mathbf{M}\}$ and $f\{c_1, \cdots, c_m \mid \overline{\mathbf{M}}\}$ be the multivariate frequency distribution functions assuming that the pixel is from $\mathbf{M}$, and from $\overline{\mathbf{M}}$, respectively. Then the *likelihood ratio* ([7,8]) at p is defined as:

$$\lambda_p(c_1, \cdots, c_m) = \frac{f\{c_1, \cdots, c_m \mid \mathbf{M}\}}{f\{c_1, \cdots, c_m \mid \overline{\mathbf{M}}\}} \ . \tag{1}$$

There are several ways to estimate $\lambda_p(c_1, \cdots, c_m)$ in (1) from discriminant analysis to Bayesian methods. Chung (2002) [5] and Chung and Keating (2002) [6] provide a detailed discussion on the likelihood ratio functions as a model applied to mineral potential mapping. For every pixel, we estimate $\lambda_p(c_1, \cdots, c_m)$. According to this model, the pixel with the largest estimate is considered as the most likely unit-area containing the undiscovered mineral deposits.

2.3 A TWO-STAGE APPROACH

Two modeling stages are proposed, namely (I) the "prediction-validation stage", and (II) the "probability estimation stage."

2.3.1 *Stage I: prediction-validation*

In prediction modeling, the most important and essential component is to carry out a "validation" procedure of the prediction results. Without a validation, the prediction models are totally useless and have hardly any scientific significance. Designing a validation procedure in *Stage I* is solely depending on the type of probability that we want to estimate in *Stage II*. Let us consider a simple validation procedure for the probability of the next discovery, so that the prediction results can provide a meaningful interpretation with respect to the next discovery.

After a prediction model is applied and the prediction results are obtained, in an application that aims at the next discovery, the proper validation should be based on the comparison between the prediction results and the unknown target pattern, i.e., the location of the next discovery. Because the target pattern, the location of next discovery, is unknown, a direct comparison is not feasible. The next best thing is to mimic the comparison by using one of the known deposit locations as if it represents the target pattern. To mimic the comparison, we must restrict the use of all the data of the known discoveries (deposits) in the study area, for instance by partitioning the dataset.

To partition all known deposits into two groups, we start by selecting a deposit. The first group consists of the selected deposit only and the second group consists of the remaining deposits. The location of the deposits in the second group is used for obtaining the prediction image. The prediction result at the location of the selected deposit in the first group is for validation and it allows estimating the "prediction-rate value." We repeat the process for each discovered deposit and obtain the corresponding prediction-rate value. The partition of the known discoveries is the corner stone of the validation techniques proposed here.

To formalize *Stage I*, we propose the following three-step procedure.

Step 1. Given the N known deposits of type D in the study area, select one of them and pretend that its location was unknown.

Step 2. Using the remaining N-1 deposits, construct a prediction model using the likelihood ratio function so that the pixel values of the prediction image generated represent relative favourability indices for undiscovered deposits of type D. Sort all the pixel values in the prediction image in increasing order. Then at each pixel, replace the pixel value by the corresponding rank divided by the total number of pixels in the study area. The revised value of the pixel with the largest favourability index (the largest pixel value in the prediction image has the value of 1 (or 100%), and the pixel with the smallest value has the value of (1/total number of pixels in the study area).

Step 3. Appraise the replaced pixel value at the deposit selected in ***Step 1*** (and not used in ***Step 2***) to verify the prediction strategy in identifying the location of the deposit. These replaced pixel values are termed "prediction rates" and they are usually reported in %.

The three-step procedure is repeated N times, once for each deposit. The N prediction rates generated in ***Step 3*** are then used to validate the results of the prediction map based on all N known deposits. The validation procedure allows also to compare models and to study the contribution of each data layer or combination of data layers to the prediction model.

2.3.2 *Stage II: probability estimation*

It consists of empirically estimating the probability of a unit area in each prediction class that the next discovery will be located within the unit. The estimate of such a probability of the next discovery requires the prediction-rate curve based on the corresponding validation procedure. In addition, it also requires further related assumptions.

Suppose that we have constructed a mineral potential map for a specific deposit type using several data layers. The original prediction pattern can contain, say, 200 elementary prediction classes each class covering a fixed number of km^2. We can select a convenient size of minimum prospective unit area, MPUA, for instance 100m x 100m. When an MPUA in a class is randomly selected for prospecting the deposits, we can estimate the probability that it will contain the new discovery. To estimate such a probability, we need two additional assumptions that: (i) at least one more undiscovered deposit of that type exists in the study area, and (ii) the current prospecting techniques have the power to find the undiscovered deposit.

Let us denote by p_c the empirical estimates of the probabilities that the corresponding classes contain the next discovery if the whole classes are to be prospected. If we take only one MPUA within a prediction class, the probability, denoted by p_δ, that the unit contains the next discovery is obtained by:

$$p_\delta = 1 - \sqrt[\delta]{1 - p_c}, \qquad (2)$$

where δ represents the number of MPUAs within the prediction class and p_δ is independent of the size of the classes.

Now, suppose that we have decided to prospect 100 MPUAs (totaling 1 km^2) in each prediction class, the probability, denoted by p_ϕ, that the next discovery will be within these 100 unit areas is obtained by:

$$p_\phi = 1 - (1 - p_\delta)^{100 \, (= \text{the number of units within the class})}, \qquad (3)$$

Theoretically speaking, the prediction-rate curve must satisfy the two conditions: (i) of being a monotone increasing function, and (ii) that the increment rate (the tangent of the curve) must be a monotone decreasing function. While p_c depends upon the size of the prediction class, p_ϕ is independent of the size of the class. Also, p_δ and p_ϕ should be monotone decreasing functions if the prediction-rate curve satisfies the two conditions above.

To further describe how to perform the probability estimation, the discussion will be continued in Section 4 with the applications to the two case studies in Canada and Korea, respectively.

3. Two case studies of mineral exploration

We will illustrate how the spatial prediction models play critical roles by using two case studies, one from the Bathurst Mining Camp in New Brunswick, Canada, [5], and the other from the Ogdong area in Korea [9]. The two case studies document two situations of different data availability and subsequent exploration levels. In both studies the likelihood ratio function was applied as discussed in Chung (2002) [5]. A computer was developed to construct the prediction models system in conjunction with PCI®'s EASI/PACE™ module [10,11].

3.1 THE BATHURST MINING CAMP CASE STUDY IN CANADA

A geological synthesis of the study area, including bedrock geology, tectonic setting, and the metallogeny of massive sulfide deposits is in an Economic Geology monograph edites by Goodfellow et al. (2002) [12]. To obtain a prediction for the Bathurst Mining Camp study, the following eight continuous data layers were used: (a) Conductivity data of three frequencies, high, medium and low; (b) Magnetic data consisting of the total fields reduced to the pole or RTP and their vertical gradient derivatives; and (c) Radiometric data of three equivalent ratios, Th/K, U/K, U/Th [13]. Several combinations of these eight layers were used to construct the prediction maps.

The study area consists of 1,647,748 pixels with resolution of 50m x 50m, covering approximately 4,120 km^2. In addition to the eight layers of geophysical data, the locations of 37 known volcanogenic massive sulfide or VMS deposits were also added to the database. Whenever we refer to an "undiscovered VMS deposit" in this example, we assume that it comes from the population of the VMS deposits where the 37 discovered VMS deposits had originated from [14]. Therefore it implies that the grades and tonnage of the undiscovered VMS deposits in the Bathurst Camp have exactly the same distribution functions (characteristics) of the 37 discovered VMS deposits.

With a spatial prediction model, the likelihood ratio function developed by Chung (2002) [5], we have constructed a mineral potential prediction map using the geophysical survey data in the Bathurst Mining Camp. The prediction map for the potential VMS mineralization based on the estimation of expression (1) is shown in Figure 3. As indicated in the legend, the purple areas have the best potential for undiscovered VMS deposits and cover 0.5% of the entire Camp, i.e., 20.6 km^2. For display purposes, unequal intervals have been used in the illustration.

Now we have to satisfy the following question: "How to use this prediction map to estimate the risk associated with exploring the area?" To assess the risk of exploring a given area, the first step is to estimate the probability that the next discovery will be made within it. The unequal intervals used in Figure 3 correspond to the study area proportions shown in the first column of Table 1, from which the probabilities for *Stage II* based on expressions (2) and (3) have been estimated. Figure 5 shows the prediction-rate curves for the Bathurst study area. The discussion of the probability estimates will be made in Section 4.

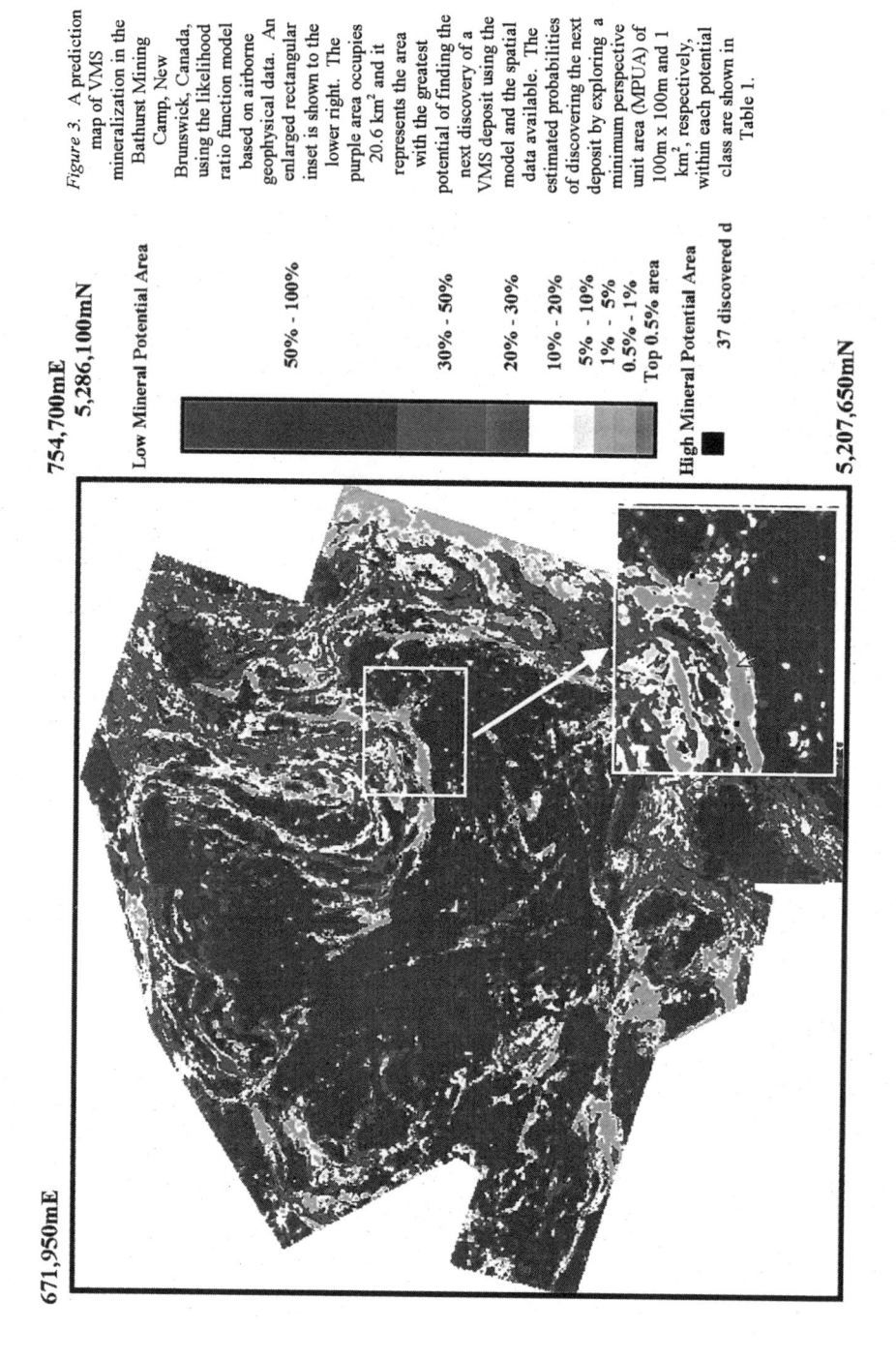

Figure 3. A prediction map of VMS mineralization in the Bathurst Mining Camp, New Brunswick, Canada, using the likelihood ratio function model based on airborne geophysical data. An enlarged rectangular inset is shown to the lower right. The purple area occupies 20.6 km² and it represents the area with the greatest potential of finding the next discovery of a VMS deposit using the model and the spatial data available. The estimated probabilities of discovering the next deposit by exploring a minimum perspective unit area (MPUA) of 100m x 100m and 1 km², respectively, within each potential class are shown in Table 1.

3.2 THE OGDONG CASE STUDY AREA IN KOREA

The next example comes from the Ogdong area in the Taeback-san Region of Korea studied by Chi et al., (2001) [9]. The study area consists of 451,870 pixels (730 by 619 pixels also with 50m x 50m resolution), covering approximately 407 km². The area is about one tenth of the Canadian study area. Geologically, the major part of the study area is composed of metasediments. Seven discovered polymetallic deposits consisting of mainly iron (Fe), lead (Pb), and zinc (Zn), occur within the thick limestone series of the Joseon Super-group and are located in the vicinity of a granite. For this example, the spatial database includes a bedrock lithology map, aeromagnetic survey data, Potassium and Thorium-equivalent Gamma-ray survey maps, and five geochemical survey data sets for Ag, Cd, Cu, Pb and Zn.

Using the same likelihood ratio function model [5,9], based on expression (1), we have constructed a mineral potential map for the poly-metallic deposits that is shown in Figure 4. As indicated in the legend, the purple and the red areas seem the most promising for polymetallic deposits and cover 1% and 4%, respectively, of the study area, i.e., 2.03 km² and 8.13 km². Even if we do not have a more definite mineral deposit model like in the Canadian study area, the characteristics of the undiscovered deposits are assumed to be identical to the seven discovered deposits shown as seven black dots in Figure 4. The corresponding prediction-rate curve is shown in Figure 5. Similarly to the previous application, Table 2 contains the probability estimates for *Stage II*, for the unequal intervals listed in the first column. The interpretation of Table 2 will be made in the next section. Here too, we have obtained a prediction map that calls for interpretation. What is the significance of the classes identified by the colors in Figures 3 and 4? The next section applies the two-stage approach to this task.

4. Application of the two-stage approach

Let us take the Canadian study area as a first example. In Figure 3 we have constructed a mineral potential map for volcanogenic massive sulfide or VMS deposits in the Bathurst Mining Camp, using several layers of geophysical data. The original prediction pattern for Figure 3 consisted of 200 prediction classes, each class covering approximately 20.6 km². For prospecting the VMS deposits in the study area, now we can consider the size of the "minimum prospective unit area", or MPUA, for example, 100m x 100m. It is roughly the area affected by a drilling operation.

Based on the validation procedure described in *Stage I*, the prediction-rate curve for the prediction pattern of Figure 3 is shown in Figure 5a. Consider a unit area in a prediction class, which is either one of the original 200 prediction classes or it consists of a combination of a number of successive original 200 prediction classes, as shown in Figure 3. In this example we use eight prediction classes, whose ranges are shown in the first column of Table 1. As mentioned earlier, to estimate the probability that a randomly selected MPUA in a class contains the new discovery, we need the two additional assumptions that: (i) at least one more undiscovered VMS deposit exists in the Bathurst Mining Camp area, and (ii) the current prospecting techniques have the power to find the undiscovered VMS deposit.

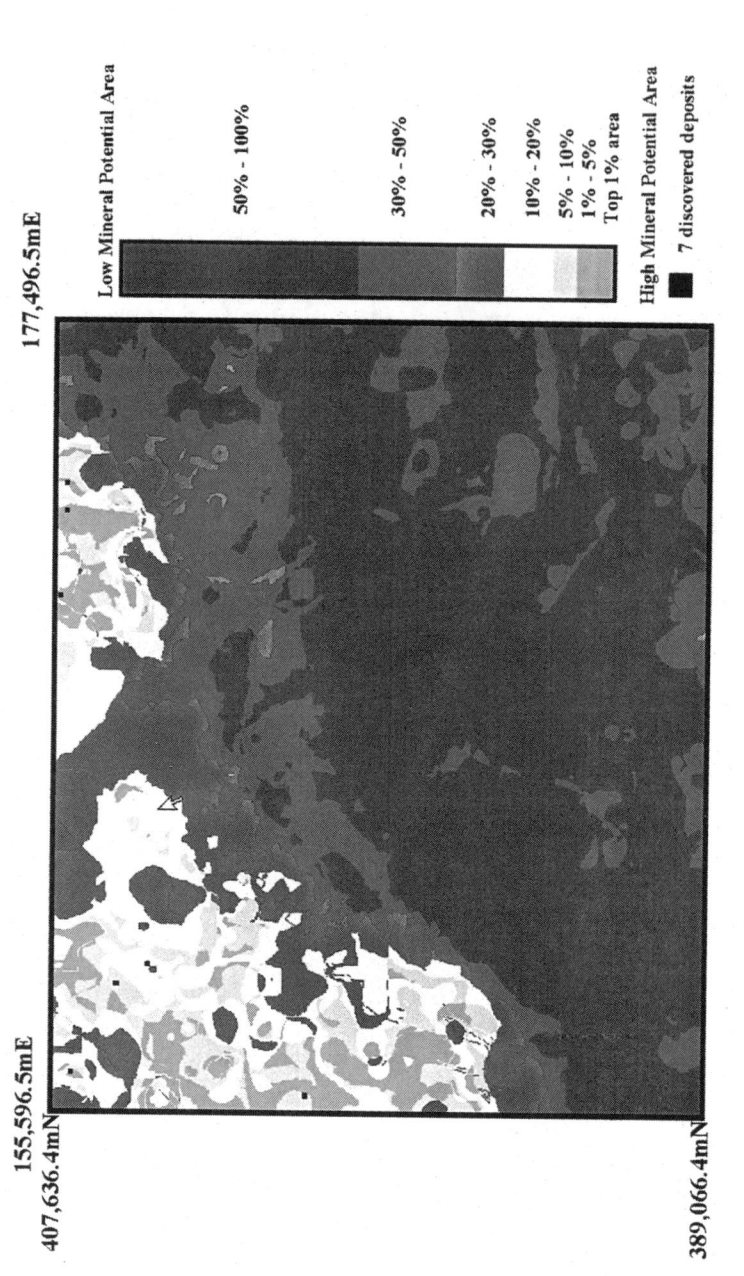

Figure 4. A prediction map of polymetallic mineralization in the Ogdong area, Korea, using the likelihood ratio function model based on lithological and geochemical survey data. Each color class covers different area proportions. The estimated probabilities of discovering the next deposit by exploring a minimum perspective unit area (MPUA) of 100m x 100m and 1 km², respectively, within each potential class are shown in Table 2.

The first column in Table 1 represents the portions of the whole study area assigned to the various prediction classes. The first two labels "Top 0.5%" and the subsequent "0.5–1%" in the column are for the most and the next most prospective classes of the original 200 classes, respectively. Each initial class covers 20.6 km^2. The next aggregated prediction class, "1-2%", consists of the next 2 classes of the original 200 classes and covers 41.2 km^2. The second column in Table 1 comes from the prediction-rate curve in Figure 5a, and is obviously based on the 37 prediction rates obtained from the validation procedure in *Stage 1*. The values in the column represent the empirical estimates of the probabilities, denoted by p_c, that the corresponding classes contain the next discovery if the whole classes are to be prospected. The corresponding p_δ values for MPUAs from expression (2) are independent of the size of the prediction classes and are shown in the third column of Table 1, assuming an area of 100m x 100m as the MPUA. Then each of the first two prediction classes consists of 2060, while the third prediction class consists of 4120 MPUAs. In reality, a more likely operational unit area in mineral exploration is 1 km^2, or 100 MPUAs. Suppose that we have decided to prospect these 100 MPUAs in each prediction class, the probability, denoted by p_ϕ, that the next discovery will be within them obtained by expression (3) is shown in the fourth column of Table 1. If the black prediction-rate curve in Figure 5a does satisfy the second condition that the increment rate must be a monotone decreasing function, the third and fourth columns in Table 1, containing the class size independent values, should also have monotone decreasing values. In the fourth column in Table 1, one (0.0021) out of the nine values is not following the monotone rule. In empirical estimation procedure, this is to be expected.

If we consider the black curve in Figure 5a, and the corresponding first and second columns of Table 1, we can see that there is a sudden drop in the percentage of deposits to be discovered, from 0.3784 to 0.0270 with 0.5% increments of the area of the highest predicted classes. Consequently, the corresponding probabilities in the fourth column reduce from 2.28% to 0.13%. It suggests that the promising prospecting area should be "the top 0.5%" area only, i.e., the highest predicted area of 20.6 km^2.

Similar considerations can be made for the Korean application. Using the same validation technique (see also [5]), we have obtained the "prediction-rate curve" for the Korean study area also shown in Figure 5a. Using the curve, we could empirically estimate the probability of the future discoveries for the classes in Figure 4. The same observations made for the Canadian study area and the corresponding Table 1, apply here for the probability values listed in Table 2, however, the Korean study area is much smaller than the Canadian area (2060 MPUAa vs. 101 MPUAs per the same 0.5% of the study area).

When we look at the gray prediction rate curve in Figure 5a for Korean study, the rate of prediction of the seven deposits for the first 10% (0 – 10%) of the study area is much lower than the subsequent 10% (10 – 20%) (one deposit vs. three deposits). The subsequent 10% (20 – 30%) predicted the remaining three deposits. Obviously, it suggests that the prediction map in Figure 4 is not reliable and the decision for prospecting should not be based on the prediction map. This unreliability is also shown in the third and fourth columns of Table 2. From the nine probabilities in fourth column, it is not possible to detect the monotone decreasing property. The next section compares the two applications in broader exploration terms.

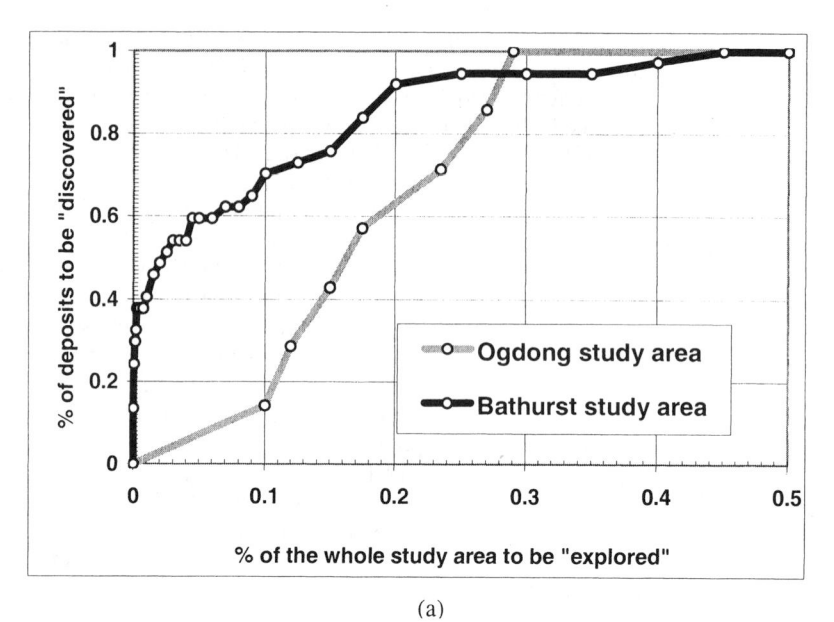

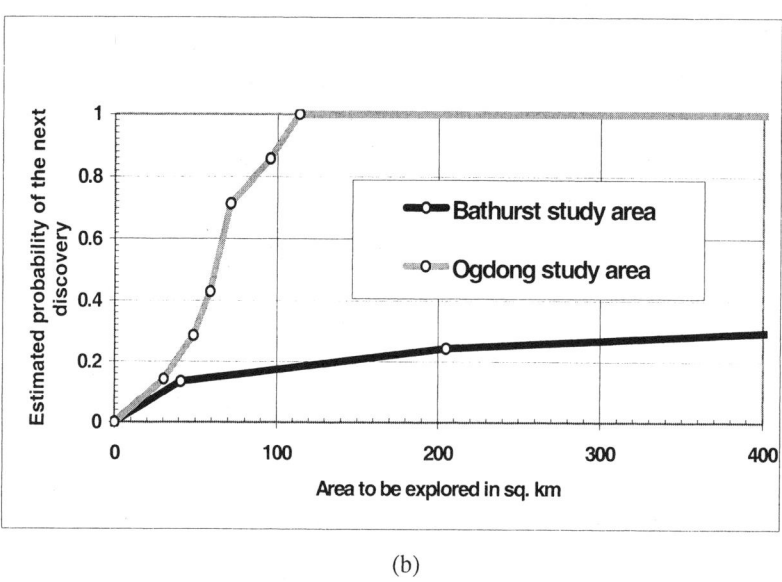

Figure 5. Two prediction rate curves for the Bathurst and the Ogdong (gray) study areas are shown with two different horizontal-axis scales. The black and gray curves represent the Bathurst and the Ogdong areas, respectively. In (a) the horizontal axis represents % of the whole study area. In (b) the horizontal axis represents actual km^2 rather than %. While the Bathurst study area is 4,126 km^2, the Ogdong study area covers 406 km^2.

Table 1. Estimates of the various probabilities discussed in *Stage II* for the prediction map in Figure 3 and the prediction-rate curve in Figure 5 for the Bathurst Mining Camp, N.B., Canada. The estimates are based on expressions (2) and (3). * Within brackets are the numbers of minimum prospective unit areas (MPUAs) of size of 100m x 100 m.

Portion of the study area considered as a desirable prospective area $1\% = 41.2$ km^2	Proportion of 37 discovered VMS deposits within the class. Estimate of p_c, that comes from Figure 5a	Estimated probability p_δ in expression (2) of the next discovery in the minimum prospective unit area (MPUA)	Estimated probability p_ϕ in expression (3) of the next discovery in 100 minimum prospective unit areas (MPUAs)
Top 0.5% (2060)*	0.3784	0.000231	0.0228
0.5 – 1% (2060)*	0.0270	0.000013	0.0013
1 – 2% (4120)*	0.0811	0.000021	0.0021
2 – 5% (12360)*	0.1081	0.000009	0.0009
5 – 10% (20600)*	0.1081	0.000005	0.0006
10 - 20% (41200)*	0.2162	0.000006	0.0006
20 – 30% (41200)*	0.0270	0.0000007	0.00006
30 – 50% (82400)*	0.0541	0.0000007	0.00006
Remaining 50%	0.0	0.0	0.0

Table 2. Estimates of various probabilities discussed in *Stage 2* for the prediction map shown in Figure 4 with the prediction rate curve in Figure 4 in the Ogdong area in Korea, based on expressions (2) and (3). *Within brackets are the numbers of minimum prospective unit areas of size of 100m x 100 m.

Portion of the study area considered as a desirable prospective area $1\% = 2.03$ km^2	Proportion of 7 discovered deposits within the class. Estimate of p_c, that comes from Figure 5a	Estimated probability p_δ in expression (2) of the next discovery in the minimum prospective unit area (MPUA)	Estimated probability p_ϕ in expression (3) of the next discovery in 100 minimum prospective unit areas (MPUAs)
Top 0.5% (101)*	0.0071	0.00007	0.007
0.5 – 1% (101)*	0.0071	0.00007	0.007
1 – 2% (203)*	0.0143	0.00007	0.007
2 – 5% (610)*	0.0429	0.00007	0.007
5 – 10% (1016)*	0.0714	0.00007	0.007
10 - 20% (2033)*	0.4286	0.00027	0.027
20 – 30% (2033)*	0.4286	0.00027	0.027
30 – 50% (4066)*	0.0	0.0	0.0
Remaining 50%	0.0	0.0	0.0

5. Comparing the two case studies

In the Canadian application, although the study area is 10 times larger than the Korean study area, we were able to delineate the 20.6 km^2 best potential area where the estimated probability (0.023%) of finding a new discovery, within the MPUA (100m x 100m). Considering the drastic 17.7 times reduction (from 0.0231% to 0.0013%) in probabilities from the best class to the second best class in the Canadian study (see the third and fourth columns and rows 1 and 2 of Table 1), it may be concluded that an exploration activity will be continued only within the best 20.6 km^2 (purple color areas in Figure 3). One of the reasons is that the spatial database for the Canadian study was the result of five years of efforts and was much better than the Korean database

In the Korean study, however, we were not able to suggest any promising prospecting area and the decision would not be clear, because (i) the probability does not change from class to class within the top 10% area (i.e., it remains at 0.007%, as shown in Table 2); and (ii) the corresponding probabilities within the "10 – 20% area" and "20 – 30% area" are more than three times higher than the previous ones. It is probably because of the spatial database characteristics. Unlike in the Canadian study, we need significantly more additional information to make an informed decision and the prediction map shown in Figure 4 with the prediction rate curve alone are not enough to make a proper decision to precede the exploration program.

Mineral exploration is a risky venture where each km^2 explored represents a cost. Reducing the area to be explored without diminishing the chances of new discoveries is a challenge. In addition to the cost of exploration, we have to consider the environmental damage that exploration itself and the subsequent stages of development and exploitation will cause. The following review provides arguments to extend the predictive modeling approach using the geoenvironmental deposit models.

6. How to develop the FF approach to use information on deposit and geoenvironmental models

This section helps in constructing the scenarios needed for predictions in terms of sustainable development. In Section 2 we have used the following proposition:

pF "p will contain an undiscovered deposit of type D".

In it the consistency of the type of mineralization and assumed geologic setting is fundamental in predictive modeling. The approach can be extended to the environmental aspects of mineral exploration. The following works have paved the way to such extension.

According to Berger and Drew (2001) [15], "mineral-deposit models are the basis for consistent resource-assessment, exploration, and environmental risk analysis methodology." For instance, the identification of VMS deposits in the Canadian example was based on such models. Furthermore, examples of volcanogenic massive sulfide deposits, VMS, are discussed by Taylor *et al.* (1995) [16], not only because they guide in mineral exploration but also because their geoenvironmental characterization models the associated environmental problems and particularly acid mine drainage in areas disturbed by surface mining and of tailing disposal. Associated are potentially toxic trace metals, such as arsenic, bismuth, cadmium, mercury, lead and antimony that are present in the deposits hosted in volcanic and sedimentary source rocks. Besides considering the geological factors that influence potential environmental effects (e.g., deposit size, host rocks, wall rock alteration, deposit trace element geochemistry, topography, physiography, mining and milling methods) they focus on the environmental signatures of the VMS deposits (e.g., drainage characteristics, metal mobility, soil and sediments prior to mining, potential environmental causes associated with mineral processing, smelting characteristics, climate and environmental geophysics). This geoenvironmental deposit model is just one example of a recent effort in compiling geoenvironmental models of mineral deposits (du Bray, 1995, [17]). Those models become essential to developing environmental scenarios. Therefore, it is illustrative to look more closely at the geoenvironmental; deposit concept. The concept is best expressed in the words of Plumley and Nash (1995, p. 1; [18]):

"Economic geologists recognize that mineral deposits can readily be classified according to similarities in their geologic characteristics (ore and gangue mineralogy, major- and trace-element geochemistry, host rock lithology, wall-rock alteration, physical aspects of ore, etc.), as well as their geologic setting ...

A next step in the process of mineral deposit modeling is the development of geology-based, geoenvironmental models of diverse mineral deposit types. Mineral deposit geology, as well as geochemical and biogeochemical processes, fundamentally control the environmental conditions that exist in naturally mineralized areas prior to mining, and conditions that result from mining and mineral processing. Other important natural controls, such as climate and anthropogenic factors (including mining and mineral processing methods) mostly modify the environmental effects controlled by mineral deposit geology and geochemical processes. Thus, deposits of a given type that have similar geologic characteristics should also have similar environmental signatures that can be quantified by pertinent field and laboratory data and summarized in a geoenvironmental model for that deposit type. Similarly, environmentally important geologic characteristics, such as the presence of an alteration type likely to produce highly acidic drainage water or an alteration type likely to help buffer acid drainage water, should also be common to most or all deposits of a given type, and thus can also be summarized in a geoenvironmental model."

It becomes obvious that such models and the corresponding spatial data layers can be used to study, predict, minimize, and possibly avoid the environmental effects of mineral exploration-exploitation and mineral-resource development. The prediction of environmental impacts is now commonly discussed in economic geology. For instance, Evans (1997, p. 36; [19]) describes how in many countries it has become mandatory for companies applying for planning permission to start a mineral operation by preparing an environmental impact statement, or EIS. The latter has to comprehensively cover effects on vegetation, climate, air quality, noise, ground and surface water, including the method of ground reclamation that will be used at the termination of the mining activity.

Appropriate scenarios of exploration development need to be anticipated to provide the supporting background to an EIS. The EIS represents information and documentation that is to be provided early in the life cycle of mining development, however, according to Ritsema (2002; [20]), to assess "the consequences of each social economic activity, such as mining minerals and coal, the full life cycle should be considered in terms of both environmental and social economic impacts. The typical asset life cycle should include exploration, appraisal, development, production and last but not least abandonment."

This assessment has a clear parallel not only in the recent developments of industrial ecology, but also in the growing application of multicriteria analysis to environmental resource planning [21]. For instance, Laniado and Beinat (1994; [22]) distinguished several phases in environmental impact assessments of development projects: *strategic* (project identification and selection), *planning* (selection of one or a few alternatives), and *operational* (selection of the final project). All those different phases are part of a development process and require predicting impacts at different levels of approximation and uncertainty as different scenarios, similarly to the stepwise mineral exploration that leads to mining development, operation and eventual decommissioning.

The similarity between industrial development and mineral resource development is obvious, so is the gradual growth of infrastructures and flows of goods, services and wastes from and to the development site. From past experience of similar mining sites and of exploration ventures, it is possible to make realistic assumptions on the environmental impact of mining from the exploration phase (strategic), through the planning phase (selection of development site), to the operational phase (actual development), to the final decommissioning (closing of operation). For each phase, an appropriate spatial database will have to be constructed in order to bring the spatial prediction problem within the same terms described in Section 2.

7. Conclusions

This contribution offers an overview of spatial prediction models not only to identify target areas for natural-resource exploration and of areas vulnerable to environmental impacts but also to provide validation techniques to estimate the probabilities of the next discoveries showing the uncertainties associated with the predictions. Previous studies by the first two authors of this contribution [1,2,3,4,5] have covered a wide range of models and estimation procedures from simple to complex. A firm theoretical motivation has been presented elsewhere and here the emphasis is maintained on practical applications and implementations covering both current scientific research topics as well as recent case studies.

Extending the approach to a more comprehensive or holistic strategy is feasible whenever we are able to construct realistic exploration or development stage scenarios before taking action. The scenarios, based on previous case studies, are to provide the location and extent of the main impacts, benefits and costs. It means that scenarios have to be developed to perform a cost-benefit analysis of the prediction-rate curves and the corresponding selected ranges of estimated probability. Similarly, scenarios can be constructed for predicting the environmental impact of the exploration/exploitation development. In Section 6 the basis for such scenario was provided.

The expected benefit (discovery of a new deposit) and the expected damage (negative impact on the environment) from the development can be estimated from similar prediction-rate curves and from the corresponding probability tables given the necessary data. In addition, the expected avoidance of damage (not affecting the environment) and the expected loss (not discovering the deposits) from outside the development area can also be estimated. The four economic values, associated with the prediction map, are critical for continuing the exploration process and the sustainable development of nonrenewable resources such as VMS deposits in the Canadian study and poly-metallic deposits in Korean study. Also, assumptions are needed on the default impact of exploration phases for a given deposit type in addition to the specific impact due to the physiography and other resources of the study area. Cost-benefit analyses can be generated from the probability estimates. The environmental impact can be considered as a cost when exploration is seen in a sustainable development perspective (see Figure 1). The prediction-rate curves and the corresponding probability tables can be dissected in terms of cost-benefit analysis and of environmental impact prediction.

It has become evident that the strength and therefore the reliability of the prediction can only be based on the prediction-rate curves and the corresponding probability tables obtained through validation. When looking into the future, only that strength and reliability allow us to make statements on the location of the future discoveries. Through this procedure, we can provide the exact assumptions that we make in estimating the probability of discovery (or of loss or of impact). The two-stage process proposed here is to select and prioritize sub-areas with different probability of discovery of mineral deposits. Given appropriate databases and scenarios, it can be extended to more holistic decision making with a sustainable development perspective. Research by the authors is in progress to meet that objective.

8. Acknowledgements

The study was partly supported by a research grant provided to the spatial Data Analysis Laboratory of the Geological Survey of Canada by PCI Geomatics Inc., Richmond Hill, Ontario, Canada. The partial support is also acknowledged of the European Commission T&MR Programme's GETS Network Project on "Geomorphology and Environmental Impact Assessment of Transportation Systems" (Contract number: ERBFMRXCT-970162). We also wish to thank No-Wook Park of KIGAM, Korea for processing the data for Korean study. Dr. Edward A. Johnson in Mojácar, Spain, has provided generous assistance for computer communications.

References

1. Chung, C.F. and Fabbri, A.G., 1993, Representation of geoscience data for information integration. Journal of Nonrenewable Resources, v.2, n. 2, p. 122-139.
2. Chung, C.F. and Fabbri, A.G., 1998, Three Bayesian prediction models for landslide hazard. Proceedings of the International Association for Mathematical Geology Annual Meeting IAMG 1998, Ischia, Italy, October 1998, p. 204-211.
3. Chung, C.F. and Fabbri, A.G., 1999, Probabilistic prediction models for landslide hazard mapping. Photogrammetric Engineering & Remote Sensing. v. 65, n. 12, p. 1389-1399.
4. Chung, C.F. and Fabbri, 2001, Prediction models for landslide hazard using a fuzzy set approach. In, M. Marchetti and V. Rivas, eds., Geomorphology and Environmental Impact Assessment, Balkema, Rotterdam, p. 31-47.
5. Chung, C.F., 2002, Use of airborne geophysical surveys for constructing mineral potential maps. In, Goodfellow, W.D., McCutcheon, S.R., and Peter Jan M., eds., Massive Sulfide Deposits of the Bathurst Mining Camp, and Northern Maine. Economic Geology Monograph 11, in press..
6. Chung, C.F. and Keating, P., 2002, Mineral potential evaluation based on airborne geophysical data. Submitted for publication to Geophysical Exploration.
7. Kshirsagar, A.M., 1972, Multivariate Analysis, Marcel Dekker Inc., New York, 534 p.

8. Cacoullos, T., 1973, Discriminant Analysis and Applications, Academic Press, New York, 434 p.
9. Chi, K.H., Park, N.W., and Chung, C.F., 2001, Spatial integration of geological data for predictive mineral mapping: a case study from the Taebaeck-san area, Korea. Proceedings of the International Association for Mathematical Geology Annual Meeting IAMG 2001, Cancun, Mexico, September 10-12, 2001, website: http://www.kgs.ukans.edu/conferences/iamg/sessions/k/papers/
10. Chung, C.F., 1998, Spatial data integration (SDI) software package for prediction models based on PCI's EASI/PACE. Unpublished computer programs and user's guide on CD-ROM.
11. PCI Geomatics, 1997, Using PCI Software, Version 6.2, Richmond Hill, Ontario, 551 p.
12. Goodfellow, W.D., McCutcheon, S.R., and Peter Jan M., eds., 2002, Massive Sulfide Deposits of the Bathurst Mining Camp, and Northern Maine. Economic Geology Monograph 11, in press.
13. Keating, P., Thomas, M.D., and Kiss, F.G., 2002, Significance of high resolution magnetic and electromagnetic surveys for exploration and geological investigations, Bathurst Mining Camp, New Brunswick, Canada. In, Goodfellow, W.D., McCutcheon, S.R., and Peter Jan M., eds., Massive Sulfide Deposits of the Bathurst Mining Camp, and Northern Maine. Economic Geology Monograph 11, in press.
14. Wright, D.F., Chung, C.F., and Leybourne, M.I., 2002, VHMS favourability mapping using kernel method analysis of stream sediment and till geochemical data in the western Bathurst Mining Camp. In, Goodfellow, W.D., McCutcheon, S.R., and Peter Jan M., eds., Massive Sulfide Deposits of the Bathurst Mining Camp, and Northern Maine. Economic Geology Monograph 11, in press.
15. Berger, B.R., and Drew, L.J., 2002, Mineral-deposit models. This volume.
16. Taylor, C.D., Zierenberg, R.A., Goldfarb, R.J., Kilburn, J.E., Seal II, R.R., and Kleinkopf, M.D., 1995, Volcanic-associated massive sulfide deposits (Models 24a-b, 28a; Singer, 1986a,b; Cox, 1986). In, Preliminary Compilation of Descriptive Geoenvironmental Mineral Deposit Models. U.S. Geological Survey Open-File report 95-831, U.S. Government Printing Office, Washington, D.C., p. 137-144.
17. du Bray, E.A., 1995, Preliminary Compilation of Descriptive Geoenvironmental Mineral Deposit Models. U.S. Geological Survey Open-File report 95-831, U.S. Government Printing Office, Washington, D.C.
18. Plumlee, G.S. and Nash, J.T., 1995, Geoenvironmental models of mineral deposits – fundamentals and applications. In, du Bray, E.A., ed., Preliminary Compilation of Descriptive Geoenvironmental Mineral Deposit Models. U.S. Geological Survey Open-File report 95-831, U.S. Government Printing Office, Washington, D.C., p. 1-9.
19. Evans, A.M., 1997, An Introduction to Economic Geology and its Environmental Impact. Oxford, Blackwell Science Ltd., 364 p.
20. Ritsema, I., 2002, Asset-life cycle in the mining industry. This volume.
21. Beinat, E., 2002, Environmental resource planning: methodology to support the decision making process. This volume.
22. Laniado, E., and Beinat, E., 1994, Decision processes and multicriteria support for environmental management: a case study. Papers and Extended Abstracts of Technical Papers for IAMG'94, Mont Tremblant, Québec, Canada, October 3-5, 1994, p. 27-31.

PART 3. RESOURCE ASSESSMENT & MANAGEMENT

Mineral deposit modelling not only allows for the assessment of mineral potential in unexplored or partially explored areas and the promise of further assistance in any follow-up exploration but most recently, provides in addition, guidance on managing environmental impacts subsequent to the exploitation of discovered resources. In the papers that follow, new deposit models are proposed that offer a broader view of the nature and genesis of mineral deposits and the potential effects such deposits have on the environment if these deposits are brought into production. Sound resource management now requires careful consideration of the risks, costs, and benefits involved in any mining operation. Both the theoretical and practical aspects need to be considered.

In their paper, Berger and Drew propose a new generation of mineral deposit models as a basis for resource assessment, mineral exploration, and environmental risk analysis. Discussions of porphyry copper and related polymetallic vein deposits are used to illustrate the value of this new generation of deposit models. Langer, Lindsay, and Knepper Jr. discuss the value of geologic information, including the geographic distribution, value and quality of aggregate resources as a basic requirement for establishing sound policies in regulating aggregate resource extraction. Langer and Arbogast review the environmental impacts of mining natural aggregates and stress that there cannot be resource development without environmental impacts. At issue are the impacts of site preparation, resource extraction, reclamation, and environmental ethics. In their paper, Drew and Berger apply a new tectonic model to undiscovered resources in the Mátra and Börzsöny-Visegrád mountains, northern Hungary. In the final paper, Dowd discusses the financial, technical, and environmental risks associated with mineral resource exploitation.

MINERAL-DEPOSIT MODELS

New Developments

B. R. BERGER[1] and L. J. DREW[2]
[1] *U.S. Geological Survey, Federal Centre MS964*
Denver, Colorado 80225, USA
[2] *U.S. Geological Survey, 954 National Centre*
Reston, Virginia 90122, USA

Abstract. Mineral-deposit models are the basis for consistent resource-assessment, exploration, and environmental risk analysis methodologies. To reduce uncertainties in these methodologies, improved predictability of deposit occurrence is essential. Advances in understanding about structure and tectonics and the geology of earthquakes, together with improved insights as to how fluid flow is coupled with active deformation, heat transport, and solute transport, provide the framework for integrating mechanical phenomena into deposit models. With this framework, it is possible, for a given deposit type, to predict where in structural systems hydrothermal systems may occur, the chances that significant concentrations of ore may be expected, and where in larger vein arrays ore bodies may be localized. Discussions of porphyry copper and related polymetallic veins illustrate the value of this new generation of deposit models.

1. Introduction

To understand mineral deposits is to understand their interdisciplinary nature. One must have an intuitive sense for the interdependent processes and mechanisms that are necessary for their formation. Mineral-deposit models are an imitation of—a pattern after—the actions of a hydrological system that result in a mineral deposit. These models are at the core of mineral-resource assessment methodologies, mineral-exploration strategies, and geoenvironmental assessments.

Because a mineral deposit is a consequence of a fluid-flow system, it is describable by a set of coupled flow equations that are operative in the context and water/rock response of the geologic framework of a hydrologic system. The purpose of this paper is to review the use and nature of models and to "read" some pages from the "book" on the flow equations, comparing the results with what is observed and drawing some implications from the comparison.

2. What is a Model?

Mineral-deposit models are systematic syntheses of descriptive and genetic information regarding a group of deposits with sufficient attributes in common to be considered of the same type [1]. Models combine all known necessary and sufficient attributes of a deposit style and attendant theory to form a "pattern" of the deposit style. Grade and tonnage data, which are included in models, and/or descriptive information may serve as a basis on which to divide a deposit style into different types. There are many different approaches to classifying mineral deposits. Most academic institutions in North America use some variation of the classification scheme of Lindgren [2]. The Geological Survey of Canada published a summary of deposit types in Canada in 1984, which was expanded in 1995. In the U.S. Geological Survey, the classification of Cox and Singer [4] is widely used.

In mineral-resource assessments, deposit models serve several purposes—(1) they guide the delineation of terranes permissive for the occurrence of a deposit type, (2) they allow the proper classification of deposits as to type, (3) they are the basis on which probability of occurrence (spatial density of deposits) models can be created, (4) they are the basis on which estimates may be made of the size and grade of undiscovered deposits, and (5) they are a framework for evaluating the environmental impacts and costs of mining.

2.1. A TYPICAL MODEL

Because porphyry-style deposits are large, they can be enormous stores of wealth. In addition, the large tonnages contained in these deposits result in substantial environmental impacts when mined. For these reasons, we have chosen to use porphyry-style copper deposits as an example of a typical model.

Porphyry-style deposits alter cubic kilometres of ground, and are low-grade magmatic-hydrothermal deposits related to felsic, commonly porphyritic, intrusions [3][5]. Primarily, the ores consist of chalcopyrite and molybdenite and are part of a set of hydrothermal alteration assemblages that can have a zoned pattern within and around an intrusive igneous core. Typically, the broadest zone of alteration, which is referred to as "propylitic", may include chlorite, epidote, albite, calcite, and pyrite. The copper ore, which is chalcopyrite with some bornite and molybdenite, is generally contained within a zone of "potassic" alteration, which consists of potassium feldspar, biotite, chlorite, and magnetite plus or minus anhydrite and andalusite. Overprinting earlier alteration assemblages may be "phyllic," "argillic," and/or "advanced-argillic" alteration zones. The phyllic zone may contain copper ores and an assemblage of sericite and potassium feldspar. The argillic zones contain one or more clay minerals, and the advanced-argillic zone may contain pyrophyllite, alunite, diaspore, and zunyite.

Porphyry-style deposits are known world-wide and are considered to have formed in consuming tectonic-plate-margin environments [6][7]. Empirically, porphyry-style deposits occur within strike-slip fault systems in volcanic- and continental-arc settings. Strike-slip faulting in overriding plates results from different plate-boundary conditions. Where the convergence is oblique, trench-linked strike-slip faults form [8], whereas the juxtaposition of similar crust on both plates (both arc crust or both continental crust) leads to buoyancy-driven indented margins and strike-slip faults [9]. Oblique convergent margins are most common, and Woodcock and Schubert [9] estimate that about 60% of present-day plate boundaries have strike-slip tectonics. Table 1 is a list of selected world-wide porphyry-style deposits.

Table 1 Examples of world-wide porphyry-style copper deposits

North America	Europe
Bagdad, Arizona, U.S.A.	Asarel, Bulgaria
Bingham Canyon, Utah, U.S.A.	Bor, Yugoslavia
Bisbee, Arizona, U.S.A.	Deva, Romania
Cananea, Mexico	Pukanec, Slovakia
Highland Valley, British Columbia, Canada	Recsk, Hungary
Island Copper, British Columbia, Canada	Rosia Poieni, Romania
La Caridad, Mexico	Skouries, Greece
Mission-Pima, Arizona, U.S.A.	
Morenci, Arizona, U.S.A.	Middle East/Asia
Santa Rita, New Mexico, U.S.A.	Boshekul, Kazakhstan
Sierrita-Esperanza, Arizona, U.S.A.	Derekoy, Turkey
Twin Buttes, Arizona, U.S.A.	Erdenet, Mongolia
Yerington, Nevada U.S.A.	Kal'makyr, Uzbekistan
	Kounrad, Kazakhstan
Central/South America	Sar Cheshmeh, Iran
Bajo de la Alumbrera, Argentina	
Cerro Colorado, Panama	South Pacific
Cerro Pantanos, Colombia	Batu Hijau, Sumbawa, Indonesia
Cerro Rico, Argentina	Dizon, Philippines
Cerro Verde-Santa Rosa, Peru	Grasberg, Irian Jaya, Indonesia
Chuquicamata, Chile	Marcopper, Philippines
El Abra, Chile	Ok Tedi, Papua-New Guinea
El Salvador, Chile	Panguna, Bougainville, Papua-New Guinea
El Teniente, Chile	Santo Tomas II, Philippines
Morococha, Peru	Yandera, Papua-New Guinea
Toquepala, Peru	

3. Problems With Current Models

The needs to reduce uncertainty and improve precision are problems in current deposit models. Significant sources of uncertainty in models are the locations of undiscovered deposits within a permissive terrane and of ore bodies within larger vein and/or alteration arrays. Generally, land-use decisions by governments are made with regard to specific tracts of land that bear little relation to a "permissive terrane." For natural-resource values to be considered equally with other competing land values and uses, the quantitative prediction of the number of undiscovered deposits and their locations is necessary. Similarly, exploration costs are significantly reduced when drilling targets are identified early in an exploration program and "continue/drop the property" types of decisions can be more easily made.

Another source of uncertainty is in the application of models. In making subjective, probabilistic estimates of the number of undiscovered deposits in a tract of land, many geologists believe that their estimates have a high uncertainty and low reproducibility. Thus, there is a need to provide improved mineral-deposit models that diminish the uncertainty and increase the precision of the estimates.

4. Improving Mineral-Deposit Models

4.1. A FRAME FOR ANALYSIS

To reduce uncertainty, it is essential to recognizse that hydrothermal mineral deposits are complex entities that are the outcomes of interdependent processes and mechanisms. Magmatic and hydrothermal fluid flow are coupled with deformation, chemical transport, and heat transport [10]. This thermal, mechanical, and hydraulic interdependence may be viewed in the context of Darcy's flow principle,

$$q = -Ksh - K_c sC_s - K_T sT, \qquad (1)$$

where q is the flow rate; K the hydraulic conductivity; K_c the diffusivity; and K_T, the thermal conductivity. The hydraulic conductivity (K) is a function of permeabilty (k), rock density (ρ), acceleration due to gravity (g), and the dynamic visocity of the fluid (μ),

$$K = k\rho g/\mu \qquad (2)$$

and depends on the nature of the porous medium. The coupled reaction equation (1) serves as the analytical frame of thinking herein.

From equation (1) we know that fluids flow from higher to lower pressures and from higher to lower temperatures. It is more probable that fluid flow will be dispersive than focussed because there are more ways in which the energy due to random molecular motion can be shared throughout a permeable region than ways in which it can be focused into one area. Focussing is a requirement for ore-body formation because, empirically, ore bodies are discrete entities within spatially larger vein and/or alteration arrays. Therefore, thermal and hydraulic phenomena are not probable mechanisms likely to focus fluid flow. Rather, mechanical (deformation) phenomena are the most probable mechanisms for focusing hydrothermal fluids sufficiently for ore body formation.

Within any stress field, in the far- (regional) or near-field, fluids flow perpendicular to the least principal stress (σ_3) and parallel with the trajectories of the maximum principal stress (σ_1) [11][12]. The coupled nature of fluid flow and deformation indicates that resolving the stresses around mineral-deposit-controlling fault systems provides a framework for predicting where within structural systems hydrothermal systems are most likely to be localised and ore bodies formed. Segall and Pollard [13] conducted experiments that elucidated the distribution of stresses around two propagating en échelon fracture tips, as shown in Figure 1. They found that maximum stress is concentrated around the fracture tips and diminishes away from them. For en echelon fractures propagating in a right-lateral sense, a right step between the fractures will result in a zone of lower stress (extension) between the fractures relative to the far-field, and stress greater than the far-field at the fracture tips and immediately outside of the stepover. For left stepovers along right-lateral fault systems, the maximum stresses are highest at the fracture tips and between the en échelon fractures and lower outside or away from the interacting fracture tips. The implication of these relations is that because fluid flow is perpendicular to σ_3 and parallel to σ_1, fluid flow will be focused at interacting, propagating fracture tips.

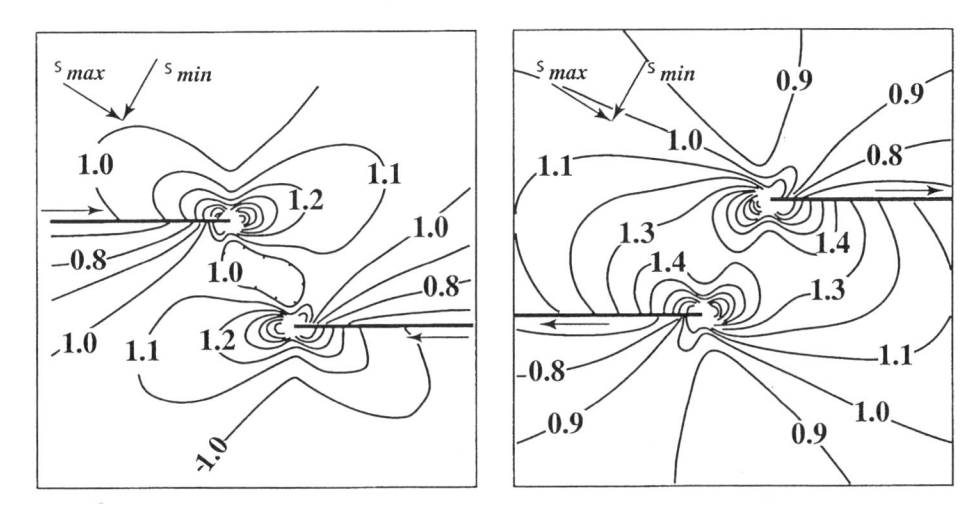

Figure 1. The state of stress around interacting right-lateral en échelon fractures for (**a**) a right stepover and (**b**) left stepover (from [13]). The contoured values are normalised by the far-field maximum principal stress; thus, background value is 1.0. The maximum and minimum far-field principal stresses are shown with heavy arrows in upper left-hand corner of each drawing.

The degree of interconnectivity of fracture networks affects the fluid flow and reactions within the fluid. If the differential stress,

$$\Delta\sigma = \sigma_1 - \sigma_2, \quad (3)$$

acting on a fracture is high (>1 MPa), then stress will be resolved primarily along the plane of the fracture, there will be little interaction between closely spaced fractures, and straight shear fractures will ensue [14]. In contrast, under low-$\Delta\sigma$ conditions (_0 Mpa), fracture tips will interact, stresses normal to the fracture trace will predominate, and curved fracture traces will ensue. Interconnected fracture networks may result from changing principal stresses and a high density of high-Δσ fractures or from conditions that result in the presence of high- and low-Δσ fractures. High interconnectivity leads to more efficient and faster heat transfer between hot rock and fluid in the fractures, thereby affecting the vapor/liquid ratio in the fluid and concentrations of solutes (discussed further in Section 7.2) [29].

4.2. WHAT IS A PORPHYRY-STYLE DEPOSIT?

To apply the principle of coupled flow equations, the definition of a porphyry copper deposit must be reexamined to recognise the role of fracture networks. Empirically, relatively higher grade porphyry-style copper deposits (_0.6% Cu) contain high- and low-Δσ fractures. In contrast, relatively lower grade deposits (_0.35% Cu) contain predominantly high-Δσ fractures. Therefore, economic porphyry-style deposits form in structural environments where high- and low-Δσ stress conditions can dominate. To reduce the uncertainty in modelling, it is important to identify structural environments conducive to forming high- and low-Δσ fractures.

5. Characteristics of Strike-Slip Fault Systems

As noted above, analysis of world-wide porphyry copper deposits shows that they form along strike-slip fault systems in convergent, consuming tectonic plate-margin environments (see also [15][16]). Strike-slip faults accommodate strain in the overriding tectonic plates, most frequently along margins with some degree of oblique convergence. Within these fault systems, the stress and strain distribution is important to the style of fracturing that occurs in the internal hydrothermal systems.

The analysis of seismograms following the 1992 Landers, California (U.S.A.), earthquakes show that seismic-wave velocities decreased as the displacement approached releasing or extensional bends along the strike-slip fault system and accelerated after crossing through stepovers along the system [17]. In addition, the analysis of interferograms of synthetic aperture radar (SAR) data obtained prior to and following the Landers earthquake indicated that the vertical component of displacement was greatest and that displacement contours were most closely spaced adjacent to releasing and constraining bends [18]. Analogous to the patterns around fracture tips in Figure 1, the SAR interferograms show the displacement contours to form semi-circular swirling patterns of displacement that increasine in diameter away from the Landers fault system. These patterns of actual strain are the same geometry as the stress patterns calculated by Segall and Pollard [13] and indicate that deformation is greatest at fracture tips in propagating en échelon strike-slip fault systems.

In 1972, there was a large magnitude earthquake along the Imperial strike-slip fault system in southern California (U.S.A.). Displacement propagated north along the fault system and stepped over to the Brawley strike-slip fault zone near Brawley, California. Johnson and Hadley [19] analysed the displacement on a number of aftershocks that occurred near Brawley during early 1975. Their analysis showed that right-lateral strike-slip motion occurred on the Brawley fault but that strain within the northern releasing bend from the northwestern sidewall of the stepover duplex onto the Brawley fault was accommodated in a much more complex way. The complexity of strain accommodation was recognised by Hill [20] to be via a network of short extensional faults separated at each end by short strike-length synthetic and antithetic shear faults. Thus, motion within extensional duplexes along strike-slip faults can take place on a mesh of faults, not simply on extensional sidewall faults that link the master strike-slip systems as implied in the results from clay model experiments (e.g., [21]).

Sibson [22] proposed that the attenuation of seismic velocities in releasing bends of strike-slip fault systems was due to the interaction of fluids and mechanics in fault zones. He further proposed that the fluids effectively would be coseismically pumped into shear fractures within fault meshes at the time of earthquakes and out of them and into tensile fractures during interseismic periods [23]. Thus, major earthquakes and aftershock events would set up cyclical pumping within fluid/rock systems, a process that might be of importance to ore formation in hydrothermal systems.

6. The Structural Localisation of Porphyry Copper Systems

Porphyry copper deposits are not randomly located within strike-slip fault systems. They are preferentially located in the releasing bends into duplexes of faults in stepovers and less commonly in positive flower structures ("pop-up" structures in zones of convergence) along strike-slip fault systems, as shown in Figure 2. Thus, there is a parallel between the coseismic distribution of stress and strain along strike-slip fault systems and the occurrence of porphyry-style deposits.

Two examples follow that illustrate the structural localisation of porphyry-style systems, Chuquicamata in South America and the Sierrita-Esperanza area in North America. However, although there are structural similarities between the examples, there are differences in the fracture networks and grades in the two examples.

6.1. CHUQUICAMATA, CHILE

Chuquicamata, located in northern Chile, is one of the largest and higher grade porphyry-style deposits in the world. It occurs along the north-striking regional trench-linked Domeyko shear zone [24]. At the onset of mineralisation (early to middle Oligocene) the sense of motion on the shear zone was right-lateral, reflecting a far-field σ_1 oriented northeast/southwest [25]. At Chuquicamata, there is a releasing bend (right-stepover) along the Domeyko shear zone and the deposit is localised within the hinge zone of this bend. The final stage of mineralisation appears to be coincident with a reorientation of the far-field stresses (refer to [24]). Lindsay et al. [24] found that the fracture networks at Chuquicamata may be separated into domains. During the earliest stage of mineralisation, it may be inferred from their data that high- and low-$\Delta\sigma$ fracturing developed throughout the deposit. The orientation of the high-$\Delta\sigma$ veins is apparently to the northeast, which would make them roughly parallel to the far-field σ_1. Pardo-Casas and Molnar [25] interpret a change in the plate convergence circa 30 Ma, which is the approximate age of quartz-sericite alteration at Chuquicamata. There was a concurrent change in the hydraulic conductivity in the southeastern part of the deposit, as may be inferred from the data of Lindsay *et al.* It is likely that the change in hydraulic conductivity reflects this change in the orientation of σ_1 rather than a synmineralisation rotation of a block of the deposit. Thus, the advent of extensive sericitic alteration is also related to this change in the far-field stress trajectories. This is discussed further in Section 7.2.

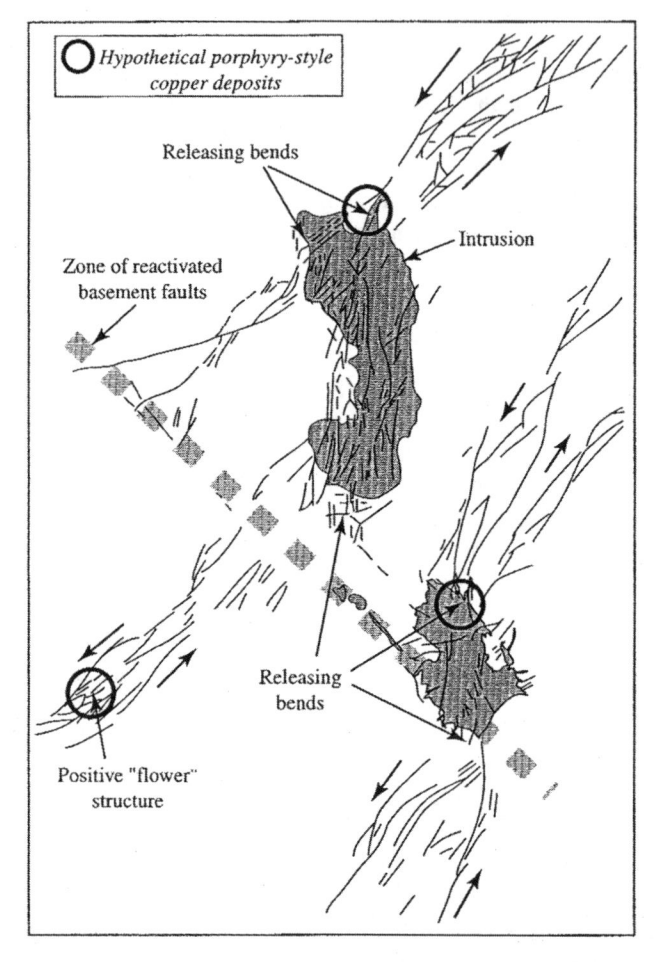

Figure 2. Left-stepover along a left-lateral strike-slip fault system that forms an extensional duplex. Hypothetical porphyry-style copper deposits are localised in the releasing bends of the duplex and within a positive flower structure along the master strike-slip fault system.

6.2. SIERRITA MOUNTAINS, ARIZONA (U.S.A.)

There are a many porphyry-style deposits in southern Arizona in the Western United States. In the Sierrita Mountains, located south of Tucson, Arizona, there are several open-pit porphyry-copper deposits presently being mined. A northwest-striking shear zone bounds the northeastern flank of the range. During the Laramide (latest Cretaceous to Eocene), motion on the shear zone was right-lateral. The Palaeocene Sierrita batholith is juxtaposed against older rocks by the fault zone and extends to the northeast, flooring an extensional duplex formed by a series of right stepovers along the shear zone.

There were at least three porphyry-copper-forming systems localised within or near the Sierrita Mountains extensional duplexes. The Sierrita-Esperanza porphyry copper/molybdenum deposit is localised in the southwestern-most releasing bend of the duplex. Stockwork, northeast-striking, vein-controlling fractures in the mine are predominantly high-$\Delta\sigma$ (B.R. Berger, unpublished data). The veins were deposited parallel to the far-field σ_1. Because of the lack of low-$\Delta\sigma$ fractures, the fracture interconnectivity is low, and the ore grade is also low relative to most producing porphyry-copper mines. The releasing zone that localised the deposit appears to have been "forced" by motion on a northeast-striking, left-lateral shear zone.

The Twin Buttes porphyry copper/copper skarn deposit is northeast of Sierrita-Esperanza along the same northeast-striking fault system that forced the stepover in the Sierrita-Esperanza mine area. The deposit occurs where the left-lateral, northeast-striking fault zone intersects a regionally extensive, northwest-striking, right-lateral fault system. The ore is localised where the left-lateral fault zone steps to the right across the right-lateral zone which forms a zone of convergence—a positive flower structure. Northwest-striking Laramide-age dykes, apophyses from the Sierrita batholith that floors the extensional duplexes, intrude Precambrian granitic rocks and Palaeozoic and Mesozoic sedimentary rocks. Ore-grade mineralisation is primarily on northwest-striking fractures and generally in the host rocks into which the Laramide-age dykes intruded.

To the north in the same mining district as the Sierrita-Esperanza and the Twin Buttes mines are the Mission-Pima-San Xavier porphyry-style deposits. These deposits are fault-dismembered remnants of a once-single deposit. They were moved from the original location by postmineralisation, low-angle detachment faulting. Kinematic indicators on the detachment surface suggest that the complex was moved southeasterly from its original location. This implies that Mission-Pima is not related to the Twin Buttes deposits as suggested by some writers (refer to [26]). It is geometrically possible that its root zone is in a releasing bend of the duplex structure to the northwest, but this is only speculation.

6.3. OTHER PORPHYRY-COPPER DEPOSITS

The empirical importance of releasing bends and positive flower structures along strike-slip faults in the localisation of porphyry-style deposits is evident world-wide. Sufficiently detailed geologic information at the proper scale is difficult to obtain for most deposits. Thus, assembling a comprehensive data base for most occurrences is difficult. Table 2 is a representative list of deposits other than those discussed above for which sufficient data are available to interpret their structural settings.

TABLE 2. The structural localisation of porphyry-style deposits

Deposit	Structural setting
United States: Bisbee, Arizona	Releasing bend, left-stepping, left-lateral fault
Ithaca Peak, Arizona	do
Red Mountain, Arizona	do
Sunnyside, Arizona	do
Twin Buttes, Arizona	Constraining bend, intersecting lateral faults
Romania: Valea Morii	Releasing bend, left-stepping, left-lateral fault
Zlatna	do
Slovakia: Banska Stiavnica	Releasing bend, left-stepping, left-lateral fault
Irian Jaya: Grasberg	Releasing bend, left-stepping, left-lateral fault

7. Discussion

The presence of high- and low-$\Delta\sigma$ fractures in higher grade porphyry-style deposits (e.g., Chuquicamata) in comparison to the presence of only high-$\Delta\sigma$ fractures in less economically valuable deposits (e.g., Sierrita-Esperanza) indicates that in spite of the analogous structural controls on localisation, special stress conditions at the site of localisation are important to ore-depositing processes in porphyry systems. Further, the preferred localisation of deposits in releasing bends and positive flower structures along strike-slip fault systems implies a necessary relation between the mechanical phenomena that characterise these sites and ore-deposition processes.

There are two characteristics of releasing bends along strike-slip fault systems that are important to porphyry-style deposit formation. One is that the steep stress and strain gradients at these effective "fracture tip" environments are advantageous to the focussing of fluid flow. The other is the nature of how stress and strain are partitioned within releasing bends.

7.1. STRESS, STRAIN, AND FLUID-FLOW FOCUSSING

There are two scales at which fluid-flow focussing is important in the genesis of porphyry-style systems. (1) From their origins to depths of _4-5 km, intermediate to siliceous magmas rise as batches elongated horizontally parallel with the far-field σ_1 and vertically (dyke-like) inside a pipelike vertical column of such batches [27]. At about 4-5 km depth, the buoyancy of the magmas is insufficient relative to its surroundings ("neutral buoyancy" [27]) to ascend further, and the batches accumulate into "magma reservoirs." These reservoirs are localised in zones of low mean stress, such as releasing bends and extensional stepovers along strike-slip faults. Within a magma reservoir, stress is concentrated, and the highest strain rates occur, along the boundaries of the reservoir including its top because of the high rheological contrast of the intrusions and the host rocks. Thus, in porphyry-style hydrothermal systems, the record of fluid flow—stockwork fracturing and wallrock alteration—frequently is in the shape of an upward converging cone surrounding a core of porphyritic intrusions. (2) The fault systems inside releasing bends in extensional duplexes are meshes of shear and tensile fractures. During the intrusion of batches of magma, the magma pressures may equal the external pressures ($P_{magma} \_ P_{external}$), and the stress regime surrounding the intrusion will be dominated by the magma. Low-$\Delta\sigma$ fracturing results as well as a high degree of interconnectivity of the fracture system in the incipient porphyry-style deposit. This style of fracturing may be intermittent during porphyry deposit formation but is most frequently observed during the earliest stages of deposit development (e.g., El Salvador, Chile [28]). During periods when $P_{external} > P_{magma}$, high-$\Delta\sigma$ shear fractures predominate. Most paragenetic studies of porphyry-style deposits have found the greatest proportion of ore minerals to occur along high-$\Delta\sigma$ fractures. We envision this to be due to two effects, one involving the greater strike length of shear fractures relative to shorter tensile fractures and the other being analogous to Sibson's [23] pump model where fluids in compressional settings flow into shear fractures during their coseismic propagation.

7.2. PARTITIONING OF STRESS AND STRAIN

Ore bodies reflect the focussing of fluid flow into restricted permeabilities; otherwise, entire vein arrays would constitute ore. Empirically, these ore bodies are within permeable zones of short strike length relative to the entire mineralised fracture network. The favorability of shorter strike-length fractures may be understood in the context of equation (1), the coupled flow relation. Heat transfer during deformation to the liquid phase in the fracture network is greater for closely spaced fractures than for widely spaced fractures [29]. Because, empirically, present-day hydrothermal systems are near their vapour-saturation condition [30], the partitioning of volatiles into the vapour phase during deformation will lead to significant temperature drops, depending on the vapour fraction produced, and the concentration of dissolved compounds in the liquid phase will increase dramatically. Thus, dense, interconnected fracture networks are conducive to the production of metal-saturated solutions during active deformation.

Because stress and stress gradients are greatest at fracture tips, "hinge" areas of releasing bends along strike-slip faults focus fluid flow, as well as the interior area of constraining bends (positive flower structures). However, because stress and strain are related, as the stress is partitioned so is the strain. We suggest that economically viable deposits form where fluid flow is mechanically constrained to sites under intermittent high strain rates such that an effective "reaction vessel" is formed. It is strain partitioning that is the biggest enemy of porphyry-style ore formation. The partitioning of strain may lead to insufficient times of fluid focus to form large tonnages of high grade ore, or even may lead to the complete failure of the fluid to be focussed at all into a restricted volume.

In addition to the decline of porphyry ore formation, when the "reaction vessel" is breached through strain partitioning, there may be several other consequences. One is that meteoric waters more easily interact with the magmatic system, leading to the greater concentrations of lead and/or zinc in the hydrothermal fluids. Veins with a polymetallic character are deposited. Intense quartz-sericite alteration has been shown isotopically in porphyry systems to be another reflection of the increased incursion of meteoric fluids into the system [31]. At Chuquicamata, the late-stage sphalerite ± galena + enargite veins associated with the zones of phyllic alteration [24] most probably reflect the effects of strain partitioning. In some deposits, the polymetallic veins cross-cut the porphyry ores (e.g., Copper Cities, Arizona), while at others, the polymetallic veins are "zoned" around the porphyry deposit (e.g., Bingham Canyon, Utah).

7.3. ORE BODIES WITHIN VEIN SYSTEMS

Vein-style ore bodies are usually discrete entities within spatially more-extensive vein systems. Why? The world's great polymetallic vein and/or replacement districts (e.g., Leadville, Colorado) are localised along strike-slip fault systems. As noted in Section 7.1, the fracture networks within these systems are made up of zones of faults under shear and under tension.

Because many polymetallic vein and replacement districts are related to intrusive centres of porphyritic stocks and dykes (e.g., Tintic, Park City, and Bingham Canyon, Utah), the veins occur in releasing bends or the hinges of extensional stepovers along strike-slip faults. Ore bodies within the veins are preferentially formed along shorter strike length (tens to hundreds of metres) tensile fractures rather than on longer strike (hundreds to thousands of metres) extensional faults. This is because large volumes of highly permeability ground lead to dispersion of flow and lower temperatures. By understanding the spatial dimensions of individual faults within fault meshes, one can better predict the probable locations of ore bodies.

8. Conclusions

The impetus for state-of-the-art models is the need to significantly reduce uncertainty in mineral-resource assessments and minerals exploration. Mineral-deposit models have traditionally been summaries of descriptive and genetic information about a specific deposit type. The emphasis with genetic information has been on geochemistry. However, mineral deposition is a dynamic process that evolves through time and space as a consequence of the coupling of thermal, mechanical, and hydraulic phenomena. By considering the interdependence of all of these phenomena, a greater understanding about ore-forming processes results, including such critical aspects as where within structural systems hydrothermal systems are localised and where within larger vein arrays ore bodies are most likely to be formed. As demonstrated for porphyry-style deposits, a new generation of deposit models is possible.

References

1. Henley, R.W. and Berger, B.R. (1993) What is an exploration model anyway?— An analysis of the cognitive development and use of models in mineral exploration, in R.V. Kirkham, W.D. Sinclair, R.I. Thorpe, and J.M. Duke (eds.), *Mineral Deposit Modeling*, Geological Association Canada Special Paper 40, pp. 41-50.

2. Lindgren, W. (1933) *Mineral Deposits*, McGraw-Hill, New York.
3. Guilbert, J.M. and Park, C.F., Jr. (1986) *The Geology of Ore Deposits*, W.H. Freeman and Company, New York.
4. Cox, D.P. and Singer, D.A. (eds.) (1986) *Mineral Deposit Models*, U.S. Geological Survey Bulletin 1693.
5. Titley, S.P. (ed.) (1982) *Advances in Geology of the Porphyry Copper Deposits*, The University of Arizona Press, Tucson.
6. Mitchell, A.H.G. and Garson, M.S. (1981) *Mineral Deposits and Global Tectonic Settings*, Academic Press, New York.
7. Sawkins, F.J. (1990) *Metal Deposits in Relation to Plate Tectonics*, second edition, Springer-Verlag, New York.
8. Woodcock, N.H. (1986) The role of strike-slip fault systems at plate boundaries, *Philosophical Transactions of the Royal Society of London* **A317**, 13-29.
9. Woodcock, N.H. and Schubert, C. (1994) Continental strike-slip tectonics, in P.L. Hancock (ed.) *Continental Deformation*, Pergamon Press, New York, pp. 251-263.
10. Ingebritsen, S.E., and Sanford, W.E. (1998) *Groundwater in Geologic Processes*, Cambridge University Press, New York.
11. Nakamura, K. (1977) Volcanoes as possible indicators of tectonic stress orientation—Principle and proposal, *Journal of Volcanology and Geothermal Research* **2**, 1-16.
12. Tsunakawa, H. (1983) Simple two-dimensional model of propagation of magma-filled cracks, *J. Volcanology and Geothermal Research* **16**, 335-343.
13. Segall, P. and Pollard, D.D. (1980) Mechanics of discontinuous faults, *Journal of Geophysical Research* **85**, 4337-4350.
14. Olson, J. and Pollard, D.D. (1989) Inferring paleostresses from natural fracture patterns: A new method, *Geology* **17**, 345-348.
15. Nishiwaki, C. (1981) Tectonic control of porphyry copper genesis in the southwestern Pacific island arc region, *Mining Geology* **31**, 131-146.
16. Nishiwaki, C. (1982) Tectonic stress and metallogenesis; primarily with reference to porphyry copper genesis, *Mining Geology* **32**, 291-304.
17. Wald, D.J. and Heaton, T.H. (1994) Spatial and temporal distribution of slip for the 1992 Landers, California, earthquake, *Bulletin of the Seismological Society of America* **84**, 668-691.
18. Massonnet, D., Rossi, M., Carmona, F., Adragna, R., Peltzer, G., Feigl, K., and Rabaute, T. (1993) The displacement field of the Landers earthquake mapped by radar interferometry, *Nature* **364**, 138-142.
19. Johnson, C.E. and Hadley, D.M. (1976) Tectonic implications of the Brawley earthquake swarm, Imperial Valley, California, January 1975, *Bulletin Seismological Society of America* **66**, 1133-1144.
20. Hill, D.P. (1977) A model for earthquake swarms, *Journal of Geophysical Research* **82**, 1347-1352.
21. Dooley, T., and McClay, K. (1997) Analog modeling of pull-apart basins, *American Association of Petroleum Geologists Bulletin* **81**, 1804-1826.
22. Sibson, R.H. (1987) Stopping of earthquake ruptures at dilational fault jogs, *Nature* **316**, 248-251.
23. Sibson, R.H. (1994) Crustal stress, faulting, and fluid flow, in J. Parnell (ed.) *Geofluids: Origin, Migration and Evolution of Fluids in Sedimentary Basins*, Geological Society of London Special Publication 78, pp. 69-84.
24. Lindsay, D.D., Zentilli, M., and Rojas de la Rivera, J. (1995) Evolution of an active ductile to brittle shear system controlling mineralization at the Chuquicamata porphyry copper deposit, northern Chile, *International Geology Review* **37**, 945-958.
25. Pardo-Casas, F. and Molnar, P. (1987) Relative motion of the Nazca (Farallon) and South American plates since late Cretaceous time, *Tectonics* **6**, 233-248.

26. Barter, C.F. and Kelly, J.L. (1982) Geology of the Twin Buttes mine, in S.R. Titley (ed.) *Advances in Geology of the Porphyry Copper Deposits*, The University of Arizona Press, Tucson, pp. 407-432.
27. Ryan, M.P. (1988) The mechanics and three-dimensional internal structure of active magmatic systems: Kilauea volcano, Hawaii, *Journal of Geophysical Research* **93**, 4213-4248.
28. Gustafson, L.B. and Hunt, J.R. (1975) The porphyry copper deposit at El Salvador, Chile, *Economic Geology* **70**, 857-912.
29. Henley, R.W. and Hughes, G.O. (in press) Excess heat effects in vein formation, unpublished manuscript.
30. Henley, R.W., Truesdell, A.H., and Barton, P.B., Jr. (1984) *Fluid-Mineral Equilibria in Hydrothermal Systems*, Society of Economic Geologists, Reviews in Economic Geology 1.
31. Sheppard, S.M.F., Nielsen, R.L., and Taylor, H.P., Jr. (1971) Hydrogen and oxygen isotope ratios in minerals from porphyry copper deposits, *Economic Geology* **66**, 515-542.

GEOLOGIC INFORMATION FOR AGGREGATE RESOURCE PLANNING

WILLIAM H. LANGER, DAVID A. LINDSEY, and DANIEL H. KNEPPER, JR.
U.S. Geological Survey
MS 973 P.O. Box 25046
Denver, Colorado 80225-0046 USA

Abstract. Construction and maintenance of the infrastructure is dependent on such raw materials as aggregate (crushed stone, sand, and gravel). Despite this dependence, urban expansion often works to the detriment of the production of those essential raw materials. The failure to plan for the protection and extraction of aggregate resources often results in increased consumer cost, environmental damage, and an adversarial relation between the aggregate industry and the community.

As an area grows, the demand for aggregate resources increases, and industries that produce these materials are established. Aggregate is a low-cost commodity, and to keep hauling costs at a minimum, the operations are located as close to the market as possible. As metropolitan areas grow, they encroach upon established aggregate operations. New residents in the vicinity of pits and quarries object to the noise, dust, and truck traffic associated with the aggregate operation. Pressure is applied to the local government to limit operation hours and truck traffic.

In addition to encroaching on established aggregate operations, urban growth commonly covers unmined aggregate resources. Frequently urban growth occurs without any consideration of the resource or an analysis of the impact of its loss. The old idea that aggregate resources can be found anywhere is false. New aggregate operations may have to be located long distances from the markets. The additional expense of the longer transport of resources must be passed on to consumers in the community. In many instances, the new deposit is of inferior quality compared with the original source, yet it is used to avoid the expense of importing high-quality material from a more-distant source.

Some governmental, including city, provincial or state, and national, agencies, have enacted regulations to help maintain access to prime aggregate resources. Although regulations have met with variable success, some policy or regulation to protect aggregate resources is worth consideration.

A basic requirement of any aggregate resource policy or regulation is the knowledge of the geographic distribution, volumes, and quality of aggregate resources. This knowledge commonly is obtained through geologic mapping and characterization of aggregate resources. Geographic Information Systems (GIS) and Decision Support Systems (DSS) provide excellent tools to help present and evaluate the information in a manner that is understandable by public decisionmakers.

1. Introduction

Infrastructure, such as highways, roads, bridges, airports, railroads, public buildings, sewage treatment, and many other facilities, is vital to the sustainability and vitality of any populated area. In many areas of rapid population growth, the infrastructure may be inadequate, and new infrastructure must be constructed to meet growing needs. In other areas, infrastructure has deteriorated to a point that extensive repair and replacement are required. Development and maintenance of the infrastructure requires large volumes of aggregate (sand, gravel, and crushed stone), which makes aggregate the number one-ranked extracted mineral resource worldwide in terms of amount and value [1].

Aggregate occurs where Mother Nature put it, not necessarily where people need it. Local resources may not have the quality requirements for some uses. Even if adequate supplies of suitable-quality aggregate resources are available, they may be difficult to obtain because of urban encroachment or other incompatible land uses.

Aggregate is a high-bulk, low-unit value commodity. Transportation is a significant part of the put-in-place cost of aggregate, and, therefore, most aggregate operations are located near populated areas. This puts aggregate producers in direct conflict with communities that have their own priorities for use of the land surface, water supply, air, and roads. In addition, aggregate operators are far outnumbered by society-at-large, and most people do not like mining. In some areas, the options for places to develop aggregate resources are extremely limited.

As available resources are consumed or preempted, the cost of maintaining or expanding the infrastructure increases and the costs are passed on to the public as higher taxes or reduced services. To provide for a continuous supply of reasonably priced, high-quality aggregate resources, the remaining resources may need to be identified and protected.

In the paper *Quarry Site Surveys in Relation to Country Planning*, Shadmon, [2] pointed out that there were many countries where opening new quarries was practically impossible. This limitation is especially true for densely populated countries where planning powers established uses of available land without paying attention to quarry needs. He called for countrywide planning to forecast long-term quarry product needs. Thirty years later, we are still calling for effective mineral-resource planning.

Land use conflicts can be divided into avoidable and unavoidable conflicts [3]. Avoidable conflicts include the careless preemption of valuable resources by allowing the land over them to be developed and the encroachment of conflicting land uses where new neighbors diminish mining activities through regulations and ordinances. Unavoidable conflicts are those that involve incompatible or unacceptable relations between aggregate mining and other natural resources with an inherent value to society.

This paper reviews several case histories to demonstrate how aggregate resources might be protected from avoidable conflicts. The case histories are summarized from published articles. The authors of those articles are cited at the beginning of each case history and included in the references at the end of the report. Readers are encouraged to consult those publications for more-detailed discussions of the issues.

Each case history is closed with a statement summarizing one specific fact of the overall history.

2. Assessments – Identification of the Resource

One necessary step in protecting aggregate resources is to identify the available resources. The 1970's was a period of increased environmental awareness. Numerous agencies began experimenting with using earth science information to assist land-use planning. As a result, a large variety of special-purpose maps derived from standard geologic maps were created. Included in this effort was the use of geologic information for the identification of aggregate resources [4].

The efforts to create special purpose maps to delineate aggregate resources apparently proved to be a worthy exercise because the same types of maps continue to be prepared today, nearly three decades later. The various methodologies for assessing aggregate resources are beyond the scope of this paper. Numerous national, state, or provincial agencies, however, have conducted mapping programs to assess and characterize aggregate resources. Descriptions of assessment techniques are included in *The Aggregate Handbook* [5], *Aggregates* [6], and in a four-part series of articles in *Rock Products* [7, 8].

Descriptions of actual assessments for parts of Australia, Brazil, Canada, France, Germany, Great Britain, Hong Kong, India, Italy, The Netherlands, Norway, Poland, South Africa, Spain, Sweden, and the United States of America can be found in such major works related to aggregate; *Aggregate Resources – A Global Perspective* [9], *Aggregates – Raw Materials' Giant* [10], and the *Proceedings from the International Symposium on Aggregates*, [11]. A study of these reports will not only provide an understanding of the many different ways to assess aggregate resources but because of a nearly 15-year time span will give an historical perspective to various assessment approaches.

To plan adequately for the protection of aggregate resources, understanding the location and character of the resources and presenting that information in a clearly understandable format that can be used by decisionmakers is vitally necessary.

3. Austria – Model to Identify Resource Extraction Areas

Identification of the distribution of aggregate resources, by itself, does not provide enough information for the protection of the resources. The physical and chemical properties of the aggregate, as well as the anticipated volumes of resources, must be understood. Another important consideration is where the resources fit within the greater environment, as well as how developing those resources might impact the environment.

A model was developed by the Geological Survey of Austria that was used to identify, characterize, and evaluate aggregate resources. The model also evaluated those resources within the broader context of environmental impacts. The model was tested by using the Mattigtal area in Upper Austria as a demonstration area [12, 13]. The model tested techniques of evaluation and took into account aggregate quality and quantity, as well as potential for unwanted impacts to the environment. Potential environmental impacts included ground water and near-surface minerals, ecology, landscape, conservation, settlement, recreation areas, climate and air quality, agriculture, and forestry. The model identified positive conservation zones for further exploitation and evaluation of resources and negative conservation zones where development should be restricted or forbidden. The model could be used to predict the impacts of aggregate extraction on other land uses, such as agriculture, forestry, recreation, waste disposal, and so forth.

To protect resources, identification of where the resources are and characterization of the resources are important.

4. East Granby – An Early Attempt at Planned Development

Planning for aggregate resources involves difficult choices. Sites with aggregate resources, particularly sand and gravel, are often good development sites. Until recently, aggregate extraction commonly was considered a final use of the land. Questions would arise as to whether it is better to develop the aggregate or use the land for some other use. Today, aggregate extraction commonly is viewed as an interim land use. But choices still need to be made about land use. Also, although aggregate is necessary to the development of an area, aggregate operations can depress the value of land in the immediate vicinity of the operations. Some of these problems can be solved through comprehensive planning.

During the 1940's, the town of East Granby, Connecticut, U.S.A., was a small, rural community. For the most part, East Granby was dependent on onsite septic system waste disposal and local wells for water supply. The town has an area of about 45 square kilometers, of which less than 25 percent was developed. During 1941, the town established a zoning commission. Many Connecticut communities, including East Granby, were severely damaged by floods in 1955, which stimulated land-use planning activities. In East Granby, zoning and subdivision were adopted during 1956, and a plan of development was adopted the following year. During the 1970's, the plan was replaced by a new plan that relied heavily on earth science information [14].

In preparing the new plan, the planning and zoning commission solicited ideas from other town agencies and from all East Granby residents by questionnaire. Three overall goals were adopted as a result of responses to the questionnaire – to maintain the rural character of the town, to provide for a balanced growth of population, and to protect the natural resources of East Granby.

Earth science information was used to prepare seven special-purpose maps. Each map is described separately in the plan. The report describes the impact of each natural factor on development and the impact of development on the natural factors. The most significant part of the plan, as related to aggregate, is that even during this early stage in the evolution of land-use planning, an area was identified for aggregate extraction. In the plan, the commission recommended creating a zone that directly related to earth products. The zone would serve two purposes - to strengthen the rights of quarry operators to continue and expand existing activities and to provide controls on the permitted extraction activities.

A comprehensive plan of growth and development can accommodate traditional land use issues, as well as offer protection to, and from, aggregate extraction.

5. Colorado - A Case of Too Little, Too Late

During the 1960's and 70's, aggregate producers were seeing their resource base threatened by legislation and urban encroachment. Some producers asked government to enact legislation to protect them from the impact of expanding development. Some governments responded by requiring assessments of aggregate resource and requiring consideration of those resources in the planning process.

Preemption of aggregate resources by encroachment and conflicting land use has affected the Denver, Colorado, U.S.A., area since the 1950's, when urban growth began to cover large deposits of prime gravel resources. The U.S. Bureau of Mines predicted that restrictive zoning, lack of general public understanding of sand and gravel occurrence and mining operations, and conflicting land uses would cause a shortage of low-cost aggregate. The aggregate industry was concerned about the trend toward preempting aggregate resources and encouraged the state legislation to take measures to protect the remaining resources [15].

During 1973, the Colorado legislature passed House Bill 1529. That act declared that:

(1) "the state's commercial mineral deposits are essential to the state's economy
(2) the populous counties of the state face a critical shortage of such deposits,
(3) such deposits should be extracted according to a rational plan, calculated to avoid waste ...and cause the least practical disruption to the ecology and quality of life of the citizens... ."

The Colorado Geological Survey prepared a folio of maps that delineated the occurrence of aggregate resources in the State's most populous areas [16, 17]. Unfortunately, House Bill (H.B.) 1529 did not succeed at protecting aggregate resources in the Denver area. In 1981, the U.S. Department of Labor pointed out that reserves continued to decline and blamed the decline on noncompliance with H.B. 1529, adverse zoning, increased demand, inadequate grain size to meet specifications, and environmental and visual concerns [15].

The last major source of local gravel, the South Platte River valley north of Denver, was being covered with preemptive land uses faster than it was being mined. Although gravel extraction had removed large amounts of the high-quality gravel, most of what remained was precluded from extraction by urbanization. By 1997, mining had been forced downstream about 9 miles to where the gravel-to-sand ratio of the deposits is about 1:4 [18]. At this location, mining the deposits is reaching economic limits. Aggregate developers are leaving the South Platte River valley to extract aggregate from better resources, even though they are located farther from the market area.

To protect aggregate resources, legislation must be enforceable.

6. California – Controlling Competition Between Aggregate Mining and Urbanization

The situation in which preemptive land use such as urban development covers aggregate resources commonly is referred to as "sterilization." Goldman [19] was one of the first people to describe the impact of urbanization on the mineral industry. The area he was describing was the State of California, U.S.A.

California has a population of about 32 million that consumes more than 150 million tonnes of construction aggregate each year. In addition, the residents are highly urbanized, well educated, politically aware, and environmentally conscious. These factors combine to make the permitting and excavation of aggregate a difficult task.

In an attempt to offset the impacts of sterilization, California enacted the Surface Mining and Reclamation Act of 1975 (SMARA), which is one of government's earliest attempts to reckon with the problem of sterilization of aggregate resources [20]. The first article of SMARA states, "The legislature hereby finds and declares that the extraction of minerals is essential to the continued economic well-being of the State and to the needs of the society, and that the reclamation of mined lands is necessary to prevent or minimize adverse effects on the environment and to protect the public health and safety."

Under SMARA, the California Division of Mines and Geology is mandated to classify specified lands within the State on the basis of mineral content. The classification is to be done "on the basis solely of geologic factors, and without regard to existing land use and land ownership." SMARA does not require that aggregate resources be permitted, but it provides decisionmakers with the information upon which to base various land-use evaluations.

SMARA uses the following six categories of Mineral Resource Zones to classify construction aggregate, ranked in order of importance: 1). economically significant measured or indicated aggregate resources, 2). economically significant inferred aggregate resources, 3). marginally economic, 4). subeconomic, 5). noneconomic, and 6). undetermined or unknown.

Having the State, an unbiased third party, prepare the classifications has distinct advantages. Accurate, objective, quantified mineral-resource data reduces the ability of special interest groups to influence the process. Having statewide standardization eliminates unfair advantages to competing mining companies.

Anecdotal evidence indicates that use of SMARA has been effective at facilitating the permitting process, increasing the life span on renewed use permits, and in the discovery and opening of new deposits.

Objective, quantified, unbiased third-party mineral-resource data improves the fair evaluation and protection of potential aggregate extraction areas.

7. Ontario –Evolution of Regulations

The environmental awareness that started in the 1970's frequently placed mineral developers and "environmentalists" in conflict. Citizens have their own ideas on how to use the land, and public opposition to aggregate mining commonly is very strong. During the 1970's, the groundwork was set for increasing loss of aggregate reserves. The citizens asked their governments to protect them from the impacts of mining. Governments responded by requiring permits or imposing regulations to control aggregate development. As discussed above, aggregate producers were seeing their resource base threatened by legislation and encroachment and also asked government for protection.

This schizophrenic approach to resource management had its problems. In some places, legislation and plans were rewritten to accommodate both concerns in a comprehensive plan. Ontario, Canada, is one such example.

Actions on environmental and planning issues that concern the mining of aggregate in Ontario, Canada extend back to 1970 with the enactment of the Niagara Escarpment Protection Act [21, 22]. Specific provisions were implemented to control extraction of aggregate from the escarpment. During the 1971, steps were taken to regulate the industry further through the Pit and Quarries Control Act (PQCA). During the late 1970's and early 80's, shortcomings of the PQCA became evident. In 1986, mineral aggregate policies were adopted by the Government of the Province as the Mineral Aggregate Resources Policy Statement (MARPS).

The goal of MARPS was to have all parts of Ontario " *** share a responsibility to identify and protect mineral aggregate resources and legally existing pits and quarries to ensure mineral aggregates are available at a reasonable cost and as close to markets as possible to meet future local, regional, and provincial needs." The statement contains a number of principles. These principles include the intent to identify and legally protect existing pits and quarries from incompatible uses, to identify and protect mineral aggregate resource areas from land uses that may be incompatible with future extraction, and to require rehabilitation that is compatible with long-term uses of the land.

During 1995, a Comprehensive Set of Policy Statements (CSPS) came into effect that amalgamated the MARPS with six other provincial policy statements. The CSPS recognizes that other non-aggregate minerals also require protection for their access and resource development in Ontario. The intent is similar to that of the MARPS.

The effectiveness of these policies for aggregate resources cannot yet be determined, although preliminary observations are that the provincial interests in aggregate are now taken into consideration in the development of planning decisions.

A comprehensive plan can accommodate protection of aggregate resources, and protection from the impacts of developing those resources.

8. Netherlands – National control - Recycling as a Major Part of the Solution

When a resource is being developed, the benefits of development are dispersed over very large areas. The benefits to the region, however, are seldom considered in the local permitting process. The community where development occurs suffers most from any adverse consequences of resource development. To complicate the issue further, if a local resource operation is denied, then higher costs are usually added. Also, impacts may be transferred outside the area where the permit was denied. For example, the decision to deny a resource extraction permit because of the unwanted increase in truck traffic ultimately results in more truck traffic in the region because of longer haul routes. Thus, more roads are subjected to truck traffic and suffer increased wear and tear; more people are subjected to accidents, noise, traffic, and other inconveniences; more fuel is consumed; and more hydrocarbons are released to the atmosphere. Any gain by the local community is often at the expense of the greater public and greater environment. The riddle is "When a political entity is evaluating whether or not to develop or improve a resource, how can we as a nation be sure that the dispersed benefits of use of that resource are adequately weighed in the final decision?" [23]. One example is the Dutch Excavation Act.

The Netherlands has the second highest population density in world. Its per capita GNP is among highest in world, and its economy is still growing. These factors create a high demand for construction materials [24]. Sand and gravel are the main sources of construction aggregate, and gravel has a very limited distribution on or near the surface. Bedrock outcrops occupy only about 1 percent of the land surface of the Netherlands. Where resource availability is limited, conflicts tend to be great.

During the 1970's, the provinces, building industry, and environmental groups wanted the national government to develop a policy on the excavation of construction materials. The aggregate extraction industry saw problems of supply because of the increased time to obtain excavation licenses. The environmental groups wanted a reduction in the use of materials through recycling and a better planning procedure to determine where extraction would be allowed.

Discussions led to the formulation of a national policy on the supply of construction materials. That policy was in place for about 10 years. Recently, the Dutch Excavation Act was adopted. The main goals of the act are to encourage economical use of construction materials and recycling, to ensure that a sufficient supply of construction materials can be excavated from within the Netherlands, to develop resources near the application, to develop aggregate extraction areas in a socially responsible manner, and to achieve a systematic approach to the extraction policy.

One of the most notable aspects of the Act is that State officials, if strictly necessary, will be authorized to force provinces to give permission for excavating a sufficient quantity of natural construction materials.

One way to ensure that aggregate resources are protected is to give the national government the authority to force provinces or states to allow extraction.

9. Geographic Information Systems – A Tool for Aggregate Resource Planning

Computers have been used as a tool for analyzing spatial relations of aggregate resource and other land-use planning for nearly 25 years. Early programs depended on analyses of data collected in a grid patterns and stored in numerical arrays. As long as different sets of data were stored in identical arrays, a computer could be used to compare, add, subtract, or otherwise manipulate data in different arrays. Data output was accomplished by using a typewriter-style printer, and to achieve a map format, characteristics of each cell were represented by single symbols or combinations of symbols.

As computers and computer displays became more elaborate, it became possible to conduct more complex analyses of data, as well as to observe the results by using computer monitors. Spatial data could be collected, stored, and analyzed as raster data (similar to data being stored in arrays) or as vectors. Robust software programs were designed to be run on large mainframe computers. This required a large investment in computer systems, as well as the need for personnel to operate the sophisticated mapping programs.

Recent improvements in technology, software, and manufacturing processes, combined with decreased costs of computer systems, has brought GIS within economic reach of many groups. The U.S. Geological Survey has a number of projects designed to manage earth science information by using standard off-the-shelf GIS. One such project is the Front Range Infrastructure Resources Project (FRIRP).

9.1 GEOGRAPHIC INFORMATION SYSTEMS

One goal of the FRIRP is to educate decisionmakers regarding the issues related to infrastructure resources. Decisionmakers need to understand that aggregate occurs only in limited areas; that the quality of the aggregate, in part, controls its use; and that aggregate quality changes from deposit to deposit. Some decisionmakers are not familiar with geology, aggregate resources, or resource maps. Consequently, visualization is a very important part of the process. GIS are valuable tools for use in outreach and visualization. The aggregate map data sets described in Section 5, have been imported into a GIS and can be displayed in combination with other digital data thus allowing users to develop a variety of scenarios, in an interactive manner. This demonstration looks at part of the FRIRP study area along the Cache la Poudre River from Ft. Collins to Greeley, Colorado.

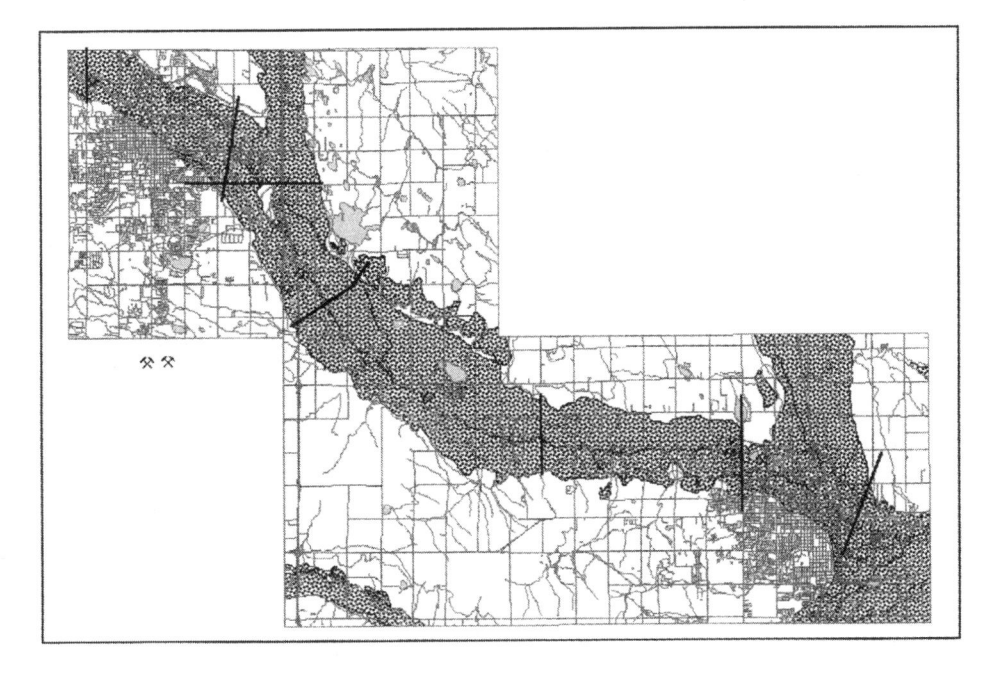

Figure 1. Computer generated image showing landforms underlain with sand and gravel. Hillshade base removed for clarity.

An artificially sun-shaded relief image created from a digital elevation model was prepared for the Cache la Poudre River area. This was combined with highway and hydrological digital line graphs (DLG) to form the base for other digital maps. The first data layer added to the base map was the location of landforms underlain by sand and gravel, shown in Figure 1.

The "hot-link" feature of the GIS was employed to demonstrate graphically some of the geologic aspects of sand and gravel resources. The hot-link feature allows you to associate an image with a map feature. For example, cross sections were prepared across the river valleys to help demonstrate that the thickness of the deposits change from place to place. The locations of the cross sections were plotted in the data base, and the cross sections were connected to the section lines by using the hot-link tool, as shown in Figure 2. Likewise, the locations of pits were plotted in the database, and a bar chart was linked to pit locations to demonstrate grain size distribution of deposits (see Figure 2).

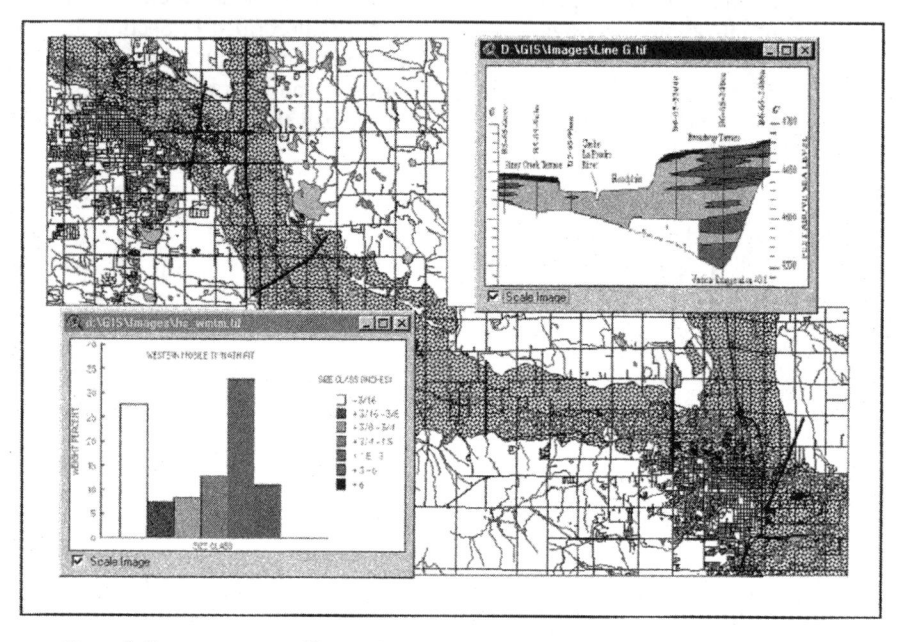

Figure 2. Computer generated image showing images of cross section and statistical chart that were linked to cross section lines and pit locations using GIS "hot link" feature.

Decisionmakers need to understand that gravel particle size affects the use of the aggregate. For example, in the United States, most cement concrete requires gravel of at least ¾-inch. Most asphalt requires particles with angular faces (obtained through crushing), which commonly requires starting with gravel of at least 1½-inch. The downstream change in particle size can clearly be demonstrated by linking photographs of pit faces to pit locations along the stream valleys, as shown in Figure 3. Gravel at the upstream end of the image (left-hand photograph) is about as large as a person's head, and gravel size particles are predominant. Gravel at the downstream end of the image (right-hand photograph) is about as large as a fist, and the deposit includes many sand lenses.

9.2 DECISION SUPPORT SYSTEMS

DSS assists decisionmakers in assessing the implications and opportunities of alternate land scenarios. The FRIRP aggregate project described above provides some of the earth science data, and the GIS provides a platform for implementing a DSS. Using a DSS, we developed a land-use scenario by adding three resource-related components to the data set - development, demand, and transportation. These components are editable and can be added or deleted for individual scenario exercises.

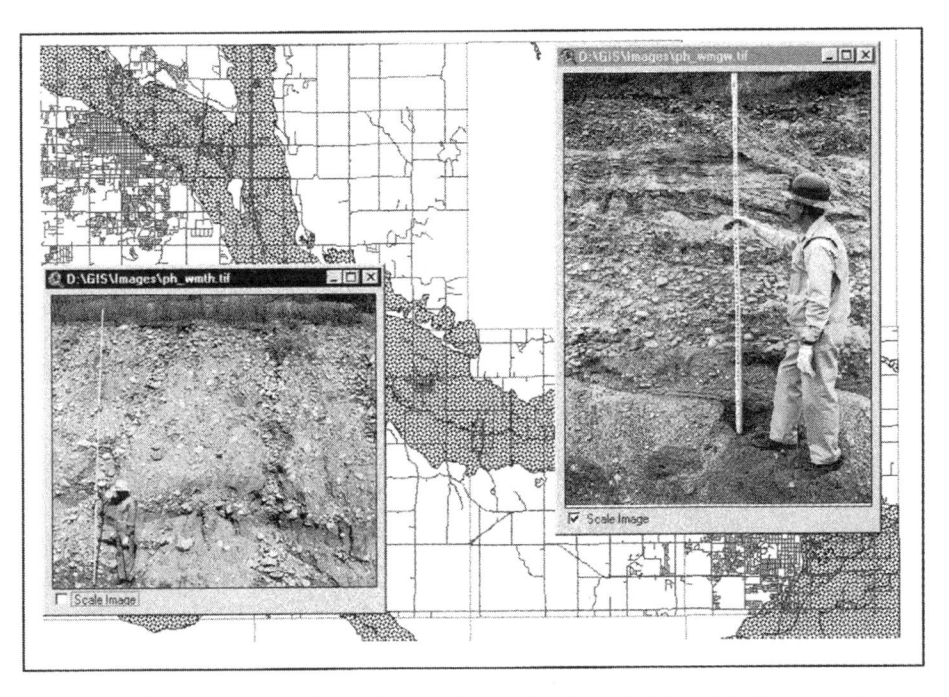

Figure 3. Computer generated image showing photographs of gravel-pit faces linked to a map of gravel pits by using GIS "hot link" feature.

Use of the DSS to estimate aggregate resource requirements for a new housing and commercial development is shown in Figure 4. First, a polygon was drawn on the map view (by using simple GIS procedures) that identified a new development area. The user is prompted to input the number of housing units per acre. The DSS then calculated the aggregate required to construct the new houses, infrastructure, and supporting commercial development. The calculations were based on the size of the development area (determined by the DSS), the number of dwellings per acre (input by the user), and an average aggregate requirement per housing unit (provided by the authors).

The DSS can also analyze aggregate resources in a new aggregate extraction area. Again, a polygon was drawn on the map to identify the resource extraction area. Resource reserves were calculated on the basis of aggregate distribution, thickness, and quality information derived from other GIS data layers. To calculate pit life, the DSS combined those data with information regarding production rates, which was input by the user. All calculations were based on formulas and data supplied by the scenario developer (the authors).

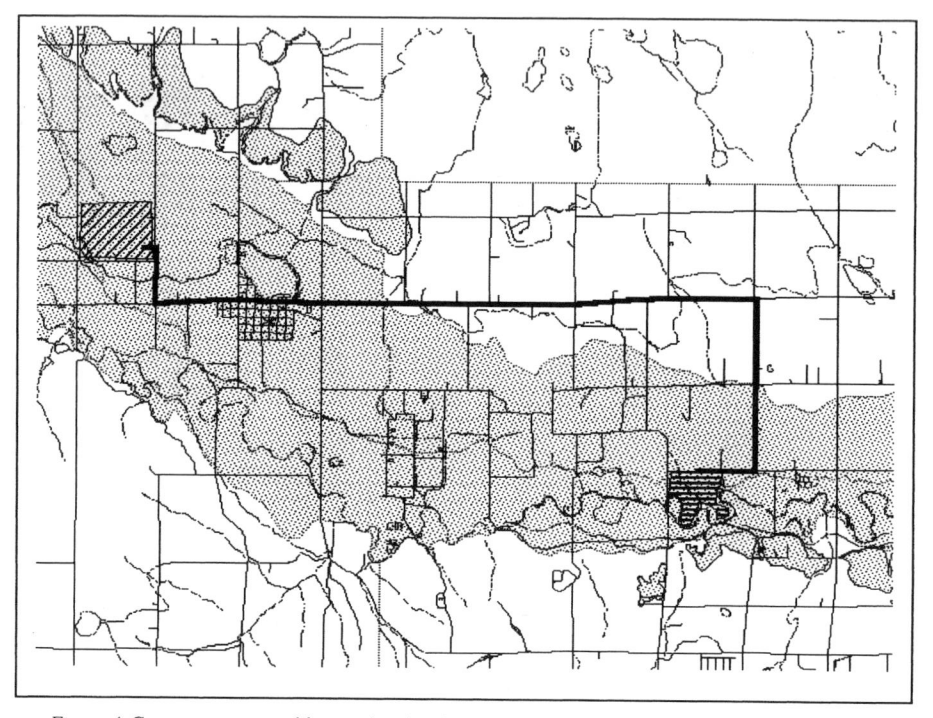

Figure 4. Computer-generated image showing three editable themes - New Development (diagonal ruled area), Extraction Area (horizontal ruled area), and Truck Transportation Route between extraction area and development area (black line).

Using the line-drawing function of the DSS, the route to transport the aggregate from the extraction area to the use area was identified. The DSS queried the "new development area" to determine aggregate needed. It calculated the cost of transporting aggregate along the indicated route, as well as such other factors such as accident exposure, fuel consumption, and emissions. These calculations were based on miles traveled (determined by the DSS), aggregate required for the new developments (determined by the DSS), and other data provided by the scenario developer (the authors).

An extremely versatile function of the DSS is a feature that allows a user to create a polygon around an area of interest and return information regarding the contents of that polygon. We created four evaluation tools for this scenario, as shown on the right side of Figure 5. Those tools can summarize the results of various land-use options regarding aggregate resources, demand, and transportation and can demonstrate the results either as numbers or as graphics. For this example, we implemented the area summary tool. By using this tool, an area can be identified on the map, and information regarding aggregate resources within that area can be returned in a graphical format.

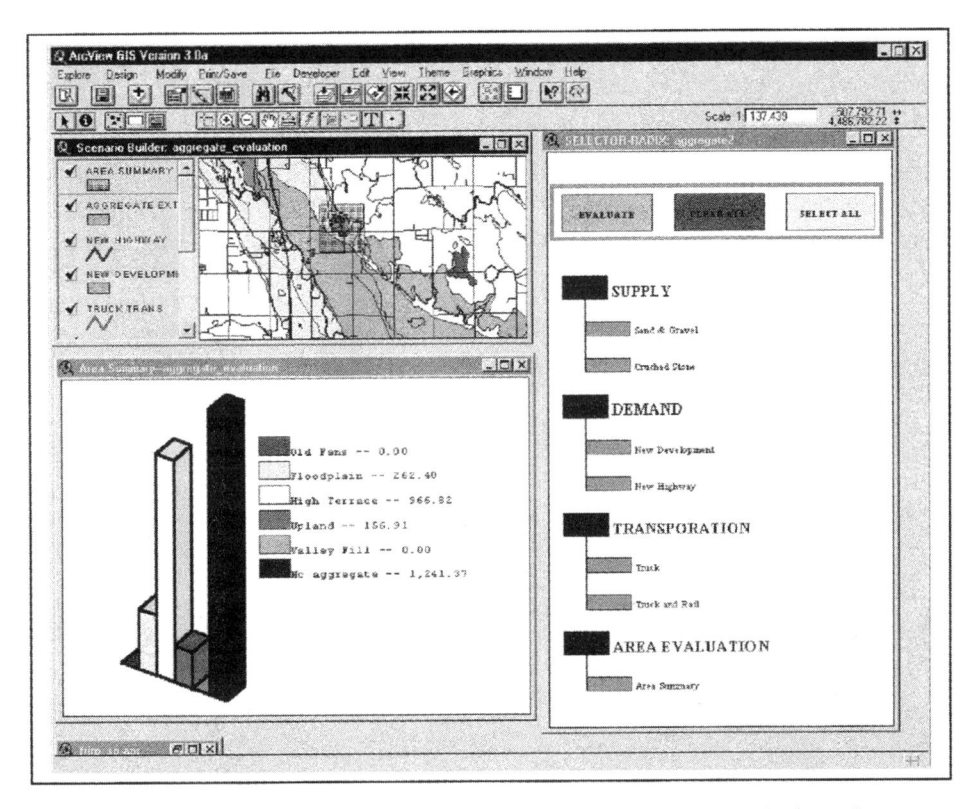

Figure 5. Computer-generated image showing summary area (upper left), area evaluation tool (right), and graphical results of evaluation (lower left).

First we zoomed in on a detailed part of the map and created a polygon around an area on the map (upper left part of Figure 5). Occurring within that polygon are four different landforms that are underlain with gravel. By enabling the area evaluation tool (right side of Figure 5), the DSS identified what types of landforms were contained within the polygon, determined how many acres each landform comprised, and displayed the results in graphical format. The DSS can query any layer of the map project, regardless of whether or not it is active, or even visible.

GIS and DSS can be used to help manage and display large amounts of spatial data for land-use decisions.

10. Summary

Mining aggregate resources will clearly be necessary well into the future. If effective planning policies are not in place, then mining may have to be relocated to areas distant from the market, thus resulting in economic and environmental consequences.

Given the current state of the art and allowing for reasonable technological progress, the use of alternate resources, and recycling, the aggregate industry is capable of maintaining a resource supply while achieving reduced environmental impacts. Major severe local aggregate shortages, however, are already occurring and will occur more frequently owing to inadequacies of present policies.

The number of factors that can be controlled to reduce the consumption of aggregate while maintaining a continuous supply of quality aggregate at a reasonable cost are finite. The consumption of local aggregate can be slowed through use of alternative resources, such as recycled concrete and asphalt pavement, and by importing aggregate from other areas. The local aggregate supply can be expanded through underground mining, technological advances, and beneficiation of aggregate. Aggregate consumption, in general, can be slowed through the modification of application designs to require less aggregate or by modifying specifications to allow the use of lower quality aggregate in certain low-end uses.

These measures only postpone the inevitable encroachment upon and sterilization of potential aggregate resources. To protect aggregate resources from avoidable conflicts, a clearly defined policy to include resource development in the comprehensive planning process should be defined. To support this policy, the distribution, volume, and quality of aggregate resources need to be determined to provide information to decisionmakers during the permitting process. GIS and DSS are tools that can help present and evaluate data in a clearly understandable format that can be used by decisionmakers. To be effective, the plan must be enforceable and should protect the aggregate resources from society and protect society from the environmental impacts of aggregate extraction.

References

1. Lüttig, G.W., 1994, Rational management of the geo-environment – A view in favour of "Geobased Planning", *in* Lüttig, G.W., ed., Aggregates -- Raw materials' giant: Report on the 2nd International Aggregate Symposium, Erlangen, p.1-34.
2. Shadmon, A., 1968, Quarry site surveys in relation to country planning, *in* International Geological Congress: Report of the Twenty-Third Session, Czechoslovakia, v. 12, p. 125-132.
3. Ahern, J.F., 1987, Planning for protection and use of sand and gravel resources in urbanizing areas, *in* Proceedings, 1987 National Symposium on Mining, Hydrology, Sedimentology, and Reclamation, University of Kentucky, Lexington, p. 363-367.
4. Crosby, E.J., Hansen, W.R., and Pendleton, J.A., 1978, Guiding development of gravel deposits and of unstable ground, *in* Robison, G.D., and Spieker, A.M., Nature to be commanded: U.S. Geological Survey Professional Paper 950, p. 28-41.
5. Barksdale, R.D., ed., 1991, The aggregate handbook: National Stone Association, individual chapters, variously numbered.
6. Smith, M.R., and Collis, L., eds., 1993, Aggregates -- Sand, gravel and crushed rock aggregates for construction purposes: Geological Society Engineering Geology Special Publication No. 9, London, The Geological Society, 339 p.
7. Timmons, B.J., 1994, Prospecting for natural aggregates; An update - Parts 1-3: Rock Products. v. 97, no. 8, p. 43-45; v. 97 no. 9, p. 23-25; v. 97 no. 10, p. 43-46, 54.

8. Timmons, B.J., 1995, Prospecting for natural aggregates; An update - Part 4: Rock Products. v. 98, no. 1, p. 31-37.
9. Bobrowski, P. T., 1998, editor, Aggregate resources – A global perspective: A.A, Balkema, Rotterdam, Netherlands, 470 p.
10. Lüttig, G.W., ed., 1994, Aggregates -- Raw materials' giant: Report on the 2nd International Aggregate Symposium, Erlangen, 346p.
11. International Association of Engineering Geology, 1984, Proceedings from the International Symposium on Aggregates, Nice, France: Bulletin of the International Association of Engineering Geology, n. 29, Paris, France, 470p.
12. Heinrich, M., Letouzé-Zezula, G., and Pirkl, H., 1994, Mineral resources vs. environmental conservation – Towards a lower conflictivity, in Lüttig, G.W., ed., Aggregates -- Raw materials' giant: Report on the 2nd International Aggregate Symposium, Erlangen, p. 289-296.
13. Stenestad, E., and Sustrac, G., 1992, The role of geoscience in planning and development, in Lumsden, G.I., 1992, ed, Geology and the environment in Western Europe: Oxford University Press, New York, p.281-301.
14. Langer, W.H., and Johnson, L.H., 1978, East Granby – A plan of development for a rural community, in Robison, G.D., and Spieker, A.M., Nature to be commanded: U.S. Geological Survey Professional Paper 950, p. 46-51.
15. Langer, W.H., 1998, Keeping ahead of encroachment – It's not child's play: Mining Engineering, v. 50, n. 8. p. 79.
16. Schwochow, S.D., Shroba, R.R., and Wicklein, P.C., 1974, Sand, gravel, and quarry aggregate resources – Colorado Front Range counties: Colorado Geological Survey Special Publication 5-A, 43 p.
17. Schwochow, S.D., Shroba, R.R., and Wicklein, P.C., 1974, Atlas of sand, gravel, and quarry aggregate resources – Colorado Front Range counties: Colorado Geological Survey Special Publication 5-B, not paginated.
18. Lindsey, D.A., Langer, W.H., and Shary, J.F., 1998, Gravel deposits of the South Platte River Valley north of Denver, Colorado – Part B – Quality of gravel deposits for aggregate: U.S. Geological Survey Open File report 98-148-B, 24p.
19. Goldman, H.B., 1959, Urbanization and the mineral industry: San Francisco, Mineral Information Service, California State Division of Mines, 11 p.
20. Beeby, D.J., Successful integration of aggregate data in land-use planning – A California case study, in Bobrowski, P. T., ed., Aggregate resources – A global perspective: A.A, Balkema, Rotterdam, Netherlands, p. 27-50.
21. Stewart, D.A., 1998, The state of aggregate resource management in Ontario, in Bobrowski, P. T., ed., Aggregate resources – A global perspective: A.A, Balkema, Rotterdam, Netherlands, p. 51-70.
22. Kelly, R.I., Rowell, D.J., and Robertson, H.M., 1998, Ontario's aggregate resources inventory program, in Bobrowski, P. T., ed., Aggregate resources – A global perspective: A.A, Balkema, Rotterdam, Netherlands, p. 421 – 438.
23. Dunn, J.R., 1983, Dispersed benefit riddle, in Ault, C.R., and Woodard, G.S., eds., Proceedings of the 18th Forum on Geology of Industrial Minerals: Indiana Geological Survey Occasional Paper 37, pp. 1-9.
24. de Jong, B., and de Mulder, E.F.J., 1998, Construction materials in the Netherlands – Resources and policy, in Bobrowski, P. T., ed., Aggregate resources – A global perspective: A.A, Balkema, Rotterdam, Netherlands, p. 203-214.

ENVIRONMENTAL IMPACTS OF MINING NATURAL AGGREGATE

William H. Langer and Belinda F. Arbogast
U.S. Geological Survey
MS 973 P.O. Box 25046
Denver, Colorado 80225-0046   USA

Abstract. Nearly every community in nearly every industrialized or industrializing country is dependent on aggregate resources (sand, gravel, and stone) to build and maintain their infrastructure. Indeed, even agrarian communities depend on well-maintained transportation systems to move produce to markets. Unfortunately, aggregate resources necessary to meet societal needs cannot be developed without causing environmental impacts.

Most environmental impacts associated with aggregate mining are benign. Extracting aggregate seldom produces acidic mine drainage or other toxic affects commonly associated with mining of metallic or energy resources. Other environmental health hazards are rare. Most of the impacts that are likely to occur are short-lived, easy to predict and easy to observe. By employing responsible operational practices and using available technology, most impacts can be controlled, mitigated or kept at tolerable levels and can be restricted to the immediate vicinity of the aggregate operation.

The most obvious environmental impact of aggregate mining is the conversion of land use, most likely from undeveloped or agricultural land use, to a (temporary) hole in the ground. This major impact is accompanied by loss of habitat, noise, dust, blasting effects, erosion, sedimentation, and changes to the visual scene.

Mining aggregate can lead to serious environmental impacts. Societal pressures can exacerbate the environmental impacts of aggregate development. In areas of high population density, resource availability, combined with conflicting land use, severely limits areas where aggregate can be developed, which can force large numbers of aggregate operations to be concentrated into small areas. Doing so can compound impacts, thus transforming what might be an innocuous nuisance under other circumstances into severe consequences. In other areas, the rush to build or update infrastructure may encourage relaxed environmental or operational controls. Under looser controls, aggregate operators may fail to follow responsible operational practices, which can result in severe environmental consequences.

The geologic characteristics of aggregate deposits (geomorphology, geometry, physical and chemical quality) play a major role in the intensity of environmental impacts generated as a result of mining. Mining deposits that are too thin or contain too much unsuitable material results in the generation of excessively large mined areas and large amounts of waste material. In addition, some geologic environments, such as active stream channels, talus slopes, and landslide-prone areas, are dynamic and respond rapidly to outside stimuli, which include aggregate mining. Some geomorphic areas and (or) ecosystems serve as habitat for rare or endangered species. Similarly, some geomorphic features are themselves rare examples of geologic phenomena. Mining aggregate might be acceptable in some of these areas but should be conducted only after careful consideration and then only with extreme prudence. Failure to do so can lead to serious, long-lasting environmental consequences, either in the vicinity of the site or even at locations distant from the site.

Mining generates a disturbed landscape. The after-mining use of the land is an important aspect of reducing environmental impacts of aggregate extraction. The development of mining provides an economic base and use of a natural resource to improve the quality of human life. Wisely restoring our environment requires a design plan and product that responds to a site's physiography, ecology, function, artistic form, and public perception. Forward-looking mining operators who employ modern technology and work within the natural restrictions can create a second use of mined-out aggregate operations that often equals or exceeds the pre-mined land use. Poor aggregate mining practices, however, commonly are accompanied by poor reclamation practices, which can worsen already existing environmental damage.

With environmental concerns, operating mines and reclaimed mine sites can no longer be considered isolated from their surroundings. Site analysis of mine works needs to go beyond site-specific information and relate to the regional context of the greater environment. Understanding design approach can turn features perceived by the public as being undesirable (mines and pits) into something perceived as being desirable.

1. Introduction

Aggregate resources (sand, gravel, and crushed stone) rank first in order of amount and value of the global extracted mineral resources [1]. Nearly every community, whether industrialized or agrarian, is dependent on aggregate resources to build and maintain their infrastructure.

Geology controls where aggregate occurs. Although aggregate tends to be widely distributed, there are large areas where it is absent. Geologic conditions also control the type and properties of the potential aggregate resource. For example, glacial deposits yield different types of gravel than marine terraces, and volcanic rocks have different properties than intrusive rocks. The properties of the aggregate determine the environmental impacts that might result from mining the resource, as well as the amount of processing and the amount of waste material generated.

Aggregate is a low-unit value, high bulk commodity. Consequently, excavation of aggregate near the point of use, which is commonly at population centers, is most economical. In these areas, conflicting land use, regulations, and citizen opposition further restrict the development of aggregate. In some areas, the options for places to develop aggregate resources are extremely limited.

Unfortunately, aggregate resources necessary to meet societal needs cannot be developed without causing environmental impacts. At one time, aggregate and other resources could be mined with little regard as to how that mining might impact the environment. Fortunately, that situation has changed in many parts of the world, and responsible operators must reckon with the environmental impacts created by aggregate mining.

When options for extracting aggregate are limited, identification of areas for extraction that are free from potentially serious environmental problems may not be possible. We may be forced to develop aggregate resources in areas that we otherwise might choose to avoid. That we understand what potential environmental impacts exist, and know how to mitigate or avoid those impacts is of utmost importance.

2. Environmental impacts from aggregate mining

2.1 DEFINITION OF ENVIRONMENTAL IMPACT

Although the scientific literature is replete with discussions about impacts from metallic and coal mining, far fewer reports describe details of environmental impacts from aggregate mining. (Some of those reports are included in symposium volumes or other comprehensive collections of individual papers, many of which are referred to in this paper.) Therefore, before beginning a discussion of environmental impacts of aggregate mining, it is beneficial to define what is meant by environmental impact. Kelk [2], defined environmental degradation or pollution as follows:

> the alteration of the environment by man through the introduction of materials which represent potential or real hazards to human health, disruption to living resources and ecological systems, [or] impairment to structures or amenity * * *

For use in this paper, Kelk's term "introduction of materials" is expanded from "introduction of materials or activities" to include aggregate mining. It is important to note that the definition includes humans, other living resources, ecological systems, and structures.

2.2 GENERAL TYPES OF IMPACTS FROM AGGREGATE MINING

We cannot obtain aggregate resources without causing some environmental disturbance. Some of the disturbance is caused directly by the mining or processing activities. The most obvious environmental impact of aggregate mining is the conversion of land use, most likely from undeveloped or agricultural land use, to a hole in the ground. This major impact may be accompanied by loss of habitat, noise, dust, vibrations, chemical spills, erosion, sedimentation, changes to the visual scene, and dereliction of the mined site. (Some references that generally describe these impacts include Barksdale [3]; Kelk [2], Smith and Collis [4], Lüttig [5], and Bobrowski [6].)

Mining may cause secondary impacts or a ripple effect. Some are obvious, such as transporting aggregate from the plant to the market frequently results in heavy traffic. Some are less obvious, such as mining in some environments may cause stream erosion. Erosion may cause loss of shade along stream banks, which, in turn, may cause loss of fish habitat.

Societal pressures can exacerbate the environmental impacts of aggregate development. In areas of high population density, resource availability, combined with conflicting land use, severely limits areas where aggregate can be developed, thus forcing large numbers of aggregate operations to be concentrated into relatively small areas. Doing so can compound impacts and transform what might be an innocuous nuisance under other circumstances into severe consequences. In other areas, the rush to build or update infrastructure may encourage relaxed controls. Under looser controls, aggregate operators may fail to follow responsible operational practices, which can result in severe environmental consequences.

2.3 NATURE OF ENVIRONMENTAL IMPACTS

One way to assess an environmental impact is to characterize its nature. The nature of an impact can be referred to by using a number of terms that include range, timing, duration, ability to predict, and ability to control. In this paper, these terms are expressed in relative values.

2.3.1 *Range of Impact*

The range of the impact refers to how large the area is that is impacted by the aggregate operation. Impacts, such as conversion of land use, are commonly (although not always) restricted to the site. Impacts, such as noise and dust, are commonly limited to the near-site area. Impacts, such as erosion, sedimentation, and changes to the visual scene, may be widespread.

2.3.2 *Timing of Impact*

The timing of the impact refers to how rapidly the impact develops. Impacts, such as conversion of land use, take place immediately. Other impacts, such as erosion, may not begin to be noticed until many years after aggregate extraction begins.

2.3.3 *Duration of Impact*

The duration of the impact refers to how long the impact lasts. The impacts associated with noise commonly last only as long as the equipment generating the noise is being operated. The impacts associated with conversion of land use commonly last until the operation is reclaimed, at which time yet another conversion of land use will occur. Other impacts, such as erosion, may last for an extended period of time

2.3.4 *Ability to Predict Impact*

The ability to predict the impact refers to how easily one can anticipate that the impact will occur and how easily one can predict the range, timing, and duration of the impact.

Predicting the range, timing and duration that results from conversion of land use is relatively easy; predicting those factors for subsequent erosion is more difficult.

2.3.5 Ability to Control Impact

The ability to control the impact refers to how easily one can avoid, minimize, or mitigate an impact. Impacts, such as dust, commonly can be avoided or minimized by using modern careful production techniques and modern technology. Impacts like erosion may be difficult to control.

The above terms often are interrelated. For example, an impact that is widespread is likely to have a long duration. Similarly, an impact that is difficult to predict is likely to be difficult to control.

3. Discussion of Impacts

Aggregate mining is divided into three distinct phases - site preparation, aggregate excavation, and aggregate processing. Each phase of mining is typified by specific activities, with each activity having the potential to create specific types of environmental impacts. This section of the report describes the impacts that result from the three phases of mining, and offers measures that can be taken to mitigate the impacts.

3.1 IMPACTS FROM SITE PREPARATION

Site preparation commonly starts with stripping a sufficient amount of overburden to access the resource. The method used depends on the type and thickness of material to be removed. Soil and partially weathered rock can be pushed aside with a bulldozer and removed with conventional loaders and haul trucks. Harder, more-consolidated material may require drilling and blasting. Organic soil commonly is separated from the overburden and stockpiled for reclamation activities. Overburden may also be used to construct berms, stockpiled, or sold. When pre-production stripping is complete, berms, haul roads, settlement ponds, processing and maintenance facilities, and other plant infrastructure are constructed by using standard building techniques.

The land use immediately preceding mining might not be the original land use. Civilizations in many parts of the world have created enormous impacts on the landscape. What we are witnessing today is only the most current nature of the landscape. The land use at any particular site may have changed many times owing to forces of nature and of our predecessors.

3.1.1 Conversion of Land Use

The most obvious environmental impact of aggregate mining is the conversion of land use, most likely from undeveloped or agricultural lands, to an aggregate operation. The impact to the site commonly is quite dramatic because open-pit mining, which is the common method of winning aggregate resources, is dramatically different from other types of land use. The mine site, however is a designed facility, and the impact is, therefore, predictable and controllable.

One method to minimize the impact of conversion of land use is to develop the best resources available. For example, quarrying a thin, flat-lying limestone bed with thick overburden will likely create more environmental impacts than quarrying a thick granitic intrusive body with thin overburden. To obtain the same amount of material, the limestone quarry will have to move more overburden, be larger in area, and result in more spoil piles than the granite quarry. Of course, the limestone might have certain properties that the granitic rock does not possess and might be developed instead of the granite for that reason.

Another means of minimizing environmental impacts is through the development of superquarries. A single huge operation at an environmentally acceptable site may be preferable to many smaller quarries at scattered locations. The concept is dependent on cheap, high-volume transport and on benefits to the local economy through revenues earned from export of aggregate [2].

Accompanying conversion of land use is a change to the visual scene (viewshed) either from the site or from locations remote from the site. The change, which can be either temporary or permanent, is a very subjective topic; what is acceptable to some people is objectionable to others. The nature of this impact depends on the topographic setting, natural ground cover and type of operations. The change to the visual scene for quarries commonly is an issue because most quarries have a very long duration. The extended duration of quarries may result in unpleasant visual impacts and semi-permanent nuisances to the local environment.

The change to the visual scene can be mitigated through sequential reclamation, buffering, and screening (including berms, tree plantings, fencing, or other landscaping techniques). Overburden and soil can be stockpiled in out of the way places. The area impacted can be predicted by using standard off-the-shelf geographic information systems with line-of-site calculation capabilities.

With most aggregate operations, conversion of land use to mining is commonly temporary. After mining has been completed, the land can be reclaimed and converted to yet another land use. In many instances, the second use is equal to or more acceptable than the original use.

3.1.2 Destruction of Habitat

Site preparation results in destruction of habitat in the actual mined area. Unless relocated, vegetation and wildlife that is not mobile is destroyed. Mobile wildlife may leave the site for other areas. Some areas of aggregate serve as habitat for rare or endangered species. In addition to habitat, aggregate mining may impact archeological sites, and in some instances, the geologic deposit may be a rare feature itself. For example, Gonggrijp [7] described how the only esker system in The Netherlands has had an extensive part of it mined for aggregate. Mining aggregate might be acceptable in some of these areas but should be conducted only after meeting all permitting requirements and then only with extreme prudence.

Habitat destruction cannot be eliminated, but can be controlled by regulations and, in some situations, by relocating selected animals or plants. In some cases, the site can be reclaimed to look similar to the original habitat after mining has ceased.

3.1.3 Erosion

The potential for erosion from site preparation occurs in moderate to steep areas where aggregate mining results in the removal of vegetation, soil cover, and changing the natural land surface slopes. Except in rare geographic settings, common engineering practices can limit erosion.

3.1.4 Landslides

Aggregate operations in, or near, the toe or head of an existing landslide deposit can remobilize the slide. Even in areas where natural factors are not conducive to slope failure, aggregate mining can cause landslides. Goswami [8] investigated landslides in Gauhati, northeastern India, and attributed the failure to aggregate mining that had disrupted the natural equilibrium of the hill slopes and their natural drainage conditions.

Aggregate operations should avoid areas of known landslides and areas with slope, aspect, and geologic conditions that are favorable for mass movement.

3.1.5 Sedimentation

Earth moving associated with site preparation may increase sedimentation in nearby streams. Sedimentation commonly can be controlled with techniques employed in normal construction, such as with catchment basins or erosion barriers.

3.1.6 Stream Flow

Aggregate mining may create impervious land that prevents infiltration, remove vegetation that normally retards runoff, or otherwise change runoff patterns. The resultant faster, higher peak runoff can result in higher peak stream flow.

Retaining the runoff in infiltration basins can mitigate the impact.

3.1.7 Ground water

Under some geologic and climatic conditions, removing vegetation and soil from the land surface can reduce evapotranspiration and ultimately increase ground water. Changes in ground-water quality in areas of sand and gravel mining [9] are attributed to the removal of soil that had been acting as a protective layer, filtering, or otherwise reducing contaminants to the ground water. The level of impact ranges depends on a number of factors, which include the thickness of material removed, the surface area involved, the total volume of the aquifer, and recharge to the aquifer.

Impacts can be mitigated by controlling recharge in aggregate operations, or by locating resource extraction areas outside of recharge areas.

3.2 IMPACTS FROM AGGREGATE EXTRACTION

A variety of methods can be used to excavate aggregate. The methods to excavate aggregate depend, in large part, on the geologic environment. Sand and gravel is commonly mined from pits or dredged from water bodies. Crushed stone is mined from quarries and requires drilling and blasting prior to excavation. The impacts are different for each type of excavation.

3.2.1 Dry Pit

In this paper, if sand and gravel mining does not penetrate the water table, then it is referred to as a "dry pit." The aggregate is dry and can be extracted by using conventional earth-moving equipment, such as bulldozers, front loaders, track hoes, and scraper graders. The equipment chosen commonly depends on the lay of the land and on operator preference.

The impacts from excavating aggregate from a dry pit are commonly easy to predict, observe, and control with standard engineering practices. Dust and noise (discussed below) are the most common impacts. Runoff patterns may be changed.

3.2.2 Wet Pit Mined Dry

In this paper if sand and gravel mining extends to a depth that penetrates the water table, then it is referred to as a "wet pit." In some geologic settings, wet pits can be made dry by dewatering the pit. This is done by collecting the groundwater in drains in the floor of the pit and pumping the water out of the pit. Construction of slurry walls or other barriers to ground-water flow around the pit may be required. After ground-water drains from the deposit, sand and gravel can be extracted by using the dry mining techniques described above.

Two potential impacts of great significance result from dry mining of sand and gravel from the saturated zone (beneath the water table) - the effect of dewatering the pit and the potential for pit capture during floods (if the pit occupies an active floodplain). Dewatering commonly lowers the water table in the vicinity of the pit and may affect the flow of nearby streams. In some geologic settings, the streamflow will decrease. In others, drained water that is returned to streams can increase streamflow. The extent of the impact depends on the type of deposit (floodplain, stream terrace, alluvial fan, marine terrace, glaciofluvial, etc.), hydraulic properties of the aggregate deposit, the thickness of water table penetrated, where the drained water is discharged, and whether or not slurry walls are used.

The impacts from dewatering a pit can be monitored by use of observation wells. In highly permeable deposits, the use slurry walls might be necessary to isolate the pit from the water table. Water removed through dewatering can be returned to nearby streams.

Cut-offs and avulsions of a stream during floods may result in capture of a wet pit located on an active floodplain. Pit capture can result in changes in channel position and can substantially alter the spatial distribution of energy and force of the stream or river [10]. The ability to predict flooding and capture of the pit largely depends on how well the hydrology and history of the adjacent stream are known.

Levees may protect floodplain pits from flooding and stream capture.

3.2.3 Wet Pit Mined Wet

In some situations where the sand and gravel pits penetrate the water table, the pit may not be able to be drained, or the operator may prefer to extract the material by using wet mining techniques. Material may be excavated by using draglines, clamshells, bucket and ladder, or hydraulic dredges.

The greatest potential impact of mining a wet pit is if the pit is located on an active floodplain. Cut-offs and avulsions of a stream during floods may result in capture of the pit, thus resulting in the impacts described above in Section 3.2.2. Again, the ability to predict flooding and capture of the pit largely depends on how well the hydrology and history of the adjacent stream are known.

Depending on the geologic and climatic conditions, the pit can act as a recharge area or a discharge area. In humid areas, precipitation can collect in the pit and recharge the ground water. In semiarid or arid climates, evaporation from water in pits can lower the water table. Monitoring wells installed around the pit and stream gauges can be used to observe effects on the water table. Levees may protect floodplain pits from flooding and stream capture.

3.2.4 In-Stream Mining

Sand and gravel can be excavated directly from stream channels or from embayments dredged off of the main channel by using draglines, clamshells, bucket and ladder, or hydraulic dredges. During times other than flooding, aggregate can be skimmed from bars in the channel or from active floodplains by using the dry mining techniques described above.

Depending on the geologic setting, in-stream mining has the potential to create very serious environmental impacts. Impacts may be particularly serious if the stream being mined is an eroding, as opposed to an aggrading stream. Mossa and Autin [11], Kondolf [12], Florsheim *et al.* [13], as well as many other authors, have described the dramatic changes to river systems where in-stream mining is being improperly managed. Impacts can be a result of extracting too much material at one site, or the combined result of many small but intensive operations [14].

The principal cause of impacts from in-stream mining is the removal of more material than the system can replenish. The major impact of removal of gravel from the stream is a change in the cross section of the stream. This, in turn, has many secondary impacts. Upstream incision can result from increased gradient, and downstream incision can result from decreased sediment load. The stream may change its course, thus causing bank erosion. Erosion may cause the removal of vegetation and overburden and the undercutting of structures. In-stream mining can also result in channel bed armoring, increases in sediment load, lowering of alluvial water tables, and stagnant low flows. All these impacts can result in major changes to aquatic and riparian habitat.

Removal of gravel from aggrading streams may not cause adverse environmental impacts. Some eroding streams underlain with large gravel layers deposited under conditions other than those prevailing at the current time may support gravel extraction with no serious environmental impacts. Jiongxin [15] described such a situation on the Hanjiang River in China where downcutting stopped when coarse bed material was reached. Similar situations exist where coarse gravels of glacial origin underlie modern stream deposits.

The best method of mitigation of the impacts of in-stream mining is prevention of the impacts. Kondolf [16] suggested a number of strategies to manage in-stream mining of aggregate. One method is to define a minimum elevation for the thalweg (the deepest part of the channel) along the river and to restrict mining to the area above this line. Another method is to estimate the annual bedload and to restrict extraction to that value or some percentage of it. Difficulties may exist, however, in realistically determining the annual bedload.

Although riverbed incision is generally assumed to be irreversible, evidence of exceptions exist. In streams that suffer from the effects of in-stream mining, the first step for mitigation is to reduce or stop the excavation. Piégay and Peiry [17] described how the Giffre River in the northern part of the French Alps rehabilitated itself following extensive extraction of gravel from the river channel. This restoration is largely due to the fact that after the amount of material allowed to be extracted from the channel was reduced, the bedload supply greatly exceeded extraction.

3.2.5 Dry Quarry

Rock quarries that do not penetrate the water table or where discharge from the water table is offset by evaporation or is otherwise insignificant are referred to as dry quarries. To produce aggregate, the rock is first drilled and blasted. The types of drills or explosives vary because of the diversity of rock types and formations used as aggregate. Blasting commonly breaks the rock into pieces suitable for crushing. If the rubble is too large, then secondary breaking may be required. The blasted material is dry and can be extracted by using conventional earth-moving equipment, such as bulldozers, front loaders, track hoes, and scraper graders. The equipment chosen commonly depends on the lay of the land and on operator preference.

The main impacts from quarrying are the affects of blasting. Poorly designed or poorly controlled blasts may cause rocks to be projected long distances from the blast site (fly rock), which is a serious hazard. Poorly designed or poorly controlled blasting can fracture the surrounding rock, thus altering the ground-water flow paths. This ill effect of blasting, if it occurs, is difficult to observe or predict. Blasting may cause ground shaking for some distance from the quarry. Ground shaking can be monitored with seismic equipment and can be limited by reducing the size of the blast or by employing time-delay blasting techniques.

The technology of rock blasting has evolved substantially, and when blasting is properly conducted, its environmental impacts should be negligible. Blasting, however receives numerous complaints from neighbors. Blasting or other noise may also frighten certain types of wildlife, although buffer areas and areas of undeveloped reserves commonly serve as habitat for some wildlife. One way to manage complaints is through education, outreach, and prompt response to complaints.

3.2.6 Wet Quarry Mined Dry

Rock quarries commonly penetrate the water table. Where this occurs, the quarries commonly are kept dry by pumping of the water. The rock is then mined by using the procedures followed in a dry quarry.

The impacts of extracting aggregate from a wet quarry mined dry are similar to those for a dry quarry (Section 3.2.5). Depending on geologic conditions, dewatering the quarry may lower the water table that surrounds the quarry. The impact of dewatering on the water table can observed with monitoring wells.

3.3 IMPACTS FROM AGGREGATE PROCESSING

Aggregate processing consists of loading rock or sand and gravel, transporting the material to the plant, crushing, screening, washing, stockpiling, and loadout. Material commonly is transported from the mining face to the plant either by conveyor or truck. If the material consists of boulders or blasted rock, then it commonly goes through a primary crusher. A conveyor then moves it to a surge pile. A gate at the bottom of the surge pile releases material at a constant feed rate to an area where it is screened and sorted by size. Depending on the type of material being processed, and on the final product, the material may be washed. After screening, sorting, and washing, if necessary, the material is moved by conveyor to stockpiles. Upon sale, the final product is loaded on trucks, railcars, or barges for transport to the final destination. For this paper, processing also includes the repair and maintenance of equipment.

Dust, noise, and a change to the visual scene are all common impacts associated with any type of earth-moving activity, which includes aggregate processing. Dust can be from a point source, such as drilling or processing equipment or as fugitive dust from blasting or haul roads. A recent concern, primarily for worker safety, is the issue of dust that contains dangerous amounts of crystalline silica or carcinogenic particulates. The impacts from dust commonly can be mitigated by use of dust suppression techniques, which includes water, dust suppression chemicals, covers on conveyors, or the use of vacuum systems and bag houses. Workers are protected from dust through the use of enclosed, air-conditioned cabs on equipment and, where necessary, the use of respirators. Worker safety commonly includes regular health screening.

Noise is generated from blasting, excavating equipment, crushing and processing equipment, and trucks. The impacts of noise can be mitigated through the use of berms, locating noisy equipment (such as crushers) away from populated areas, limiting blast sizes, use of conveyors instead of trucks for in-pit movement of materials, and limiting the hours of operation. Workers are protected from noise through the use of enclosed, air-conditioned cabs on equipment and, where necessary, the use of hearing protectors. Worker safety commonly includes regular health screening.

Maintenance of equipment may result in the accidental release of fuel, solvents, or other chemicals. Accidental spillage can be controlled by limiting the amount or type of chemicals on hand and by careful operating and safety training and procedures.

4. Reclamation

Reclamation commonly is considered to be the start of the end of environmental impacts from mining. The development of mining provides an economic base and use of a natural resource to improve the quality of human life. Equally important, properly reclaimed land can also improve the quality of life. Wisely shaping mined out land requires a design plan and product that responds to a site's physiography, ecology, function, artistic form, and public perception.

Examining selected sites for their landscape design suggested nine approaches for mining reclamation. The oldest design approach around is nature itself. Humans may sometimes do more damage going into an area in the attempt to repair it. Given enough geologic time, a small site scale, and stable adjacent ecosystems, disturbed areas recover without mankind's input. Visual screens and buffer zones conceal the facility in a camouflage approach. Typically, earth berms, fences, and plantings are used to disguise the mining facility. A restoration design approach attempts to restore the land to its original landscape character. Rehabilitation targets social or economic benefits by reusing the site for public amenities, most often in urban centers with large populations. A mitigation approach attempts to protect the environment and to return mined areas to use with scientific input. Recognizing the limited supply of mineral resources and encouraging recycling efforts are steps in a renewable resource approach. An educative design approach effectively communicates mining information through outreach, land stewardship, and community service. Mine sites used for art show a celebration of beauty and experience-abstract geology. The last design approach combines art and science in a human/nature ecosystem termed "integration."

Site analysis of mine works needs to go beyond site-specific information and relate to the regional context of the greater landscape. Understanding design approach can turn undesirable features (mines and pits) into something perceived as desirable by the public.

In her study of landfill and sewage treatment, Engler [18] discussed eight approaches to designing waste landscapes. Mining generates a disturbed landscape that many consider waste until reclaimed, and Engler's terminology appears to be adaptable to reclaimed mine sites with minor reinterpretation (table 1).

Table 1. Design approaches to reclaiming mine sites

ENGLER'S LIST	AUTHORS' LIST	DESCRIPTION
	Natural	Allow nature to reclaim site with no, or minimal, human influence.
Camouflage	Camouflage	Conceal a mining facility with visual screens and buffers.
Restoration	Restoration	Return land to its approximate original condition.
Recycling	Rehabilitation	Use site for public amenities.
Mitigation	Mitigation	Repair mined-out site that has experienced extensive damage from human or natural causes.
Sustainable	Renewable resource	Use site to recycle man-made or natural resources.
Educative	Education	Communicate resource information through outreach.
Celebrative	Art	Treat site as work of beauty and unique experience.
Integrative	Integration	Combination of approaches integrating art and science.

One approach that Engler did not mention is nature itself. Although a combination of the above approaches is most often applied, examination of the specific categories with examples is still useful.

4.1 NATURAL

Wait long enough and no matter how great the disturbance, nature works to regenerate with or without the benefit of man. Some areas devastated from fire, landslide, volcanic eruption, or quarrying manage to recover well without human intervention. Therefore, a conscious natural design approach may be one of hands off. For example, the moist climate, dense vegetation, and remoteness of areas like Alaska are likely places for some mined-out pits to be passively reclaimed by nature. Heavy equipment brought in to recontour these old sites may do far more damage to the existing ground cover and surface soil than the benefit gained.

In other areas, long-term natural recovery may not bring about the specific changes people find desirable. Few, if any, people living near Appalachian coal mining sites would want to wait 30 years for hardwood seedlings to sprout. Studying nature's ability to heal is one way scientists and designers can learn new techniques for reclamation by taking maximum advantage of natural geological and biological processes.

Natural seedfall for restoration of cottonwood (*Populus* sp.) and willow (*Salix* sp.) was tested in a sand and gravel pit near Fort Collins, Colorado, U.S.A. Controlled flooding was used to simulate historic spring flooding conditions along the Poudre River to establish vegetation [19]. With additional timed floodings, clearing of undesirable exotic saltcedar (*Tamarix chinensis*) seedlings was successful. The pit is an example of seminative riparian vegetation being used to reclaim a site with little human involvement and cost.

4.2 CAMOUFLAGE

Camouflage uses visual screens and buffer zones to conceal the mining or mined out facility and to provide barriers to sound, dust, and noise. Austin [20] calls using landscape skills "merely to provide a cosmetic touch to an otherwise ravaged landscape, an exercise akin to putting lipstick on a pig." This may have been true in the past when reclamation took a backseat to exploration, but in some instances, a minimal approach is justified. In the past, an immediate response by the industry was to use fences, earth berms, and plantings (small-scale features) to disguise the activity from residential areas. The design solution was frequently associated just with the site perimeter. Camouflage commonly made use of linear, uniform rows of quick-growing plant species. Wide buffer zones were frequently abandoned in the interest of cost. Today, progressive aggregate companies recognize the value of including landscape architects early in the planning stage. For example, a long-term approach to camouflage may involve the use of quick-growing plants as part of a matrix that contains the slower growing native species matched to grow in the overburden and spoil material. Another consideration is the profound effect vegetation has on water control (infiltration and erosion).

4.3 RESTORATION

Returning the land exactly to its original condition is a restorative approach. Mining is considered to be a temporary activity that leaves a disturbed area that many think should be returned to pre-mining biological conditions. Restoration to the original condition is seldom possible because we do not currently have the level of information and skill required to return ecosystems exactly to their original structure. In addition, the new land is environmentally unstable, and exotic species invade disturbed sites. Many native organisms do not return or fill the same ecological niche. Instead of returning an area to its original condition, a more-realistic approach is to approximate the new habitat as closely as possible to its original function and to recapture the landscape character.

4.4 REHABILITATION

A rehabilitative approach reclaims mined-out land for public amenities with social or economic benefits. Beginning in 1904, Butchart Gardens in British Columbia, Canada, reclaimed 50 acres of an exhausted limestone quarry to a premier botanical garden. The city hall of Hagen, Germany, is located on the site of an inactive quarry [21]. Many of the natural rock outcroppings are left for visual impact, and quarry rock is used on the interior and exterior of the building. Golf courses have been built over abandoned quarries near many U.S. cities. Construction of townhouses, shopping centers, or industrial parks are other examples of a rehabilitative approach.

Near Mombassa, on the coast of Kenya, an abandoned quarry illustrates a more-comprehensive rehabilitation plan. Once barren land with almost no underground water, it is covered with grass and trees. Rene Haller, a Swiss agronomist, introduced agriforestry, animal farming (including cattle, sheep, oryx, tilapia, and crocodiles), and tourism to the wasted landscape [22].

4.5 MITIGATION

Some mined areas have undergone major negative environmental changes after mining ceased. A mitigation approach uses scientific input to protect the environment and to return these mined areas to beneficial uses. For example, dumping of contaminated debris from the U.S. Department of Defense and Atomic Energy Commission sites into the Weldon Spring Quarry (near St. Louis, Missouri, U.S.A.) went on for nearly 30 years [23]. Ground-water contamination was spreading towards well fields that supplied homes and industries throughout the area. A quarry cleanup of the bulk waste began in 1989 under the U.S. Environmental Protection Agency (EPA) Superfund Program and the State of Missouri. High-quality clay soil is required to construct the permanent disposal facility and to make it impervious to water. Nearly 2 million cubic yards of clay was excavated from more than 200 acres of land in the nearby Weldon Spring Conservation Area. In this case, the cleanup of one mine site requires the construction of another.

Nature can also degrade mined lands. During a 35-year period, Cooley Gravel Co. extracted more than 26 million tons of aggregate from a site along the South Platte River, Colorado, U.S.A. Floods in 1965 and 1973 breached levees and changed river channels, which caused catastrophic impact on the land. Cooley reclaimed the land and donated 425 acres to the city of Littleton. The design made use of native seed mixes, incorporated trails, fishing along the South Platte River, and educational tours at the Carson Nature Center. Today, together with adjacent land dedicated by Littleton, South Platte Park is one of the largest wildlife parks within city limits in the United States.

4.6 RENEWABLE RESOURCE

Mined-out land can be a source of or a place to process renewable resources. One of the most dramatic effects that man has had on the ecosystem relates to loss of wetlands. The reclamation of gravel mining pits at the Farm (Boulder Valley, Colorado, USA) to wetlands is replacing renewable resources. During the period from 1780 to 1990, Colorado lost about half of its wetlands [24]. The Farm design incorporates oxbow lakes and more than 39,000 plants native to prairie wetlands in a hundred-acre site [25].

For some experts, a sustainable relationship with the earth will only come about by controlling growth, reducing our consumption of goods, and preserving diverse landscapes. A tremendous amount of demolition debris is buried in landfills and becomes a wasted resource. Some mined-out areas are contributing to the reuse effort by serving as locations for recycling of concrete, macadam, glass, and other resources. Recognizing the limited supply of mineral resources and encouraging recycling efforts are beneficial steps toward sustainability. England and Wales have established planning guidelines that include reducing the proportion of stone removed from land from the current 83 percent to 68 percent by 2006 [26].

4.7 EDUCATION

In the educative approach, one tries to effectively communicate mining information through outreach so that citizens can make informed choices about future land use. Europeans tend to be ahead of the United States in this regard, perhaps, owing in part to their limited available land. They also focus on aesthetics as much as the functional after use. For example, scientists in the United Kingdom have understood the importance of gravel pits for bird habitat since the 1930's and 40's.

An example of educational outreach is in Albuquerque, New Mexico, U.S.A., where Western Mobile replaced the grass lawn at its corporate headquarters with a xeriscape garden. The ground cover, parking lot, pavement, signage, and building make use of aggregate and, together with the garden, help educate and promote their products. The garden requires about 10 to 20 percent of the water previously needed and is open to the public.

4.8 ART

An artistic approach is one where the site is celebrated as a work of beauty and unique experiences. Engler [18] categorized the approach as celebrative; people become fully aware of the connection between the production of an item and their everyday lives. One pioneer in the earthworks-as-art movement was Robert Smithson. Smithson proposed "Art can become a resource that mediates between the ecologist and the industrialist" [27]. Sites examined for their artistic design approach include Parc des Buttes-Chaumont, Aexoni Quarry, and Smithson's *Broken Circle*.

The public Parc des Buttes-Chaumont in Paris (c.1864-69) was built upon old quarried limestone and gypsum pits, abandoned gallows, sanitary sewage dump, and a mass grave. After the reclamation, Parisians could stroll "aimlessly while observing the city's changing physical and social structure" and appreciate "artistic urban accomplishments" [28].

A sculpted quarry at Aexoni, Greece, celebrates music and dance while acknowledging the natural landscape. The quarry expresses the Greek philosophy of the unity of all things. Although it provides a stage for performing arts and exhibitions, the sculptured quarry illustrates the artistic and rehabilitative approaches. Regional plant species were planted, and sculptural forms relate to adjacent rock formations. The floor plane and backdrop of the stage design are an impression of excavation-a cave [29].

In 1971, Smithson used a sand pit and body of water to create a circular jetty and canal entitled *Broken Circle*. The project was planned in The Netherlands as part of an international art exhibition. The symmetrical landform is about 140 feet in diameter and suggests yin and yang, thus inviting human passage. The earthwork also evokes images of dikes and polders that are the backbone of the Dutch landscape.

4.9 INTEGRATION

The combination of art and science in a "human-nature ecosystem where work and leisure coexist" is an integrated approach [18].

Quarry Cove, on the Oregon coast, U.S.A., is a quarry that has been converted into a man-made tidal zone fed and nourished by wave action [30]. Quarry Cove, which was developed by the Bureau of Land Management, provides a variety of wildlife habitats and is expected to have species diversity comparable to a natural tidal pool within 5 to 10 years. Visitors (the site is wheel-chair accessible) can view nature taking its course as marine life invades the area. The cove is an exciting example of an exhausted site becoming a "natural" biological laboratory with community outreach.

On a larger architectural scale is a design by sculptor Michael Heizer. Entitled *Effigy Tumuli,* the project is sited on land disturbed by coal mining. The reclamation plan celebrates the region's history of Indian burial mounds with earth shapes or tumuli; water strider, frog, turtle, snake, and catfish [31]. Located on a sandstone bluff above the Illinois River, the 150-acre outdoor project was a cooperative team effort by Ottawa Silica Company, a nonprofit organization, and State and Federal Governments [32]. The artist minimized expensive earth moving by studying the existing mine-site topography for hidden forms. The project included the treatment of millions of gallons of acid water, neutralizing acid spoil, and seeding with wildflower and grass. Although this instance concerns a coal mine, pit and quarry reclamation can be tailored to include landform sculptures, as demonstrated Section 4.8.

5. Environmental Ethics

There is an economic cost to limiting environmental impacts while developing aggregate resources. Economic costs can be in the form of increased costs for land acquisition, equipment, processing, or transportation. In some areas, the economic costs may add an overwhelming burden to the cost of developing the resource. If so, then compromises between the cost of extracting aggregate and the environmental impacts of extraction might have to be made.

What constitutes an overwhelming burden is relative. In part, it depends on the wealth of an area and the ability of the citizenry in the area to pay the added costs. Less economically developed areas have less economic wealth to prevent or address environmental impacts. In some cases, the incentive to address environmental issues may be lacking. In less economically developed areas, a few environmental impacts might appear to be a small price to pay for what can be gained through developing aggregate resources to build an infrastructure.

This report can help guide compromises between the economics of extracting aggregate and the environmental impacts of extraction. It should be clear from reading the environmental impacts described above that some impacts are restricted to the site, short term, easy to predict, and easy to control. These commonly would be the best impacts to consider for compromise. Impacts that affect health and safety or impacts that are long lasting and far-reaching would be unlikely candidates for compromise.

Economically developed areas commonly have funding available to address environmental issues. Even so, citizens in some economically developed areas may feel that the best way to avoid environmental impacts is to avoid development of resources within their jurisdiction. This philosophy is many hundreds of years old. Pliney the Elder, a Roman naturalist who died at the eruption of Vesuvius during A.D. 79, wrote that Roman citizens felt mining, although an appropriate activity for conquered lands, was not an appropriate activity in the homeland [33].

The question of ethics arises when one recognizes that aggregate resources have to be developed somewhere. When a wealthy area excludes development of aggregate resources within its jurisdiction, it is depending on obtaining resources from some other area. If the area providing the aggregate is significantly less wealthy, then it likely will have less capability to address environmental issues. To add to the burden, per capita consumption of aggregate commonly increases with wealth. The wealthy area may demand more aggregate from a poor area than the poor area demands for its own needs. Not only must the citizens in the less economically developed area suffer their own environmental problems, but they also must suffer the environmental impacts for those in the more wealthy area. The overall affect may be one of not preventing environmental impacts, but of exacerbating and relocating them.

Sometimes, areas without large wealth (or even wealthy areas) may want to be suppliers of resources. If an area has large supplies of high-quality aggregate resources, then it may be able to develop superquarries or other types of aggregate operations and export those resources to areas that are unwilling or unable to develop their own resources. It is, however, important to include environmental safeguards, and pass the cost of those safeguards on to the consumer.

6. Summary

Most environmental impacts associated with aggregate mining are benign. Extracting aggregate seldom produces acidic mine drainage or other toxic affects commonly associated with mining of metallic or energy resources. Other environmental health hazards are rare. Most of the impacts that are likely to occur are short lived, easy to predict, and easy to observe. By employing responsible operational practices and using available technology, most impacts can be controlled, mitigated, or kept at tolerable levels and can be restricted to the immediate vicinity of the aggregate operation.

There are, however, a number of situations in which mining aggregate can lead to serious environmental impacts. The rush to build quickly and cheaply with whatever materials come to hand can result in severe environmental consequences. Mining in areas of high population density can force large numbers of aggregate operations to be concentrated into relatively small areas and can exacerbate the environmental impacts of aggregate development. Societal pressures may encourage relaxed controls, and aggregate operators may fail to follow responsible operational practices.

Some geologic environments, such as active stream channels and landslide-prone areas, are dynamic and respond rapidly to outside stimuli, which include aggregate mining. Mining thin deposits or deposits that contain large amounts of unsuitable material results in the generation of large mined areas and large amounts of waste material. Some areas and (or) ecosystems serve as habitat for rare or endangered species, and any activity in these areas may be detrimental. Similarly, some geomorphic features are themselves rare examples of geologic phenomena. Mining aggregate might be acceptable but should be conducted only after careful consideration and then only with extreme prudence.

Important operational conditions that affect environmental impacts are the location and type of the mine, mining techniques, processing techniques, the type and effectiveness of regulations, and the enlightenment of the operator.

Planners are using landscape architects to help enhance the native character of reclaimed sites by fitting the built landscape into the natural design of the surrounding area. By discovering the true landscape and understanding it, reclamation planning and design can concentrate on ecology and development, research and technology, culture and nature, science, and art. The data may not say what choices to make, but they can help with wise options.

Perhaps the most effective means to minimize the impact of conversion of land use is to develop the best resources available and to implement a second land use that is compatible with nature and acceptable to the public.

References

1. Lüttig, G.W., 1994, Rational management of the geo-environment – A view in favour of "Geobased Planning", in Lüttig, G.W., ed Aggregates -- Raw materials' giant: Report on the 2nd International Aggregate Symposium, Erlangen, p.1-34.
2. Kelk, B., 1992, Natural resources in the geological environment, in Lumsden, G.I., 1992, ed. Geology and the environment in Western Europe: Oxford University Press, New York, p. 34 – 138.
3. Barksdale, R.D., ed., 1991, The aggregate handbook: National Stone Association, individual chapters, variously numbered.
4. Smith, M.R., and Collis, L., eds., 1993, Aggregates -- Sand, gravel and crushed rock aggregates for construction purposes: Geological Society Engineering Geology Special Publication No. 9, London, The Geological Society, 339 p.
5. Lüttig, G.W., ed., 1994a, Aggregates -- Raw materials' giant: Report on the 2nd International Aggregate Symposium, Erlangen, 346p.
6. Bobrowski, P.T., ed., 1998, Aggregate resources - A global perspective: A.A. Balkema, Rotterdam, Netherlands, 470p.
7. Gonggrijp, G.P., 1994, Aggregates extraction and Earth-Science Conservation, in L✝ttig, G.W., ed., Aggregates -- Raw materials' giant: Report on the 2nd International Aggregate Symposium, Erlangen, p. 215-226.
8. Goswami, S.C., 1984, Quarrying of aggregates in and around Gauhati – Impact on the Environment: Bulletin of the International Association of Engineering Geology, n. 29, p. 265-268.
9. Hatva, T., 1994, Effect of gravel extraction on groundwater, in Soveri, J., and Suokko, T., eds., Future groundwater resources at risk, Proceedings of the Helsinki Conference): IAHS Publication no. 222, p. 427-434.

10. Graf, W.L., 1979, Mining and channel response: Annals of the Association of American Geographers, v. 69, n. 2, p. 262-275.
11. Mossa, J., aqnd Autin, W.J., 1998, Geologic and geographic aspects of sand and gravel production in Louisiana, *in* Bobrowski, P. T., ed., Aggregate resources – A global perspective: A.A, Balkema, Rotterdam, Netherlands, p. 439 - 464.
12. Kondolf, G.M., 1997, Hungry water – Effects of dams and gravel mining on river channels: Environmental Management, v. 21, n. 4. p. 533-551.
13. Florsheim, J., Goodwin, P., and Marcus, L., 1998, Geomorphic effects of gravel extraction in the Russian River, California, *in* Bobrowski, P. T., ed., Aggregate resources – A global perspective: A.A, Balkema, Rotterdam, Netherlands, p. 87-100.
14. Rowan, J.S., and Kitetu, J.J., 1998, Assessing the environmental impacts of sand harvesting from Kenyan rivers, *in* Bobrowski, P. T., ed., Aggregate resources – A global perspective: A.A, Balkema, Rotterdam, Netherlands, p. 331-354.
15. Jiongxin, Xu, 1996, Underlying gravel layers in a large sand bed river and their influence on downstream-dam channel adjustment: Geomorphology, v. 17, n. 4, p. 351-360.
16. Kondolf, G.M., 1998, Environmental effects of aggregate extraction from river channels and floodplains, *in* Bobrowski, P. T., ed., Aggregate resources – A global perspective: A.A, Balkema, Rotterdam, Netherlands, p. 113 – 130.
17. Piégay, H., and Peiry, J.L., 1997, Long profile evolution of a mountain stream in relation to gravel load management – Example of the Middle Griffe River (French Alps): Environmental Management, v. 21, n. 6, p. 909-919.
18. Engler, Mira, 1995, Waste Landscapes: Permissible Metaphors in Landscape Architecture: Landscape Journal, v. 14, no. 1, p. 11-25.
19. Gladwin, D.N. and J.E. Roelle, 1997, Evaluate Habitat Restoration for the WREN Surface Mine near Fort Collins, Colorado: US Geological Survey, Midcontinent Ecological Science Center. [Online]: *available at* http://www.mesc.nbs.gov/WREN-surface-mine
20. Austin, Peter, 1995, Unlimited restoration: Landscape Design, no. 238, p. 26-28.
21. Dietrich, Norman L., 1990, European Rehabilitation Projects Reflect Cultural and Regional Diversity: Rock Products, v. 93, no. 2, p. 45-47.
22. Myers, Norman, 1990, Miracle in a lifeless pit; quarry restoration in Kenya: Whole Earth Review, no. 66, p. 98.
23. Department of Energy, Weldon Spring Site Remedial Action Project, 1996, Cleanup of the Weldon Spring Quarry. [Online]: *available at* http://www.em.doe.gov/wssrap/quarry
24. Dahl, T.E. 1990. *Wetland Losses in the United States 1780's to 1980's*. US Department of the Interior, Fish and Wildlife Service, Washington D.C., 21 p.
25. Leccese, Michael, 1996, Little Marsh on the Prairie: Landscape Architecture, v. 86, no. 7, p. 50-55.
26. Richardson, Gordon, 1995, Selling land by the tonne: Landscape Design, no. 238, p. 41-44.
27. Holt, Nancy, ed., 1979, The Writings of Robert Smithson: New York University Press, New York.
28. Meyer, Elizabeth K., 1991, The Public Park as Avante-Garde (Landscape) Architecture: A Comparative Interpretation of Two Parisian Parks, Parc de la Villette (1983-1990) and Parc des Buttes-Chaumont (1864-1867): Landscape Journal, v. 10, no. 1, p. 16-26.
29. Golanda, Nella, 1994, Aexoni Quarry, Glyfada, Attica, Greece, *in* Michael Lancaster, The New European Landscape: Butterworth-Heinemann Ltd., Oxford, 162 p.
30. Thompson, J. William, 1996, Taming the Tide: Landscape Architecture, v. 86, no. 5, p. 74-81, 100-102.
31. Massie, Sue, 1985, Timeless Healing at Buffalo Rock: Landscape Architecture, v. 75, no. 3, p. 70-71.
32. Illinois Department of Conservation, 1997, Ancient Art Form Recalled. [Online]: *available at* http://www.iit.edu/~travel/efftxt
33. Timmons, B.J., 1990, Aggregates evaluation in a recreationally oriented state, *in* Martin, J.A., compiler, Proceedings of the 16th Annual Forum on the Geology of Industrial Minerals: Missouri Division of Geology and Land Survey Special Publication no. 7, p. 29-32.

APPLICATION OF THE PORPHYRY COPPER/POLYMETALLIC VEIN KIN DEPOSIT SYSTEM TO MINERAL-RESOURCE ASSESSMENT IN THE MÁTRA MOUNTAINS, NORTHERN HUNGARY

L.J. DREW[1] and B.R. BERGER[2]
[1] U.S. Geological Survey
954 National Center
Reston, Virginia 20192
[2] U.S. Geological Survey
Federal Center MS 964
Denver, Colorado 80225

Abstract. The application of the tectonic model for the porphyry copper/polymetallic vein kin-deposit system, proposed by the authors and used to assess the undiscovered metallic mineral resources of northern Hungary, is illustrated here for the Mátra Mountains, northern Hungary. This model is based on the evolution of strain features (duplexes and flower structures) developed in the strike-slip fault systems in continental crust above a subducting plate and the localization of mineral deposits within them.

The application of this model relied on the integration of data from the literature, satellite images, and geologic maps. During the time of the mineralization (15-16 Ma), the regional-scale tectonic framework was dominated by a right-lateral, pull-apart structure cored by the Mátra Mountains volcanic complex. While synthesizing these data, it was found that the tectonic elements identified on the satellite images could be closely associated with the location of the vein mineralization discovered in the Mátra Mountains. Displays are presented that illustrate how the individual tectonic and geologic elements (faults, volcanic rocks, and sedimentary basins), as well as the relation of the mineralized structures to the complex array of extensional and shear faults that are created inside of a strike-slip fault duplex, are associated.

1. Introduction

The application of the tectonic model for the porphyry copper/polymetallic vein kin-deposit system proposed by [1,2,3] and used to assess the undiscovered metallic mineral resources in northern Hungary [4] is illustrated. This model is based on empirical descriptive models from field data [5], model and theoretical studies of the behavior of strike-slip fault systems [6,7,8,9,10,11], and studies of heat dissipation and mechanics associated with intrusive rocks [12,13]. The kernel of this model is derived from observations that these kin deposits occur in close spatial and temporal associations in the strike-slip fault systems and, most commonly, in those zones where stress is transferred from one fault to another; that is, in duplexes. The members of this system of kin deposits range from those formed in an initial magmatic phase to those that are derived subsequently from mixed meteoric/magmatic inputs; that is, from porphyry copper deposits to polymetallic veins.

The purpose of this paper is to summarize the existing model; to contribute additional elements, such as the extensional-shear vein mesh network; and to illustrate the use of the refined model in mineral-resource assessment with an example from the Mátra Mountains of northern Hungary.

2. The Model

The region of application of the porphyry copper/polymetallic kin-deposit system is the brittle continental crust. The sites within this crust that are the potential locations of the mineral deposits that belong to this system are situated in strike-slip fault systems inboard from a collision orogen; this is, in that section of continental crust above the subducting plate associated with the collision. These locations are determined by the creation of strain features that release the stress in the far field.

As discussed in [14], stress is released through the development of strain features in the continental crust above the subducting plate in a continent-to-continent collision, as shown in Figure 1. Indent-linked strike-slip faults are the far field strain features caused

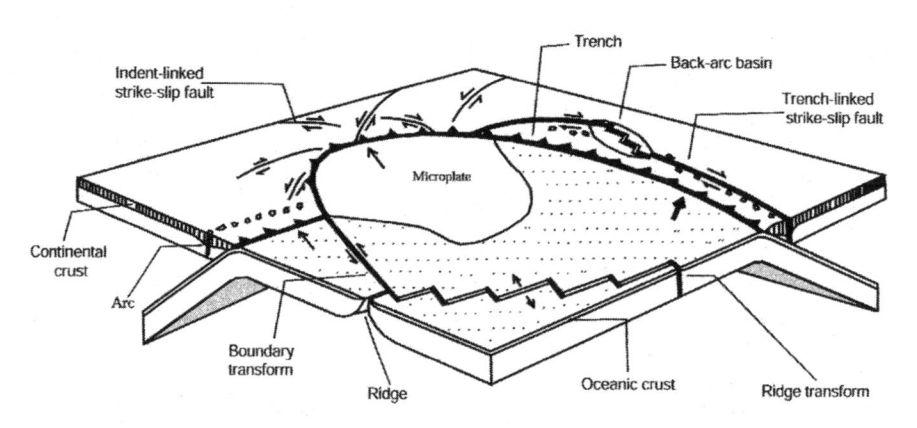

Figure 1. Plate tectonic setting of major classes of strike-slip faults [14].

by such collisions; thus, the maximum principal stress (σ_1) is in the plane of the Earth's surface. Within the principal deformation zones (PDZ) of these faults systems, extensional and compressional strain features developed that can localize magmas at shallow crustal levels. The flow of hydrothermal fluids is similarly controlled by the same stress field and localized in the resulting strain features [9]. In addition to the focusing of magma and fluids in the crust, sedimentary basins and pop-up structures are created as stress is dissipated in the same PDZ [15].

173

The association between the creation of mineral deposits in the porphyry copper/ polymetallic kin-deposit system and strain features created in continental crust located above a subducting plate in a continent-to-continent collision is illustrated in Figure 2. The angle of collision (Figure 2A) is oblique and shown to generate a right-lateral strike-slip fault system behind the volcanic arc. In this fault system, stress is relayed from one fault tip to another, thereby creating strain features known as duplexes. The duplexes shown in Figure 2 have left and right senses of direction of stress transfer. The lower duplex has a right-stepping sense and is extensional, whereas the upper duplex has a left-stepping sense and is, therefore, compressional [15].

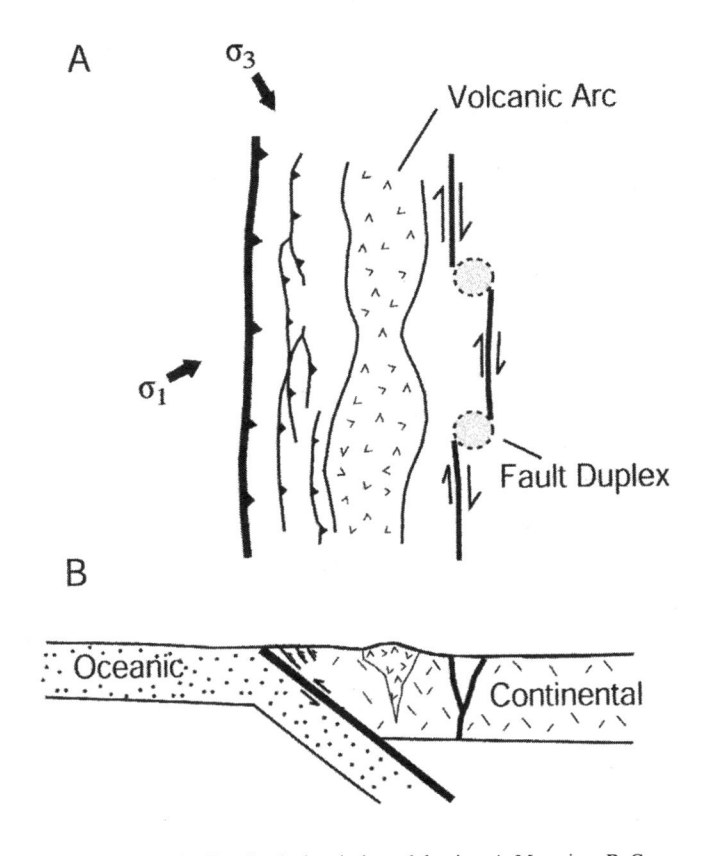

Figure 2. Location of strike-slip duplex during subduction. A, Map view; B, Cross section.

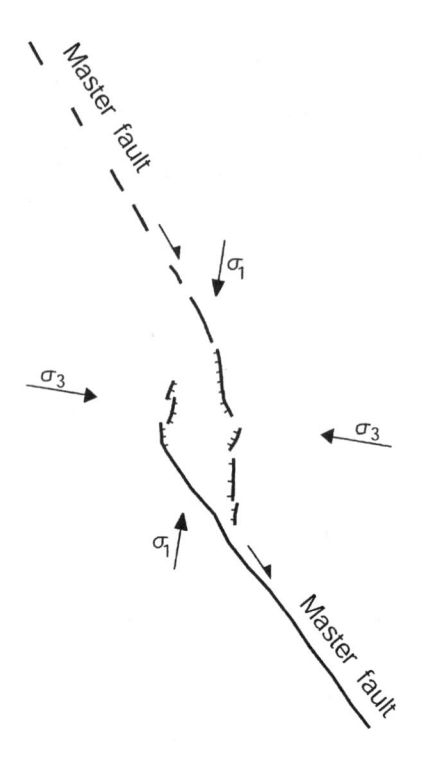

Figure 3. Elements of an extensional strike-slip fault duplex. Modified from [6].

Although both types of duplexes can provide the necessary focusing mechanism to control the assent of magma so that a porphyry copper deposit can form and vein deposits can be deposited by hydrothermal fluids, we will consider here only the extensional duplex, as shown in Figure 3. The far-field stress was resolved into right-lateral movement in a strike-slip fault system. The extensional duplex was created by local inhomogeneities in the crust, such as a previously developed basement structure and as a strain feature in which stress was dissipated as it was being relayed from the northern master fault to the southern master fault. The surface expression associated with such an extensional duplex is usually a sedimentary basin into which locally derived detritus has been shed. Plutonic stocks frequently intrude into various positions in such a duplex, often at its corners; sometimes the entire duplex will be filled with volcanic and associated intrusive rocks.

Porphyry copper deposits are regularly found to occur at the top of plutonic stocks that have intruded into the corners of extensional strike-slip fault duplexes [3]. Genetically related (kin) polymetallic veins are usually found to be deposited subsequently to the deposition of the porphyry deposit(s) in faults that were developed as stress was relayed across the duplex [3].

The model that we have proposed and apply here describes an initial phase of mineralization that begins in the crustal plate above the subducting plate as far-field stress is dissipated in strike-slip fault duplexes. As stress is transferred from one master fault tip to another across a duplex, zones of tensional fracturing are created that provide channelways for magma to rise to shallow levels in the crust [7]. With emplacement of magma in the duplex structure, as shown in Figure 4, the surrounding wallrock is elevated to mesothermal levels, in Figure 5, thus creating a local ductile environment. In this local environment, the high differential stress in the far field ($\sigma_1 > \sigma_3$) is temporarily nullified (see Figure 4); that is, the differential stress ($\Delta\sigma = \sigma_1 - \sigma_3$) is very low or zero. This is the necessary condition for the formation of a porphyry copper deposit. Here, the magmatic stock may be held in place long enough for the necessary mesothermal reactions (carapace development, hydrofracturing, and focusing of hydrothermal fluid flow) to occur. In a few words, we can say that the local ductile environment created by the introduction of magmatic heat into a duplex forms a self-sealing chemical reaction containment vessel.

With mesothermal temperatures and low differential stress, the rocks fracture when the hydraulic pressure under the carapace exceeds the confining pressure. With repeated sealing through mineral precipitation and fracturing, an interconnected stockwork of veins is developed. Silica, potassium feldspar, and copper sulfide minerals precipitate in the low-pressure environment that exists after a fracturing event.

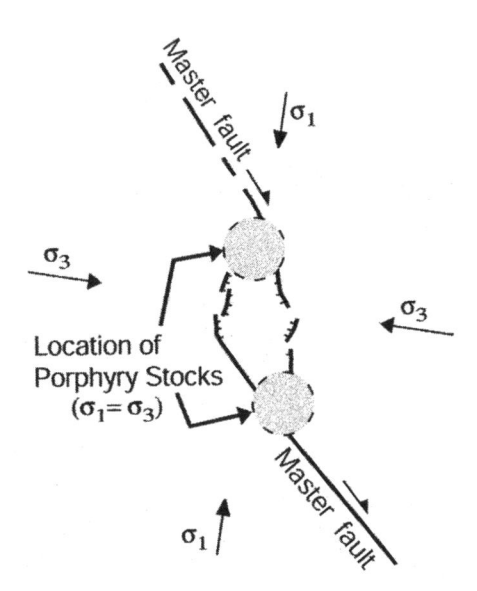

Figure 4. Position in a strike-slip fault duplex favorable for the intrusion of magmatic stocks and porphyry copper deposition (initial phase of mineral deposition.

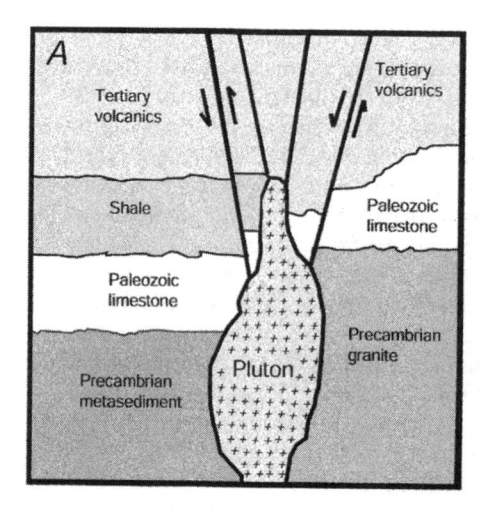

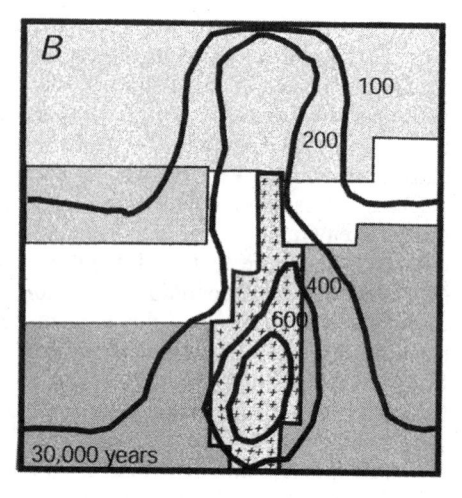

Figure 5. Model of the distribution of magmatic heat associated with a plutonic stock and isotherms for temperatures after 30,000 years of heat dissipation [

When new batches of magma are no longer emplaced into the magma chamber and heat dissipates in the stock and surrounding wallrock, the likelihood for throughgoing brittle fracturing increases as the far-field stress regains structural dominance over the rock volume (σ_1 again dominates, or $\Delta\sigma \gg 0$); that is, as the thermal environment becomes retrograde, the effect of the stress in the far field is reestablished, and throughgoing brittle fracturing dominates, as shown in Figure 6.

The polymetallic veins that often crosscut and (or) are intimately associated with porphyry copper deposits are deposited in the brittle fractures that develop after strain is partitioned and accommodated within the duplex and also along segments of the master faults. The mechanism in which these veins are deposited is a network of extension and shear fractures (a "mesh") that are created inside of the extensional fault duplex, as shown in Figure 7. Vein deposition in segments of a master fault is controlled by the same type of mechanism as deposition inside a duplex, as when a lozenge-shaped graben is created in a PDZ after a P-shear fracture connects two R-fractures [16]. Our understanding of this extension-shear mesh inside of a duplex is provided by [9] and originates from a study of the mechanism that controls the intrusion of dikes in a tectonic stress field [6]. The extension-shear mesh was initially discussed by [9] in the context of the properties of earthquakes and the associated flow of fluid inside of a strike-slip fault duplex; he later adapted these ideas to the deposition of gold veins [10].

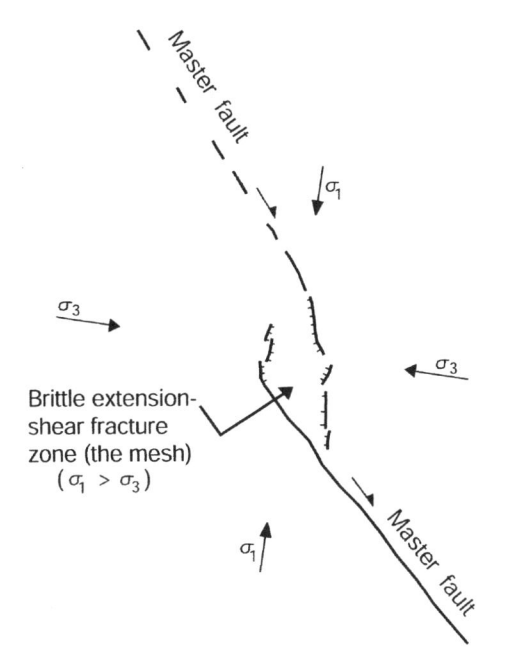

Figure 6. Location in a strike-slip fault duplex favorable for development of the extension-shear mesh and polymetallic vein deposition (second phase of mineral deposition).

The model that we have proposed incorporates this extension-shear mesh as the mechanism for the deposition of the polymetallic veins that were deposited under the influence of brittle fracturing, that is, subsequent to the reestablishment of the effect of the stress in the far field.

Under the influence of this brittle fracture, the containment vessel in which the porphyry copper system was active is breached, and the hydrothermal system is then open to nonmagmatic sources of zinc, lead, and other components that have been leached from the rocks that surround the intrusive complex by the incoming meteoric waters. Once these constituents are entrained in the hydrothermal system, they enter the extension-shear mesh and are deposited as vein material under the conditions of the repeated cycling of the

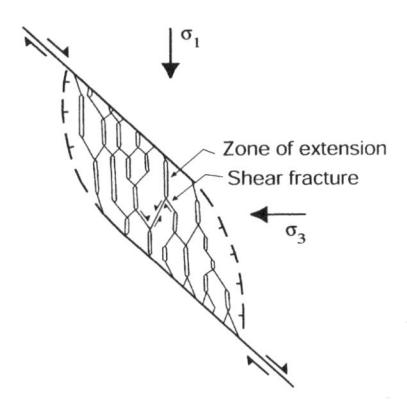

Figure 7. The extension-shear mesh of brittle fracture within a strike-slip duplex. Polymetallic veins are deposited along the interconnected network of zones of extension connected by shear fractures [9].

seismic pumping system triggered by earthquakes. The extension-shear mesh in the interior of the duplex should be considered to be a self-sealing chemical reaction containment vessel for the polymetallic veins just as the local ductile environment created by the intruding magmatic stock is for the porphyry copper deposits at the corners of the duplex.

3. Examples of the Porphyry Copper/Polymetallic Vein Kin-Deposit Model

Well-documented examples of the porphyry copper/polymetallic vein kin-deposit system that illustrate the initial phase of porphyry development followed by the development of polymetallic veins are found within the inner Carpathian magmatic arc in the Apuseni Mountains, Romania [17,18,19,20]. For example, in the Zlatna region, as shown in the cross in Figure 8, where low-grade porphyry copper deposits have been emplaced, these deposits are often cut by polymetallic veins that have grades that range from 5- to 7-percent combined zinc and lead (I. Berbeleac, Prospectiuni S.A., oral commun., May 1997). Also in the same region, the economic polymetallic veins that occur in the Hanes deposits, shown in Figure 9, are in the footwall (more-compressive) section of complex positive flower structures. Flower structures are the common strain features (configuration of horse blocks) that develop in a duplex. They take their name from the flowerlike arrangement of horse blocks found when a cross section is constructed for a duplex.

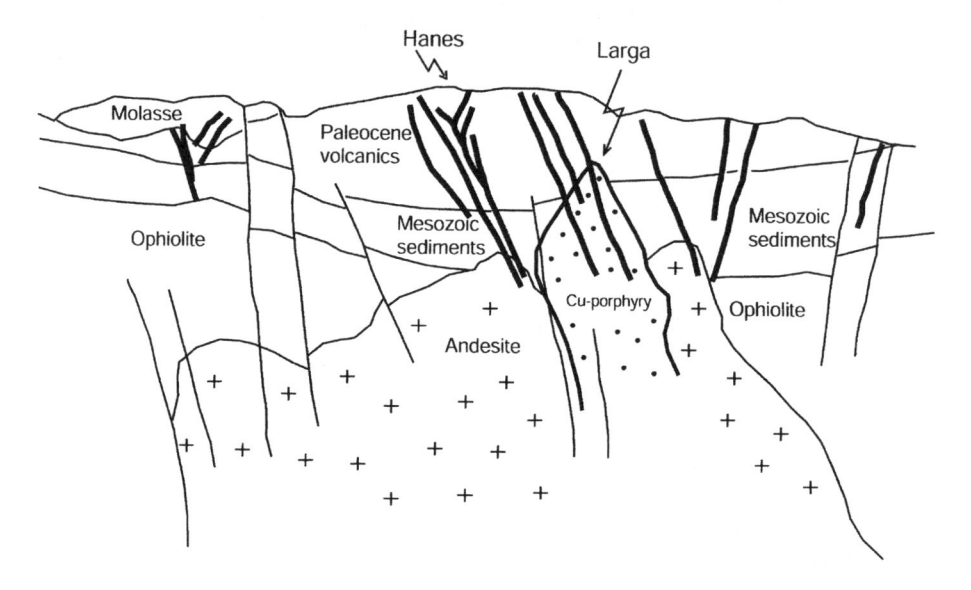

Figure 8. Cross section through the Larga porphyry copper and Hanes polymetallic vein deposits, Zlatna region, Romania [17].

The flower structure at the Hanes polymetallic vein deposit is highly evolved because the structure has been filled with intruding andesite. Only the overall form of the flower structure and two horse blocks remain of the original structure. The intruding structure, and the footwall (southwestern side) is interleaved with horses of marl beds

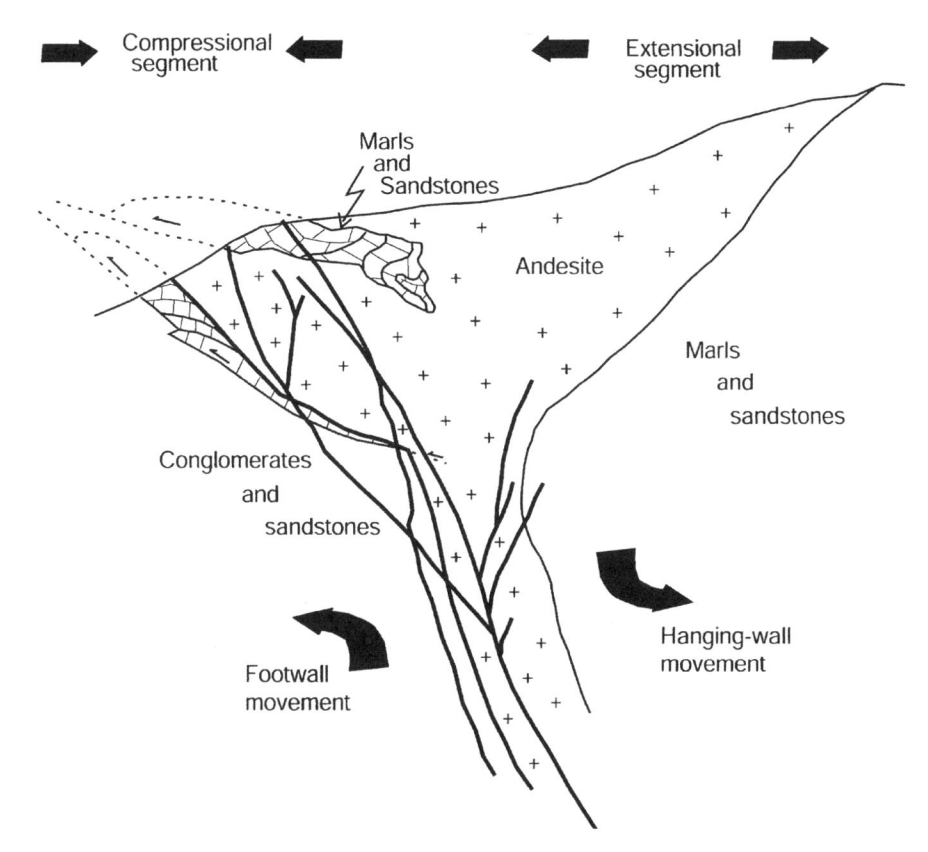

Figure 9. Interpretation of the flower structure that hosts the polymetallic veins at the Hanes deposit, Zlatna district, Romania (see Figure 8 for location). Based on data from [21].

that were thrust upward from lower in the stratigraphic section. Andesite completely filled the hanging-wall segment (northeastern side) of this flower

At Hanes, then the intimate relation between the strain features created by the strike-slip faults (flower structure) and the intruding andesite is obvious—the andesite body takes the form of the flower structure with a dike-like root and a dome-shaped top (see Figure 9). Subsequent to intrusion of the andesite along the fault system, throughgoing brittle fracturing occurred in the andesite, and polymetallic veins were deposited in the southwestern side of the flower structure in a network of veins interpreted as being elements of an extensional-shear mesh.

4. Mátra Mountains, Northern Hungary

The study area chosen for an application of the porphyry copper/polymetallic vein kin-deposit system is the Mátra Mountains, northern Hungary. These mountains are a Middle Miocene volcanic complex (Badenian age; shown in Figure 10) located in the inner Carpathian magmatic arc about 85 km northwest of Budapest. The volcanic rocks in the Mátra Mountains volcanic complex are calc-alkaline affinity and are principally andesitic in composition with some dacites and rhyolites [22]. In a summary of the chronology of the Neogene volcanism of the Carpathian-Pannonian region, the ages of the volcanic rock in the Mátra Mountains were estimated by [23] to range from 13.7 to 16.0 Ma.

On the basis of data from [24,25,26], this volcanic complex is interpreted to have originated in a right-lateral and right stepping strike-slip fault duplex, as shown in Figure 11. Today, this volcanic complex is the core of a much larger sedimentary basin filled with Badenian and Sarmatian age detritus [24,25]. This basin is only one of a large number of sedimentary basins that formed as the Carpatho-Pannonian escape structure (an approximately 150,000-km^2 region) moved eastward during the Late Oligocene and throughout the Miocene periods as a result of the continent-to-continent collision that formed the Alps to the west of the Carpatho-Pannonian region.

One of the master faults along which the far-field stress developed in this collision was dissipated in the Carpatho-Pannonian continental plate in the vicinity of the Mátra Mountains and is located in the Etes trough (see Figure 11). The second master fault to which this stress was relayed is located in the middle of the Csernat block (see Figure 11). As stress was relayed from the northern master fault to the southern master fault, a strain feature (duplex) developed that initially filled with the Mátra Mountains volcanic rocks. During the initial stage of tectonic development, porphyry copper deposits may have formed, but subsequent tectonic activity (strain partitioning along the numerous detachment and other large faults in the Carpatho-Pannonian plate of northern Hungary) removed any evidence that they were formed in the initial configuration of the duplex.

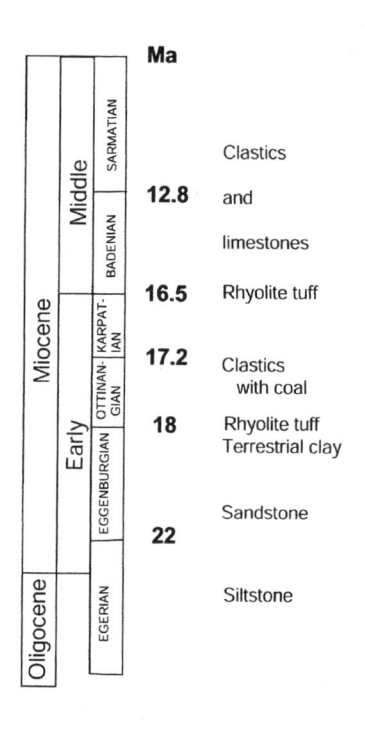

Figure 10. Stratigraphic column of northern Hungarian Paleocene-Miocene basins [27,28,29].

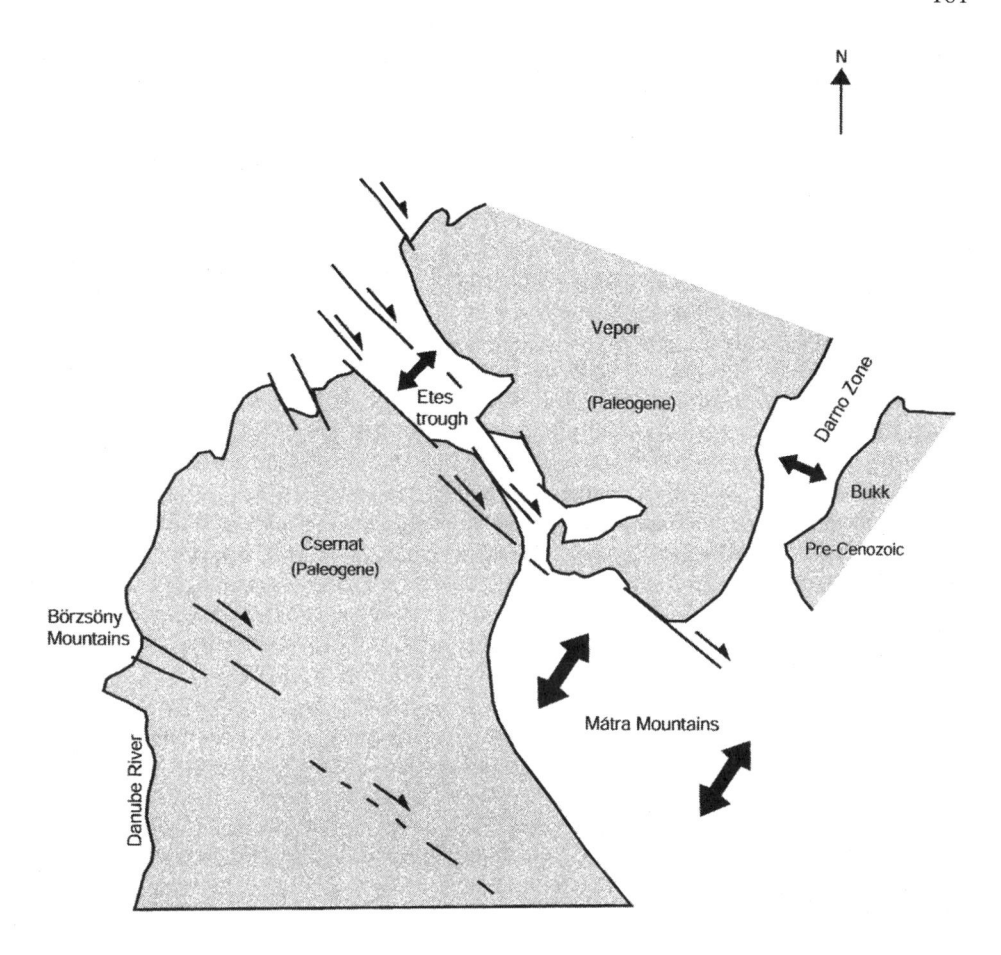

Figure 11. Extension of the continental crust in northern Hungary during the Early Badenian. Modified from [24,25,26].

In striking contrast, many polymetallic veins were deposited in a system of brittle faults that formed when the Mátra Mountains volcanic complex filled the initial duplex, as shown in Figure 12; [4]. Intense prospecting in the Mátra Mountains has resulted in the discovery of many polymetallic veins, the Gyöngyösoroszi mining district being the largest with 4.8 million t of proved reserves of lead, zinc, and silver ores [30]. The polymetallic veins in the mining district are hosted in an extension-shear mesh network. The horse blocks of rock that host these veins are arranged in the typical flower structure configuration.

As the right-lateral strike-slip fault system continued to evolve, sedimentary basins marginal to the Mátra Mountains volcanic complex developed—the Zagyva trough to the west of the complex, as shown in Figure 12, and other basins to the east and south. When [24] interpreted the seismic sections across the Zagyva trough, they discovered half-graben structures in a transfer fault system of the duplex. This conclusion is consistent with the idea that the Mátra Mountains-Zagyva trough complex was created in an extensional duplex between two master faults. The subsequent development of this strike-slip fault duplex from having an initial magmatic phase to becoming a "cold" sedimentary basin provides important determinants in the application of the porphyry copper/polymetallic vein-kin deposit system to estimate the probability that members of this family of deposits remain to be discovered in the study area.

5. The Assessment

In May 1997, a team of geologists from the Geologic Institute of Hungary and the U.S. Geological Survey met in Budapest to assess the undiscovered mineral resources of a study area in northern Hungary of which the Mátra Mountains was a subregion [4].

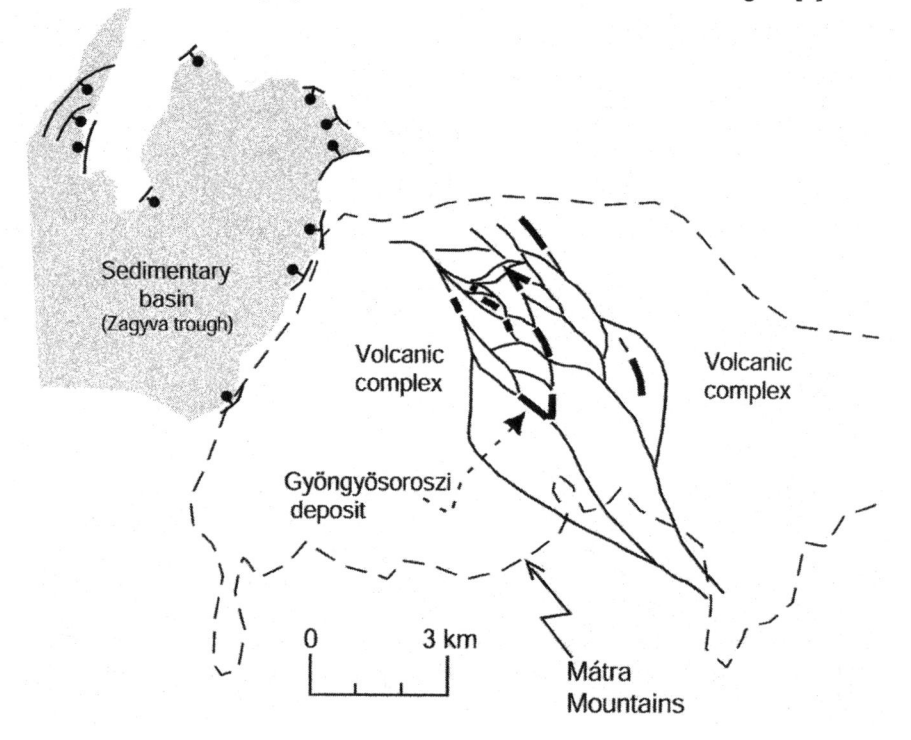

Figure 12. Polymetallic veins located in an extension-shear mesh in the Mátra Mountains, northern Hungary. Dashed line, assessment area; bold lines, polymetallic vein deposits in the extensional zones of the extension-shear mesh system (figure 7; data from [31]); solid lines, faults identified from satellite imagery and interpreted to be the shear faults in the mesh; and bars and balls, normal faults.

The goal of this assessment was to estimate the number of undiscovered deposits for the deposit types in the porphyry copper/polymetallic vein kin-deposit system, which, in this assessment, included skarn deposits. The subjective technique that was used in this assessment required each geologist to transduce into probabilistic statements his/her sense of the magnitude and uncertainty associated with these types of yet-to-be-discovered mineral deposits. The individual assessments were cast by secret ballot and resolved into a consensus in group discussion.

The Mátra Mountains are a volcanic complex that was supplied by at least six volcanic centers. The assessment area is shown in Figure 12. Within the approximately 100-km^2 area, a collection of strike-slip faults that host polymetallic veins occurs. The collection of northwest- to southeast-trending faults that cut across these mountains were located by combining data from a detailed map [31] and photolinear analysis of a satellite image of the area.

During the assessment of the Mátra Mountains, the following data were considered to be most relevant for predicting the occurrence of undiscovered polymetallic vein deposits: the basic geology is permissive (six or more volcanic centers), hydrothermal alteration is widely distributed, one or more producing polymetallic vein districts (Gyöngyösoroszi) has been discovered, and the inner Carpathian arc hosts many polymetallic vein deposits (nearby in Slovakia and Romania). The consensus estimate for the inventory of undiscovered polymetallic veins in the Mátra Mountains is a 90-percent probability of 4 deposits, a 50-percent probability of 5 deposits, a 10-percent probability of 6 deposits, and a 1-percent probability of 10 deposits. On the basis of this distribution, the expected number of undiscovered deposits is 5.08 which was estimated by weighting the individual probabilities associated with the regions and the probability of occurrence of each number of deposits [32].

This estimate of 5.08 deposits is for the occurrence of deposits distributed in size, as shown in Figure 13. Inspection of this figure reveals that the Gyöngyösoroszi deposit is an exceptionally large polymetallic vein deposit with reserves of 4.8 million t of ore. This tonnage is at the upper end of the

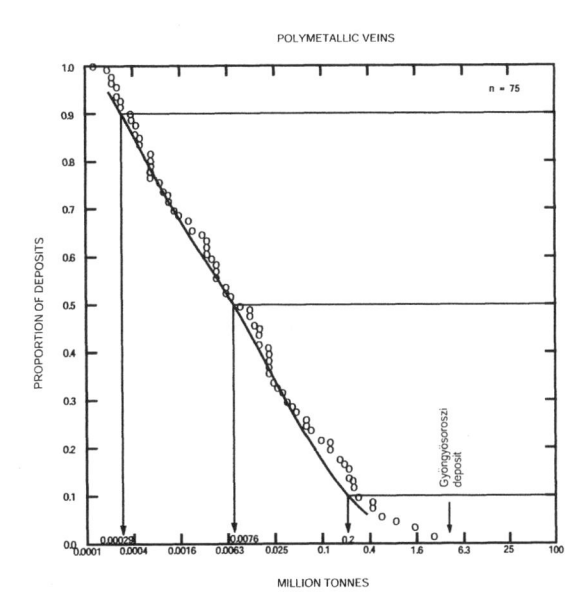

Figure 13. Cumulative distribution for the size of polymetallic veins. Modified from [33].

observed range of sizes for this type of deposit. The mean size of these deposits is 7,600 t, which is 632 times smaller than that of the Gyöngyösoroszi deposit. The assessment of the expected 5.08 deposits remaining to be discovered is associated with this mean, the distribution of which is shown in Figure 13. Therefore, the team's assessment for ore-grade material "at the mean" is equal to 5.08 x 7,600 = 38,680 t. This expectation is also a small fraction of the amount of ore produced from the Gyöngyösoroszi deposit.

In addition to an assessment for polymetallic vein deposits, the assessment team concluded that the probability of porphyry copper deposits remaining to be discovered in the Mátra Mountains volcanic complex is nontrivial. The general basis for this determination was that although no porphyry copper deposits have been discovered, three silicified areas have been mapped, and polymetallic vein deposits have been discovered. The observed silicification could be a manifestation of a porphyry copper system and (or) the occurrence of polymetallic vein deposits. After considering the field data and exploration history, the assessment team reached a consensus that no porphyry copper deposits occur in the Mátra Mountains volcanic complex at the 90- and 50-percent probability levels, that one would occur at 10-percent level, and that two would occur at the 1-percent level. The expected number of undiscovered porphyry copper deposits was computed to be 0.38 [32]. Configuration of the polymetallic veins in the Mátra Mountains (see Figure 12), vis-à-vis the subsequent down-to-the-basin faulting, shows that the corners of the strike-slip duplex have been down faulted into the sedimentary basins. Thus, either the porphyry copper deposits that were formed have been down faulted into the sedimentary basins or the preferred sites for their formation were removed before they could form as depicted in the graphical displays shown in Figures 4 and 5.

6. Aggregate Resource Estimates

The Mark3 simulation software [32] was used to estimate the inventory of undiscovered mineral resources contained in the expectation (statistical) that 0.38 porphyry copper and 5.08 polymetallic vein deposits remain to be discovered in the Mátra Mountains. The summary statistics for the tonnage of contained metal and ore in these undiscovered deposits is shown in table 1.

The mean tonnage of ore material from both deposit types is estimated to 65 million t. The individual mean-aggregate metal tonnages are estimated to be 340,000 t of copper, 42,000 t of lead, 28,000 t of zinc, 2,300 t of molybdenum, 600 t of silver, and 28 t of gold. The table also lists the companion tonnages for each metal at three probability levels in the aggregate metal distribution.

7. Conclusions

The tectonic model for the occurrence of deposits in the porphyry copper/ polymetallic-kin deposit system proposed by [1,2,3] and used to assess the undiscovered mineral resources of a study area in northern Hungary [4] has been illustrated in detail in an assessment of the undiscovered mineral resources of the Mátra Mountains, northern

Hungary. The model was used to show that the probability for the occurrence of

Table 1. Summary statistics for the tonnage of contained metal and ore in the undiscovered deposits in the Mátra Mountains, northern Hungary

Metal or ore	Tonnage			Mean
	90 percent	50 percent	10 percent	
Copper	1.5	140	1.0×10^6	340,000
Silver	17	180	1,100	600
Zinc	330	7,200	140,000	28,000
Lead	1,900	18,000	94,000	42,000
Gold	0.0060	0.45	81	28
Molybdenum	0.0	0.0	3,600	2,300
Ore material	33,000	580,000	200×10^6	65×10^6

polymetallic veins is high and that of porphyry copper deposits is low. This conclusion is based on a synthesis of tectonic data that show that the tectonic evolution of the Mátra Mountains migrated from a initial center of volcanism situated in a extensional strike-slip fault duplex to a vastly expanded cold sedimentary basin. By using these data and the model, it was concluded that either the porphyry copper deposits which may have formed were subsequently down faulted into sedimentary basins as the strike-slip faults converted the region from a volcanic center to a cold sedimentary basin by strain partitioning or the preferred sites for the formation of porphyry copper deposits (the corners of the strike-slip duplex) were down faulted before porphyry formation.

References

1. Berger, B.R. and Drew, L. J. (1997) Role of strike-slip duplexes in localization of volcanoes, related intrusions, and epithermal ore deposits, *1997 Annual Meeting of the Geological Society of America*, Abstracts with Programs, A329-A360.
2. Berger, B.R., Drew, L. J., and Singer, D.A., (1999) Quantifying mineral-deposit models for resource assessment, *Geologica Hungarica, v. 24, p. 41-54.*
3. Berger, B.R., and Drew, L. J., (in press) Mineral-deposit models—New developments, in Andrea Fabbri (ed.), *title unknown*, Kluwer Academic Publishers, Dordrecht.
4. Drew, L.J., Berger, B.R., Bawiec, W.J., Sutphin, D. M., Csirik, G., Korpás, L., Vető-Akos, E., Odor, L., and Kiss, J. (1999) Mineral-resource assessment of the Mátra and Börzsöny-Visgrád Mountains, north Hungary, *Geologica Hungarica*, v. 24, p 79-96.
5. Cox, D.P., and Singer, D. A. (eds.) (1986) *Mineral Deposits Models*, U.S. Geological Survey Bulletin 1693.
6. Hill, D.P. (1977) A model for earthquake swarms, *Journal of Geophysical Research* 82, 1347-1352.
7. Segall, P. and Pollard, D.D. (1980) Mechanics of discontinuous faults, *Journal of Geophysical Research* 86, 4337-4350.
8. Sibson, R.H. (1985) Stopping of earthquake ruptures at dilational fault jogs, *Nature* 316, 248-251.
9. Sibson, R.H. (1986) Earthquakes and lineament infrastructure, *Transactions of the Royal Philosophical Society of London* A-317, 63-79.
10. Sibson, R.H. (1989) *Structure and Mechanics of Fault Zones in Relation to Fault-Hosted Mineralization* Australian Mining Foundation, Glenside.
11. Sibson, R.H. (1994) Crustal stress, faulting and fluid flow, in J. Parnell (ed.), *Geofluids—Origin,*

Migration and Evolutions of Fluids in Sedimentary Basins, Geological Society Special Publication **78**, pp. 69-84.

12. Norton, D.L. (1982) Fluid and heat transport phenomena typical of copper-bearing pluton environments, in Titley, S.R. (ed.), *Advances in the Geology of the Porphyry Copper Deposits, Southwestern North America*, University of Arizona Press, Tucson, pp. 59-72.
13. Sonder, L.J. and England, P.C. (1989) Effects of temperature-dependent rheology on large-scale continental extension, *Journal of Geophysical Research* **94**, 7603-7619.
14. Woodcock, N.H. (1986) The role of strike-slip fault systems at plate boundaries, *Transactions of the Royal Philosophical Society of London* **A-317**, 13-29.
15. Sylvester, A.G. (1988) Strike-slip faults, *Geological Society of America Bulletin* **100**, 1666-1703.
16. Davison, I. (1994) Linked fault systems—Extensional, strike-slip and contractional, in P. L. Hancock (ed.) *Continental Deformation*, Pergamon Press, Oxford, pp. 121-142.
17. Borcos, M. (1994) Volcanicity/metallogeny in the south Apuseni Mts (Metalliferi Mts), *in* Borcos, M., and Vlad, S. (eds.), *Plate tectonics and metallogeny in the east Carpathians and Apuseni Mts--June 7-19, 1994--Field trip guide* Geological Institute of Romania, IGCP Project No. 356, p. 32-38.
18. Berbeleac, I., Iliescu, D., Andrei, J., Ciuculescu, O., and Ciuculescu, R. (1995) Relationships between alterations, porphyry copper-gold and base metal-gold hydrothermal vein mineralizations in Tertiary intrusions, Talagiu area, Zarand Mountains, Romania: *Journal of Mineral Deposits* **76**, 31-39.
19. Berbeleac, I., Popa, T., Marian, I., Iliescu, D., and Costea, Cr. (1995) Neogene porphyry copper-gold, gold, and gold bearing epithermal deposits in the South Apuseni Mountains, Romania, 1995, *Congress of the Carpathian-Balcan Geological Association*, 15th, Proceedings, 665-670.
20. Mitchell, A.H.G. (1996) Distribution and genesis of some epizonal Zn-Pb and Au provinces in the Carpathian-Balkan region, *Transactions of the Institution of Mining and Metallurgy*, B127-B138.
21. Borcos, M. Gheorghita, I., Bostinescu, S., and Maties, P. (1962) Consideratii asupra unor manifestari magmatice Neogene cu character linear in Muntii Metaliferi si sasupra structurii aparatului vulvanic Hanes, *Dari de Seama ale Sedintelor* **49**,19-30.
22. Szabó, C., Harangi, S., and Csontos, L. (1992) Review of Neogene and Quaternary volcanism of the Carpathian-Pannonian region, *Tectonophysics* **208**, 243-256.
23. Pécskay, Z., Lexa, J., Szakács, A., Balogh, Kad., Seghedi, I., Konecny, V., Kovács, M., Márton, E., Kaliciak, M., Széky-Fux, Póka, T., Gyarmati, P., Edelstein, O., Rosu, E. and Zev, B. (1995) Space and time distribution of Neogene-Quaternary volcanism in the Carpatho-Pannonian Region, in H. Downes and D. Vaselli (eds), *Neogene and Related Magmatism in the Carpatho-Balcanian Region*, Acta Vulcanologica **7**, pp. 15-28.
24. Tari, G., Horváth, F., and Rumpler, J. (1992) Styles of extension in the Pannonian Basin, *Tectonophysics* **208**, 203-219.
25. Márton, E., and Fodor, L. (1995) Combination of paleomagnetic and stress data--A case study from North Hungary, *Tectonophysics* **242**, 99-114.
26. Peresson, H., and Decker, K. (1997) Far-field effects of Late Miocene subduction in the Eastern Carpathians--E-W compression and inversion of structures in the Alpine-Carpathian-Pannonian region, *Tectonics* **16**, 38-56.
27. Balogh, K. and others (1984) Genetic aspects of the Recsk mineralized complex, *Bulletin of the Hungarian Geological Society* **114**, 335-348.
28. Hámor, G. (1985) Geology of the Nógrád-Cserhát area, *Geologica Hungarica* **22**.
29. Vass, D., Konečný, V., and Sesfara, J. (1979) *Geology of Ipel'skwá Kotlina (Depression) and Krupinská Planina Mts.* Geologický Ústav Dionýza Štúra, Bratislava.
30. Bartók, A. and Nagy, I. (1992) *Magyarország érchordozó ásványi nyersanyagai, színes és feketefémérc vagyona*, Központi Földtani Hivatal, Budapest.
31. Varga, G., Csillagne-Teplánszky, E., and, Félegyházi, Z. (1975), *Geology of the Mátra Mountains*, Annals of the Hungarian Geological Institute, Budapest.
32. Root, D.H., Menzie, W.D., and Scott, W.A., Jr. (1992) Computer Monte Carlo simulation in quantitative resource estimation, *Nonrenewable Resources* **1**, 125-138.
33. Bliss, J.D., and Cox, D.P. (1986) Grade and tonnage model for polymetallic veins, in D.P. Cox and D.A. Singer (eds.), *Mineral Deposit Models* U.S. Geological Survey Bulletin 1693, pp. 125-129.

THE ASSESSMENT AND ANALYSIS OF FINANCIAL, TECHNICAL AND ENVIRONMENTAL RISK IN MINERAL RESOURCE EXPLOITATION

P.A. DOWD
Department of Mining and Mineral Engineering
University of Leeds
Leeds, LS2 9JT
United Kingdom

Abstract. This lecture explores the various types of risk that arise in the exploitation of mineral resources and describes ways in which they can be identified, assessed, quantified and managed. The emphasis is on the quantification of risk and on the use of quantitative methods for the visualization of risk.

1. Introduction

Realistic quantification of risk requires adequate models of the processes that give rise to risk. This lecture covers the integrated use of the following technologies:

- Correlated, multivariate Monte Carlo Simulation
- Geostatistical simulation
- Geographical Information Systems (GIS)
- Knowledge-Based Systems (KBS)
- Numerical modelling of risk processes,

to conduct realistic risk analyses that incorporate the complex inter-relationships among financial, technical, natural and environmental variables.

Although, in general, risk can be quantified it is important to recognise that quantification is only one way of assessing and reporting risk and that numerical approaches to risk assessment are limited. In general, risk is perceived and this perception must be addressed when conducting risk analyses and reporting the results, especially in public forums. The modelling and visualisation capabilities of an integrated use of the technologies described here provides a means of dealing with perception in risk analysis.

2. Risk

All decisions that involve risk also carry a benefit, otherwise they would not be considered, and the consequent risk would not be incurred. This is true whether the decision is largely qualitative (e.g., health and safety) or largely quantitative (e.g., financial or environmental). In dealing with concepts such as risk and benefit it is important to separate aspects that are empirical and can therefore be quantified from those that are value judgments and cannot be quantified. Table 1 lists the four empirical components of decisions together with the characteristic value judgment corresponding to each of these components. The definitions of these factors and judgments follows Lowrance [1].

Table 1. Empirical components of decisions and corresponding value judgments (after Lowrance [1])

Empirical factor	Value judgment
Risk	Safety
Efficacy	Benefit
Monetary cost	Social cost
Distribution	Equity/fairness of distribution

Risk is a measure of the probability and severity of adverse effects whereas **Safety** is the degree to which risks are judged acceptable.
Efficacy is a measure of the probability and intensity of beneficial effects whereas **Benefit** is the degree to which efficacies are judged acceptable.
Monetary cost is the measurable amount (direct and indirect) paid for implementing the decision; **Social cost** is the perceived cost of intangibles such as deterioration in the environment or the quality of life.
Distribution is the manner in which risks, efficacies and costs are borne; **Equity of the distribution** is a judgment of the appropriateness of the distribution. Equity may involve a judgment of the fitness of a specified element of a project to bear the risks and costs allocated to it; more generally it may be a judgment of fairness and/or social justice. Risks are, for example, distributed amongst partners in a project, over segments of a project, over time, over geographical areas and over sections of the population and these distributions may be more important than the magnitude of the risk.

The distinction between the empirical and the judgmental was made by Handler [2]: "The *estimation* of risk is a scientific question and, therefore, a legitimate activity of scientists in federal agencies, in universities and the National Research Council. The *acceptability* of a given level of risk, however, is a political question, to be determined in the political arena." In the applications described in this lecture 'political' refers to all forms of subjective decision-making, including company and shareholder policies and attitudes to technical, financial and environmental risk.

The components and corresponding value judgments can be related to different periods of time. For example, project operating costs must be viewed in the short term by the mining company whereas it may be necessary for shareholders and the general community to take a long term view of benefits.

No minerals industry project is completely risk-free or its success completely certain: there is always a non-zero risk of failure. As risk cannot be eliminated the best that can be done is to reduce it to a minimum (generally determined by cost) and then manage the residual risk. A general strategy for dealing with risk thus comprises the following elements:

- identify the risk
- assess and quantify the risk
- determine the minimum acceptable level of risk
- reduce the risk to a minimum
- manage the residual risk

The acceptability of risk is generally determined by a risk-benefit analysis: is the benefit worth the residual risk? (e.g., "Do the alleged risks to public health and to ecosystems from the emissions of coal-fired electric power plants warrant the increase in the price of energy necessitated by scrubbers to reduce those emissions?" [2]; Are adverse, short-term environmental impacts worth the long-term economic benefits generated by mining projects?). This trade-off approach is simply a recognition that economic progress entails a cost and that the former must always be balanced against the latter. A cost-benefit analysis compares the cost of a project to the value of the benefits generated. A cost-effective analysis recognises that resources are limited and seeks to identify how to conduct a project, or meet an objective, for the least cost.

The diversity and complexity of the risks involved in the exploitation of mineral resources make effective assessment and management very difficult. This lecture explores the use of computer-based technologies in the quantification, assessment, visualisation, perception and management of risk associated with minerals industry projects.

2.1 RISK IN MINERALS INDUSTRY PROJECTS

Minerals industry projects are designed on the basis of variables that are subject to extreme uncertainty. This uncertainty arises because of the nature of the variables and the cost of obtaining information about them. Financial variables, for example, are inherently unpredictable: forecast mineral prices are based on historical data and assessments of future markets. Geological variables can only be assessed and quantified on the basis of sparse drilling and sampling programmes; even at the (relatively advanced) mine planning stage, the grades of 4m x 4m x 5m selective mining units may be estimated from the grades of samples taken from drillholes on a 20m x 30m grid. Geotechnical design is based on the geotechnical properties of sparse samples often not even collected for the purpose at hand. In many cases environmental risk may involve assessing the likelihood of an event (e.g. contamination of the water table by tailings dam leakage) that has not occurred on the basis of variables that have not been directly measured (e.g. fluid flow travel paths from the tailings dam through the sub-strata to the water table).

Risk in minerals industry projects arises under a number of headings, the most important of which are financial, technical and environmental. Some authors would add legal risk to this list, but legal risk generally arises only as a consequence of risk in one or more of these three areas. Similarly, sovereign and political risks (including expropriation and retrospective reassignment of mineral rights) could, in general, be included as part of financial risk.

The degree and relative importance of each type of risk depends on the type and location of the project. In general, underground mining incurs a higher technical and financial risk than surface mining; mine development in a developing country may carry higher sovereign and political (and, therefore, financial) risks than it does in a developed country.

Whilst it is possible to delineate the areas of risk as financial, technical and environmental, it is difficult and misleading to separate these areas as they are highly inter-related. Take, for example, groundwater problems in surface mining. The inflow of groundwater in mining operations poses a risk in all three areas:

- Financial (e.g. pumping costs; delays in production)
- Technical (e.g. reduced slope stability; efficient means of pumping)
- Environmental (e.g. pollution of water courses)

In addition, meeting legislative and planning requirements on groundwater control requires technical solutions that, in turn, incur further costs.

Throughout the world environmental considerations are now the major constraint on the exploitation of mineral resources. Responsible extraction requires a comprehensive environmental impact assessment together with extensive, integrated risk analyses and evaluations of remediation measures.

The key to valid risk analysis is complete integration of all risks within specific categories and between categories. This approach is widely recognised in legislation (e.g. Integrated Pollution Control under the 1990 UK Environmental Protection Act and the EU Integrated Pollution Prevention and Control Directive) and by industry bodies and professional associations (e.g., the UK Engineering Council *Guidelines on Environmental Issues* [3]).

2.2 ASSESSING RISK

Risk analysis is the identification of the possible outcomes of decisions. Risk assessment is the quantification of risk in terms of probability (or likelihood) and size of outcome. Risk can be assessed in a number of ways from the completely analytical to the completely synthetic and the assessment can vary from subjective to objective. One common way of identifying the risks associated with an activity and assessing their impact is by the use of assessment matrices. Matrix methods are best illustrated by the most well known, the Leopold matrix, summarised in Figure 1 for an application to environmental impact assessment.

Each column in the matrix represents a particular type of activity that will have an environmental impact; each row represents a form of the impact. For example, types of activity might include transport and blasting; forms of impact might include vibration, dust and noise.

It is a small step to extend this simple recording procedure to include the probability of occurrence of each impact. Ultimately, quantified risk analysis requires an estimate of the likelihood, or probability, of an event occurring. It may be argued that in the case of true uncertainty it is not possible to determine probabilities. This is a simplistic view of probability, however, and, in the context of most of the assessments required in the minerals industry, an incorrect one. What is required is the generation of possible states of nature based on process models followed by an assessment of the likelihood of particular events occurring, given these states of nature.

INSTRUCTIONS

1. Identify all actions or activities (located across the top of the matrix) that are part of the proposed project.

2. Identify all environmental impacts of activities.

3. Under each of the proposed actions place a slash at the intersection with each impact if that impact is possible.

4. Having completed the matrix, in the upper left-hand corner of each box with a slash place a number from 1 to 10, which indicates the MAGNITUDE of the possible impact; 10 represents the greatest magnitude of impact and 1, the least (no zeroes). Before each number place + if the impact would be beneficial. In the lower right-hand corner of the box place a number from 1 to 10, which indicates the IMPORTANCE of the possible impact (e.g., regional vs. local); 10 represents the greatest importance and 1, the least (no zeroes).

5. The text that accompanies the matrix should be a discussion of the significant impacts of the activities (i.e., of those columns and rows with large numbers of boxes marked and individual boxes with the larger numbers).

SAMPLE MATRIX
Type of activity

		a	b	c	d	e
Impact	a		2/1		/	8/5
	b		7/2	8/8	3/1	9/7

Figure 1: Instructions for setting up a Leopold matrix for an application to environmental impact assessment (after Leopold et al [4]).

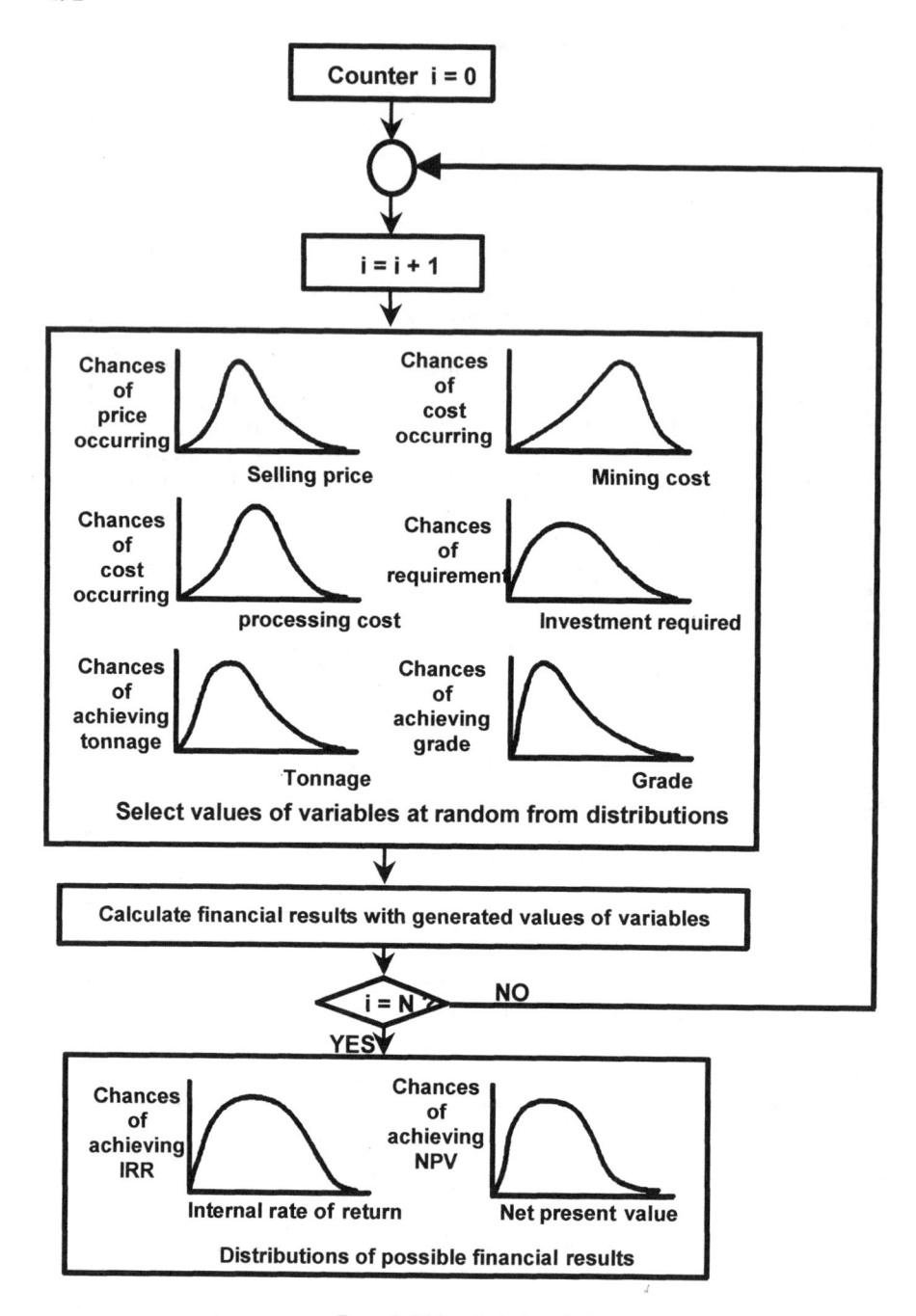

Figure 2. Risk analysis flow chart

3. Risk Analysis and Assessment by Monte Carlo Simulation

In its most general form simulation is a method of mimicking reality. Physical simulation is well understood as, for example, in the use of flight simulators for training pilots and, within the minerals industry, computer-based, virtual reality simulators for training oil rig workers and even open-pit truck operators. Additional examples are bench-scale pilot plants of mineral processing operations and laboratory models of ventilation systems and of underground roadways.

Probabilistic simulation that uses the well-known Monte Carlo techniques is widely used in operational research applications. The method involves setting up a model in the form of a histogram that represents all possible (or at least all foreseen) values of a variable and selecting values at random from the model. It has found application in many aspects of mine planning (e.g., optimal truck despatching) and has found growing acceptance as a means of financial risk analysis [5]. Financial criteria, such as net present value or internal rate of return, are calculated on the basis of uncertain values of technical and economic variables. These uncertainties are the major contributors to the high financial risk associated with mining projects. In view of these uncertainties it is usually meaningless to report a single value of an estimated financial criterion, such as net present value, unless it is accompanied by a statement of the reliability of the estimate. One way of assessing this reliability is to assess the sensitivity of the calculated value of the financial criterion to simulated changes in the values of the variables from which it is calculated. There are two general ways of doing this:

- Take one variable at a time and change its values systematically. For each value of the variable calculate the corresponding value of the financial criterion and record the variations in this value. This is known as sensitivity analysis.

- Allow all variables to change their values simultaneously in such a way that:

  - any correlations among variables are reproduced in their values
  - the values of each variable are selected at random from a histogram that either summarises the pattern of values that has occurred in the past or that models projected future values of the variable.

  This approach is known as risk analysis.

Standard risk analysis simulates the uncertainties in the values of economic and technical variables and assesses the effects of these uncertainties on calculated criteria. The simplest case of multivariable simulation is for uncorrelated variables in which the values of each variable are selected from their respective distributions independently of all other variables. In mining applications, however, many variables are highly correlated (e.g., grade and tonnage) and any realistic risk analysis must include inter-relationships between variables and/or serial correlations in the values of a variable (e.g. prices over time). In the general case a value of each relevant variable is generated from its associated distribution in such a way that any serial correlations and/or correlations between variables are reproduced in the values. An evaluation is then conducted on the basis of these values. This procedure is then repeated a set number of times (typically 50 to 100), each repetition yielding a value of one or more specified criteria. The simulation procedure is illustrated for financial risk analysis by the flowchart in Figure 2 in which values of six random variables are chosen at random from their respective distributions and an internal rate of return (IRR) and net present value (NPV) of the project are calculated for each combination of the values of the six variables. The process is repeated N times and N IRR and NPV are obtained.

These statistics provide measures of reliability, such as the probability of achieving a stated value, the range of possible values or the probability of not achieving set goals. In general, the outcome from a risk analysis is a set of simulated values that are best summarized in the form of a histogram. The mean, or expected value, μ, and the variance, σ^2, of the simulated financial results can be used to calculate μ/σ, known as the risk per unit of value or the risk factor. This risk factor is commonly multiplied by 100 and expressed as a percentage.

In principle, any technical, geological or financial parameter can be treated as a random variable for risk analysis: mill recovery, annual tonnage, grade; stripping ratio, metal or mineral recovered, price, gross revenue, operating costs, capital expenditure, cash flow, discount rate, reinvestment rate, pit slopes and groundwater inflow. Any number of these parameters can be selected as random variables for risk analysis, and different distributions for each variable can be specified for each year of operation. Relationships between variables can be imposed by correlation coefficients between all pairs of variables.

An example of the output from a risk simulation is shown in Figure 3. Risk analysis results in the form shown in the Figure allow more-specific questions to be answered. For example, "what is the probability that a net present value of at least £10.5 million will not be achieved by this project?" or "what is the most likely range of net present values for the project?".

There are, however, fundamental differences in the manner in which variability and uncertainty arise in technical and environmental variables that make a simple application of Monte Carlo simulation techniques inapplicable:

- Variability in the values of geological, geomechanical, petrophysical, environmental and other natural variables cannot be quantified by histograms and simple correlation coefficients; it is complex, spatial, rarely random, and is controlled and influenced by diverse factors many of which have complex inter-relationships
- Many variables are descriptive and non-quantifiable (e.g., the presence of particular rock types or geological units)

- Some events for which the risk assessment is made are so rare that it is not possible to establish a distribution model from a statistical analysis of historical data; in extreme cases the event may not yet have occurred

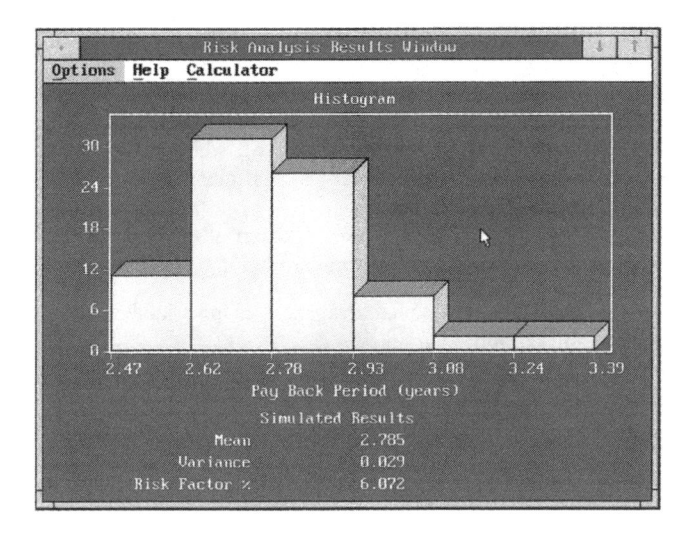

Figure 3. Example of risk analysis output

The first two of these problems relate to the spatial simulation of natural variables (states of nature). The third relates to an event that may occur if a certain combination of values of these variables is encountered or because of technical decisions made on the basis of inadequate knowledge of the states of nature. It is usually possible to establish the conditions under which the event will occur, and it may then be possible to simulate states of nature to assess the likelihood that such conditions will occur. For example, given a mining operation in specified geological conditions what is the likelihood that there will be an inflow of groundwater of such volume that a proposed or existing pumping system will be unable to cope? Or what is the probability that the volume of water to be discharged from the operation will exceed planning restrictions? By simulating geological structures and then simulating the progress of the workings, it is possible to assess the frequency with which points of inflow, such as fault zones, fractures or permeable beds, are encountered. The effects of encountering these points of inflow can be assessed by combining the simulated geological model with a groundwater flow model.

3.1 GEOSTATISTICAL SIMULATION

The values of geological, geotechnical and environmental variables are only known at sampled locations. Accurate field measurements of these variables at a single location is difficult and measurement at all locations is impossible because of cost and access constraints. Such variables lend themselves to a spatial statistical approach.

Spatial variability can be quantified in the form of variograms [6] or correlograms which describe the manner in which the values of a variable (e.g., grades, seam thickness, impurities, porosity, transmissivity, Rock Quality Designation) change or vary in the three-dimensional space of the mineral deposit. Variograms are essentially the spatial equivalent of the simple autocorrelation coefficient generalized to include correlations between the values of the variable over all distances and in all directions. This quantification includes all relevant causes of variability at all relevant scales. Spatial correlation between variables (e.g., porosity and solids, silver and lead grades) can be quantified in the form of cross-variograms [6], which are essentially the three-dimensional spatial equivalent of the correlation coefficient between values of pairs of variables at all distances and in all directions.

Geostatistical simulation [6,7,8] is a generalization of the concepts of Monte Carlo simulation to include three-dimensional spatial correlation and is one in which:

- At sampled locations the simulated values of each variable are the same as the measured values of those variables.
- All simulated values of a given variable have the same spatial relationships observed in the data values (spatial correlation).
- All simulated values of any pair of variables have the same spatial inter-relationships observed in the data values (spatial cross-correlation).
- The histograms of the simulated values of all variables are the same as those observed for the data.

The methods can be extended to most descriptive or qualitative variables simply by defining the variables in terms of presence/absence at sampled and simulation locations [9]. When natural, physical structures are a significant source of variability and/or exert a significant controlling influence on other variables (e.g., geological controls on mineralization, lithostratigraphic controls on porosity and permeability; rock types and rock properties may be significant factors in the physical distribution of grade), they, or at least their effects, must be included in the simulation. In some cases the modelling of categorical or descriptive variables may be an intermediate stage that provides a means of accurately modelling a quantitative variable (e.g., gold grades associated with quartz veins); in other cases they may be the object of simulation (e.g., flow zones for the prediction of groundwater flows).

Geostatistical simulation [9, 10, 11] is now widely used and accepted as a method of generating stochastic models of mineral deposits, hydrocarbon reservoirs and geological structures that can then be subjected to various operational procedures [12, 13]. Many applications are described in the literature using one or more of the available range of methods [7].

Figure 4 shows the simulation of quartz veins and gold grades on horizontal sections through a gold orebody. In this application the ability to discriminate between ore and waste depends on the ability to predict quartz vein locations. Images such as these could be taken as possible "realities", and the ability to predict reality could be assessed by sampling the image and attempting to predict and map quartz vein occurrence from the sample information. These ideas can readily be extended to other areas, such as the ability to predict geological hazards in longwall coal mining or the ability to predict groundwater flow on the basis of a given set of geological and hydraulic data. By repeating simulations and generating additional images risk can be assessed; for example, what is the minimum number of drillholes required in order to be 95% certain that all geological hazards will be predicted? For a given drilling grid what is the likelihood that they will not be detected?

Once a mineral deposit or geological region has been simulated, it can then be subjected to any number of simulated operational activities, and possible outcomes can be assessed. For example, a simulated hydrogeological image can be subjected to a groundwater flow modeller that, together with a model of an evolving mining operation, can be used to assess the likely flows of water over the life of the mine. A mining operation can be designed and scheduled on the basis of sampled values and the result compared with a design based on the entire image ("reality"). Fracture networks and faults can be generated to assess the probability of water flows and subsequent contamination. Rock properties can be simulated to provide models on which the behaviour of buried wastes over time can be assessed. Simulated models of porosity and permeability can be used to assess environmental risks associated with the tailings dams.

The steps in a geostatistical risk analysis for an open-pit gold mining project are summarized in Figure 5. Gold grades within the orebody are simulated on the smallest relevant scale, which, in this case, is 5m vertical drill core lengths on a 1m x 1m horizontal grid. Gold grades on all other scales are simulated by averaging the simulated core grades within specified volumes (e.g., 5m x 5m x 5m selective mining units). The in-situ reserves are then calculated on the basis of the ore that can be recovered within each planning block by selective mining (i.e., on the basis of the selective mining unit). Costs and the gold price are obtained from the corresponding distributions of values and are applied to the recoverable tonnages and grades within each planning block to produce a revenue block model. An optimal open-pit design algorithm (e.g., Lerchs-Grossmann [14]) is applied to the revenue block model to determine the optimal open pit. The blocks that constitute the optimal open pit are scheduled (e.g., by dynamic programming [15]) so as to yield the sequence of cut-off grades and production rates that maximises the net present value of the operation. A total of N revenue block models and corresponding open pits are generated for N different combinations of gold price and costs selected from the corresponding distributions of those variables. The whole process is repeated M times by generating M simulated orebodies. Each pit yields a minable, recoverable reserve, and each schedule yields an optimal net present value, internal rate of return, payback period and mine life, production rates and cut-off grades.

Geostatistically simulated orebodies allow an assessment of technical and operational risks such as slope stability and risk of failure in open pits; the ability to discriminate ore and waste by different blasting procedures and the associated ability to locate and follow trends in the mineralization, such as quartz veining and fracturing; the effects of different slopes and haul roads [16] on reserves and production schedules in open pits; the physical extent of an open pit and its effects on surface operations and layout; and the ability to predict geological hazards in underground mining.

198

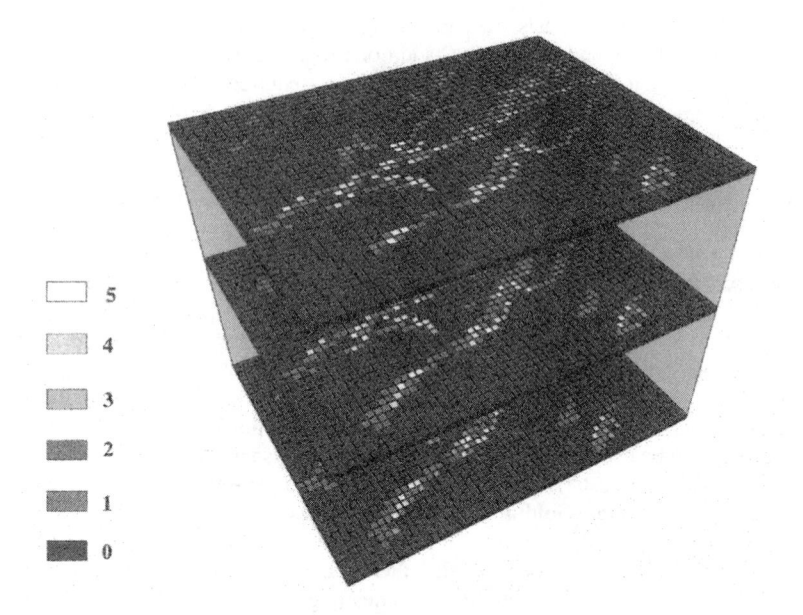

Figure 4. Simulated horizontal sections through a gold ore-body on which quartz veins and gold grades have been co-simulated.

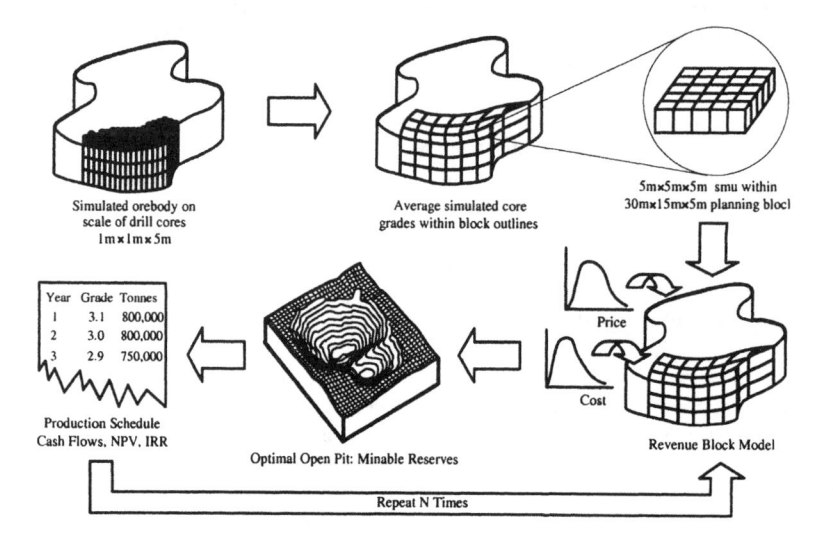

Figure 5. Summary of steps in a technical and financial risk analysis framework in an open-pit gold mining operation

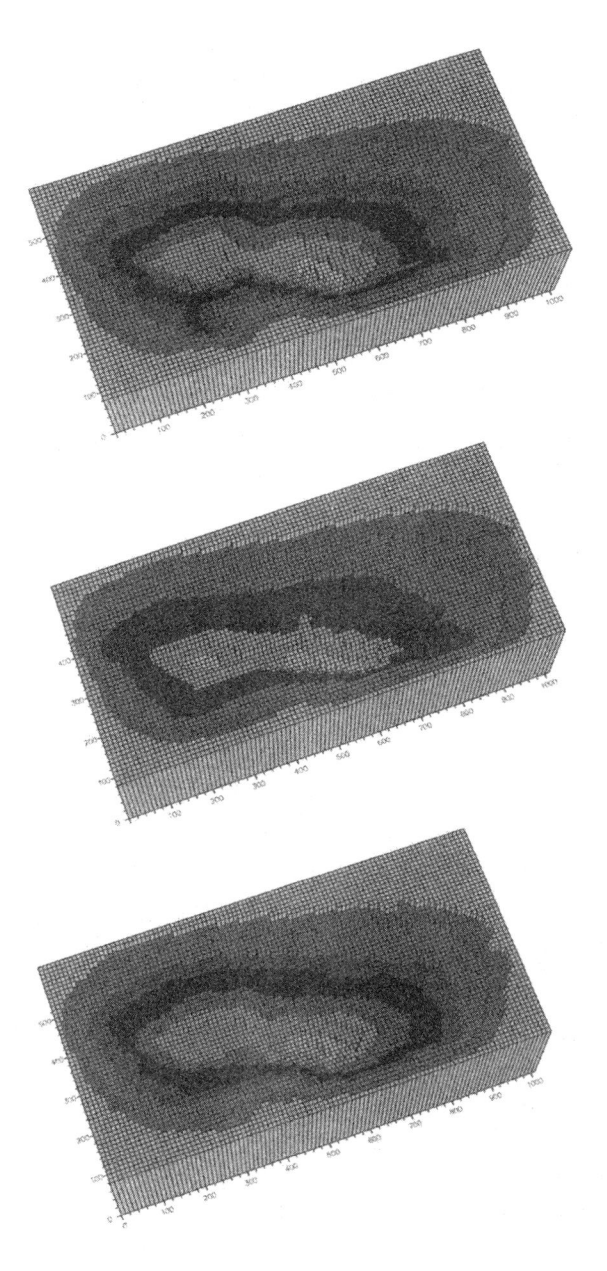

Figure 6. Three optimal open-pit designs each corresponding to a simulated orebody of the type shown in Figure 4.

All grades and tonnages used in feasibility studies and mine planning must be estimated from the available samples. In general, the denser the drilling grid, the less will be the uncertainty associated with estimated grades and tonnages and consequently the less the risk associated with feasibility and planning results. The risk associated with different drilling grids can be assessed by "drilling" a simulated orebody on a given grid and then using these simulated grade values to estimate block grades. The real, simulated block grades are, of course, known and can be compared with the estimated values. Measures, such as the variance of the difference between real and estimated block grades for different drilling grids, can be plotted against the cost of implementing the associated drilling grid; in this way the reduction in risk can be expressed as a function of achieving that reduction.

Any realistic risk analysis must be based on block tonnages and grades estimated from "samples" obtained by "drilling" the simulated orebody on the grid envisaged for mine planning and design. In this way the uncertainty attached to the use of relatively sparse information, as well as the cost of obtaining that information, can be included in the risk analysis. To base the risk analysis on the complete simulated orebody would be equivalent to planning the mining operation with perfect knowledge of the mineralisation.

As an example, an open pit could be designed for the simulated orebody (grades and geology) shown in Figure 4. Figure 6 shows three open pit designs each of which is optimal (maximum profit as determined by the Lerchs-Grossmann algorithm [14]) for a specific simulation of the orebody. Each of these pit designs effectively determines a minable reserve, and several simulations would allow a highly sophisticated assessment of the risk associated with the ore reserve and the consequent financial risk. By incorporating petrophysical variables, such as rock properties, in the simulations pit designs such as these could be used to assess other forms of technical risk.

4. Geographical Information Systems

The term Geographical Information System (GIS) is variously used to refer to any computer technology used to manipulate geographical data. One definition [17] states that a GIS is "a system for capturing, storing, checking, manipulating, analysing and displaying data which are spatially referenced to the Earth." In this lecture GIS is taken to mean a software package that incorporates one or more geographic databases and a facility for processing the information (usually in the form of maps) stored in the database(s). In addition, a GIS may include a facility for spatial analysis. The fundamental components of a GIS are shown in Figure 7.

The aim of posing and answering the generic questions is to allow the user to understand and detect what is happening now and to model what might happen in the future. The extension of the spatial analysis component to probabilistic spatial analysis provides a means of realistically assessing and incorporating the technical and environmental risks associated with minerals industry projects. It could also provide assistance in the management and monitoring of risk.

For minerals industry applications effective risk analysis requires the ability to integrate widely disparate types of data and to display them in a meaningful and realistic manner that allows all interactions to be assessed. GIS provide a means of data integration and interrogation together with a visualization capability that aids analysis and interpretation. GIS can be also be used for the assessment and management phases of risk analysis.

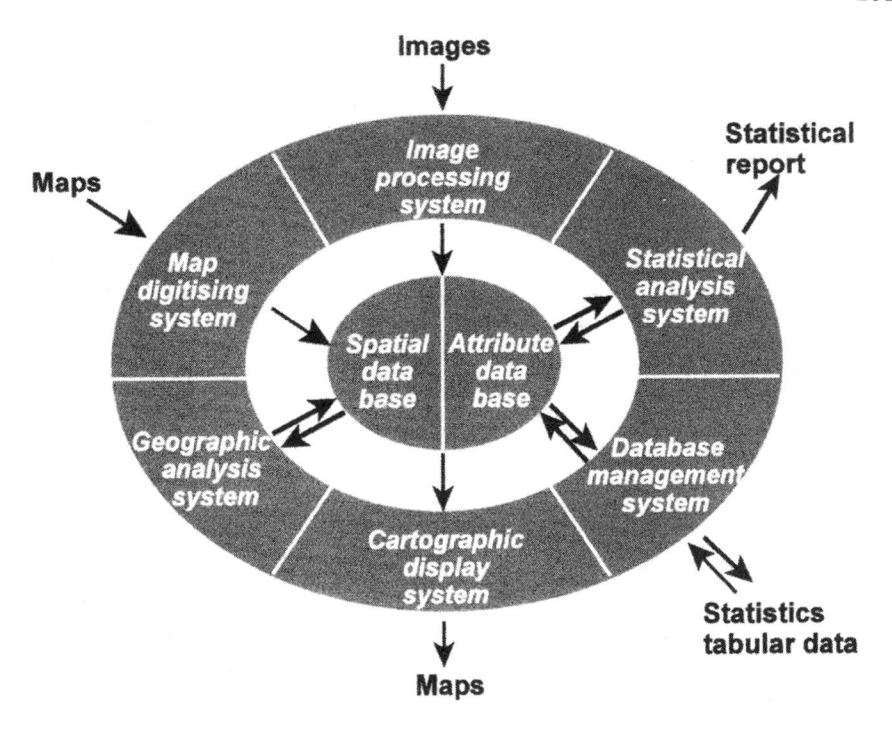

Figure 7. Fundamental components of a GIS (after Eastman [18])

Table 2 lists the generic questions, identified by Rhind [19], that a GIS should be able to answer.

Table 2 Generic questions that a GIS should be able to answer Source: Rhind [19]

Question	Task type
What is at ….?	Inventory, Monitoring, Locating
Where is the best route from … to …?	Route finding
Where is ….. true [or not true]?	Inventory, Monitoring
What has changed since ….?	Inventory, Monitoring
What spatial pattern(s) exist(s)?	Spatial analysis
What if …..?	Modelling

Figure 8 exemplifies how overlays in a GIS can be used in the planning of a minerals industry project, in this case a quarry. The pit shape overlays a geological interpretation map that, in turn, overlays a local land use map and a vegetation and habitat map. Many other maps could be added to the overlays, including a groundwater assessment for each stage of the pit development.

Quarry/pit design

Simulated geology

Land use map

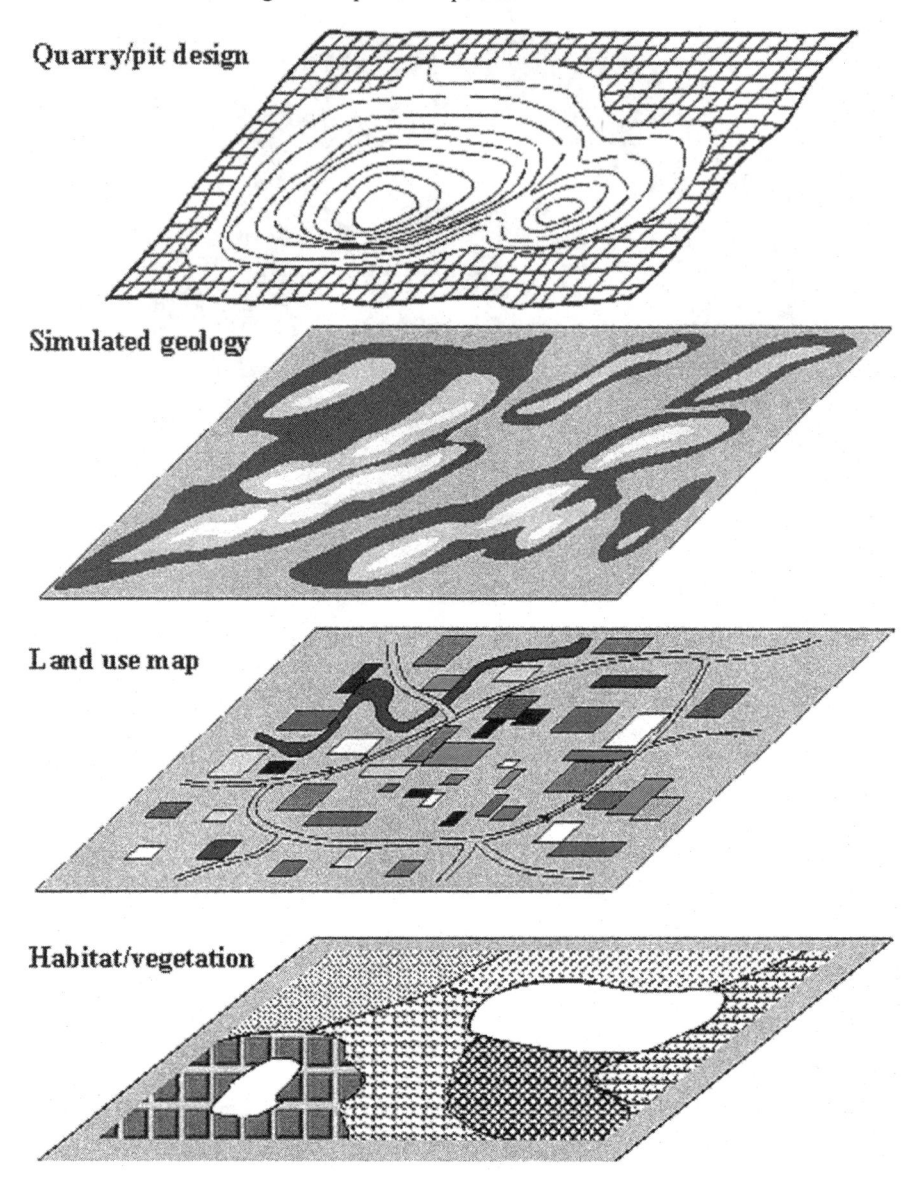

Habitat/vegetation

Figure 8. Schematic of map overlays in GIS display of surface mining project

Figure 9 shows how a sequence of overlays might be used to assess various technical and financial risks associated with an open pit project: slope stability, ore reserves and production schedules. Each of the overlay operations illustrated generates a new map that can be used to identify coincident features (e.g., all low slope areas in high grade zones) or to assess impacts (e.g., the additional dilution caused by low pit slopes in high-grade zones).

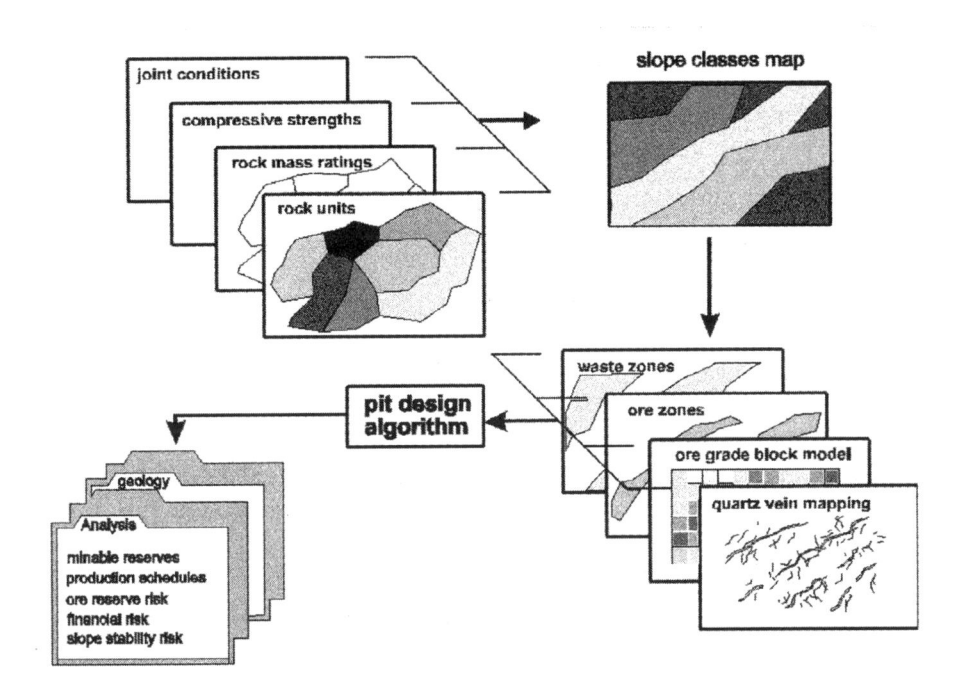

Figure 9. Use of GIS map overlays to assess technical and financial risks associated with an open-pit project.

4.1 GIS AND GEOSTATISTICAL SIMULATION

In mineral industry applications many of the maps used in GIS analyses are only estimates subject to considerable uncertainty. Prior to mining, the geological map is an estimate, or a possible image of reality, that is based on sparse sampling. Similarly the technical pit design is based on the uncertain geological map, sparse geotechnical data and forecast economic factors. The effect of the evolving pit on groundwater flow is also a prediction subject to uncertainty. Geostatistical simulation provides a means of generating alternative images of geological, geomechanical, petrophysical and flow structures. By incorporating these alternative images into the GIS presentation the possible range of effects on technical design, environmental impact and financial performance can be assessed. GIS combines a powerful, two- and three-dimensional visualization tool - for display of such items as terrains, groundwater flows, orebody structures and locations - with an interactive interrogation facility (cf. Table 2). Geostatistical simulation adds the spatial modelling and risk assessment components.

The combined approach provides an integrated risk monitoring and management system and a means of visualizing, and hence aiding the perception of, risk.

5. Knowledge Based Systems (KBS)

A Knowledge based system, or expert system, is an intelligent computer program that uses stored knowledge together with inference procedures to simulate experts' thinking to solve problems in specific domains. The essential components of a KBS are a knowledge base, an inference engine, a user interface and an external interface.

The objectives of using a KBS include substituting for unavailable human expertise, assimilating the knowledge and experience of several human experts, training new experts, providing requisite expertise on projects that do not attract, or retain, experts and providing expertise to projects that cannot afford human experts.

KBS have particular application in the environmental impact assessment of the exploitation of mineral resources and the consequent assessment of mitigation measures. Serious impacts not only cause damage to receivers but may also incur financial penalties on mine operators ranging from fines to mine closure. The recognition of the adverse socio-political and financial effects of pollution, together with the enforcement of measures for compliance with environmental legislation, has meant that Environmental Impact Assessments (EIA) have become an essential component of all mining projects. An EIA allows environmental problems to be identified, assessed and predicted and provides a means of deciding on any necessary mitigation measures in the event of serious adverse impacts.

Although a great deal of expertise has been accumulated over the past three decades there are still many problems in the EIA process which affect the quality of Environmental Impact Statements (EIS) and environmental risk assessments. The severity of these problems can sometimes lead to project failure and/or significant deterioration of the surrounding environment. The most significant of these problems are:

- inexperienced personnel involved in the EIA;
- poor quality EIS due to the inexperience;
- time required for an acceptable EIA;
- staff required for the EIA;
- cost.

In a recent survey of UK EIS, the Institute of Environmental Assessment [20] concluded that the quality of a significant proportion of EIS is poor. The main problems occur in the process of prediction and evaluation of an environmental impact. KBS provide a means of addressing these problems by compensating for the lack of available human expertise and accumulating expertise as it is acquired. When combined with simulation and GIS it becomes part of a powerful integrated tool for comprehensive risk analysis in the exploitation of mineral resources.

5.1 KNOWLEDGE BASED GEOGRAPHICAL INFORMATION SYSTEM

Such systems combine the capabilities of KBS and GIS and have enormous potential in the assessment of environmental risk and of the effectiveness of mitigation measures.

Conventional modelling methods are based on point processes and thus cannot deal adequately with environmental impacts from mineral projects that occur over relatively large areas often containing significant variations in elevation. This three-dimensional aspect of environmental impacts can be modelled and analysed by combining appropriate numerical models of pollutant generation and dispersal with the spatial analysis capabilities of GIS. One such approach [21, 22] is:

- Adapt appropriate numerical models to the generation and dispersal of minesite pollutants
- Use a GIS to combine the numerical models with a geographical model of the surface mine and surrounding areas
- Use the GIS to identify impacts on environmentally sensitive areas surrounding the minesite
- Use a KBS to assist in the impact analysis and mitigation of impacts.
- analyse the results of the environmental impact modelling, presented as geographic images, against various criteria

One example of KBS/GIS is the APPRAISAL-Environmental system [21, 22] which has three basic components:

- A mine model database that includes a relational and a spatial database
- A knowledge-based system (E2) for primary, qualitative environmental impact assessment
- A GIS for secondary, quantitative environmental impact modelling.

The three sub-systems are closely integrated and form a powerful environmental impact assessment system for surface mining and quarrying projects. The general architecture of the APPRAISAL-Environmental system is shown in Figure 10.

A mine model must be constructed before the environmental impact assessment can be conducted. This model contains all relevant geographic information about the proposed mine site, historical meteorological information in the area around the site and mine design information. The mine model is implemented in a relational database system and a GIS.

In the system shown in Figure 10 the user would normally begin by running the primary EIA system (E2) to do a qualitative assessment of the environmental impact of the mining project. The E2 system takes information from the user and the mine model database. If the outcome of the primary assessment session is serious, then the user can use the secondary EIA system (in this application using the OPENDUST module for assessing dust impact) to investigate the impact further in a quantitative way. The secondary EIA system uses information from the primary EIA session and the mine model database system. Finally, the system presents the qualitative assessment outcome in a comprehensive environmental impact statement. The quantitative modelling and analysis results are presented as a sequence of GIS images.

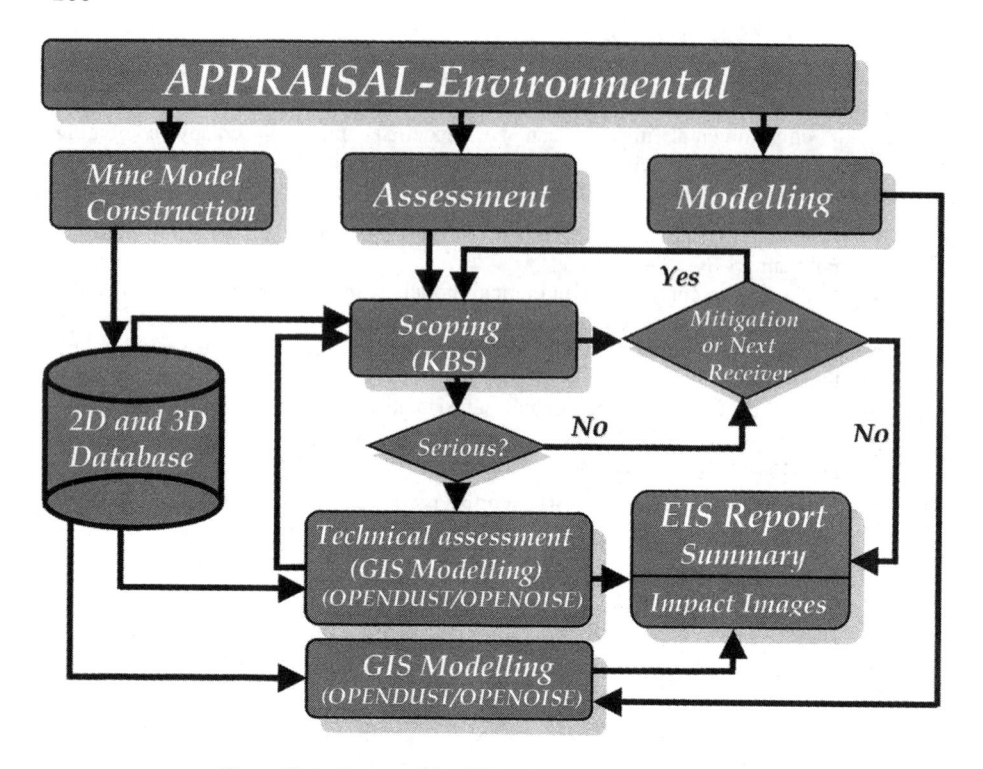

Figure 10. Architecture of the APPRAISAL-Environmental system

6. Numerical modelling

Numerical methods are a useful means of modelling the processes that generate specific states of nature. In minerals industry risk analysis the most appropriate applications are to the generation and dispersal of pollutants, (e.g., airborne dust, noise, vibration, contaminant trajectories in water and/or rock). Dust is used here for the sake of a specific example.

At surface mining operations dust measurements are normally the concentration and deposition at ground level. The concentration of the dust is a short-term indicator of the dust impact that varies with time, while dust deposition is a long-term or chronic process. Dust generation and dispersal is a function of topsoil characteristics, geological conditions, meteorological conditions, hydrogeology, mining activities, working cycles, vegetation and natural and artificial barriers.

The Industrial Source Complex (ISC) model [22,23] is a short-term, point-source dust dispersion model. This point source model can also be used for area or line dust source dispersion modelling by dividing the area or line source into a number of reasonably small segments that can be approximated by point sources. A raster-based GIS allows every dust source to be approximated as a point or a set of points that significantly simplifies dust dispersion modelling when there is a variety of types of dust sources as is usually the case at surface mine sites.

The hourly ground-level concentration at a downwind distance x and crosswind distance y is given by [22, 23]:

$$c(x,y) = \frac{KQ}{\pi \bar{u}(h) \sigma_y \sigma_x} e^{-\frac{1}{2}\left(\frac{y}{\sigma_y}\right)^2} \{Vertical\ term\}\ \{Decay\ term\} \quad (1)$$

where
- Q is the dust emission rate (g.s^{-1});
- K is a scaling coefficient to convert calculated concentrations into the desired units (default value of 1 x 106 for Q in g.s^{-1} and concentration in µg^{-3})
- σ_y & σ_x are the standard deviations of the lateral and vertical concentration distributions respectively (m)
- $\bar{u}(h)$ is the mean wind speed (m.s^{-1}) at source height h.

The vertical term is added to the equation to take account of the effects of source elevation, plume rise, limited mixing in the vertical plane, and the gravitational settling and dry deposition of larger dust particles (particles with diameters of greater than 20µm). The decay term is used to account for pollutant removal by physical or chemical processes. For the modelling of dust dispersion at surface mines, the decay term can be ignored because, normally, no measures are taken to remove the dust from the atmosphere. For large particulates the effects of gravitational settling and dry deposition are very important factors in model construction. The vertical term [22, 24] is given by:

$$VT = \frac{\Phi_n}{2} \left\{ \sum_{i=0}^{\infty} \left[\gamma_n^i e^{-\frac{1}{2}\left(\frac{2iH_m - H + (V_{sn} x/\bar{u}(h))}{\sigma_z}\right)^2} + \gamma_n^{i+1} e^{-\frac{1}{2}\left(\frac{2iH_m + H - (V_{sn} x/\bar{u}(h))}{\sigma_z}\right)^2} \right] + \sum_{i=1}^{\infty} \left[\gamma_n^i e^{-\frac{1}{2}\left(\frac{2iH_m + H - (V_{sn} x/\bar{u}(h))}{\sigma_z}\right)^2} + \gamma_n^{i+1} e^{-\frac{1}{2}\left(\frac{2iH_m - H + (V_{sn} x/\bar{u}(h))}{\sigma_z}\right)^2} \right] \right\} \quad (2)$$

where
- Φ_n is the mass fraction of dust particles in the nth settling velocity category,
- γ_n is the reflection coefficient of dust particles in the nth settling velocity category,
- V_{sn} is the settling velocity of dust particles in the nth settling velocity category.

7. Integrated systems

Simulation, KBS, GIS and numerical modelling can be combined to form an integrated system for risk assessment and analysis [21, 22]. An example of outputs from such a system is shown in Figures 11 and 12. Figure 11 is a GIS display of a proposed mine site together with colour-coded land use of the surrounding area. Figure 12 is a GIS display of likely dust generation and dispersion generated by numerical modelling and prediction of dust from planned sources and activities on the minesite. Further impact analyses and the assessment of mitigation measures can be conducted with the assistance of an integrated KBS.

The complete integration of geostatistical simulation within such systems is still under development but, when completed, will provide the basis for comprehensive spatial risk analysis and assessment. This approach will provide realistic and meaningful quantitative and qualitative solutions to risk assessment for the exploitation of mineral resources and for applications in related areas, such as the design, assessment and monitoring of underground waste disposal repositories and landfill sites.

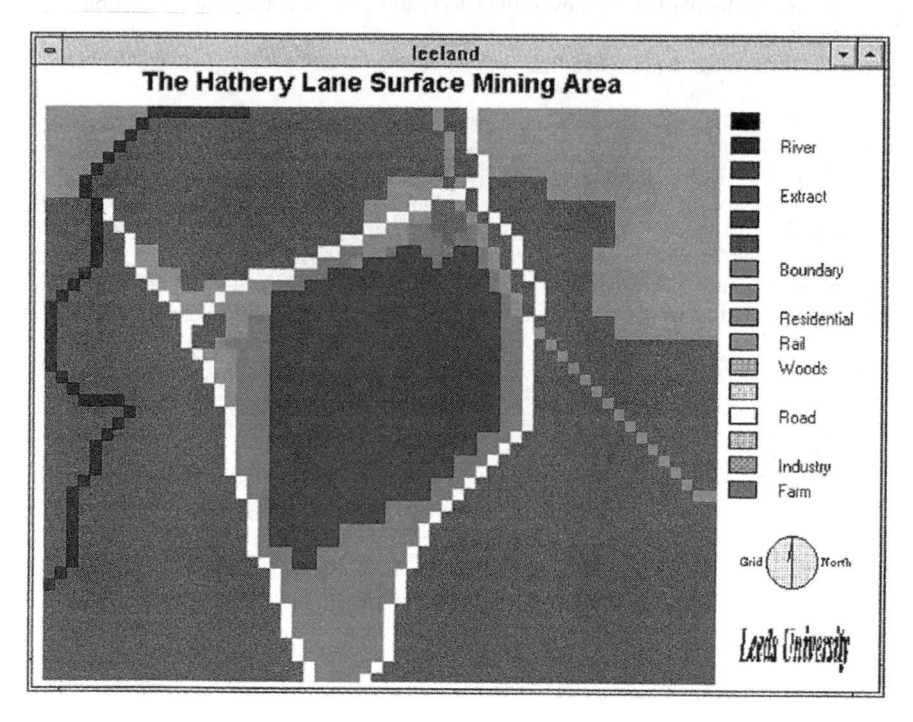

Figure 11. Proposed mine site

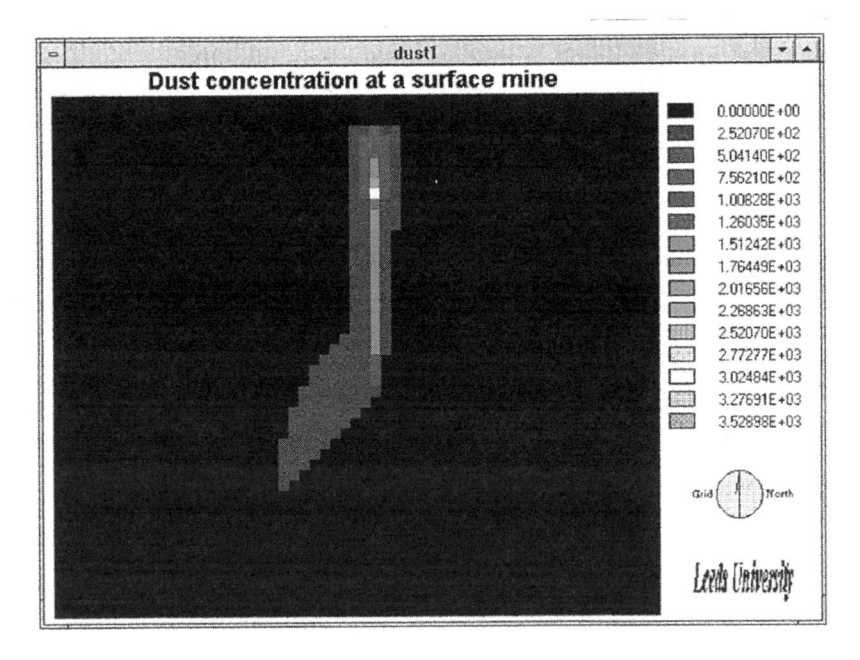

Figure 12. Dust dispersion image

8. Risk perception

Quantified risk may not be a sufficient, or even an acceptable, criterion for decision-making. In practice, many decisions are made on the basis of the perception of risk by project managers and the general public. Whilst probabilistic risk analysis reflects the real world it is only one way of thinking about that world. A probabilistic risk analysis may, rightly and realistically, indicate a negligible risk associated with a particular event, but this may be irreconcilable with the public or management perception of the safety or acceptability of the event. The chances of being knocked down by a car are probably much greater than the chances of environmental damage to a house in the vicinity of mine workings but the public perception is generally the reverse. The acceptability of risk is subjective and in many cases no amount of quantitative evidence to the contrary will alter the perception. Handler [2] provided the example of acceptability of risk in relation to nuclear power: "Are the estimated risks of nuclear power plants too great to be acceptable; are they more or less acceptable than those associated with coal combustion? Is a very small probability of a large catastrophe more or less acceptable than a much larger probability, indeed almost a guarantee of a small number of casualties annually?"

A large element of any planning enquiry related to mining applications is concerned with risk analysis (an examination and identification of the available choices and their consequences), risk assessment (evaluation and comparison of the risks arising from the various choices) and risk perception. The calculation of probabilities and the use of sophisticated methods, such as simulation and decision theory, may lead to appropriate inputs for a cost-benefit analysis of the mining project. People who live in the vicinity of the proposed project will conduct their own, often intuitive and subjective, cost-benefit analysis. Differences arise because both sides use different, and often irreconcilable, scales of values in their analyses. A board of directors may similarly be unconvinced by an "optimum" exploration or investment strategy yielded by a decision theory analysis; mining companies take particular views on all forms of risk and the perceptions of these risks are embodied in company philosophies, attitudes and policies.

Table 3, summarises the ways in which qualitative risk is perceived and the considerations that influence safety judgments (i.e., the extent to which risks are deemed to be acceptable). Although the table was originally designed for hazards it can be applied equally well to the perception of all the types of risk discussed in this paper.

9. Conclusions

It is unrealistic to expect that probabilistic risk analysis, no matter how sophisticated and accurate it may become, will ever be able to settle all debates about the acceptability of risk. The scientific assessment of risk, however, has an important role to play in quantifying risk and communicating expressions of risk in meaningful ways.

Table 3. The public perception of the qualitative risk (Source: Lowrance [1])

Risk assumed voluntarily	Risk borne involuntarily
Effect immediate	Effect delayed
No alternative available	Many alternatives available
Risk known with certainty	Risk not known
Exposure is essential	Exposure is a luxury
Encountered occupationally	Encountered non-occupationally
Common hazard	'Dread' hazard
Affects average people	Affects especially sensitive people
Will be used as intended	Likely to be misused
Consequences reversible	Consequences irreversible

The proper use of probabilistic risk analysis must recognise that risks have causes, and it is thus not sufficient simply to generate probabilities from a supposed model based on summaries of historical data. It is essential to understand the nature of the events that cause failure and to incorporate these events and their associated uncertainties into the risk models as drivers of the simulation procedures used for risk analysis. Geostatistical simulation provides a means of simulating physical structures and numerical variables. GIS provides a means of integrating raw data and the output from geostatistical simulations; it also provides a means of visualising and communicating the effects and nature of risk.

The methods outlined in this paper provide a sophisticated means of establishing expressions of risk and arriving at correct answers. Whilst they do not establish acceptability they can assist in reaching agreed conclusions about risk.

References

1. Lowrance, W.W. (1976) Of acceptable risk. William Kaufman, California.
2. Handler, P. (1979) Some comments on risk assessment. National Research Council Current Issues and Studies, Annual Report, National Academy of Sciences, Washington DC, USA.
3. Engineering Council Guidelines on Environmental Issues, September 1994, The Engineering Council, 10 Maltravers Street, London, WC2R 3ER
4. Leopold, L.B., Clarke, F.E., Hanshaw, B.B. and Balsley, J.R. (1971) A procedure for evaluating environmental impact. USA Department of the Interior Geological Survey Circular 645.
5. Dowd, P.A. (1994) Risk assessment in reserve estimation and open-pit planning. Trans. Instn. Min. Metall. (Sect. A: Min. industry), 103, A148-154.
6. Journel, A.G. and Huijbregts, C. (1978) Mining Geostatistics. Academic Press, New York.
7. Dowd, P.A. (1992) A review of recent developments in geostatistics. Computers and Geosciences, 17, (10), 1481-1500.
8. Journel, A.G. and Alabert, F. (1989) Non-gaussian data expansion in the earth sciences. Terra Nova, 1, 123-134.
9. Dowd, P.A. (1994) Geological controls in the geostatistical simulation of hydrocarbon reservoirs. The Arabian Journal for Science and Engineering, 19, (2B), 237-247.
10. Journel, A.G. and Alabert, F. (1990) New method for reservoir mapping. JPT February, 212-218.
11. Journel, A.G. and Isaaks, E.H. (1984) Conditional indicator simulation : application to a Saskatchewan uranium deposit. Mathematical Geology, 16, (7), 685-718.
12. Dowd, P.A. and David, M. (1976) Planning from estimates: sensitivity of mine production schedules to estimation methods. In Advanced Geostatistics in the Mining Industry NATO ASI Series C: Mathematical and Physical Sciences, vol 24. M. Guarascio, M. David, and C. Huijbregts, eds. (D. Reidel Pub. Co. Dordrecht), 163-183.
13. David, M. Dowd, P.A. and Korobov, S. (1974) Forecasting departure from planning in open pit design and grade control. In 12th Symposium on the application of computers and operations research in the mineral industries (APCOM), 2 . (Golden, Colo: Colorado School of Mines), F131-F142
14. Onur, A.H. and Dowd, P.A. (1993) Open pit optimization - Part 1: optimal open pit design. Trans. Instn. Min. Metall. (Sect. A: Mining Industry), 102, A95 - A104.
15. Dowd, P.A. (1976) Application of dynamic and stochastic programming to optimize cut-off grades and production rates. Trans. Instn. Min. Metall. (Section A: Mining industry) 85, A22-31.
16. Dowd, P.A. and Onur, A.H. (1993) Open pit optimization - Part 2: production scheduling and inclusion of roadways. Trans. Instn. Min. Metall. (Sect. A: Mining Industry), 102, A105-13.
17. Department of the Environment (1987). Handling geographic information. HMSO, London.
18. Eastman, J.R. (1992) IDRISI User's Guide. Clark University, Massachusetts, USA.
19. Rhind, D.W. (1990) Global databases and GIS. in M.J. Foster. and P.J. Shand, (1990) The Association for Geographic Information yearbook 1990. Taylor, Francis and Miles Arnold, London, 218-223.
20. Broadbent, C.P., Coppin, N.J. and Brown, C.W. (1996). Environmental impact assessments and the mining industry. Mineral Resources Engineering, 5 (3), 189-207.
21. Dowd, P.A. and Li, S. (1996) Development of a knowledge-based geographical information system for environmental impact assessment of surface mining projects. Proc. Minerals, Metals and the Environment II. Institution of Mining and Metallurgy, London. 65-83
22. Li, S. and Dowd, P.A. (1998) Integrated computer system for environmental impact assessment of surface mining and quarrying projects. Proc. 27th International Symposium on Computer Applications in the Minerals Industries (APCOM), Institution of Mining and Metallurgy, London. 125-136.
23. Cramer, H.E. (1986) Industrial Source Complex (ISC) Dispersion Model User's Guide, H E Cramer Co Inc., Salt Lake City, USA.
24. Dumbauld, R.K. and Bjorklund, J.R. (1975) NASA/MSFC Multilayer Diffusion Models and Computer Programs , Version 5, National Aeronautics and Space Administration, Washington, D.C., USA.

PART 4. RESOURCE POLICIES AND SUSTAINABLE DEVELOPMENT

For sustainable development, the availability of mineral resources and the environmental impact of their extraction are important considerations.

Mineral resources are non-renewable and consequently, depletion of most mined commodities is inevitable. Despite this fact, international regulations are lacking with respect to the rational use of mineral commodities with the result that many of these resources are being wasted. Furthermore, these resources are unevenly distributed worldwide. Sustainability is possible provided rational use is made of current mineral production and replacement of some commodities by other commodities. With proper resource management, the future impact of shortages can be lessened, thereby prolonging the lifetime of many commodities. Global resource assessments of the most important mineral commodities will serve to ensure an adequate mineral supply while retaining effective stewardship of resources in the future.

Numerous instances of poor mining practices in the past decade have greatly increased public awareness of environmental impacts of extraction of minerals and hydrocarbons. In many cases, this has led to a negative attitude towards mining, even though it is obvious that humankind will need minerals and hydrocarbons to maintain sustainable development for society as a whole. In response to the current situation, mining companies are adopting new techniques in environmental and resource planning. Scientists can contribute to these improved practices by focusing attention on the life cycle of resource extraction. The role of science is essential in all phases of mining, starting with deposit modelling, mineral exploration, resource extraction, waste disposal, and finally, remediation and restoration of the natural environment.

The paper by Shields proposes a collaborative resource management strategy based on a hierarchical model. Beinat's contribution introduces decision support methods for environmentally sound resource planning. In the short paper by Olivero, the implementation of the ISO 14001 industrial standard for a small industry is described. The paper by Ritsema provides examples of how information technology (IT) can contribute to the management of a mine asset-lifecycle. Finally, the paper by Feoli reviews the state-of-the-art in industrial ecology and bioremediation.

A HIERARCHICAL MODEL OF COLLABORATIVE RESOURCE MANAGEMENT

D.J. SHIELDS
U.S. Department of Agriculture, Forest Service
Rocky Mountain Research Station
240 W. Prospect St., Ft. Collins, CO, USA 80526

Abstract. This contribution describes a quantitative tool designed to support resource management decision-making. The Hierarchical Model of Resource Management is based on the assumption that the publics' objectives for resource management are a function of their values and further, that influence flows down, whereas information with respect to the state of the world and the consequences of fulfilling objectives flows back up. The purpose of the Hierarchical Model is to facilitate the public participation process. It can be quantified with either a measurement approach or a preference approach. The measurement model provides a means for linking values to behaviors and has been derived from the traditional values-attitudes-behaviors framework widely accepted in social psychology and consumer behavior literature. Data are collected via survey and analyzed using a variety of mathematical and statistical methods, including structural equations, clustering and descriptive statistics. The preference model provides a means for linking objectives for natural resource management to choices among potential implementation strategies. The approach is based on decision theoretic methods that are widely used to evaluate alternatives involving multiple objectives. This approach follows a path from strategic objectives to fundamental and means objectives, to preference functions, ending with preference ordering and choice. The measurement and preference paths run in parallel, but have significant differences that are explained in the text. Finally, the Hierarchical model is placed within a recursive decision framework that is based on the principles of adaptive management. The range of information provided by the two approaches for quantifying the Hierarchical Model provide information that will be essential if public preferences are to incorporated into resource management decision making.

1. Introduction

This contribution describes a model of resource management that explicitly incorporates the interests and desires of the public in the decision making process.

Since the 1970's, interest in environmental quality has grown steadily. In many countries, prior paternalistic approaches to resource management are gradually being replaced by more participatory processes in response to public demand and legal mandates. Resource management decisions that were once within the exclusive purview of one or a few individuals must now be made with substantial public input. Public interaction heretofore has taken many forms, including written comments, public hearings, meetings, workshops, open houses, interpretive programs, surveys, informal contacts, and collaborative processes. All of these are useful activities; however, they seldom lead to formal statements of objectives for natural resources. Nor do they facilitate direct participation in the decision process.

Effective public participation in resource management, whether direct or indirect, requires open communication and information sharing. Managers need to know what the public actually wants and the public needs to understand the current state of the world and the management alternatives open to them. The model presented here offers a relatively straight-forward approach for determining objectives for resource management. The objectives can be linked to measures of system condition, the status of which can be predicted under alternative management schemes. It is then possible to describe the potential consequences for socio-economic and biophysical systems of fulfilling any given set of objectives.

Most modern societies are diverse, with the result that there is seldom universal agreement as to the appropriate course of action. Choices need to be made. One approach is to seek consensus among all the participants in the decision process. The information that the model provides on public preferences can be used as a starting point for negotiations. It can also be used as input to formal choice models in game theoretic or axiomatic form. Finally, because natural resource management is of necessity an ongoing process, the model is placed within a recursive decision framework.

The conceptual resource management model is presented in Section 2 of this paper. Next, methods for quantifying the theoretical construct are described. Section 3.1 introduces a values-objectives-attitudes-behaviors model, describes the related survey instrument and recommends several methods for analyzing survey results. This is followed by a discussion of a quantitative method for linking objectives to indicators of system condition to behavioral choice in Section 3.2. In Section 4, the model is placed within the context of adaptive management. Conclusions are presented in Section 5.

2. The Hierarchical Model of Collaborative Resource Management

The design of the Hierarchical Model of Collaborative Resource Management (hereafter the HM) is based on the pragmatic view that resource management is, by definition, a human action motivated by human desires. The philosophical perspective is anthropogenic. All values, including those dealing with the environment, are human derived [1]; however, humans are capable of finding value in nature [2,3]. Ultimately, human values drive the environmental and resource goal setting processes that inform resource management decision making. This perspective allows for a comprehensive conception of human well-being, one that embraces what might otherwise be thought of as irrational, e.g., non-self motivated, behaviors and choices. The survival of species and the needs of future generations, in addition to a desire for jobs, income and commodities, become legitimate concerns, and in the process influence decisions and actions.

In the HM, decisions about how resources are managed follow as a consequence of societal values in general and those associated with nature in particular. Societal values, and related objectives for natural resources, translate into alternative management actions. It is possible to estimate the consequences of implementing an alternative by forecasting the status of measures or indicators of current and projected environmental, social and economic system condition. Moreover, since the measures are linked directly to the objectives, information is also available on the potential consequences of fulfilling specific objectives. Thus, the HM takes the form of a complex systems problem, i.e., one that is hierarchical and ordered and embodies control and information flows [4].

In this section, a brief introduction to hierarchy theory is followed by a description of the conceptual structure of the hierarchical model. The layers of the hierarchical model are then discussed in more detail.

2.1 HIERARCHY THEORY

Hierarchy theory states that all hierarchies, whether ecological or economic, are complex systems composed of partially ordered sets with an asymmetrical relationship among their vertical elements [5]. The relationship is asymmetrical in that direct control tends to move in only one direction, from top to bottom, while feedback tends to be in the form of information rather than control.

Any level in a hierarchy exerts some control over all lower levels with which it communicates [6]. However, communication is nearly always limited in vertical extent to one or two layers because signals are filtered in hierarchies [7]. Within level linkages in hierarchies tend to be stronger than those between levels, and higher levels behave at lower frequencies (longer cycle time) than do the levels below them. Hierarchies may be nested or non-nested. A nested hierarchy is one in which higher levels completely contain lower ones, e.g., ecosphere, ecosystem, organism, and cell [8]. Non-nested hierarchies are made up of ranked structures of distinct members. A hierarchy composed of both socio-economic and biotic components would be considered to be non-nested because its apical layer does not contain all attributes of lower layers, nor can it be derived from them. Nonetheless, higher levels in non-nested hierarchies still constrain the behavior of those below them.

Feedback in hierarchies can affect the status of higher levels, but the relationship is not like a closed-loop control system or a closed information feedback system [9]. Rather, feedback is often in the form of communication and is thus indeterminate, especially when it is complex or incessant [10]. Such relationships are more rigorously and appropriately described by the stochastic control principles of cybernetics [11]. Cybernetic feedback mechanisms have already been recognized as a beneficial component of environmental monitoring systems [12].

Regardless of its indefinite character, the importance of feedback in socio-economic systems is receiving increased recognition and scrutiny. Arrow *et al.* [13] argued that reforms are needed in the way signals are provided to resource users. According to the authors, feedback must clearly identify the risks and consequences of economic activity in a way that is easily observed, with low probability of misinterpretation. Risser [14] added that, to be effective, such signals (he was referring to measures of environmental condition) must also be compatible with the perceived needs of decision-makers. These views are consistent with the emphasis now being placed on development of indicators of sustainability [see e.g., 15,16].

2.2 THE HIERARCHICAL MODEL

Three decades ago Cooper [17] pointed out that decisions about natural resource management take place at a level of integration higher than that of the ecosystem itself.

"Systems, social and biological, exist at many levels of integration. The social and economic system within which resource management decisions are made represents a higher level of integration than that of the ecosystem, narrowly defined.... The usefulness of the level of integration concept is that the goal of any given level of integration is manifested at the level above, while its mechanism is derived from the next lower level... Thus the goal of a resource management system is generation of maximum sustained outputs of human satisfaction over time, a goal that has meaning only when expressed at the level of integration represented by the human social system."

The HM is based on Cooper's principle that resource management goals make sense only within the context of the human social system. Consistent with that view, basic human values (sometimes called held values) are placed over the cultural, institutional, and economic framework within which societal goals and objectives are communicated, over assigned value measures, and over the actions that can impact the bio-physical system (Figure 1). Typical of hierarchies, control flows from top to bottom; information flows in the opposite direction. Basic held values are assumed to determine the social objectives that drive land management decisions; implementation of those decisions leads to biophysical and social impacts; and, in a fully functioning application of this model, information about those impacts would be passed back up through the levels of the hierarchy. The model's layers, as well as current and potential connections between them, are discussed below.

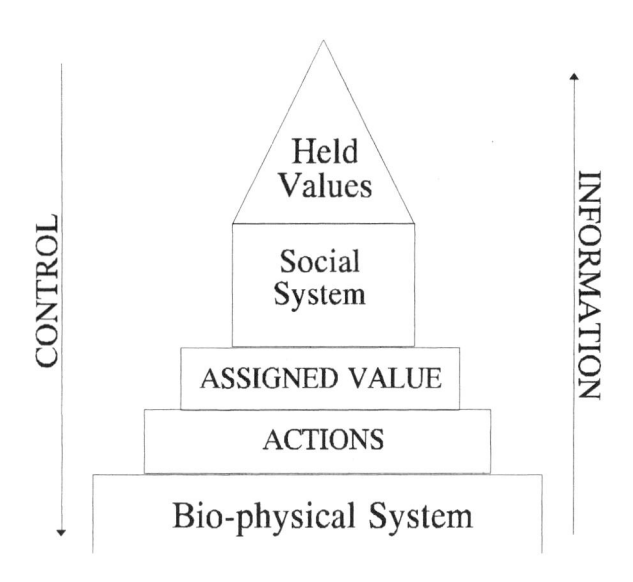

Figure 1. The Hierarchical Model of Collaborative Resource Management

2.2.1. *Values*

Western society has struggled with the concept of value at least since the time of Plato and Aristotle. More recently, psychologists, sociologists, and economists have joined philosophers in proffering definitions of value. Rokeach [18], for example, explained it as "an enduring belief [concerning] desirable modes of conduct or desirable end states of existence." Brown [19] stated that value is "an enduring conception of the preferable which influences choice and action."

Both Rokeach and Brown were referring to held, or fundamental, values. These comprise highest-order qualities that motivate all ensuing lower-order preferences and, consequently, decisions and actions. The set of possible held values is finite and includes such welfare indicators as generosity, responsibility, fairness, freedom, beauty, and friendship [19].

Each person has their own set of held values; the sets are not necessarily identical, although they may overlap. Individuals tend to have relatively few held values (perhaps 10 or 15 at most) and also tend to order them, i.e., certain values are given precedence and emphasis over others [20]. It is possible to identify held values through psychometric tests or carefully worded survey instruments [21].

Held values are extremely resistant to change without immense outside perturbation. Nonetheless, such large-scale disturbances have been known to happen. New values may be acquired or old ones abandoned, as in the case of religious conversion. In less extreme situations, existing values may be given increased or decreased emphasis, to be shifted to a higher or lower place in the overall value hierarchy [22].

There are numerous ways to classify values, including subscribership, objects at issue, nature of benefit, etc., (Rescher 1969). Subscribership relates to who holds the value, not what the value is per se. In the case of resource management, a relevant question might be "What value set does a specific user group espouse?". Objects at issue classifies values according to the objects to which they apply, i.e., values (the value assigned to some tangible object such as a car), environmental values, personal values, group values, and societal values. Clearly environmental values will be important for resource management, but other categories such as personal values may be influential as well.

Nature of benefit refers to the qualitative outcome or state of the world that will obtain if a value is realized. The categories include material and physical, economic, moral, social, aesthetic, etc., and relate to human wants and needs. Conflicts occur, in part, because groups or individuals espouse values that lead to different, and sometimes mutually exclusive, benefits.

In addition to the foregoing, values can be differentiated with respect to their level of application, i.e., to individuals, groups, or society as a whole. Because specific values may be distributed more or less widely among the population, the value sets of individuals, and groups differ. Values tend to be fairly consistent within groups, but differ to some degree among groups. For example, environmentalists have historically fallen into two general categories defined by different value sets, one is conservation oriented and the other preservation oriented.

2.2.2. Social System

Social values are derived from held values; together they provide the ethical foundation for social institutions and the rationale for governmental, corporate, and individual decisions. A social system and institutions may be exclusively reflective of the values held by groups dominant in society or may reflect accommodations to groups with differing or conflicting value sets. In either case, as Wilson [23, p.32] pointed out,

"At the supra-individual level, institutions espouse specific established preferences in accord with which men are motivated, relationships are established, or organizations function. Some of these values are clearly articulated in creeds, moral codes, legal systems, charters, declarations, treaties, and the like; other remain inchoate, faintly apprehended, only rarely or partially articulated."

Societies hold social values with regard to equity before the law, economic justice, and the environment, plus other public values less relevant in a resource context. Environmental values are the source of the strategic objectives people hold for natural resources and so have been of particular interest to those concerned with resource management decision making [24,25,26]. Whether stated or unstated, strategic objectives guide decision making because, as Keeney [27] wrote, "Quite simply, deciding what is important requires [held] value judgements."

The social system and its related institutions evolve out of held values and are ranked below them in the HM. By definition the system will change more frequently than the value set. The same is true for the objectives espoused by society; they are more likely to change or be reprioritized than either values or social constructs. The availability of new information has the potential to focus attention on an objective and result either in its being accorded increased importance or its reformulation to reflect the enhanced level of understanding.

The HM is designed to facilitate the sharing of information on the current state of the world and the expected state of the world, given the implementation of management strategies designed to fulfill a set of objectives. As will be discussed in more detail in Section 3.2, this is done by linking objectives directly to measures of social system and ecosystem condition. Such information can assist groups and individuals in determining whether the predicted or actual impacts associated with fulfilling their objectives are acceptable.

2.2.3. Assigned Values

The public policy decision process is one in which held values, beliefs and knowledge inform a cognitive valuation process that results in the assigning of relative value to alternative policies or states of the world [28]. In the HM, the primary connections from society to the biophysical system are through this intervening level of assigned values. Decision makers chose from among alternative actions based on a comparison of their relative values.

In western societies, value is often expressed in terms of a money metric; the local currency acts as a proxy for how strongly some object or state of the world is desired. In the case of resource management, decisions are often based on a comparison of expected net present value (the sum of all costs and benefits in monetary terms, brought to a single point in time). There are problems with using a purely economic measure of relative valuation. Prices are determined in markets, but if the market is not competitive the observed price may not truly represent the public's preferences. The process of discounting, i.e., expressing costs and benefits in terms of today's currency, simultaneously discounts the interests of future generations and thus has serious intergenerational equity implications. Perhaps most importantly, not every value can or should be monetized.

Decision makers have historically relied too heavily upon monetized values as a way to compare resource management alternatives. These decision processes can benefit from the addition of information on how implementation of various management alternatives would affect different group's, or individual's, perceptions of their own well-being. Assuming rationality, individuals make choices so as to maximize their well-being. Those choices are based on a comparison of alternatives. Preferences, which are derived from objectives, drive the comparison process. There is general agreement that people's objectives and preferences for specific resource management schemes are a function of their value sets [29]. This is simply the downward flow of the HM; fundamental or held values lead to social and personal objectives, which lead to assigned value expressed as preferences, which lead to actions and ultimately impacts on social and biophysical systems.

2.2.4. Actions, Potential Impacts, and Information Flow

Resource management involves making decisions, such as whether to grant permits for a new mine, or cut an area of forest, or reduce damage to hiking trails. The issues in question are rarely clear cut and simplistic. Frequently, each situation could be handled a number of different ways. In the case of the new mine, there could be decisions related to engineering design, size, and reclamation methods. In the case of timber, a choice must be made regarding logging method; for example, clear cut versus shelter wood. Also recreation impacts can be ameliorated in different ways; e.g., by limiting the number of hikers allowed on the trail at any given time, or by building trails that can withstand more intensive use. Each alternative would result in a set of impacts on the social, economic and environmental systems in question.

The HM facilitates the flow of information regarding the effects of alternative choices to decision makers, be they land managers or members of the general public. This information is presented in two forms. First, the current state of the world is measured and reported as baseline data. Second, the impacts on socio-economic and biophysical systems can be predicted for each proposed management action and changes in system condition calculated. Comparisons of the range and magnitude of impacts across alternatives can then be made.

This type of comparison is useful because it helps both governments and members of the public understand the short- and long-term consequences of implementing various management alternatives. Actual and predicted impacts may be different from or more intense than what was expected. In some instances, understanding the broader implications of decisions may lead to a change in the value assigned to a specific activity or situation. This in turn could lead to a reprioritization of objectives, which might imply that a different choice would be made. The ability to facilitate learning and adaptation through the upward flow of relevant information is one of the most important features of the HM.

3. Quantifying the Hierarchical Model

The hierarchical model is a theoretical construct that hypothesizes a relationship among values, social institutions, assigned value, personal or agency actions, and their associated environmental impacts. Although the construct is realistic, it cannot be applied directly to resource management problems in the absence of practical quantitative tools. To bridge the gap between theory and the reality of management decision making, two alternative implementation approaches have been developed. The measurement model investigates the pathway from values to actions and behaviors. The preference model uses a decision theoretic approach to link objectives to assigned values (preferences) and thus to choices among alternative actions. The relationship between the theoretical construct, and the measurement and preference models is shown in Figure 2. Both quantitative models are described in more detail below.

3.1 THE MEASUREMENT MODEL

The measurement model provides a means for linking values to behaviors and has been derived from the traditional values-attitudes-behaviors framework widely accepted in social psychology and consumer behavior literature [see e.g., 30,31]. Those models omit objectives; however, most behavior is, in reality, goal directed [32,33]. Thus, the framework has been revised to incorporate objectives, which have been demonstrated to mediate the relationship between values and attitudes [34]. This leads to a values-objectives-attitudes-behaviors construct.

The measurement model mirrors the hierarchical model. As is the case in the theoretical model, the uppermost level is values. These are assumed to provide the motivation for objectives, attitude formation, and behavior. However, the focus here is on values with respect to the environment, rather than the full suite of values that could potentially be held by an individual. Objectives reside at the second level of the measurement model. Objectives reflect and are tempered by the existing social, institutional, cultural milieu within which an individual lives, in addition to being derived from the individual's value set. This model does not specifically incorporate a variable for assigned valuation, but rather its precursor, attitudes. And finally, actions are represented by behaviors, which can take the form of personal behaviors or, at the agency level, management decisions. The actions represent tangible evidence of attitude.

Theoretical Model	Measurement Model	Preference Model
Held Values	Values	Strategic Objectives
↓	↓	↓
Social System	Objectives	Fundament and Means Objectives
↓	↓	↓
	Attitudes	
↓		
Assigned Value		Preference functions
		↑ ← Alternative Actions
↓	↓	↓
Actions	Behaviors	Preference Ordering and Choice

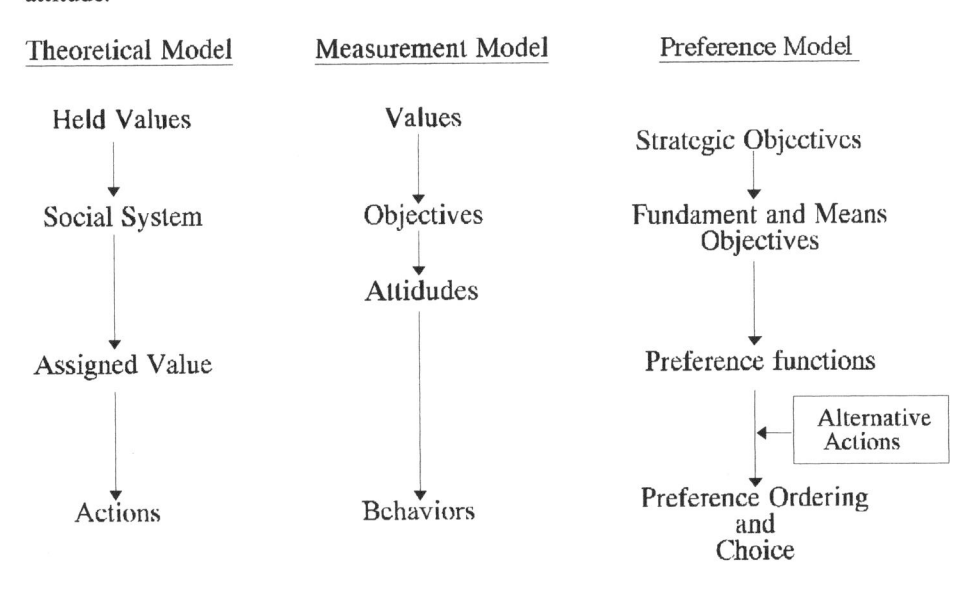

Figure 2. Quantifying the Hierarchical Model of Collaborative Resource Management

A survey instrument has been developed to elicit information from individuals about their values, objectives, attitudes, and behaviors with respect to natural resources. Survey results can be explored using descriptive statistics, cluster analysis and structural equations models, each of which is briefly considered in Section 3.1.5. Prior to that the four parts of the survey instrument are introduced.

3.1.1. *Values*
Values have long been recognized as a means to better understand society and culture based on the premise that values play an important role in human behavior. Values

tend to be context specific. Thus, a value measurement scale is needed that has been designed to be relevant in the natural resource management context. The Consumption of Public Lands value scale, developed by Martin *et al.* [35], represents a value system that encompasses two underlying dimensions concerning the environmental and resource related social and economic issues. The first value dimension is socially responsible individual values, which encompasses values related to individual actions. The second value dimension is socially responsible management values. Here the focus is on those values that would inform the actions of land managers. This section of the survey comprises 25 Likert scale questions (1=strongly disagree – to - 7=strongly agree) divided between the two value dimensions.

3.1.2. *Objectives*
Prevailing community objectives generally make up basic conditions for achieving legitimate land use policy [36]. Consistent with the view that stakeholder involvement is essential to effective natural resource management, both the measurement and preference models are designed to incorporate stakeholder group or individual objectives. By comparing objectives (i.e., across individuals, firms, agencies, etc.), it is possible to identify those which are held in common and those which are unique to special interests. Often significant overlap among objectives sets exists, which presents opportunities for consensus building among stakeholders.

Information on objectives can be organized in the form of an objectives hierarchy (Figure 3), which is a tree-like representation of an individual's or group's objectives [36,37]. An objective is a statement of what one desires to achieve and is characterized by having a context (in this instance, natural resources), an object (a action alternative) and a direction of preference [27].

```
┌─────────────────────┐
│ Strategic Objectives│
└─────────────────────┘
  Goals Derived from Held Values
              ↓
┌─────────────────────┐
│Fundamental Objectives│
└─────────────────────┘
     Situation Specific Goals
              ↓
┌─────────────────────┐
│  Means Objectives   │
└─────────────────────┘
  Means to attain fund. obj's linked to
      measures of attainment (x_j)
              ↓
┌─────────────────────┐
│Preference Functions (for i)│
└─────────────────────┘
     Defined over measures (x_j)
┌──────────┐                 │
│ Action   │                 │
│Alternatives├──────────→    │
│   (k)    │                 │
└──────────┘                 ↓
┌─────────────────────┐
│ Preference Ordering │
└─────────────────────┘
   (of k by i as function of x_j)
```

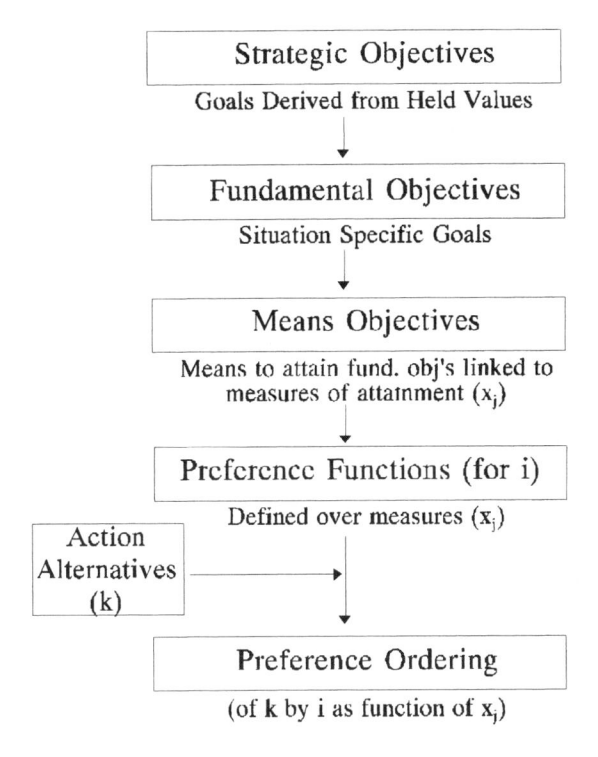

Figure 3. An Objectives Hierarchy with Preference Functions and Ordering

Strategic objectives reside at the highest level of the hierarchy. These are value-based objectives and are general in nature. Next are the main branches of the objectives hierarchy, which describe fundamental objectives. These are situation specific applications of the strategic objectives. In the process of building the hierarchy, the broad, fundamental objectives are clarified through the addition of ever more specific sub-objectives. These so-called means objectives are intended as measures of the degree to which the fundamental objectives have been achieved[1]. Objectives hierarchies have been built for many public policy applications [39,40,41].

As noted above, objectives hierarchies are situation specific. For this reason, the same objectives questions cannot be used in every applcation of the measurement model. Rather the questions are tailored to whatever problem or issue is being investigated. The tailoring is accomplished by first defining a decision context and then eliciting objectives hierarchies from a range of interested groups or individuals. These hierarchies are then merged to create a comprehensive hierarchy that includes everyone's objectives. The means-level objectives are then rephrased as declarative statements and included in the survey instrument as rating scale type questions (1=not at all important - to - 7=very important.)

---

[1] The balance of Figure 3 will be discussed in Section 3.2.

3.1.3. *Attitudes*

An attitude is an ".. organization of beliefs around an object or situation that predisposes one to respond in a a given manner" [42]. The focus in the measurement model will be on an individuals attitudes about certain behaviors related to natural resources. These could be individual behaviors, such as recreational activities like fishing or mountain biking, or the management behaviors, such as land set reservation for wilderness or implementation of environmental regulations. Attitudes are influenced both by the individual's underlying values and by their objectives. Objectives can be thought of as outcomes toward which people hold positive attitudes.

Attitudes can be organized into a hierarchy, in which each descending level represents a decreasing degree of abstraction, similar to the manner in which objectives are organized. The level of specificity for attitude scales in the survey should be equivalent to that for objectives and behaviors questions [43,44]. The attitude scales are presented in the survey instrument as statements requiring a Likert rating scale type response (1=strongly disagree – to – 7=strongly agree). In addition, composite scales listing alternative individual activities or management actions (depending upon the application involved) can be added. Respondents are asked to rate each activity (or action) on a scale of 1 to 7 in terms of suitability (1=not at all suitable – to – 7=highly suitable) and potential for causing damage (1=no damage – to – 7=severe damage).

3.1.4. *Behaviors*

The final section of a measures model survey addresses actual behaviors associated with either public usage of natural resources or alternative management actions that could be implemented. If the survey concerns individual usage, then questions could address how often resources are used (1=never – to – 7=very frequently), in what capacity they are used (e.g., 1=private use, 2=as licensed guide or permittee, etc), and in what social situations (1=alone, 2=with family,3=with a small group of friends, etc). If the survey addresses management actions, then questions might ask respondents to choose between alternative actions, given specific circumstances. According to the theory of reasoned actions, attitudes will predict compatible behaviors only under certain conditions [45]. First, a highly reliable index of behaviors must be used. Second, attitudes and behaviors must be specified at the same level of generality.

3.1.5. *Analysis of Survey Results*

The results of a measurement model survey are in the form of an $n \times m$ matrix of numeric, rating scale responses to the values, objectives, attitudes, and behaviors questions (responses to n questions by m respondants), plus categorical responses to the demographic questions. A survey response provides information about the relative importance of various values, i.e., whether the value embodied in the value statement is not at all important (Likert rating of 1) or very important to that person or group (rating of 7). Similarly, most individuals have multiple objectives for natural resource management, a few of which they deem vital (rating of 7), others merely desirable (rating of 5), and some unimportant (rating of 2 or less). Analogous logic follows for attitudes and behaviors questions.

Responses can also be analyzed individually or in sets. The survey-response data for an individual or group provides a fairly complete description of their views with respect to natural resources. Sets of survey results can be analyzed using a variety of statistical and quantitative methods. Only three will be considered here: descriptive statistics, cluster analysis and structural equations modeling.

Descriptive Statistics. The purpose of descriptive statistics is to summarize and describe a data set[2]. If survey results are available for the entire population, summary measures are termed parameters. If, as is much more likely, the survey was completed by a subset of the total population, i.e., if we are dealing with a sample, then the summary measures are termed statistics.

The two most common summary statistics are measures of central tendency and measures of dispersion. Measures of central tendency include the arithmetic mean, the median and the mode. The arithmetic mean of the data is simply the sum of the numbers in the data set divided by the number of numbers. In the case of survey responses, the mean for any specific question represents the "average response" over all people participating in the survey. The median is the middle value in the data set; half the responses are above (or equal to) the median, and half are below (or equal). The mode is the most frequently observed value of the measurements in the data set.

The mean is frequently used because it is easy to calculate and has convenient mathematical properties. However, the median may be preferable to the mean, if the data set includes a few extreme observations that would strongly influence the value of the mean. The mode is particularly useful in situations where it differs from the mean, e.g., when one response value occurs often enough to be noteworthy, but the mean does not indicate its importance.

Measures of dispersion describe the variability of the data. The most significant measures of average deviation are variance and standard deviation. The variance is the arithmetic mean of the squared deviations of the measurements from their mean. The standard deviation is the positive square root of the variance. The more dispersion there is in a set of responses, the larger standard deviation will be. Conversely, if every response to a specific survey question equals the mean, the standard deviation will be zero. If the variance around the mean is large, then the mean would be a misleading characterization of the full data set (because many of the responses could fall far above and below the mean.)

Cluster Analysis. Cluster analysis is a method for partitioning data sets (or points), or objects, into groups based on the similarity of their characteristics[3]. Profiles of those data grouped in the same cluster should be similar to each other, and should be distinctly different from the profiles of data sets grouped into other clusters. Cluster analysis is a three stage process: partitioning, interpretation, and profiling. Partitioning involves developing the clusters themselves. Interpretation is a process of understanding the nature of the cluster and its characteristics. Profiling means describing the cluster and identifying how it differs from other clusters.

---

[2] See [46, 47] for an introduction to descriptive statistical techniques and calculation procedures.
[3] See [48,49,50] for an introduction to cluster analysis and calculation procedures.

Clusters exhibit the properties of internal cohesion and external isolation, i.e., data sets or objects within a cluster are close to each other and distant from other data sets or objects not in the cluster. As a result, similarity measures (either proximity or distance) are typically used to define clusters. The most commonly used measure of similarity is the Euclidean distance. The simple Euclidean distance between two points with coordinates $[X_1, Y_1]$ and $[X_2, Y_2]$ respectively, is the length of the hypotenuse of the right triangle formed with the third point $[X_2, Y_1]$ (or $[X_1, Y_2]$ for that matter.) There are numerous other, more complex, distance measures. Similarly, there are a multitude of procedures for determining clusters, each of which attempts to maximize the differences between clusters relative to the variation within clusters. Some methods group all observations into a single cluster and then start splitting off dissimilar observations. Others start with n clusters (where n = the number of observations) and then start grouping similar observations together. The most appropriate method will depend upon the form and type of data being clustered.

When applied to measurement model survey results, the goal of clustering is to identify and group together those responses that are most similar to each other. Although an individual's responses will reflect his or her life experiences, background, and social circumstance, it is reasonable to expect some degree of consistency among the responses for selected subsets of the population. Responses can be clustered on the full survey data set, or on each separate survey section (values, objectives, attitudes, and behaviors). In the latter case, cluster membership can be tracked to determine whether individuals holding similar values espouse similar objectives and attitudes, and partake in similar activities.

There is a great deal of art involved in the science of cluster analysis. Assuming the methods have been applied correctly, clusters are not "right or wrong, true or false," but rather meaningful or not meaningful. This is why the second phase of the process, interpretation, is so vital. The analyst must examine the cluster, its members and their similarities and dissimilarities, and tease out logical reasons for why these observations should be grouped together, and what it means when they are. Once that has been done, characterization is a straight forward application of descriptive statistics.

Structural Equations Modelling. Structural equations modelling is used to assess theoretical models by analyzing the order and strength of the relationships among a set of variables[4]. It is by far the most complex of the three analytical methods presented here and a comprehensive description is beyond the scope of this contribution. Nonetheless, it is an extremely useful approach for analyzing survey results and so worthy of a brief introduction. Structural equations modelling is normally undertaken as a two step process, factor analysis followed by confirmatory structural modelling. The factor analysis specifies the relationships of the observed measures to their underlying constructs, with the constructs allowed to intercorrelate freely. The structural model then specifies and quantifies the causal relations among the constructs. Model output is in the form of standardized path coefficients and related goodness of fit measures.

---

[4] See [51] for an introduction to factor analysis, and [52,53] for an introduction to structural equations modelling and calculation techniques.

In the case of HM survey results, maximum likelihood factor analysis with Varimax rotation is used to identify the underlying latent factors associated with values, objectives, attitudes and behaviors. The structural model is then built using the factors associated with the four measurement model sections. This type of analysis is useful for cluster comparison because both the factors themselves and the relationships among them will differ depending upon the set of responses used in the analysis.

3.2. THE PREFERENCE MODEL

The preference model provides a means for linking objectives for natural resource management to choices among potential implementation strategies. The approach is based on decision theoretic methods that are widely used to evaluate alternatives involving multiple objectives [see e.g., 37,38]. As is shown in Figure 2, this approach follows a path from strategic objectives to fundamental and means objectives, to preference functions, ending with preference ordering and choice.

This path parallels the measurement model, but differs from it in a number of significant ways. First, values are not explicitly addressed, but rather are assumed to be the basis for strategic objectives. Second, objectives are quantified in preference functions. Preference refers to the act of preferring, i.e., valuing or holding some action, thing, or state of the world, in high regard or esteem. The preference function, then, is a situation specific application of values. Its purpose is to quantify an individual's or group's objectives with regard to the action, thing or state in terms of a relevant set of indices or measures of social, economic and ecosystem condition. Third, given that alternative actions will generate different indicator or measure levels, it is possible to calculate a score for each action, thing or state by solving the preference function with the relevant measures data. Finally, processes of individual and group choice can be investigated by using the scores as inputs to voting, game theoretic, and equity models.

3.2.1. *Linking Objectives to Measurable Attributes*
Objectives hierarchies are integral to the preference model (see introduction in Section 3.1.2 and Figure 3). Some social scientists [37,38] have recommended focusing on the objectives of one or a few decision makers. Objectives important to a wide variety of stakeholders could be considered; however, in the final analysis, only those deemed relevant by the decision maker would be included in the hierarchy. This approach is inappropriate for collaborative resource management, a fundamental principle of which is community-of-interest involvement in decision making. An alternative approach is to elicit information about objectives from all relevant stakeholders and interest groups. The individual objectives are then merged into a single combined hierarchy that incorporates every objective deemed relevant by at least one person or group participating in the process.

The final step in building an objectives hierarchy is selection of measures (sometimes referred to as indicators, attributes or variables) to be associated with the lowest level objectives, in this case the means objectives. The purpose of the measures (the set of x_j's, where j is a means objective) is to describe the degree to which the associated objective is being met.

As previously noted, resource management actions can have a variety of impacts on socio-economic and biophysical systems. The type and severity of the impacts will vary depending upon which management alternative is implemented. As a result, levels of individual measures will tend to vary across alternatives. Typically, the set of values for each measure (a single x_j considered over every alternative) is normalized, with the worst case value assigned a zero and the best case value assigned a 1. Note that what is the worst outcome and what the best will depend upon the perspective of the person choosing.

Whether biophysical or social, measures are gauges of system position or condition. To be meaningful in the context of the HM, they must have the potential to vary as resource management strategies are varied, i.e., to have different values in different circumstances. The definition of a measure should be internally consistent; however, there is no need for all the measures to be comparable or defined with the same terms. For example, one measure could be valued in terms of the local currency and another in terms of weight per unit.

Keeney [27] identifies three types of measures: natural, constructed, and proxy. Natural measures have a commonly accepted interpretation and are obviously relevant to the problem at hand. If there are no natural measures, or if the ones available are hampered with value judgments or misleading connotations, then it may be necessary to construct a measure. Constructed measures often take the form of descriptions of alternative states of nature, each of which is assigned a score. Proxy measures are used only when there are no natural measures and constructed measures are deemed inadequate. In cases where objectives are complex and multifaceted, proxy measures often have multiple components.

The measures need to be technically and economically feasible, cost effective and preferably unambiguous. A technically feasible measure can be determined using currently available scientific methods and equipment. An economically feasible measure is one that is affordable, i.e., within the scope of the budget. A measure is cost effective if the value of the information provided thereby is greater than the cost of collecting it. Ambiguity presents a somewhat more complex problem. It is essential that the level of the measure be unambiguous; however, interpretation of the meaning of that level may be open to debate, particularly in cases where meaning is dependent upon world view.

3.2.2. *Preference Functions*

Recall from Section 2.2.3. that rational individuals are assumed to act so as to maximize their well-being. They do so by making choices that are consistent with their preferences.. Preferences over multiple objectives can be quantified as a function written in terms of the set of measures introduced above: a function U_i (where i is the group or individual whose preferences are being modelled) defined on x_j. The functional relationship may be ordinal or cardinal, i.e., capable of ranking alternatives so as to preserve the order of preference, or capable of ranking so as to reflect the relative degree of preference. The preference relationship may be termed a value function or a multiattribute utility function[5], where the terms value and utility connote human judgment and perception of well-being.

When the alternative outcomes of an event, and the respective probabilities, are known the quantitative statement of preferences is termed a value function. The term utility (or expected utility) is reserved for situations where outcomes are risky or uncertain. Under uncertainty, neither the alternative outcomes nor their respective probabilities are known.

Utility functions can take a number of forms, the most common of which are the additive, multilinear and multiplicative. Selection from among these is based upon the degree of conformance of individual or group preferences to standard independence conditions [27]. For the sake of simplicity, only the additive utility function will be considered here. It is applicable when all measures exhibit what is termed additive independence [54,55].

$$U_i(A_k) = 3_{i=1-n}\ w_{ij}\ u_{ij}(x_j) \qquad (1)$$

Where: $U_i(A_k)$ = the value, or utility, generated for individual or group i by implementation of alternative k.
w_{ij} = the weight placed by individual or group i on means objective j.
$u_{ij}(x_j)$ = the value, or utility, generated for group or individual i by fulfillment of means objective j.

The additive expected utility function does exactly what the name suggests; it sums up the utility generated by each constituent part of the function. In this case, the constituent parts are the single measure functions, $u_{ij}(x_j)$. Just as people prioritize and order their values, they rank order their objectives. Seldom is every means objective, or fundamental object, equally important. Rather, fulfilling some will be deemed more essential than fulfilling others. The weights are intended to reflect that prioritization. Weights are also indicative of the range of measure j in the problem at hand, relative to its feasible range; less weight should be given to measures that vary over only a limited portion of their feasible range.

Resource management problems often involve events that are risky and uncertain. While uncertainty cannot be modelled, the form of the single measure functions, $u_{ij}(x_j)$, can be chosen to reflect risk attitude, the willingness to accept, or perhaps even the desire to seek out, risk.. As a result, U_i will reflect overall risk attitude. Some individuals and groups are very risk averse, unwilling to take chances; others, such as mountain climbers, find risk exhilarating. The numerous models of risk attitude and forms of single measure functions will not be described here. For a more complete discussion of the topic, see [56,57].

The preference function U_i can be solved with the set of normalized measures (x_j's) associated with a specific alternative (A_k). When this is done over all the alternatives, the result is a vector of scores. Scores can be thought of as non-monetary quantitative expressions of how individual or group i values alternative states of world. When score

---

[5] Beinat, this volume, discusses value functions.

vectors are calculated for multiple individuals, the result is a matrix of expected utility or value estimates.

3.2.3. *Preference Ordering and Choice*

Ordering of alternatives is straightforward in the individual case. Again assuming rational behavior, an individual will rank-order alternative resource management schemes according to the utility or well-being generated by each. The alternative that is expected to generate the greatest amount of utility, the highest level of well-being, will be the one ranked highest and thus the one chosen for implementation.

Choice is more complex in situations where multiple interests need to be considered. Each individual and group holds a unique set of objectives. It is seldom possible to maximize every objective simultaneously, due in part to financial, human, and resource constraints. Moreover, some objectives are mutually exclusive. For example, it is not possible to maximize in the same place both the amount of timber cut and the amount of habitat reserved for a species that requires a large area of closed-canopy forest. For these reasons, resource management alternatives typically differ in terms of which objectives are given primary emphasis, and choice among alternatives becomes an issue of choice between objectives. Given that individuals have preferences over the set of possible objectives, the management decision can become one of choosing whose objectives will be fulfilled, of whose preferences matter most. This almost inevitably leads to conflicts.

No failure proof method of choosing among alternatives exists that is acceptable in a democratic society. However, once objectives hierarchies and utility score matrices are known, the choice problem reduces to one of finding a compromise solution through negotiation or some kind of arbitration. This is still a difficult task because likely compromise solutions depend not only on stakeholder's utility, but also on other factors such as relative bargaining power and differential levels of knowledge.

Assuming that all relevant points of view and conflicting interests are adequately represented, a choice that provides a socially acceptable tradeoff among these interests could be considered satisfactory. In reality, there exist only more or less satisfactory approximations to this ideal. In this section, three methods for making a socially acceptable choice are introduced: negotiated outcomes based on game theoretic approaches, political solutions based on voting, and fair solutions based on axioms of equity (Figure 4). Each of these quantitative methods utilizes the scores in the preference matrix to model the process of conflict resolution.

Negotiated Outcomes. Many decisions are reached through a process of bargaining or negotiation. The decision may be mundane, e.g., how long a child will be allowed to watch television, or momentous, e.g., what should be the contents of an international treaty on environmental protection. But in each case the negotiation process can be modelled as a game. Game theory is concerned with decision making in situations where participants can affect each other through their actions and as a result use strategic behavior to enhance their position.

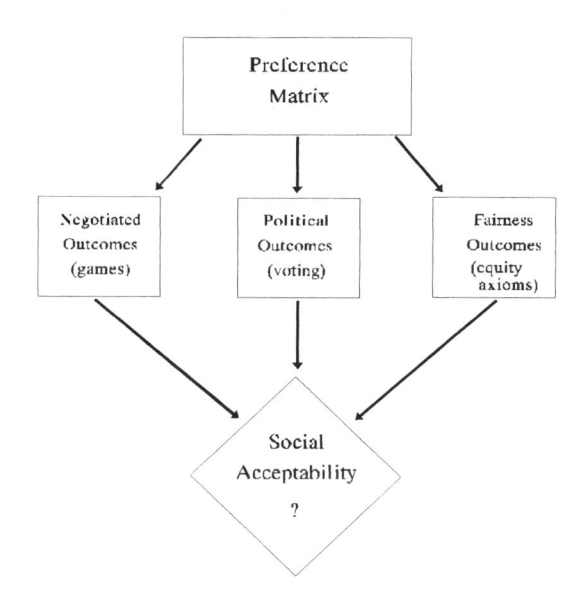

Figure 4. Using the Preference Matrix to Identify Socially Acceptable Alternatives

Games have many forms, only one of which will be discussed here, cooperative games in the characteristic function form[6]. A cooperative game is one in which the players can enter into binding, enforceable agreements with each other prior to commencing play. The characteristic function of a game with n participants (hereafter called players) is defined on a set of all subsets (coalitions) of the set of players P and specifies, for each coalition $K \subset P$, what the members of this coalition would jointly achieve if they cooperated among themselves but did not cooperate with the remaining players. The winning coalition gets to select the alternative to be implemented, which will be the one that maximizes the utility of its members.

Probably the most popular solution to a game in the characteristic function form has been proposed by Shapley and is called the Shapley value (hereafter the SV). The SV is a vector of n components, where n is the number of players in the game. The components represent the payoffs to each player, which in the resource management problem is reported in terms of utility.

The SV assumes that a player's utility is transferable, although utility as such is not transferred since utility is a derivative concept. Rather, there is assumed to exist an infinitely divisible, real, desirable commodity, i.e., money or something that acts like money, for which player's utilities are linear. Trades of the commodity will yield increases and decreases in utilities that sum to zero across all coalition members.

In a resource management problem, transferable utility could work as follows. Assume that a management alternative that generates exactly the utility levels of the SV has not been formulated. The alternative generating a utility vector closest to the SV

---

[6] See [58,59] for an introduction to game theory, [60] for a discussion of the Shapley value, and [61] for application of the Shapley value to resource management problems.

could be considered as a potential management choice. The differences in utilities generated by the "close" alternative and the utilities hypothetically achievable under the game theoretic solution to the problem indicate where there is potential for utility transfer. Stakeholders with higher utility could compensate others with lower utility by transferring to them money or commodities, and in that manner convince them to accept the "close" alternative. For example, an oil company might agree to pay for improvements at a wildlife refuge to make oil exploration activities acceptable to environmental groups.

There is, however, a less traditional way to look at transferable utility. As discussed previously, the utility function quantifies an individual's or group's preferences and risk aversion with respect to the measures associated with each means objective. The utility estimates depend upon the levels of each measure, which in turn are a function of the management action taken. Different actions have different impacts and so generate different levels of utility. Assuming that participants are willing to make at least some tradeoffs among social and environmental objectives, it may be possible to design a new management strategy that will result in measure levels that yield utilities closer to the SV. That new alternative would presumably be acceptable to the winning coalition and thus could be considered as a compromise solution to the management problem.

Political Outcomes. In democracies, voting is a popular method for making choices. Participants in resource management decision making could be endowed with votes. Each may be given one vote, or some may be given more than one to reflect their larger constituencies, greater political influence or geographic proximity to the resource in question. In such cases, vote allocation itself may become controversial. In addition, decisions about who is allowed to vote at all, regardless of how many votes they are given, can also be very controversial. Nonetheless, more and more natural resource decisions are being made through a voting process. For example, in some parts of the United States it is possible for citizens to place referenda items on general election ballots. Voters might be to accept or reject a specific resource management approach. The citizens of Montana voted to ban cyanide heap leach mining in their state through just such a process. (Implementation has been delayed through litigation.)

There are a number of interesting theories that can be used to define solutions to voting games, including one-stage (single vote) and two-stage (primary and runoff) For simplicity, the discussion here will be restricted to one-stage procedures in which each participant is endowed with one vote[7].

Consider a model were a joint group preference among alternatives is decided by voting. Any coalition of players controlling a majority of votes can impose their preference on the decision. That is A_i is preferred to A_j by a simple majority of players. An alternative that is preferred to any other in pairwise majority comparisons is called a Condorcet winner. Not every voting game has a Condorcet winner, but existence is guaranteed if the preferences of all stakeholders are restricted to the special class of so-called single peaked preferences. Preferences of this type have a unique, most desired

---

[7] See [62] for an introduction to voting models and [63] for an introduction to the mathematics of voting procedures and [64] for an application of voting to natural resource management problems.

point, and the choice of any other point is less preferred. Preferences would not be single peaked if two alternatives were equally and most desirable, i.e., ties for highest ranking are not permitted.

A Condorcet winner, if it exists, is strategy proof; each stakeholder maximizes his utility by truthfully reporting his preferences. It is also core stable, meaning that no winning coalition can find a voting strategy that would increase the utility of coalition members. However, in cases where there is not a majority alternative the Condorcet method can give inconsistent results. An alternative is rank order voting. Here voters assign scores to each alternative, with more preferred alternatives receiving higher scores. The point total for an alternative equals the sum of the scores assigned to it by all participants. The alternative with the highest net score is the one chosen. The Borda voting method is an example of this type of process. Rank order voting does not require single peaked preferences, but it does have another drawback. The majority alternative, i.e., that which would receive a strict majority of votes in pairwise comparisons, may not be chosen the winner.

Fair Outcomes. Resource projects that generate net gains for some groups or individuals, or for specific geopolitical entities, while simultaneously generating net losses for other groups, individuals or geopolitical units, have become increasingly controversial. This, despite continued demand for resources and a need for jobs and income generated by the extractive industries. One source of controversy is the perceived fairness of relative distributions of benefits and costs stemming from a resource management decision. To the degree that the distribution is uneven it may be perceived as inequitable.

Equity can be described with many words, e.g., just, fair, or equal. All these terms are related, but they are not synonymous and each has specific connotations. A just outcome is one that is based on principles of justice, as defined by a set of absolute, universal rules on individual needs, rights, and duties. A fair outcome is one that is based on broadly accepted rules of conduct, as defined by social norms. Equal outcomes would be those in which resources, access, opportunity, or net benefits are distributed to each individual or segment of the population in equal amounts.

Alternative definitions of equity can take axiomatic form[8]. Decision rules derived from axioms of equity can be used to rank alternative states of the world in terms of their relative fairness across groups and individuals. These types of decision rules require interpersonal comparisons of utility (ICU). It is generally agreed that individual preferences should not be used to generate a socially preferred ranking of alternatives. However, ICU is an acceptable way to investigate intensity of preference. Resource management can be a contentious process; not least because stakeholder's objectives are often incompatible. Objectives hierarchies, multiattribute utility functions, and cardinal utility estimates, can provide insights as to which groups or individuals might respond in a strongly negative (or positive) fashion to the selection of a specific management alternative.

---

[8] See [65,66] for an introduction to equity theory; see [67] for an application of axioms of equity to resource management problems.

A social justice axiom will rank order management alternatives (from most to least preferred) in terms of their consistency with a set of principles of justice. An axiom based on fairness orders alternatives in a manner consistent with stated social norms. And an axiom based on equalizing utility can be used to determine which alternative would minimize differences in the levels of satisfaction among groups or individuals. Assuming a group's or individual's philosophical perspective is known, their opinion about the relative equity of alternative management actions can be inferred from the rankings predicted by the appropriate equity axiom.

Although economists are loath to make ICU, stakeholders do it all the time. They tend to have strong opinions about who would gain unfairly, from their perspective, should a specific management alternative be implemented. The advantage of using axioms of equity in resource management is that they enable resource managers to address equity in an analytical fashion, including predicting stakeholder reactions to various possible choices. The limitation is that they do not provide a decision rule, unless the highest goal of the decision maker is consistency with some version of equity.

4. Utilizing the Hierarchical Model of Collaborative Resource Management in Decision Making

Resource management is a challenging process in which social values and expectations interact with the biological limitations, physical structures, economic realities, and technical capabilities. Decisions must often be made with limited knowledge of environmental and earth resource system structures, processes, interactions and components. As a result, every alternative management action has aspects that are risky and outcomes are inherently uncertain.

Adaptive management (AM) has been proposed as an approach for dealing with these complexities and uncertainties [68,69]. It is based on five basic concepts: 1) resource management is an experiment; 2) uncertainty is inevitable; 3) objectives need to be quantified; 4) current understanding of system operation (or lack thereof) is acknowledged; and 5) feedback, evaluation, and redirection are essential [70]. The HM supports AM by providing both information on objectives and a structured means for feedback and redirection. In this section, the HM is placed in the AM framework.

4.1. THE ADAPTIVE MANAGEMENT MODEL

Adaptive management is based on a fundamentally different set of ideas than those of traditional natural resource management. Surprises are expected because we do not and cannot fully understand natural systems. Therefore, adaptive management requires that uncertainty be confronted and dealt with a priori, rather than being ignored. This is accomplished in part by viewing resource management as an experiment. The purpose of science is then to try to remove uncertainty and limit risk by first posing and testing hypotheses, and second, learning from the results. Essentially, AM is a process of learning while doing.

Perfect knowledge is not a requirement of AM, but quantitatively explicit hypotheses about system structures and processes, clear management objectives, and a set of targeted actions are necessary [71,72]. AM then determines whether the goals have been achieved and assesses the validity of the initial assumptions of the experiment. The review and validation process provides the information required for any necessary redefinition of the conceptual model, reprioritization of objectives, and the design of new management actions.

4.2. THE HIERARCHICAL MODEL IN ADAPTIVE MANAGEMENT

The AM process is illustrated in Figure 5 as a recursive system. The cycle is initiated by the existence of a resource management issue. This could be the need to decide whether to grant a mine operating permit, respond to the presence of increased risk of wildfire in a specific area, or provide habitat to an endangered plant or animal species. The next step is to develop a suite of alternative management responses, which could range from no action to significant activity.

Each management alternative is examined in detail. The costs and technical parameters, as well as the social and environmental impacts are estimated. A series of questions are then posed. Is the action technically feasible, i.e., is it possible to implement the action given currently available technology? Is the action economically feasible, i.e., are costs within the allowed budget? Will the impacts on biophysical systems positively or negative affect environmental sustainability? Will the impacts on social systems positively or negatively affect community stability, i.e., is the predicted outcome consistent with sustainable development? In each case, if the answer is no, the alternative is modified and the tests reapplied.

The public is brought into the decision process through the use of the HM. All relevant stakeholders are identified. Objectives are elicited and linked to measures of system condition. Preference functions are constructed. Preferences over the set of sustainable and feasible alternatives are calculated and the socially acceptable alternatives identified. Alternatives that are deemed unacceptable can be revised or dropped from consideration. The result will be a set of technically and economically feasible, socially and environmentally sustainable, and socially acceptable management alternatives. A decision model, such as the multicriteria approach recommended by Beinat [this volume], can then be used to make a final selection from among the viable alternatives.

One of the most important features of the AM paradigm is the feedback loop. Once a management action has been implemented, systems are monitored. Information on system condition is passed back to decision makers, scientists and the public. The information can be presented in the context of the public's stated objectives for socio-economic and biophysical systems. Finally, a process of evaluation, adaptation, and revision is instituted, and through which emerging natural resource issues can be identified. The cycle then begins again.

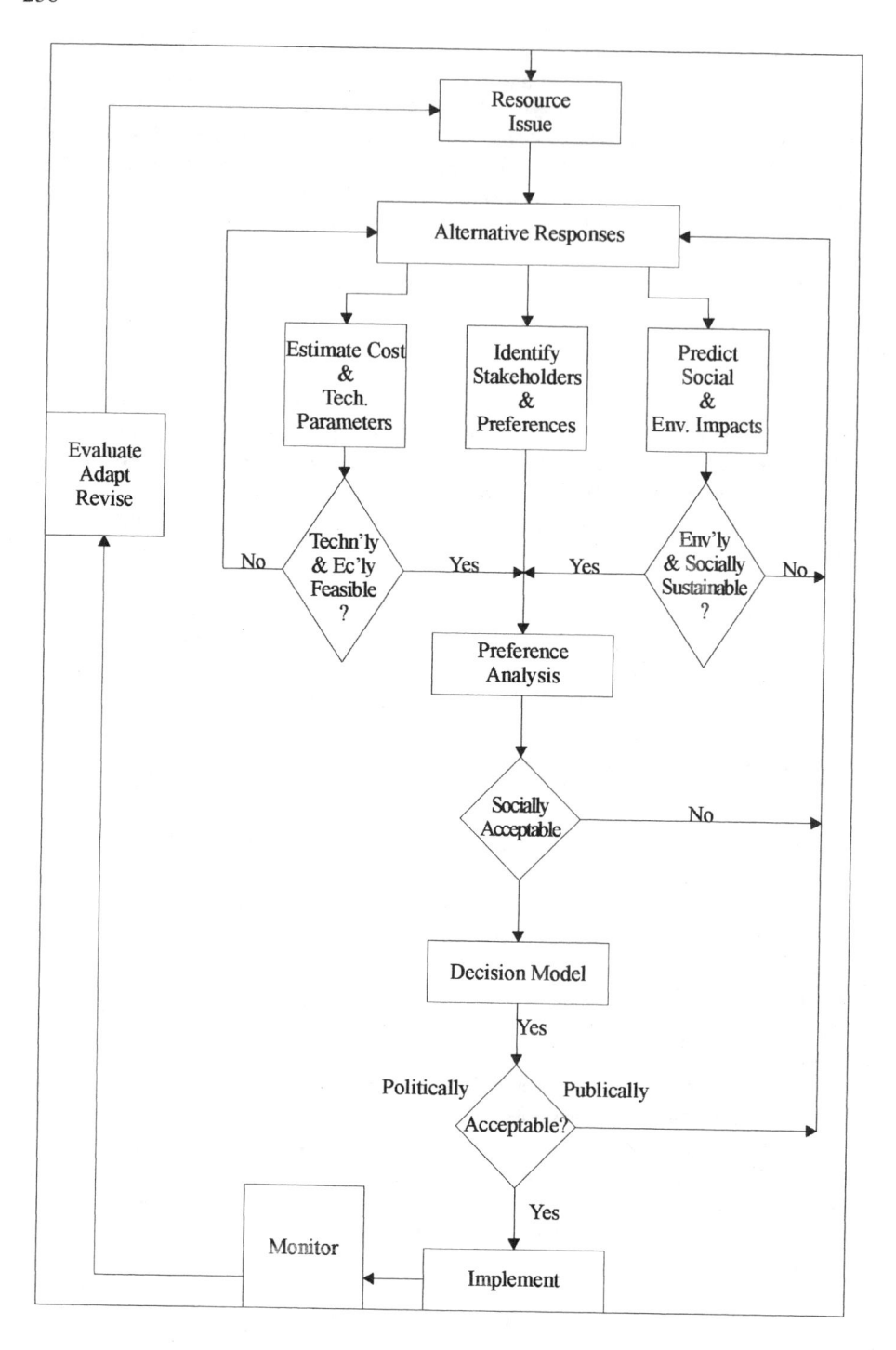

Figure 5. The Hierarchical Model of Collaborative Resource Management in Adaptive Management

5. Conclusions

It has been argued [41] that stakeholders in a planning process should be identified and consulted about their values and objectives at the very beginning of the planning process. Furthermore, they should be participants in the formulation of alternative management strategies, rather than being asked to react to alternatives selected by the agency. It has been demonstrated that, unless communities themselves are involved in rural development programs with some degree of permanence, the programs are unlikely to succeed [73]. Natural resource management programs in rural areas are tightly linked to local economies, and the success of both can depend upon the information feedback systems between them. However, local community values are sometimes in conflict with natural resource planning objectives and/or national social objectives [74]. This is particularly common in cases where costs (negative impacts) of management are born locally, but the benefits are national in scope.

The purpose of the Hierarchical Model of Collaborative Resource Management presented here is to facilitate the public participation process. The measurement model offers an approach for studying stakeholder values, objectives, attitudes, and behaviors with respect to natural resources. The preference model provides a framework for objectives to indicators of ecosystem and community status. In addition, it is possible to investigate how stakeholder groups perceive themselves to be differentially affected by alternative management plans through application of the utility models. This range of information is essential to the successful implementation of adaptive management.

References

1. Santayana, George. (1896) The Sense of Beauty, Scribner's Sons, New York.
2. Rolston, H., III. (1994) *Conserving Natural Value*. Columbia University Press, New York.
3. Brown, T.C. and Peterson, G.L. (1994) A political-economic perspective on sustained ecosystem management, in L.F. DeBano (ed.) *Sustainable ecological systems: implementing an ecological approach to land management*, Gen. Tech. Rep. RM-247, USDA Forest Service, Fort Collins, CO, pp. 228-235.
4. Shields, D.J. and Mitchell, J.E.(in press) *A Hierarchical Systems Model of Ecosystem Management*, Gen. Tech. Rep. RM-xxx, USDA Forest Service, Fort Collins, CO.
5. O'Neill, R.V., DeAngelis, D.L., Waide, J.B., and Allen, T.H.F. (1986) *A Hierarchical Concept of Ecosystems*, Princeton Univ. Press, Princeton, NJ.
6. Allen, T.F.H., O'Neill, R.V., and Hoekstra. T.W. (1984) *Interlevel Relations in Ecological Research and Management: some working principles from hierarchy theory*, Gen. Tech. Rep. RM-110, USDA Forest Service, Fort Collins, CO.
7. Waide, J.B., and Webster, J.R. (1976) Engineering systems analysis: applications to ecosystems, in B.C. Patten (ed.), *Systems Analysis and Simulation in Ecology. Vol. IV*, Academic Press, New York, pp. 329-371.
8. Allen, T.F.H., and Hoekstra, T.W. (1992) *Toward a unified ecology*, Columbia Univ. Press, New York.
9. Forrester, J.W. (1961) *Industrial Dynamics*, MIT Press, Cambridge, MA:
10. Shannon, C.E, and Weaver, W. (1949) *The Mathematical Theory of Communications*, Univ. of Illinois Press, Urbana, IL.
11. Ashby, W. R. (1956) *An introduction to Cybernetics*, Chapman and Hall, Ltd., London.
12. Fairweather, P.G. (1993) Links between ecology and ecophilosophy, ethics and the requirements of environmental management, *Australian J. of Ecology* **18**, 3-19.
13. Arrow, K., Bolin, B., Costanza, R., Dasgupta, P., Folke, C., Holling, C.S., Jansson, B., Levin, S., Mäler, K., Perrings, C., and Piementel, D. (1995) Economic growth, carrying capacity, and the environment, *Science* **268**, 520-521.

14. Risser, P.G. (1996) Decision-makers must lead in defining some environmental science, *Ecological Applications* **6**, 24-26.
15. Shields, D.J. (1998) Nonrenewable resources in economic, social and environmental sustainability, *Nonrenewable Resources* **7**, 251-262.
16. Moldan, B. and Billharz, S. (eds.) (1997) *SCOPE Report 58*, International Council of Scientific Unions, Paris.
17. Cooper, C.F. (1969) Ecosystem models in watershed management, in G.M. Van Dyne (ed.), *The Ecosystem Concept in Natural Resource Management*, Academic Press, New York, pp. 309-324.
18. Rokeach, M. (1973) *The Nature of Human Values*, The Free Press, New York.
19. Brown, T.C. (1984) The concept of value in resource allocation, *Land Economics* **60**, 231-246.
20. Boulding, K.E., and Lundstedt, S.B. (1988) Value concepts and justifications, in G.L. Peterson, B.L. Driver and R. Gregory (eds.) *Amenity resource valuation*, Venture Publishing, State College, PA, pp. 13-22.
21. Hetherington, J., Daniel, T.C., and Brown, T.C. (1994) Anything goes means everything stays: the perils of uncritical pluralism in the study of ecosystem values, *Society and Natural Resources* **7**, 535-546.
22. Rescher, N. (1969) *Introduction to Value Theory*, Prentice Hall, Englewood Cliffs, NJ.
23. Wilson, B. (1988) Values and society. In B. Almond and B. Wilson (eds.), *Values*, Humanities Press International, Inc., Atlantic Highlands, NJ, pp. 31-46.
24. Craik, K. (1978) The environmental dispositions of environmental decision makers, *The Annals Amer. Acad. Polit. and Social Science* **389**, 87-94.
25. Andrews, R.N. and Waits, M.J. (1980) Theory and methods of environmental values research, *Interdisciplinary Science Reviews* **5**, 71-78.
26. Bengston, D.S. (1994) Changing forest values and ecosystem management, *Society and Natural Resources* **7**, 515-533.
27. Keeney, R.L. (1992) *Value-focused Thinking*, Harvard University Press, Cambridge, MT.
28. Stoll, J.R., and Gregory, R. (1988) Overview, in G.L. Peterson, B.L. Driver and R. Gregory (eds.) *Amenity resource valuation*, Venture Publishing, State College, PA:, pp. 3-6.
29. Steel, B.S.; List, P., and Schindler, B. (1994) Conflicting values about federal forests:a comparison of national and Oregon publics, *Society and Natural Resources* **7**, 137-153.
30. Homer, P. M. and Kahle, L.R. (1988) A structural equation test of the Value-Attitude-Behavior hierarchy, *J. Personality and Social Psychology* **54**, 638-46.
31. Manzer, L. L. and Miller, S.J. (1982) An examination of Value-Attitude structure in the study of donor behavior, *Advances in Consumer Research* **10**, 204-6.
32. Huffman, C. and Houston, M. (1993) Goal-oriented experiences in the development of knowledge, *J. Consumer Research* **20**, 190-207.
33. Park, C.W. and Smith, D.C. (1989) Product-level choice : a top-down and bottom-up process?, *J. Consumer Research* **16**, 289-299.
34. Martin, I.M., Bender, H.W., Martin, W.E., and Shields, D.J. (in press) The impact of goals on the "values-attitudes-behaviors" framework, *Decision Line*.
35. Martin, I. M., Bender, H.W., Martin, W.E, and Shields, D.J. (1997) The development of a values scale: consumption of public lands, Colorado School of Mines, Division of Economics and Business, Working Paper.
36. Caldwell, L. K. (1990) Landscape, law and public policy: conditions for an ecological perspective, *Landscape Ecology*, **5**, 3-8.
37. Keeney, R.L. and Raiffa, H. (1976) *Decisions with Multiple Objectives*, Wiley and Sons, New York.
38. von Winterfeldt, D. and Edwards, W. (1986) *Decision analysis and behavioral research*, Cambridge University Press, Cambridge.
39. Keeney, R.L. (1988) Structuring objectives for problems of public interest, *Operations Research*, **36** 396-405.
40. Keeney, R.L., von Winterfeldt, D., and Eppel, T. (1990) Eliciting public values for complex policy decisions, *Management Science* **36**, 1011-1030.
41. Gregory, R., and Keeney, R. (1994) Creating policy alternatives using stakeholder values, *Management Science* **40**, 1035-1048.
42. Fishbein, M. and Ajzen, I. (1974) Attitudes toward objects as predictors of single and multiple behavior criteria, *Psychological Review* **81**, 59-74.
43. Ajzen, I. And Fishbein, M. (1997) Attitude-behavior relations: a theoretical analysis and review of empirical research, *Psychological Bulletin* **84**, 888-918.
44. Eagley, A. H. and Chaiken, S. (1993) *The Psychology of Attitudes*, Harcourt Brace Jovanovich College Publishers, Orlando, FL.

45. Ajzen, I. (1987) Attitudes, traits, and actions: dispositional prediction of behavior in personality and social psychology, in L. Berkowitz (ed.), *Advances in Experimental Social Psychology*, Academic Press, San Diego, CA, pp. 1-63.
46. Mansfield, E. (1987) *Statistics for Business and Economics*, Norton, New York.
47. Christensen, H.B. (1992) *Introduction to Statistics*, Saunders College Publishing, New York.
48. Hair, Jr., J.F., Anderson, R.E. and Tatham, R.L. (1987) *Multivariate Data Analysis: with readings*, Macmillan Publishing, New York.
49. Jobson, J.D. (1991) *Applied Multivariate Data Analysis, Volume II: Categorical Multivariate Methods*, Springer-Verlag, New York.
50. Everitt, B.S. (1993) *Cluster Analysis*, Edward Arnold, London.
51. Gorsuch, R.L. (1974) *Factor Analysis*, Saunders Co., Philadephia, PA.
52. Jöreskog, K.G. (1978) Structural analysis of sovariance and correlation matrices, *Psychometrika* **43**, 433-477.
53. Jöreskog, K.G., Sörbom, D. (1993) *Lisrel 8: Structural Equation Modeling with the SIMPLIS Command Language*, Scientific software International, Chicago, IL.
54. Fishburn, P.C. (1965) Independence in utility theory with whole product sets, *Operations Research* **13**, 28-43.
55. Fishburn, P.C. (1970) *Utility Theory for Decision Making*, Wiley, New York.
56. Howard, R.A. (1988) Decision analysis: practice and promise, *Management Science* **34**, 679-695.
57. Harvey, C. (1990) Structured prescriptive models of risk attitudes, *Management Science* **36**, 1479-1501.
58. Colman, A.M. (1995) *Game Theory and its Applications*, Butterworth, Oxford.
59. Bierman, H.S. and Fernandez, L. (1993) *Game Theory with Economic Applications*, Addison Wesley, New York.
60. Shapley, L.S. (1969) in La Decision, Aggregation et Dynamique, in G. Guilbaud (ed.), *Colloque Internationaux du Centre de la Recherche Scientifique no. 171*, Editions Centre National de la Recherche Scientific, Paris.
61. Shields, D.J., Tolwinski, B., and Kent, B.M. (1999) Models for conflict resolution in ecosystem management, *Socio-Economic Planning Scinces* **33**, 61-84.
62. Nurmi, H. (1987) *Comparing Voting Systems*, D.Reidel, Dordrecht.
63. Fishburn, P.C. (1984) *Review of Economic Studies* **51**, 683-692.
64. Martin, W., Shields, D.J., Tolwinski, B. and Kent, B. (1996) An application of social choice theory to U.S.D.A. Forest Service decision making, *J. Policy Modelling* **18**, 603-621.
65. Young, H.P. (1994) *Equity: in theory and practice*, Princeton University Press, Princeton.
66. Hausman, D. and McPherson, M. (1996) *Economic Analysis and Moral Philosophy*, Cambridge University Press, Cambridge.
67. Shields, D.J. (1997) An axiomatic approach to equity in natural resource management, Dissertation, Colorado State University, Ft. Collins, CO.
68. Walters, C. (1986) *Adaptive Management of Renewable Resources*, MacMillan and Co., New York.
69. Lee, K.N. (1993) *Compass and Gyroscope*, Island Press, Washington, D.C.
70. Everett, R., Oliver, C., Saveland, P., Hessburg, P., Diaz, N., and Irwin, L. (1993) Adaptive ecosystem management, in M. Jensen and P Bourgeren (eds.) *Ecosystem Management: Principles and Applications*, USDA Forest Service, Pacific Northwest Research Station, Portland, OR, pp. 351-355.
72. Walters, C.J. and Holling, C.S. (1990) Large-scale management experiments and learning by doing, *Ecology* **71**, 2060-2068.
73. Thomas, M.G. (1990) *Forest Resource Strategies for Rural Development*, U.S. Department of Agriculture, Washington, DC.
74. Galston, W. A. and Baehler, K.J. (1995) *Rural development in the United States: connecting theory, practice, and possibilities*, Island Press, Washington, D.C:

ENVIRONMENTAL AND RESOURCE PLANNING: METHODOLOGIES TO SUPPORT THE DECISION MAKING PROCESS

Euro BEINAT
Institute for Environmental Studies
Vrije Universiteit Amsterdam
de Boelelaan 1115, 1081 Amsterdam, The Netherlands

Abstract. This chapter illustrates two types of evaluation techniques for environmental and resource planning: economic valuations and multicriteria analysis. A general scheme for the evaluation is first introduced as a reference for the techniques described in the chapter: Benefit-Cost Analysis (BCA), Cost-Effectiveness Analysis (CEA) and Multicriteria Analysis (MCA). These techniques are among the most applied in environmental and resource planning. The chapter first describes BCA and CEA within a general framework of welfare economics and efficient allocation of resources. It then describes the main features of the methods and the most frequently used techniques to associate monetary values to environmental resources. The usefulness, applicability and limits of these techniques are discussed. Multicriteria analysis is then introduced as a representative of a class of methods that do not require monetary valuations. The chapter focuses on a particular MCA technique, value functions, and describes the main features of the methodology and the differences compared to BCA and CEA. Instead of describing the technicalities of value functions, the chapter focus is on the most critical application issues, which determine the added value of MCA in practice.

1. Introduction

Economic valuations support the analysis and comparison of projects, plans and policies in economic terms. The main assumption is that society benefits by an efficient allocation of scarce resources, which occurs when the difference between the benefits and costs of a given activity is maximised. In economic evaluations it is assumed that individual preferences are recorded in the market place (or would be recorded if there were a market), so that the price of goods and services are signals of people desires [23]. Therefore, by correctly measuring the monetary value of costs and benefits of a project[1] it is possible to assess whether or not society would benefit from engaging resources in this activity. Benefit-Cost Analysis (BCA) is a technique that supports this evaluation. It requires the analysis of the impacts of a project, their time distribution, and their classification into costs and benefits. The unit prices of impacts are then estimated and

---

[1] To simplify the exposition, the term project will include plans and policies unless otherwise stated.

discounted, so that a unique common denominator can be used to rate the plan. Although the use of BCA is widespread, there are objective difficulties in estimating the monetary value of many impacts. Also, the assumptions implied in the methodology can be criticised and many authors cast doubts on its applicability to environmental problems.

Cost-Effectiveness Analysis (CEA) bypasses some of these criticisms by restricting economic valuations to costs and benefits for which a clear monetary value exists. In combination with setting goals for other impacts (such as, health effects), CEA can be used to assess the degree to which a project meets some environmental goals in an economically sound way.

Multi-Criteria Analysis (MCA) does not attempt to rate all impacts of a project in monetary terms. Instead, the focus is on the exploration of the multiple objectives of the actors involved in the decision (e.g., policy makers, public authorities, stakeholders). A project is attractive if it is able to satisfy the multiple objectives of the actors involved. MCA includes techniques for making trade-offs transparent and for assessing the pros-and-cons of a project from multiple perspectives.

The chapter first provides a brief illustration of BCA. It lists the main valuation methods and the pros and cons of the methodology. CEA is introduced as a bridge between BCA and multicriteria analysis. The main part of the chapter is devoted to MCA, and in particular to the value functions approach.

2. A General Scheme for the Evaluation

The evaluation of projects, plans or policies follows a general scheme which is illustrated in Figure 1. This scheme, with slight variations, is used in many fields ranging from management decisions [21] to environmental assessments (cf., [36, 16]) and economic valuations [10][2].

This scheme shows four main phases: identification, analysis, assessment, and implementation. The choice of the methodology to support this process (BCA, CEA or MCA, or other) reflects conviction and assumptions, which have consequences on the process itself. For instance, BCA offers limited insights into equity issues and conflicts. Thus, applying BCA alone often means that public participation and stakeholders inputs are not encouraged. BCA is more suited for top-down evaluations, such as to assess government investments. MCA, on the other hand, requires direct input from actors in the decision process, thus implying an open, flexible agenda.

This type of distinctions have practical consequences reflected in: the set of actors involved (few vs. many); the role of the actors (decision makers and spectators vs. actors

---

[2] This scheme is useful to streamline the exposition. However, there are many intricate links between the steps of the analysis, so that it is risky to assume that the decision-making process follows a rational and linear analysis path. Lindblom and Woodhouse (1993), [20], for instance, describe the policy making process as a "primeval soup" rather than an organised step-wise process. Morgan and Henrion (1990), [22], maintain that a linear approach to analysis is the fingerprint of inexperienced analysts. The practice of decision support and project analysis demonstrated that the scheme is a valid conceptual guideline and can serve to organise the tools for the analysis. However, it makes little practical sense to separate out and isolate stages and analyse them in turn. The feedback between steps is where the learning-cycle takes place, which is one of the main added values of an analytical approach to decision aid.

and players); the issues considered (technical vs. social/psychological); the type of subjective input to the process (reflected in market preferences vs. expressed directly by people) and the feature of the optimal outcome (economically efficient vs. socially acceptable).

There are other process differences implied by methodological choices. For instance, a common stage to all methodologies is the identification of the projects to be evaluated and of their impacts. In BCA there is a clear distinction between project identification, and evaluation. MCA, on the other hand, is rapidly shifting the focus from an evaluation tool towards a methodology that supports also the generation of alternatives (cf. the concept of value-focused thinking in [17]).

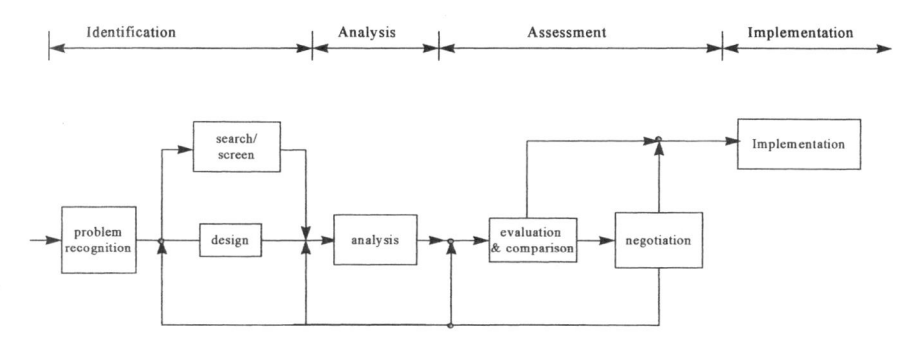

Figure 1. A scheme for project evaluation (adapted from [21]).

3. Benefit-Cost Analysis and Cost-Effectiveness Analysis[3]

3.1 TYPE OF IMPACTS

Impact analysis is the preliminary step for economic valuations. Impacts should include on-site and off-site impacts as well as those for which there is a market price and those for which there is not.

Table 1 illustrates four broad classes of impacts for an application to the management of mangrove forests in the Philippines [14]. Those included in (d) are the most difficult to analyse, because there is no obvious market value for impacts, and because the boundaries of the system are not easy to define. Therefore these impacts are affected by two sources of uncertainty: uncertainty on the magnitude of the impacts and uncertainty on the monetary value of the impacts.

---

[3] This section is based on Dixon *et al.* (1995), [10], and Pierce and Turner (1990), [26].

Table 1. Goods and services for mangrove forest management (adapted from [10], page 27; goods and services analysed in [14]).

	On-site	Off-site
Market values are available	(a) wood, charcoal, on site crabs, leaves, tannins	(b) offshore fish and shellfish, water
Market values are not available	(c) medicinal resources shoreline protection and flood mitigation	(d) shoreline protection and flood mitigation, biodiversity, uptake of carbon dioxide

3.2 THE BENEFIT-COST CRITERIA

Three main criteria are commonly used to rate projects in terms of their costs and benefits [16]: Net Present Value (NPV), Internal Rate of Return (IRR) and Benefit/Cost Ratio (B/CR). The NPV is the balance between the discounted benefits and costs. The higher is the NPV, the better is the project because the benefits overcompensate the costs. The IRR is the interest rate that results in equality of discounted benefits and costs. If the IRR is higher than a reference interest rate (e.g., the funds "cost"), then the project is financially attractive. The B/CR is the ratio of discounted costs and benefits, and is it a derivative of the NPV criterion. A B/CR higher than 1 means that the project is economically attractive. The mathematical rules of the three criteria are the following:

$$\text{NPV} = \sum_{t=0}^{n} \frac{B_t - C_t}{(1+r)^t} \quad (1)$$

$$\text{IRR} \xrightarrow{\text{the particular value of r such that}} : \sum_{t=0}^{n} \frac{B_t - C_t}{(1+r)^t} = 0 \quad (2)$$

$$\text{B/CR} = \frac{\sum_{t=0}^{n} \frac{B_t}{(1+r)^t}}{\sum_{t=0}^{n} \frac{C_t}{(1+r)^t}} \quad (3)$$

where B_t and C_t are the costs and benefits at time "t", respectively, "n" is the time horizon for the evaluation, and "r" is the discount rate.

The discount rate is used to make comparable costs and benefits, which occur at different points in time. Discount rates arise for two main reasons: the time-dependence of preferences, and the productivity of capitals [26]. People tend to trade-off between present and future consumption: a cost or a benefit matters more if it is experienced now rather than in the future. In addition, a capital, which is not diverted for consumption now, will be able to yield a higher level of consumption later. Thus, the *impatience* for the present consumption and the productivity of capital, or both, give rise to a discount rate, which takes into account the current and future values of costs and benefits in the decision rule.

3.3 THE MONETARY VALUE OF ENVIRONMENTAL IMPACTS

The literature on valuation techniques is extremely voluminous. This section will merely attempt at listing some of the most used techniques for valuing environmental goods and services. Complete overviews and detailed analyses can be found in [26], [10], and [8]. A sample of valuation techniques is provided in

1. Techniques, which use market prices to value impacts. These techniques use current prices of goods and services affected by a project to value the cost and benefits of the project. Techniques in this class are: changes in productivity, cost of illness, opportunity costs, replacement costs, shadow projects, etc.
2. Techniques in which a surrogate (indirect) market price is used. These methods use information on goods and services traded in markets to value the non-market goods under consideration. An example is the travel cost method.
3. Contingent valuation methods. When real or surrogate markets are not available, the value of environmental goods can be approximated by asking individuals for their willingness to pay for an additional quantity of an environmental good, or for the willingness to accept a compensation for the loss of an environmental good.
4. Hedonic value methods. When the market value of a good embeds an environmental component (for instance the value of a house with a nice view), the value of this component can be extracted by comparing the market value of goods, which are equivalent in all attributes except for the environmental attribute. An example of this technique is the property value approach.

3.4 APPLICABILITY AND LIMITS TO BCA

Successful economic development depends on the rational use of environmental resources and on minimising the negative impacts of development projects. BCA can be very useful to promote an efficient use of resources, which are coherent with socio-economic priorities. Especially when there is a range of alternative investment opportunities (e.g., investments in health care, conservation, or infrastructure), the assessment of the economic efficiency of the investment is of paramount importance to maximise the welfare contribution from the resources employed.

Table 2. A summary of economic valuation techniques.

Technique	Description	Comments
Change in productivity	Assesses the changes in the physical level of an economic variable, which has a market price (e.g., crops). The difference with and without the project is computed and valued.	Assumes that the effects of the project are "small" and do not change the market prices of the variable.
Cost of illness	Uses a dose-effect curve to relate environmental quality and health (e.g., indoor air pollution and respiratory diseases). Costs are the sum of medical costs and loss of earnings related to environmental factors.	Best applied for short, non-chronic illnesses. Availability of credible dose-effect functions is crucial. Disregards minor effects (pain, stress, etc.). When applied to chronic effects, or to the loss of life, it raises ethical and moral issues.
Opportunity costs	Cost of preservation is measured by the forgone income from other uses of the resource (e.g. preserving a forest instead of harvesting timber).	Uses foregone benefits as proxy estimates of the value of a resource, which are not estimated directly.
Replacement costs (relocation costs)	The benefits of measures which prevent a damage are approximated by the costs that would be needed to repair the damage (e.g., the costs or relocating a water well due to water pollution).	Damage needs to be measurable. Does not give an indication of the value of environmental protection per se.
Shadow projects	Costs of the loss of environmental goods/services are equal to the costs of a project, which would provide the same amount of goods/services (e.g., the destruction of a wetland and the restoration of another elsewhere as a compensation).	Assumes substitutability of environmental resources. The original level of the environmental goods/services was desirable.
Travel costs	The value associated to a recreational site is approximated by the expenditures of users to travel to and visit the site.	Used to value recreational goods/services. Option and existence values are not included: they may be relevant for unique sites (e.g. historical sites). It is based on assumptions of visitors' behaviour.
Contingent valuation	By means of interviews, requires people to assess the willingness to pay (e.g., for more green areas, lower noise levels) or the willingness to accept compensation (e.g., for the loss of green areas, for higher noise levels).	General applicability, but affected by information biases, difficulties in valuing hypothetical situations, strategic behaviours and, in general, by the survey approach.
Hedonic valuation	Prices of environmental goods are extracted from the total price of goods, which includes an environmental attribute (e.g., the value of noise by comparing similar houses in quite and noisy surroundings).	Requires extensive data to make it possible to separate the effect on price of a single variable.

There are also many difficulties related to the use of BCA for environmental and resource assessments. First, the economic valuation of environmental, social, cultural and health effects is often difficult, sometimes impossible, and almost always questionable [24]. In some cases the valuation may be extremely controversial so as to cast doubts on the validity of the whole BCA. For instance, when applied to assess the value of life (the value of a *statistical life*, not that of a given individual), the results usually show significant differences for people of different ages, race, sex, and income level (cf., [13]). These results can hardly be included in the evaluation of projects without raising moral and ethical issues. Social differences, though unfair, are a matter of fact. However, including them as a structural part of the evaluation implicitly means to accept and preserve them.

Another criticism to BCA is that there is no mention of the distribution of costs and benefits. Efficiency is measured disregarding to whom costs and befits accrue, and disregarding the desirability of the current distribution of income. The result of the analysis, though being economically efficient, could be socially inadmissible and determine social conflicts, which are hidden in the economical evaluation. Even when people are involved directly in the process, such as in contingent valuation methods, they are a source of information, and there is no attempt to initiate a negotiation process to resolve conflicts over distribution effects [24].

In economic evaluations it is well acknowledged that the value of money changes during the time: a dollar now has a different value that a dollar next year. In BCA, a positive discount rate means that actual costs and benefits are weighted more than future costs and benefits. Since many environmental effects appear after several years, the discounting process may result into an underestimation of future consequences and allow the implementation of unacceptable (in the future) projects. Several solutions have been proposed to avoid this effect, such as the use of low or even zero discount rates. Dixon *et al.* (1995), [10], maintain that these suggestions are ill conceived, as they disregard the facts that: (1) people prefer a certain amount of money today than tomorrow, and (2) alternative investment opportunities exist that would yield a larger sum in the future. Thus, eliminating discounting would result in a decrease of social welfare. Nevertheless, using a discount rate does imply the inter-temporal allocation of resources and affects intergenerational equity. It makes sense, therefore, to consider these effects independently of a discount rate, to assess their acceptability in addition to the economic efficiency of a project.

From a more philosophical standpoint, Sagoff (1988), [31], maintains that in BCA there is confusion between material interests and human values. Sagoff points out that the solution of environmental problems can hardly be found by assessing prices for environmental resources, but rather through cultural evolution and political approaches. The use of the money rod seems appropriate for evaluating material interests, where citizen act as consumers. However, when the evaluation concerns the quality of the environment and the quality of life the evaluation shifts to convictions, principles and cultural values, which do not easily lend themselves to monetary evaluations. Citizens do not act as plain consumers in these cases.

There are now two main approaches to BCA. Some analysts prefer to use it for economic evaluations, avoiding forced monetary quantification when there is no clear background for them. However, in the last two decades efforts increased to develop sophisticated techniques able to overcome the shortcomings of valuation techniques. The other tendency is thus to justify the use of BCA by increasing the validity of the evaluation techniques. At present it is unclear which of these two approaches will prevail in the future. Dixon *et al.* (1995), [10], distinguish between generally applicable valuation techniques (those for which there are market prices), selectively applicable, and potentially applicable techniques. This distinction implicitly suggests guidelines for the use of BCA.

3.5 COST-EFFECTIVENESS ANALYSIS

One of the common alternatives to full BCA is Cost-Effectiveness Analysis (CEA). CEA is used when costs or benefits cannot be expressed, or it is not convenient to express them, in monetary terms. The solution is to fix some targets for the aspects, which are not measurable in monetary terms, and to apply the monetary calculations to the remaining terms. For example, let us consider the evaluation of a project to improve urban air quality in a highly polluted city. Two natural objectives of the project are the maximisation of air quality and the minimisation of action costs. If air quality cannot be evaluated in monetary terms, a possible solution is to fix target levels for air pollutants and minimise the costs for achieving this level. This can be written as:

$$\min \sum_{t=0}^{T} \frac{C_t}{(1+r)^t} \tag{4}$$

$$s.t. \ Q \geq Q^*$$

where the costs (C) are minimised under the constraint that air quality (Q) is at least equal to a minimum level (Q*)[4].

This is the approach through environmental standards. Some of the aspects disappear from the objective function of BCA and become constraints. Standards for air pollution, water contamination etc., can be seen as a pragmatic solution to the impossibility of determining sensible economic valuations for environmental problems. However, some inconveniences arise also with these techniques. Figure 2 shows three pollution abatement solutions (A, B and C). Solution B is discharged since it clearly exceeds the standard (Std).

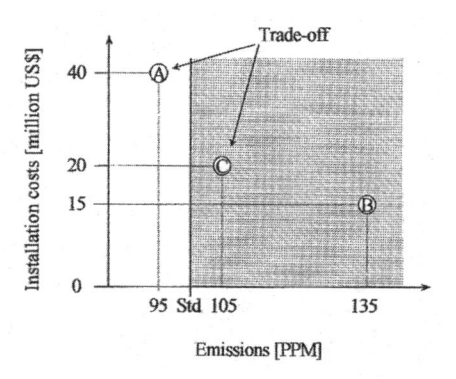

Figure 2. Example of evaluations with CEA.

---

[4] The variable t stays for time, T for the time horizon of the assessment, and r for the interest rate.

The emission levels of A and C are very similar, but A is twice as expensive as C. A mechanistic application of CEA would discard C. A closer analysis may suggest that the trade-off between saving 20 million US$ and exceeding the standard of 5% may make more sense.

Characteristic of CEA is the fact that conflicts are hidden behind the technical evaluations. In some circumstances, sub-optimal solutions are highlighted determining choices that could be improved rather easily by introducing some flexibility. In any case, CEA does not offer any aid to solve the trade-offs which may be emerge during the evaluation. Multicriteria analysis may be used for this purpose.

4. Multicriteria Analysis

The need for developing a methodological platform on which different evaluation perspectives can be considered is one of the main reasons for the growing interest in Multicriteria Analysis (MCA) (cf., [25]; [5]; and [7]). MCA has evolved from a technical procedure for the selection of the best alternative from a set of competing options, to a range of decision-aid techniques. MCA now supports the structuring of a decision problem, the exploration of the concerns of decision actors, the evaluation of alternatives under different perspectives, and the analysis of their robustness against uncertainty. At present, MCA comprises a wide set of tools, but MCA is especially a way of approaching complex decision problems. In particular:

- it makes it possible to analyse the trade-offs between different objectives and concerns, thus supporting the analysis of the pros-and-cons of different options in a transparent way;
- the effects of alternative options can be presented in a variety of forms, such as monetary units, physical units, qualitative judgements, etc.
- it offers a framework for the exploration of the objectives and concerns of decision actors, making it possible to understand and justify the main issues involved in a decision and the interests of the actors involved ;
- it makes it possible to consider the different positions of different decision actors, thus identifying and addressing potential conflicts at an early stage of the decision process;
- it offers the possibility of analysing the sensitivity and robustness of different choices against the effects of uncertainty and on the basis of different future scenarios.

There are various differences between MCA and BCA. First of all, in MCA a solution is never good or bad in absolute terms. It is such only in relation to the revealed objectives of the people who have the role and responsibility of making the decision or to influence the process. It makes little sense to carry out the analysis in isolation from the interests and tastes of the people involved. Doing MCA means to interact with decision actors, and explore their concerns and preferences. The result of MCA is by definition subjective. More so, MCA is meant to shed light into subjectivity and help people combine facts and values to reach a better decision. In addition, although MCA and BCA are both based on the subjective values of people, BCA does this indirectly, using market interactions as the vehicle for expressing people's tastes and preferences. Prices are assumed to reflect these preferences. MCA requires a direct involvement of decision actors and the subjective information in MCA is that of those involved in the assessment. BCA links choices to market preferences; MCA links choices to the preferences of individuals.

4.1 THE MAIN STEPS IN MCA

The attractiveness of a project for an actor (or group of actors) depends on many factors, some of which are hard to identify and to pin down in precise terms. However, it is usually possible to identify some key aspects, which are at the basis of the evaluation perspective of an actor, such as the main objectives and concerns of the actor. The attractiveness of a policy can then be assessed in terms of the degree to which the policy meets the revealed objectives, and it responds to the revealed concerns. In operational terms, this can be supported by a multicriteria approach. There are numerous techniques for MCA. The value-function method, in particular, is simple and transparent and has been applied successfully in many circumstances (cf., [5]; [17]; and [35])[5]. The value-function method is based on three main phases:

1. A structuring phase. Structuring means to explore, identify and make explicit the objectives and concerns of an actor. This requires interviews, brainstorming sessions, etc., usually carried out by an analyst together with the actor(s) involved. The result of a structuring exercise is the identification of objectives and concerns and the impact descriptors, which eventually allow the actors to state whether the project outcomes are attractive or unattractive from their perspective.
2. An impact-analysis phase. Tools for impact analysis are well developed and widely used (cf., [33] and [36]). When spatial information is involved, Geographical Information Systems are often the platform on which environmental models are built and impacts are calculated (cf., [32]).
3. A value-analysis phase. The attractiveness of a project depends on the stated objectives and concerns of the actors and on the impacts of the project. The value analysis phase serves to state the degree to which individual impacts meet individual objectives/concerns, and to assess the relative importance of different objectives/concerns. This is the more technical part of multicriteria analysis.

Going through these three phases means to build a *value model*. In the simplest possible terms, the components of the value model are shown in Figure 3.

---

[5] An overview of alternative approaches to MCA is in Bana e Costa (1990 [1]) and Climaco (1997 [9]).

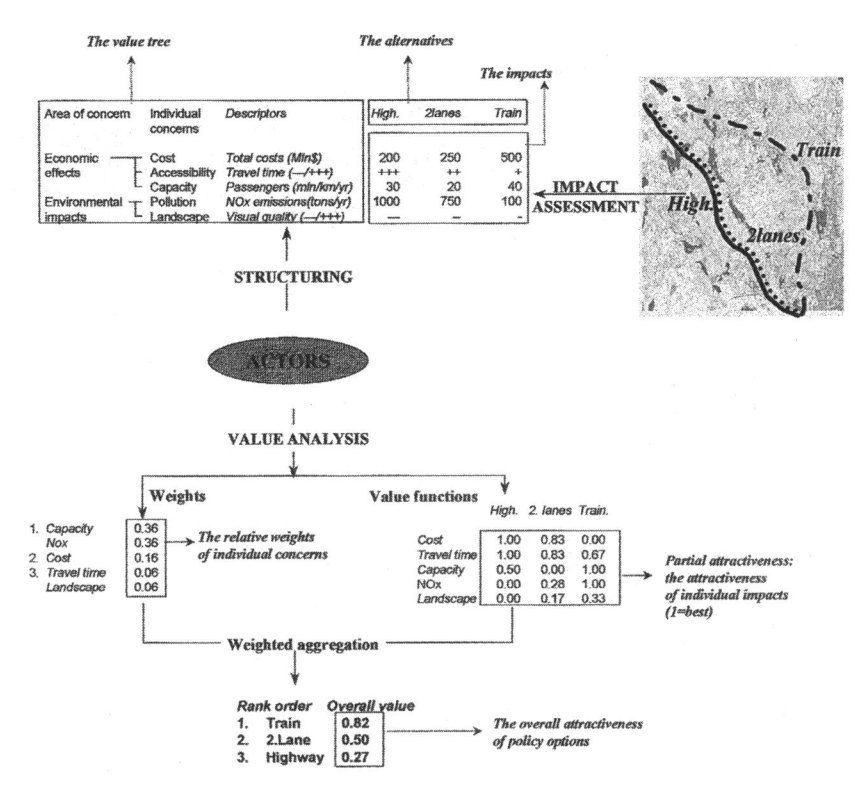

Figure 3. Example of value model applied to the comparison of three alternatives.

This model serves to help the actors understand the issues at stake, to analyse the options available in a systematic way, and to search for alternative solutions that could better satisfy the objectives/concerns revealed (cf., [17]). The structuring phase leads to an organised representation of the decision problem, which identifies the concerns, the descriptors that make the concerns operational, and the impacts of the policy alternatives. This information is organised in an impact table. Value functions are then applied to assess the attractiveness of individual impacts: the higher the value attached to an impact, the better the impact. The weights allow for the aggregation of individual attractiveness scores. This leads to an overall attractiveness score associated to each alternative, which is the basis to provide recommendations for decision.

5. Problem Structuring

Decision support techniques, any analytical approach in general, can be applied when there is a clear definition and a suitable representation of the problem. In technical or engineering projects this is often the case at the beginning of the analysis. In many other

cases, such as for problems that include social and environmental aspects, this is not the case. Structuring means to find a suitable, workable, analytical representation to complex, ill-defined problems. Most environmental management problems have these characteristics in the sense that (cf., [19]):
- There is no definite, unique and clear-cut formulation of an environmental management problem.
- There is no stopping-rule, in the sense that there is no obvious end to the search of better, more suitable solutions.
- Solutions are not true or false, but rather acceptable, good, suitable, etc. to those who are involved in the policy-making process.

Problem structuring is the process that starts from a cloud of concepts, concerns and ideas and leads to an analytical model suitable for the analysis. Amongst the techniques available to support the structuring phase (cf., [28]), the *cognitive mapping technique* (Eden, 1988) can be very useful in combination with MCA. Cognitive maps are built during interview sessions and can serve to disentangle concerns, objectives, cause-effect relationships etc. from the cloud of concepts and ideas which usually emerge at the early stages of the analysis (cf., [3]).

Figure 4 shows a cognitive map used to structure the analysis of environmental management strategies for an industrial area in Northern Italy [15]. The map is the result of several interactions. At the end of this process it was possible to identify:
1. the fundamental objectives and concerns of the analysis (grey boxes);
2. the variables which were beyond the control of the policy makers (ellipsis);
3. the decision areas, that are the actions which could be undertaken to achieve the objectives/ concerns under scenario hypothesis (shaded boxes).

When a cognitive map is used in combination with MCA, one important result of the structuring activity in the identification of a family of fundamental concerns. They can be further organised in a *value tree*, which distinguishes between the areas of concern of the actor (e.g., social effects, environmental impacts), the specific concerns (e.g., noise impacts, landscape modifications) and the descriptors selected by the actor to measure the impacts of interest (e.g. number of houses exposed to noise levels above a threshold) (see the value tree in Figure 3).

5.1 OBJECTIVES AND CONCERNS

A concern is any aspect within a specific context that actors consider relevant for evaluating the attractiveness of a project. Both *stated* objectives and *active* characteristics of a project are termed concerns. As an example (cf., [2]), let us consider the Betuweroute high-speed railway project, which will connect Rotterdam with Germany. One characteristic of the project is its costs. However, costs are not a concern for all actors.

255

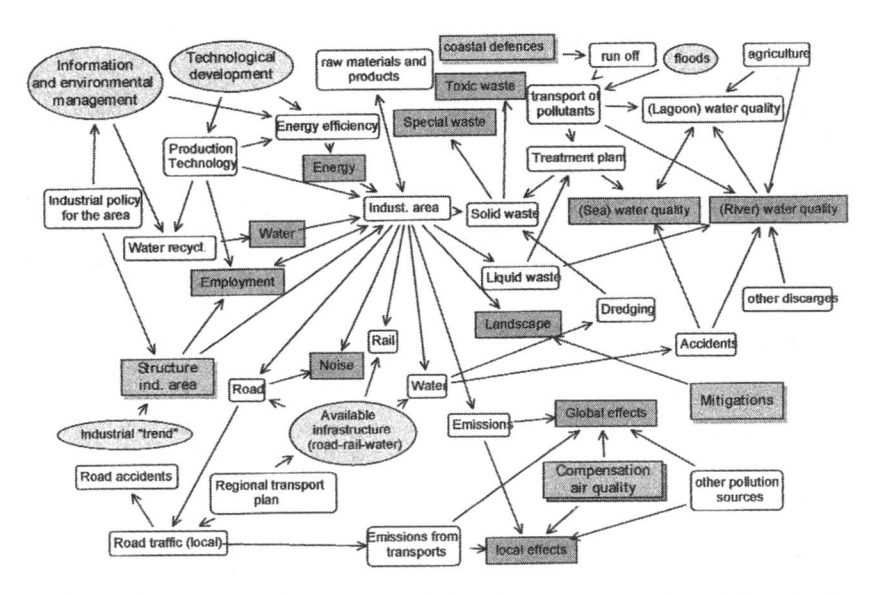

Figure 4. Example of a cognitive map used to structure the assessment of environmental impacts of development strategies for an industrial area in Northern Italy [15]. The grey boxes are the fundamental objectives and concerns of the analysis; the ellipses are the variables beyond the control of the policy makers (scenario variables); and the shaded boxes are the decision areas, or the actions which could be undertaken to achieve the objectives/ concerns of the actors. The white boxes indicate intermediate processes.

While the national authorities (who pay for the project) want to minimise costs, the municipalities involved in the routing decision may well disregard costs. Thus, *to minimise costs* is a stated objective for the Ministry, but it is not a relevant characteristic of the project for the municipalities. As another example, let us consider the problem of improving the road network in the region of Lisbon. When exploring the meaning of this general objective (to improve quality of the regional road network), two fundamental concerns emerged: *to improve the accessibility* in the region at a *reasonable cost*, which correspond to the objectives *maximise accessibility* and *minimise costs*. The construction of new roads, junctions and interfaces between existing and new roads was proposed to achieve these objectives. However, the characteristics of these projects were such that new concerns arose, such as environmental effects and temporary effects during the construction phase. Each of these broad areas of concern was further decomposed in a few more specific concerns: for instance, *noise pollution, air pollution, watershed effects*, and *land use effects*.

An objective can emerge without reference to any explicit characteristic of the actions considered. However, the analysis of the characteristics of the actions can contribute to reveal hidden objectives or other types of implicit concerns.

An objective/concern is fundamental if the projects can be ranked with respect to that concern irrespective of their impacts on any other concern (ordinal independence). A stricter form of independence is cardinal independence, which is valid when the difference of attractiveness between two actions does not depend on other concerns.

Let us consider an example: suppose we want to evaluate the attractiveness of the project in terms of noise pollution, and in particular in terms of "peak noise levels during the night" and "average noise level in 24 hours". These concerns are ordinally independent: whatever the average noise, low peaks during the night are always preferred to high peaks (and vice versa). Cardinal independence, on the contrary, is doubtful. The peaks and the average contribute to a decrease of environmental quality, but their combination is likely to produce a more than additive deterioration. In other words, it is hard to say if a given increase in the noise average produces a large or small deterioration of environmental quality without knowing the peaks during the night. These two concerns are not fundamental. A good candidate would be a combined noise concern, but the combination should not be additive!

5.2 DESCRIPTORS

A descriptor, or criterion, is an ordered set of plausible impact levels in terms of a fundamental concern. It is intended to serve as a basis to describe, as much as possible objectively, the impacts of projects with respect to that FC. A descriptor can be *direct*, or *natural*, (its levels directly reflects effects; such as the number of people affected by respiratory diseases), *indirect* (an indicator of causes more than effects; such as the concentration of air pollutants), or *constructed* (a finite set of reference levels, such as a finite number of plausible health effects of pollution). Moreover, a descriptor can be *quantitative* or *qualitative* (or even *descriptive* or *pictorial*), and *continuous* or *discrete*.

Table 3 shows two examples of descriptors: X_1 is quantitative and continuous; X_2 is constructed and qualitative. They are used to make operational two concerns for the evaluation of alternative layouts for a new railway line in the city of Lisbon. Figure 5 shows a pictorial descriptor used to assess the visual impacts of different interventions on a river basin in Chile [24].

Defining impact descriptors is, in itself, an important activity, where several pitfalls can occur. For example, several alternative descriptors can be chosen to describe the impacts with respect to a given concern, but it is essential to choose only one descriptor for each concern to avoid redundancy and double counting. When discussing descriptors it often happens that hidden concerns are revealed, and the value tree should be restructured, making this stage a recursive learning process.

Table 3. Examples of continuous and constructed descriptors (reproduced and adapted from [4]).

Fundamental concern	Descriptors
FC_1: Impact of the new line on the urban structure	FC_1 is described by the "length in meters of the line on viaduct or in open trench, where the line crosses consolidated urban areas" $X_1 = \{\text{meters} \mid 0 \leq x \leq 1200\}$
FC_2: Effects of the construction of the new line on the railway service to the port of Lisbon	FC_2 is described by a five level scale defined by different possibilities of splitting up the construction of the new line into track-sections, while keeping (or not keeping) in operation the old railway line $X_2 = \{++, +, 0, -, --\}$ where: ++ It is possible to split up the construction of the new line in all its track-sections, while keeping in operation the old line + It is possible to split up the construction in only two track-sections, while keeping in operation the old line o It is impossible to split up the construction, while keeping in operation the old line - It is possible to split up the construction in only two track-sections, but it is impossible to keep in operation the old line -- It is impossible either to split up the construction, or to keep in operation the old line

Figure 5. Pictorial descriptor (reproduced and adapted from [24], page 214). The pictures show the visual effects of changing the water level in a river and the consequences for the visual quality of a waterfall. Level 1 corresponds to the highest flow reduction, while Level 4 is the current situation without flow reductions.

5.3 IMPACT ANALYSIS

Guidelines and checklists for impact assessment are widely available (cf., for instance, the practical guidance documents of the European Commission; [11]). It is important to recognise that the impacts used in MCA are not a general list of the consequences of a project. They reflect the explicit concerns expressed by an actor, in the sense that only impacts that are useful to "trigger" the value system of an actor are useful for the evaluation of a project.

Figure 6 shows the impacts of the Betuweroute freight railway (cf., [6]) for two municipalities involved in the routing decisions. The impacts, computed on a GIS platform, are the main impacts considered as relevant by both municipalities to assess the attractiveness of the railway in their territory. Other municipalities, the regional, and the national authorities employed different descriptors for the assessment of the same railway.

As mentioned earlier, the search for objectives/concerns and the analysis of impacts are strongly interrelated. Some objectives/concerns may emerge only because some specific impacts are expected to occur. They may be even "discovered" during the analysis, in the sense that they were not considered as a general trait of the evaluator's perspective. At the same time, impacts are the identity card of a policy option only if they relate to the objectives/concerns of somebody (an individual, organisation, or even society as a whole).

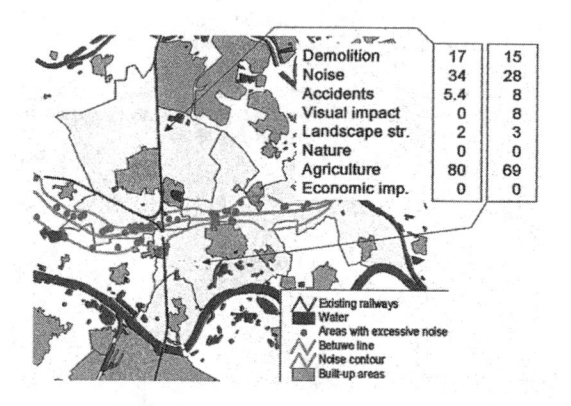

Figure 6. Impact analysis for the Betuweroute freight railway. The two columns of scores are the impacts associated to two municipalities shown in light grey in the picture.

6. Value Analysis

Impact analysis serves to estimate, more or less precisely, the effects of a project. The result is a list of impacts, which describe the consequences of the project in terms of what the actors care about. However, this is meant to be a factual, objective analysis. Impact analysis does not answer questions such as: Does an increase in accessibility of 10% justify the costs of a transport infrastructure? Is a land take of 500 ha acceptable? Is a reduction of 5% in air emissions a large or small improvement? Value analysis serves to analyse the attractiveness of impacts and to provide an answer to these questions.

By requiring the policy actors to evaluate a series of hypothetical questions, the evaluation strategy of the actors can be revealed, understood and checked for coherence. Value functions are a mathematical representation of this strategy, which, albeit in a simplified form, reflects the preferences and tastes of the actors involved.

The list of impacts can thus be translated into a list of attractiveness scores: the higher is the value, the better is the impact. The overall attractiveness of a project depends on these values and on the weight associated to individual objectives/concerns. Through a weighted scheme, an overall value score can be associated with a project, reflecting its overall attractiveness.

Given an alternative A and its impacts $a_1, a_2, ..., a_n$ on the n descriptors, the overall attractiveness of A is computed as:

$$V(A) = \sum_{i=1}^{n} v_i(a_i) \cdot w_i \tag{5}$$

where: $v_i(a_i)$ is the value function attached to the i-th concern and w_i the weight of the i-th concern.

6.1 VALUE FUNCTIONS

Figure 7 shows four examples of value functions, representing different judgement strategies in various decision situations (see [5]). The simplest value function is shown in Figure 7a and is used to appraise job offers. The value of salary increases with the salary level, following the concave shape commonly accepted in economics: a 1,000$ increase makes a larger improvement in the case of low salaries.

Figure 7b shows a radically different strategy applied to the evaluation of the pollution emissions of an industrial plant. This curve distinguishes only between good and bad situations and does not highlight intermediate values. Any level below the threshold is as good as any other.

Figure 7c shows a mixed strategy for the evaluation of the same pollution levels. Below a certain concentration the values are high, representing good environmental quality. After a certain level there is a sharp decrease in values, which corresponds to a rapid increase in environmental deterioration. The final part of the curve is again flat, indicating saturation of effects.

Figure 7d describes a value function for the evaluation of the number of parking places in a city centre. The parking spaces are described qualitatively in a five-points scale. The function shows that the current situation is associated to a low value compared to the best solution of increasing the number of parking spaces and introducing parking fees simultaneously. A low number of parking places has adverse effects on commercial activities, while a high number of parking places increases the risk of congestion and decreases the quality of the urban area, if parking fees are not introduces to limit the demand.

The curve in Figure 7a is a pure representation of personal values. If 40,000$ and 80,000$ are the actual salary and the goal salary respectively, then the value function specifies how close/far each job offer is to/from the objective and the status quo. A different evaluator could have expressed different values; for instance, highlighting more salary increases for low salaries (a more concave value function). The curve in Figure 7b is likely to represent concerns that go beyond the adverse effects of pollution; for instance, the risk of fines or legal actions for exceeding allowed emissions. In this case, the pollution level is a proxy for money or for the good image of the company. Pollution levels *per se* are irrelevant. Figure 7c shows another case in which pollution is a proxy for another concern: in this case, the effects of pollution on humans or animals. Similarly, the curve in Figure 7d involves indirect concerns such as the effects on commercial activities and urban quality. In all these three cases, the value judgement is combined with factual information: for instance, on the existence of fines, on the adverse effects of pollution or on the traffic levels at which congestion occurs.

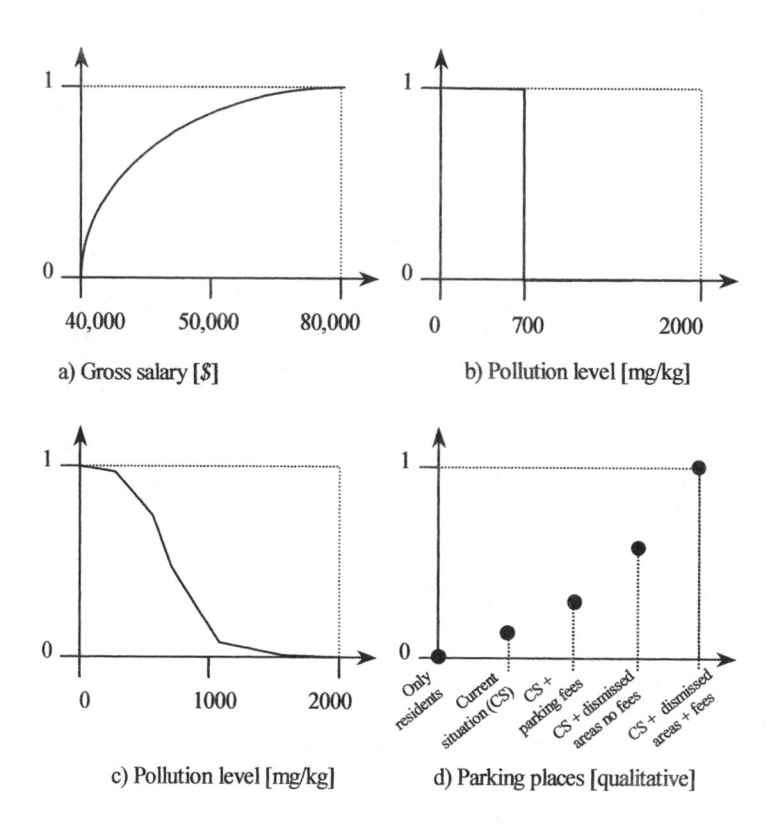

Figure 7. Examples of value functions (reproduced and adapted from [5]).

Value functions can be applied to numerical descriptors, such as those in Figure 7a-c, and qualitative descriptors, such as that in Figure 7d. The value is a dimensionless score: a value of 1 indicates the best available performance and a high objective achievement, while a value of zero indicates the worst performance. With a value function, the decision variables - for example, pollution levels - are not analysed for their face value but for their meaning. This makes explicit what counts for the decision (the value) compared with what is measured to support the decision (the pollution).

Techniques for value function assessment can be classified taking into account (see [5] for details):

1. The **assessment strategy**, which refers to the items on which the assessor is asked to reveal his/her preferences. Two strategies are used: decomposed scaling, which

requires the assessor to evaluate and compare different levels of a descriptor, and holistic scaling, which estimates the value functions indirectly starting from the preferences of the assessor for hypothetical alternatives.
2. The **information available**, which relates to the degree to which the assessor is able to specify preferences numerically. The assessor may be able to specify: only the direction of preferences (increasing or decreasing curve), or the shape (e.g., concave), or interval point estimates (e.g., the value of this impact is between 0.4 and 0.6), or point estimates (the value of this impact is 0.5), or the full curve.
3. The **assessment techniques**, which include the question-answer protocols used to assess information and the methods used to estimate the model parameters when they are only indirectly linked with the information assessed.
4. The **assessment procedures**, which are organised and structured sequences of strategies and techniques to assess the value function model from the information available.

6.2 WEIGHTS

The notion of relative importance, the so-called weights, is perhaps the most fundamental notion in MCA. It is important to stress that there is no sense, and it is theoretically incorrect, to specify weights outside of the context of the evaluation model, that is without reference to the mathematical rule which is used to aggregate partial attractiveness into an overall attractiveness value (the weighted sum, in this case). In an additive model of values, in all compensatory[6] approaches in general, the *scaling constants* (weights) have no absolute or intrinsic meaning. It is meaningless to derive them without reference to the impact ranges. Keeney (1992; pp. 147-148 [17]) clearly explains this critical mistake:

> *"There is one mistake that is very commonly made in prioritizing objectives. Unfortunately, this mistake is sometimes the basis for poor decisionmaking. It is always a basis for poor information. As an illustration, consider an air pollution problem where the concerns are air pollution concentrations and the costs of regulating air pollution emissions. Administrators, regulators, and members of the public are asked questions such as "In this air pollution problem, which is more important, costs or pollutant concentrations?" Almost anyone will answer such a question. They will even answer when asked how much more important the "more important" objective is. For instance, a respondent might state that pollutant concentrations are three times as important as costs. While the sentiment of this statement may make sense, it is completely useless for understanding values or for building a model of values. Does it mean, for example, that lowering pollutant concentrations in a metropolitan area by one part per billion would be worth the cost of $2 billion? The likely answer is 'of course not.' Indeed, this answer would probably come from the respondent who had just stated that pollutant concentrations were three times as important as costs. When asked to clarify the apparent discrepancy, he or she would naturally state that the decrease in air pollution was very small, only one part in a billion, and the cost was a very large $2 billion.*

---

[6] Compensatory means that a "bad" performance in one aspect can balanced out by a "good" performance on another. Other methods, such as the Electre methods, do not assume compensation (see [34]).

The point should now be clear. It is necessary to know how much the change in air pollution concentrations will be and how much the costs of regulation will be in order to logically discuss and quantify the relative importance of the two objectives.

This error is significant for two reasons. First, it doesn't really afford the in-depth appraisal of values that should be done in important decision situations. If we are talking about the effects on the public health of pollutant concentrations and billion-dollar expenditures, I personally don't want some administrator to give two minutes of thought to the matter and state that pollutant concentrations are three times as important as costs. Second, such judgements are often elicited from the public, concerned groups, or legislators. Then decisionmakers use these indications of relative importance in inappropriate ways. {...}

If the value tradeoffs are done properly and address the question of how much of one specific attribute is worth how much of another specific attribute, the insights from the analysis are greatly increased and the likelihood of misuse of those judgements is greatly decreased."

Unfortunately, there are several popular weighting procedures in which the assessment of weights to be used in an additive model is made by directly comparing criteria in terms of their relative *importance*. Correct weights must be assessed with reference to the impact ranges. This is what is done in the *trade-off* procedure [18] and in *swing weighting* [35], so as in the MACBETH weighting procedure [4].

Swings are applied between two impact levels, for instance the extremes of the descriptor ranges. These extremes may corresponds to the best and worst impact on each descriptor, or they can be selected to represent specific reference points (e.g., a target value, or a constraint). Figure 8 illustrates the series of swings for the example of Figure 3. The most convenient swing is that of *capacity* and NO_x. Swinging these two levels from their worst to their best level produces the largest improvements. The improvement obtained by swinging *capacity* and that obtained by swinging NO_x are equivalent. The second swing is *costs*, and the third, in this case, is that of *landscape* and *travel time*.

The sequence of swings corresponds to a qualitative assessment of weights (a weight ranking). In this case: $w_{capacity} = w_{NOx} > w_{costs} > w_{travel-time} = w_{landscape}$. These qualitative weights can already be sufficient to compare the alternatives under evaluation.

Let us consider a simpler case with three alternatives (A_1, A_2 and A_3) and three descriptors for which the weights are: $w_1 > w_2 > w_3$. The dotted area in Figure 9 shows all possible combinations of w_1, w_2 and w_3 consistent with the weight ranking. The vertexes of the area are the extreme weights (1, 0, 0), (0.5, 0.5, 0) and (0.333, 0.333, 0.333), where weights are assumed to add up to one. It is easy to demonstrate that any rank order of the alternatives corresponding to an intermediate point is included in the rankings corresponding to the extreme weights.

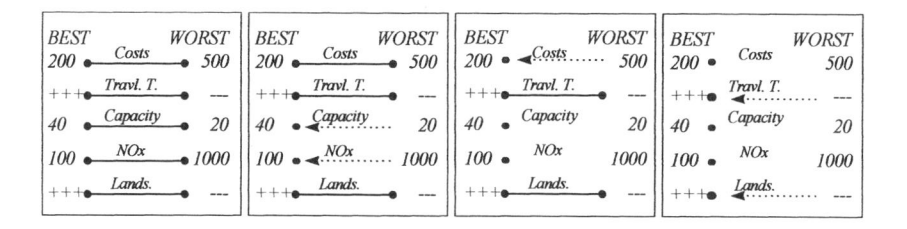

Figure 8. The swing technique.

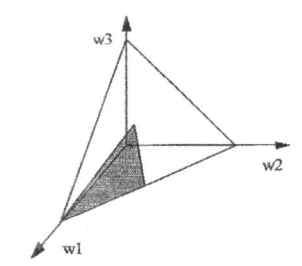

Figure 9. Graphical representation of a set of three weights.

If, for instance, the extreme weights correspond the following rankings: $A_1 > A_2 > A_3$, $A_1 > A_3 > A_2$ and $A_1 > A_2 > A_3$, then any other consistent set of weights would lead to $A_1 > A_2, A_3$. It can thus be concluded that A_1 is preferred to A_2 and A_3. This partial outcome is only based on a qualitative ranking of criterion importance and does not require additional assumptions. To distinguish between A_2 and A_3 it is necessary to gather further details on the weights.

Quantitative weights can be obtained by rating the swings in Figure 8, and assigning to each swing a numerical value that quantifies the improvement associated to a swing (see [5], for a collection of methods for weight assessment). It important to recognise that meaningful numerical statements are difficult to obtain and a certain degree of erratic behaviour seems unavoidable.

The pair-wise comparison method has been developed to simplify the assessment. The idea is to compare each criterion with each other and specify an importance ratio[7]. Table 4 shows a simple matrix.

---

[7] It is worthwhile stressing again that importance is actually interpreted in terms of trade-offs.

Table 4. Pair-wise comparison matrix for tree criteria

	C_1	C_2	C_3
C_1	-	3	5
C_2	1/3	-	3
C_3	1/5	1/3	-

Each cell in the table shows the importance ratio between the row concern and the column concern. For instance, cell (1,2) shows that C_1 is about three times as important as C_2. The set of weights that better represent this information is computed mathematically (for instance, as the maximum eigenvector of this matrix, cf., [30]). In this case: $w_1=0.6$, $w_2=0.3$, $w_3=0.1$. This method requires the criteria to be analysed in pairs, reducing significantly the difficulty of the task. However, the process becomes long and complicated if the number of concerns increases beyond a certain level (more than 10).

6.3 SENSITIVITY ANALYSIS

Estimates involved in project assessment may be affected by substantial errors (e.g., what will be the number of jobs - directly or indirectly - created by a transport infrastructure?). People may also hesitate about their preferences and priorities, which will be reflected in uncertain value function models (e.g., does it make sense to spend 1M$ to prevent on average 100 car accidents per year?). Sensitivity and robustness analysis are used to analyse the credibility and stability of the decision when uncertainty and information gaps are present.

Sensitivity analysis explores the effects of changing data and model parameters in a "limited" surrounding of a solution [16, 27]. Robustness analysis takes a more radical approach (cf., [29]). Starting from the consideration that many assumptions and numerical estimates are somewhat arbitrary, robustness analysis performs a systematic analysis of a large set of variations, which are considered as plausible in the context in question. None of these approaches reduces uncertainty or solves the uncertainty problem. However, if a solution is stable in the face of these tests, the reliability of the outcome substantially increases.

Table 5 shows an example of uncertainty on impacts scores (+/- 25%). The table shows the probability of each alternative to be ranked first, second and third. To obtain this result, the impact matrix is first randomly modified within the uncertainty interval and then the alternatives are ranked with the weighted sum. This process is repeated a large number of times (one thousand in this example).

The alternative Train is ranked first in 99% of the cases and second in 1% of the cases. Within the 25% uncertainty level is seems reasonable to conclude that the alternative Train ranks first. In other cases this analysis leads to uncertain rankings and more stable results can only be obtained by increasing the accuracy of impact measures. The same analysis can be computed for the weights or any other parameter of the value function model.

Table 5. Impact uncertainty for the test case.

	Train	2.Lan	High
Position 1	0.99	0.01	0.00
Position 2	0.01	0.89	0.11
Position 3	0.00	0.11	0.89

A more radical approach is shown in Table 6, which computes the minimum weight change that produces a specified ranking. The initial situation shows the initial set of weights and the corresponding ranking. The final situation shows the closest set of weights that would make Train and Highway equally attractive. This can only be achieved by totally modifying the weights and their rank order, implicitly discarding the possibility of rank reversal between Train and Highway.

Table 6. Weight combination uncertainty analysis

Initial			Final		
1. Capacity	0.357	1. Train	1. Cost	0.389	1. 2.lanes
2. NOx	0.357	2. 2.lanes	2. Capacity	0.214	2. High
3. Cost	0.157	3. High	3. NOx	0.186	Train
4. Time	0.065		4. Landscape	0.112	
5. Landscape	0.065		5. Time	0.100	

Sensitivity analysis plays a key role in reducing *ex-post* the requirements for precise weights and data. This is especially important for the weights because there is always a degree of fuzziness in translating human preferences and beliefs into numbers. The availability of sophisticated sensitivity techniques allows, to some extent, the use of uncertain estimations since it is possible to test if this uncertainty is relevant.

7. Conclusions

The basic conviction underlying a Multi Criteria decision Aid (MCA) approach is that the explicit introduction of several criteria, each representing a fundamental concern (e.g., economic effects, social effects, environmental deterioration) is a better path for robust decision-making for multidimensional and ill-defined problems. In contrast to more classical approaches, the MCA framework facilitates learning about the problem and about the alternative actions, by enabling the actors to think about their values and preferences from several points of view (see [2]).

It can be argued that a multidimensional problem can be re-formulated by using a *single-criterion approach*, provided a single criterion which incorporates all concerns can be found (such as in BCA). A superficial view of MCA techniques may suggest this. However, value function models, when properly constructed and used, are tools for exploring the interplay between criteria. The justification of the usefulness of an MCA approach thus requires more than a reference to the multidimensional nature of complex real situations.

Nevertheless, BCA, CEA and MCA are not competitive methods. The debate on conventional monetary evaluations and multidimensional assessments increasingly recognises that these approaches are complementary. The results of BCA are a useful input to the decision process, provided the limitations of the technique are recognised. MCA tools can support the analysis from a broader perspective, but it makes little sense to disregard the economic dimension.

References

1. Bana e Costa, C.A. (1990) *Readings in Multiple Criteria Decision Aid*, Springer, Berlin.
2. Bana e Costa, C.A., E. Beinat (1998) Assessment and evaluation of policy options, in E. Beinat (Ed.) *A methodology for policy analysis and spatial conflicts in transport policies*, Research Report no. R98/8, Institute for Environmental Studies, Amsterdam: 29-44.
3. Bana e Costa, C.A., L. Ensslin, A.P. Costa (1998) Structuring the process of choosing a rice variety at the South of Brazil, in E. Beinat, P. Nijkamp (1998) *Multicriteria analysis for land-use management*, Kluwer, Dordrecht: 33-46.
4. Bana e Costa, C.A., Vansnick J.C. (1997) A theoretical framework for measuring attractiveness by a categorical based evaluation technique (MACBETH), in J. Clímaco (Ed.), *Multicriteria Analysis*, Springer-Verlag, Berlin: 15-24.
5. Beinat, E. (1997) *Value functions for environmental management*, Kluwer, Dordrecht.
6. Beinat, E. (1998) Case study: the Betuweroute freight railway, Beinat (Ed.) *A methodology for policy analysis and spatial conflicts in transport policies*, Research Report no. R 6/98, Institute for Environmental Studies, Amsterdam:133-153.
7. Beinat, E., P. Nijkamp (1998) *Multicriteria analysis for land-use management*, Kluwer, Dordrecht.
8. Brent, R.J. (1996) *Applied cost-benefit analysis*, Edward Elgar, Cheltenham.
9. Climaco, J. (1997) *Multicriteria Analysis*, Springer, Berlin.
10. Dixon, J.A., L. Fallon Scura, R.A. Carpenter, P.B. Sherman (1995) *Economic analysis of environmental impacts*, Earthscan, London.
11. EC - European Commission (1996) *Environmental Impact Assessment: guidance on Scoping*, http://europa.eu.int/en/dg11/eia/scop-en.htm (accessed August 1998).
12. Eden, C. (1988) Cognitive mapping, *European Journal of Operational Research*, 36:1-13.
13. Freeman, A.M.III (1979) *The benefits of environmental improvements: theory and practice*, The John Hopkins University Press, Baltimore.
14. Gilbert, A., R. Janssen (1998) *Use of environmental functions to communicate the values of a mangrove ecosystem under different management regimes*, Ecological Economics, 25, 323-346.
15. IAL (1998) The industrial policy for the Aussa-Corno area: an application of strategic EIA (in Italian), Post-graduate course in Environmental Impact Assessment, IAL-University of Udine, Udine, Italy.
16. Janssen, R. (1992) *Multiobjective decision support for environmental management*, Kluwer, Dordrecht.
17. Keeney, R.L. (1992) *Value-Focused Thinking*, Harvard University Press, Cambridge.
18. Keeney, R.L., H. Raiffa (1976) *Decisions With Multiple Objectives: Preferences and Value Tradeoffs*, Wiley, New York.
19. Khisty, C.J. (1997) *The dilemma of bounded rationality and unbounded uncertainty in transport planning: how can adductive inference help?*, Paper no. 970295, Transportation Research Board, 76th Annual Meeting, January 12-16, 1997, Washington, D.C.
20. Lindblom, C.E., J. Woodhouse (1993) *The policy making process*, Prentice Hall, New Jersey.
21. Mintzberg, H., D. Raisinghani, A. Théorét (1976) The structure of unstructured decision process, *Administrative Science Quarterly*, 21:246-275.
22. Morgan, M.G., M. Henrion (1990) *Uncertainty: a Guide to Dealing With Uncertainty in Quantitative Risk and Policy Analysis*, Cambridge University Press, New York
23. Munda, G. (1995) *Multicriteria evaluation in a fuzzy environment*, Physica-Verlag, Heidelberg.

24. Nardini, A. (1998) Improving decision-making for land-use management: key ideas for an integrated approach based on MCA negotiation forums, in E. Beinat, P. Nijkamp (1998) *Multicriteria analysis for land-use management*, Kluwer, Dordrecht: 199-225.
25. Nijkamp, P., P. Rietveld (1986) Multiobjective decision analysis in regional economics, in P. Nijkamp (Ed.) *Handbook of Regional and Urban Economics*, North Holland, Amsterdam: 493-541.
26. Pierce, D., K. Turner (1990) *Economics of natural resources and the environment*, Harvester Wheatsheaf, London.
27. Rios Insua, D. (1990) *Sensitivity Analysis in Multiobjective Decision Making*, Springer, Berlin.
28. Rosenhead, J. (1989) *Rational analysis for a problematic world*, Wiley, Chichester.
29. Roy, B., D. Bouyssou (1984) Comparison of a multiattribute utility and an outranking model applied to a nuclear power plant siting example, in Y.Y. Haimes, V. Chankong (Eds.) *Decision Making With Multiple Objectives*, Springer, Berlin: 482-494.
30. Saaty, T.L. (1990) How to make a decision: the Analytic Hierarchy Process, *European Journal of Operational Research*: 48: 9-26.
31. Sagoff, M. (1988) *The economy of the earth*, Cambridge University Press, Cambridge.
32. Scholten, H., Stilwell, J.C.H. (1990, Eds.) *Geographical Information Systems and urban and regional planning*, Kluwer, Dordrecht.
33. Thérivel, R., R.M. Partidário (1996) *The practice of strategic environmental assessment*, Earthscan, London.
34. Vincke, Ph. (1992) Multicriteria Decision-Aid, Wiley, Chichester.
35. von Winterfeldt, D., W. Edwards (1986) *Decision Analysis and Behavioral Research*, Cambridge University Press, Cambridge.
36. Wathern, P. (1988) Environmental Impact Assessment, Routledge, London.

IMPLEMENTATION OF ENVIRONMENTAL SYSTEMS IN COMPLIANCE WITH ISO 14001 STANDARD IN THE CONTEXT OF SMEs IN NORTHWESTERN ITALY

SERGIO OLIVERO
Quality System Auditor (Cepas, Italian National Register)
CONSIEL
Via Meravigli 12/14 – 20123 Milano - Italy

Abstract. This paper discusses the problems related to the implementation of Environmental Systems in compliance with ISO 14001 standard in the context of Small and Medium Enterprises (SME) in nortwestern Italy.

Results are here presented of a survey performed in the period January - June 1998 on a sample of industries and breeding firms in the Province of Cuneo. The survey aimed at the determination of the basic requirements for building an *ISO 14001-compliant* Environmental System. It presents the main implementation issues together with possible operational choices.

Many Quality Systems were implemented in compliance with the ISO 9001 standard, that is today well known, and useful to participate in Tenders. This paper points out how the ISO 9001 can be considered as a basis on which to build an integrated system able to comply with the ISO 14001 requirements in terms of organisational structure, documental management, and assessment & audit criteria.

1. Introduction

This paper describes the problems related to the implementation of Environmental Systems in compliance with the ISO 14001 standard in the context of Small and Medium Enterprises (SME) in nortwestern Italy, through an analisys performed in the Province of Cuneo, near the French border. The Province is a densely populated area, with many small towns, a strong agricultural production, a variety of Small and Medium Enterprises, and one of the Italian Natural Parks named Argentera.

Results are presented of a survey performed in the period January - June 1998 on a sample of industries and breeding firms in the Province of Cuneo. Handicraft and farming firms were not taken into consideration both because their size can hardly match the ISO System requirements, and also as a consequence of today's still statistically immature farmers' attitude in the field of new organisation models. The survey aimed at the determination of the basic requirements for the building of an ISO 14001-compliant Environmental System. The main implementation issues are presented together with possible operational choices.

A description is provided of the structure of Italian Laws on Environmental subjects resulting from European Directives and the present laws. The main critical issues in application with respect to SMEs and the enforcement bodies are pointed out.

Many Quality Systems were implemented in compliance with the ISO 9001 standard, that is today well known and useful to participate in tenders. This paper points out how ISO 9001 can be considered as a basis on which to build an integrated system able to comply with the ISO 14001 requirements, in terms of organisational structure, documental management, and assessment and audit criteria.

2. The Province of Cuneo

The Italian territory is subdivided into twenty Regions. Each of them is further divided into Provinces. The widest in Italy is the Province of Cuneo, in southern Piemonte Region, near the French border. It covers 6902 square Kilometres, and has 547,234 inhabitants, 250 country towns with independent Councils, a strong agricultural, breeding and an industrial production. The numbers of industries and handicraft firms are about 1,000 and 18,000, respectively [2].

3. The Laws in Italy

Italian Central and Regional Administrations have aknowledged the need to provide Enterprises with new normative tools able to effectively operate for environmental safety and security's sake: many laws were issued, sometimes contradictory and often difficult to understand, with a wealth of obligations but often without the definition of clear application criteria.

The framework of a typical Italian Law is characterised by a high rate of complexity, that is mainly a consequence of a process of continuous step-by-step change. For instance, a law issued in 1991 is modified by another law in 1993, but the change only affects some articles, while the other ones remain in force. Some years later -say in 1995- another change is introduced by another law, and this process goes on endlessly, while some circulars are issued by the Ministries with the respective interpretations. In order to be sure of being fully compliant with the Law, one must take into consideration a structure made up of four parts. The first one is obviously the text of the Law under examination. The second one contains the references to other laws explicitly declared in the text. The third and fourth ones must be filled with the references to all other applicable laws and circulars issued by the Ministries, respectively. In order to be sure of a full compliance, the other applicable laws and the circulars must be found: in fact, the list of the laws in force, with a document correlation tree is not systematically made available to enterprises by the Enforcement Bodies. There is no systematic review of the existing lists of the laws in force. Furthermore, there are many Enforcement Bodies: the Industrial Accident Agency, the Regional Environmental Agency, the Health Department, the Fire Department, the Town Councils, the Judiciary and the Police. The ordinary man has sometimes problems in correctly identifying the laws that he has to observe and the enforcement Bodies involved. The language used is juridical, and interpretation problems may arise. These facts, together with the common lack of application criteria and the unavailability of know-how for effective self assessment, can lead to a widespread fear of "forgetting something", and this feeling often leads to extreme caution, chiefly in the "dangerous" field of the environment. In other words, today in northwestern Italy the **driving force** of organisational change can be identified in the compulsory rules issued by the government (and not in an enhanced environmental culture of SMEs), but these rules are often hardly understandable, and data mining is required to know who must do what. The Council of the Province of Cuneo has brought into operation a Waste Management Program [3] [4] with the aim of helping SMEs to reach full compliance with the environmental laws in force. It should be obtained through an integrated management of the Environmental processes, by the creation of four geographical Basins provided with up-to-date technology and by the implementation of a management system in compliance with the ISO 9001 and ISO 14001 Standards.

4. The Survey

A sample of industries and breeding firms active in the Province of Cuneo was analysed.

4.1 COMPOSITION OF THE INDUSTRY SAMPLE

Machinery	4
Paper	2
Wood	2
Building material	2
Food	2
Services	4
Dairy	2

4.2 COMPOSITION OF THE BREEDING-FIRM SAMPLE

Cattle	3
Swine	5
Sheep	1

4.3 THE REQUIREMENTS

The executive managers were asked to express their requirements: their answers were listed with respect to specific classes of aggregation and processed according to a logical framework that proved its effectiveness in the past in the context of the implementation of ISO 9001-compliant Quality Systems. No firms wanted to be expressly quoted, but this requirement does not seem to negatively influence the results of the analysis, thanks to the consequent total lack of inconsistency between actual operations and answers.

The classes of aggregation of the requirements are shown with an identifying number ranging from 1 to 10. Inside each class a number of sub-classes are defined, and they are numbered in a sequence with a root corresponding to the number of its class. Here the original numeration of the classes is maintained.

1. Initial Self Assessment
    1.1 Clear mandate to Self Assessment Team
    1.2 Theoretical reference framework
    1.3 Operational criteria
    1.4 Compatibility with existing power balances (who is assessing whom)
    1.5 Translation of methodologies into criteria respectful of Italian culture

2. Continuous Improvement
    2.1 Analytical tools
    2.2 Process description (e.g., SADT-IDEF0)
    2.3 Internal indicators with consistent data bases
    2.4 Process effectiveness indicators
    2.5 Tools for process capability determination

3. Preventive Approach
    3.1 Continuous improvement tools
    3.2 Matching of process effectiveness and internal indicators
    3.3 Workflow technicalities (e.g., IDEF3)
    3.4 Systematic review of the corrective actions in progress
    3.5 Simulation tools

4. Organisation
    4.1 Plan, programs, goals and targets (what to do)
    4.2 Responsibilities and mandate (who is doing what)
    4.3 Procedures (how to do what)
    4.4 Resource allocation
    4.5 Communication flow
    4.6 Functional, inter-functional and Team links

5. Training
    5.1 Communication tools
    5.2 Information
    5.3 Top Management training
    5.4 Middle Management training
    5.5 Employee training
    5.6 Training update
    5.7 Applicability of know-how
    5.8 Creation of a new culture (compliance to law as a layer for improvement)

6. Document Management
    6.1 Traceability
    6.2 Reachability and identification
    6.3 Periodic review

6.4 Availability in adequate places
6.5 Elimination in case of obsolescence
6.6 Classification
6.7 Availability of history of all actions and performances
6.8 No "sea of paper" effect

7. Auditing
7.1 Internal procedure
7.2 Strong mandate to internal auditor
7.3 Elimination of the root causes of non-compliances
7.4 Completely independent auditing functions
7.5 Auditing consistent methodologies for any specific market sector
7.6 Specific process know-how
7.7 Confidentiality (criminal trial risk for environmental non-compliances)

8. Communication
8.1 Internal communication policy
8.2 External communication policy
8.3 What is to be declared
8.4 What must not be declared

9. Fear of penalties

10. Quantitative description of advantages of ISO 14001
10.1 Great Customers' order
10.2 Insurance premium
10.3 Requirement in Tenders

It must be remarked that the purpose of this contribution is not to provide the grounds for an exhaustive list of the evidences resulting from the survey. The survey was oriented to the definition of operational tools able to contribute to the effectiveness of consultancy when the growth of an organisation System is the concern. However, the information considered seems to contribute to a sharper understanding of the implementation problems.

After the analytical description of the requirements, industry and breeding firms' representatives were asked to choose the three most critical sub-classes or classes, in order of importance. Every choice was then given the values +3, +2 or +1 with reference to the class/subclass involved and the order of priority. The twelve most critical classes/subclasses resulting from this processing are listed below starting from the most crucial:

9.	*Fear of penalties*
8.3	*What is to be declared*
8.2	*External communication strategy*
10.	*Quantitative description of the advantages of ISO 14001*
7.6	*Specific process know-how*
1.4	*Compatibility with existing power balances*
5.4	*Middle Management training*

5.7   *Applicability of know-how*
2.5   *Tools for process capability determination*
4.3   *Procedures*
3.3   *Workflow technicalities*
6.7   *Availability of history of all actions and performances*

4.4. ISO 14001 AREAS OF IMPLEMENTATION

Therefore, when implementing an ISO 14001-compliant Environmental System six Areas of Implementation can be identified:

Knowledge of the Rules
Business
Audit Effectiveness
Organisation and Power
Training
Process Know-How

4.4.1. Knowledge of the Rules

The choice of the ISO 14001 certification is rarely the result of a full awareness of the potentials of this standard, but should be considered as the search of a tool useful for the reengineering of the organisation with the aim of full compliance with the laws on Environment. In other words, the persons in charge of SMEs know that the capability to effectively manage environmentally-related activities represents an unavoidable step in their tasks, but they are also aware that today in Italy it is really difficult to get all the data in terms of applicable laws (with reference to their specific market sector), of correct interpretation of the laws in force and of the identification of the enforcement bodies involved. As a consequence of this lack of thorough and reliable information, the general attitude of managers is influenced by the fear of penalties and by an obsessive interest in defining what is to be declared and what must not be declared. The only answer to this need of information is to assure the availability of reliable data, together with application criteria tailored upon the specific market sectors and with self assessment and internal auditing technicalities and methodologies. With reference to the market sectors of Wooden Furniture, Town Water Treatment and Paper-mill, an analysis is being performed according to the following steps: data collection (EU Directives, Italian National and Regional Laws, Circulars and Judiciary's Sentences), systematic filing of the collected data, review and upgrade. As soon as the data mining is fully operational, a plan of systematic communication will be implemented with a first sample of enterprises.

4.4.2. Business

The cost of the implementation of an ISO 14001-compliant environmental system must be clearly evaluated with reference to the size of the enterprises and the specific market sector [1]. The growth in the number of *ISO 9000-Series Standard* Certificates in Italy is a consequence of the new vendor-rating criteria of the main Italian contractors and of the evolution of the procedures in public tenders. Thanks to these two drivers, the number of

ISO 9000-compliant Quality Systems has grown steadily in the last years, up to about 10,000 [1].

The implementation of an ISO 14001-compliant environmental system can easily take advantage of the existence of a Quality System in compliance with the ISO 9001/2/3 Standard, expeciallly in case of conformity to the ISO 9001. The statement of a Policy, the organisation structure, the review methodologies and preventive approaches, and the internal auditing procedures oriented to continuous improvement can be adjusted to the ISO 14001 requirements by the "tuning" of the existing Quality System with reduced and quantifiable costs. Last but not least, the top managers are still aware of the reasons for the ISO 9000 choice and can therefore be more easily persuaded to move towards the ISO 14001 compliance.

4.4.3. *Audit Effectiveness*

The chances of ISO (9000 and 14001) rely on the capability to really contribute to the effectiveness of the processes through new organisation criteria and methodologies. One of the most powerful tools for process improvement is audit effectiveness: by the identification of the non-compliances with respect to a chosen framework, an SME can establish corrective actions and move to excellence step-by-step [1].

The main requirements for audit effectiveness are the Auditors' specific process know-how and the availability of the history of all actions and perfomances: the former can be attested by Certification Bodies of Personnel (e.g., CEPAS in Italy); the latter through process workflow description and electronic data bases. The use of SADT-IDEF0 and IDEF3 representation languages is proving effective, provided that human-interface protocols are defined with particular regards to simplified drawing tools and new methodologies to establish indicators.

4.4.4. *Organisation and Training*

The upgrade and continuous improvement of the organisation can be obtained mainly by a new training approach aiming both at building new managing criteria and at making all employees aware of the need of a change in their way of working. With reference to a plan being currently developed and whose operational output was scheduled for January 1999, the following steps are taken into consideration:

- Choice of the market sectors
- Analysis of the training requirements with reference to each chosen market sector and to the organisation levels of the people involved
- Definition of the training contents
- Definition of the didactical units
- Engineering of the training paths (two main areas: functional/specialistic and management)
- Planning of the training modules
- Development of field options (stages focused upon the test of effectiveness of the training paths)

4.4.5. *Process Know-how*

After the clear identification of the requirements both of the ISO-14001 compliant Environmental System and of the laws in force, all implementation issues can prove effective only if a concurrent re-engineering of the organisation of the enterprise is performed. The Self Assessment of the enterprise with respect to a consistent framework (e.g., the Model proposed by the European Foundation for Quality Management - EFQM) should be performed, possibly on the existent layer of a Quality System in compliance with the ISO 9001 Standard (and its VISION 2000 upgrade) [1]. The EMAS approach shall obviously be taken into consideration, together with high-profile technicalities for the measurement of the satisfaction of all Stakeholders involved in the environmental process: the enterprise itself, the customers, the suppliers, the local Administration and the Enforcement Bodies.

5. Conclusions

In northwestern Italy laws and compulsory constraints are the major driving force towards the implementation of environmental systems: the complexity and number of obligations force SMEs to build a new kind of organisation able to define what to do with reference to their specific market sector and geographical location, and how to do it in time and with full awareness. Many SMEs have an ISO 9001/2/3 certified Quality System, that can build a layer for an Environmental System in compliance with the ISO 14001 Standard. The development of new application tools and assessing criteria can contribute to the evolution of the organisation of SMEs. This growth in know-how can be accomplished by the joint work of Industries and Research Centres, through an effective match between data and processing methodologies.

The partnership of Research Centres and Industries is still at an early stage, with negative impact on the integration of the Province of Cuneo in the area of the so called "Provençal Belt" (Catalonia, Languedoc, Provence and Italian Western Alps). Through the development of excellence cases and a systematic strategy of benchmarking the level of awareness about the need of a new approach to applied research can reach the critical threshold and create new ways to environmental security.

References

1. Olivero, S. (1997) I Sistemi di Gestione Ambientale e la Norma ISO 14001, Unpublished Turin University Training Course.
2. Unione Industriali della Provincia di Cuneo (1998) Relazione sull'Attività Associativa 1997, Unione Industriali della Provincia di Cuneo, Cuneo, Italy.
3. Agenzia Regionale per la Protezione Ambientale del Piemonte (1998) Conferenza sulla Situazione Regionale della Prevenzione e Tutela Ambientale, 2nd, Agenzia Regionale per la Protezione Ambientale del Piemonte, Turin, Italy.
4. Provincia di Cuneo (1998) Sintesi del Progetto del Programma Provinciale Gestione Rifiuti, Provincia di Cuneo, Cuneo, Italy.

ASSET LIFE-CYCLE IN THE MINING INDUSTRY
How to improve economic and environmental decision-making by applying ICT.

I. L. RITSEMA
TNO-NITG, Netherlands Institute of Applied Geoscience
P.O. Box 80015, 3508 TA, Utrecht, The Netherlands

Abstract. Understanding the interactions between earth systems and social economic systems is of crucial importance to assess, manage and steer towards sustainable environmental and social economic systems. This is especially true for social economic activities exploiting the earth systems as a natural resource, such as mining of natural resources.

In assessing the consequences of each social economic activity, such as mining minerals or coal, the full life cycle should be considered in terms of both environmental and social economic impacts. The typical asset life cycle should include exploration, appraisal, development, production and last but not least abandonment. Typically the asset life-cycles of minerals or coal mines are finite lasting up to tens of years (being both storage resources), while the asset life-cycle of groundwater production could be infinite and in equilibrium, while its is recharged (being a sustainable resource). Impacts of all relevant mining scenarios should be assessed and should be compared to the desired state of the earth and social economic systems in each phase of the life cycle, optimizing decision-making. In early phases of the life cycle, impacts can only roughly be estimated and decision parameters will contain large uncertainties (including opportunities and risks), while in later phases more accurate estimates can be made and decisions will be more straightforward.

Most mining companies are very good in assessing economic and social impact indicators, but to a lesser extent the environmental impact indicators. Economic impact is normally expressed in yearly cash flows and results in indicators such as net present values (NPV), return on investment (ROI), etc. Social impact includes employment and health indicators, etc. Earth systems impact is assessed for each of the following four interactions between these and socioeconomic systems:

- geo-resources, where earth material reserves are extracted (such as minerals, coal, water,);
- geo-space, where materials are stored in the subsurface (such as waste, pipelines, buildings),;
- geo-risks, which are a risks to social economic activities (such as land subsidence, floods);
- geo-environment, where the earth systems are disturbed (such as soil and water pollution).

Each type of interactions has to be assessed and expressed in relevant indicator values, the first two expressed as potential benefits or values and the last two as potential risks or losses. The main economic and environmental aspects of asset life cycle management are discussed, illustrated with examples from similar industries.

The conclusions are that there is a need for a systematic and integrated approach in assessing environmental and social economic impacts throughout the asset life cycle of mining operations. It is very important for companies to co-operate with all stakeholders, such as governmental ministries and surveys, regional and local authorities. Therefore a fast, clear and sound communication is essential.

In this paper examples will be given how Information and Communication Technology (ICT) products can greatly facilitate these new business requirements. Examples of the use of telecommunication, data management, applications and decision support systems using WWW technology are shown.

1. Introduction

Natural resource production companies, in particular the mining industry, are operating under more and more severe economic and environmental conditions. Shareholders have an increasing number of options in the global economy to invest or divest their shares urging the mining companies to have a clear view on the economic value of a particular asset throughout its lifecycle. On the other hand new environmental regulations aimed at a sustainable environment also urges for a clear and systematic insight in the impact mining activities have on environmental systems throughout the asset lifecycle and thereafter. Therefore economic and environmental variables or indicators play a vital role in each major decision in the asset lifecycle of a ore prospect or mine. As there are many stakeholders (mining company, governmental authorities, public, etc.) participating in the decision process, this process should be as transparent as possible and at same time still being effective [2].

As will be illustrated by examples, Information and Communication Technology (ICT) can definitively help to achieve this.

2. Asset Lifecycle in the Mining Industry

The asset lifecycle of a finite natural resource can be schematized as in Figure 1. Also for mineral exploitation it consists of the following phases:
- Exploration (discovering natural resources resulting in assessed ore prospects)
- Appraisal (assessing natural resources volumes resulting in assessed ore reserves)
- Development (design and construction of production facilities resulting assessed mine reserves)
- Production (production operations and enhancement developments: assessed open mine reserves)
- Abandonment (decommissioning of facilities and waste management: assessed closed mines)

279

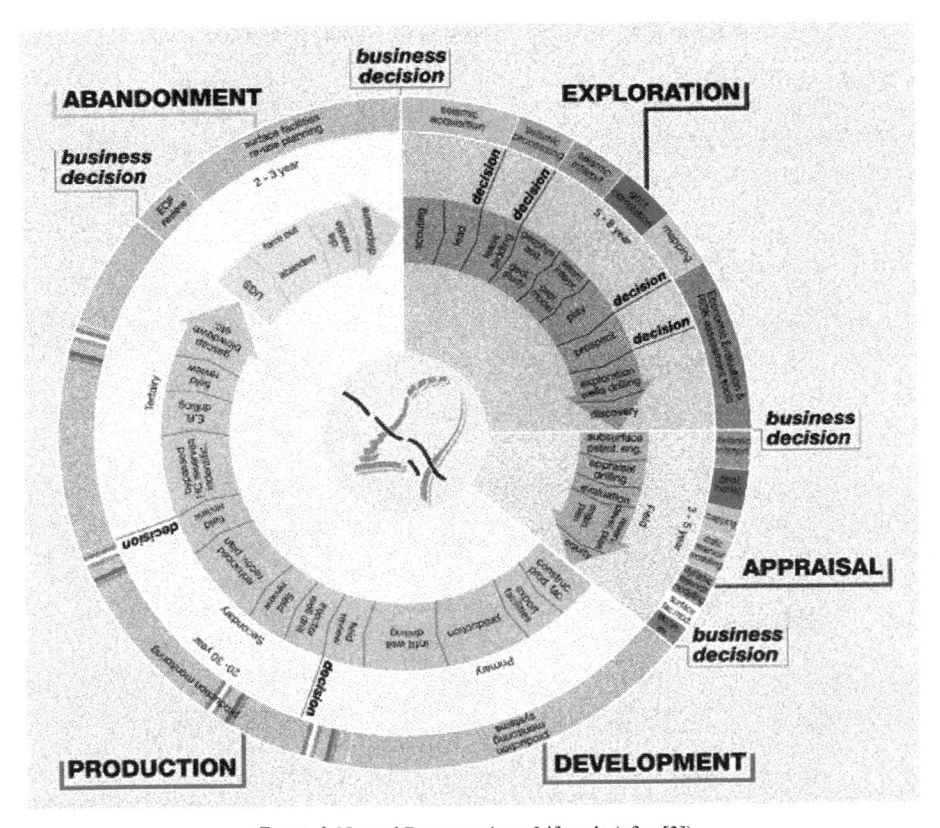

Figure 1. Natural Resource Asset Lifecycle (after [3]).

The lifecycle of natural resource assets (and also ores) is very long, typically tens of years. Especially the production phase is long and is normally for management purposes divided into yearly cycles of planning and asset value assessment.

3. Decision process in the Asset Lifecycle of Mineral Prospects or Mines

Between each phase of the lifecycle important decisions are taken to proceed (or not) and which implementation scenario of the next phase is the best. These are strategic, commercial business decisions for the mining company and political, but also financial decisions for governmental bodies. Assessing the value of the natural resources means evaluating the social economic (costs versus benefits) and environmental (negative and positive impacts) variables. These values should be expressed including their probabilities of occurrence (<1 in the exploration phase and ca. 1 for the other phases) and uncertainties (>100%, before the first phases and decreasing, ca. 40%, 15% and < 5% for the others). This information, created at the operational level, is aggregated and integrated for the decision-makers.

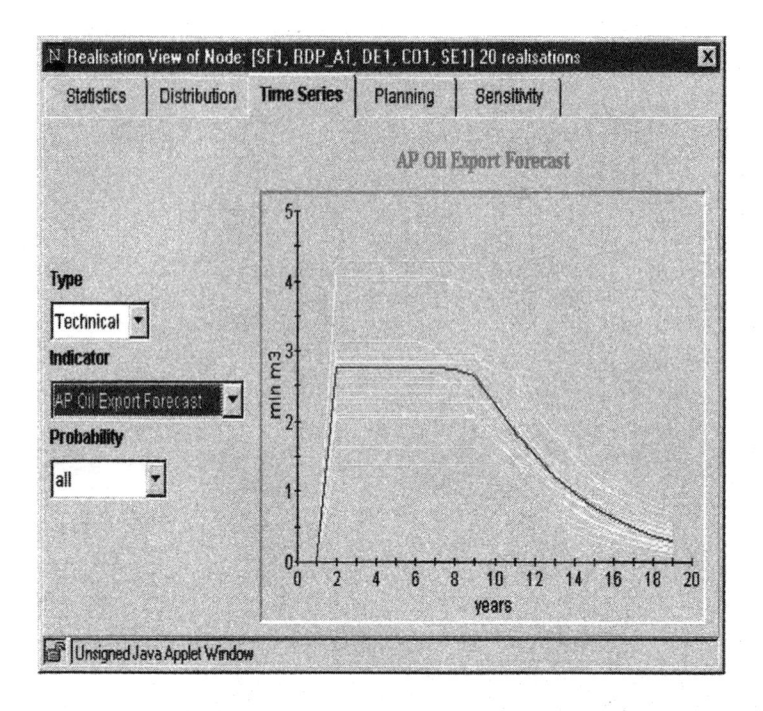

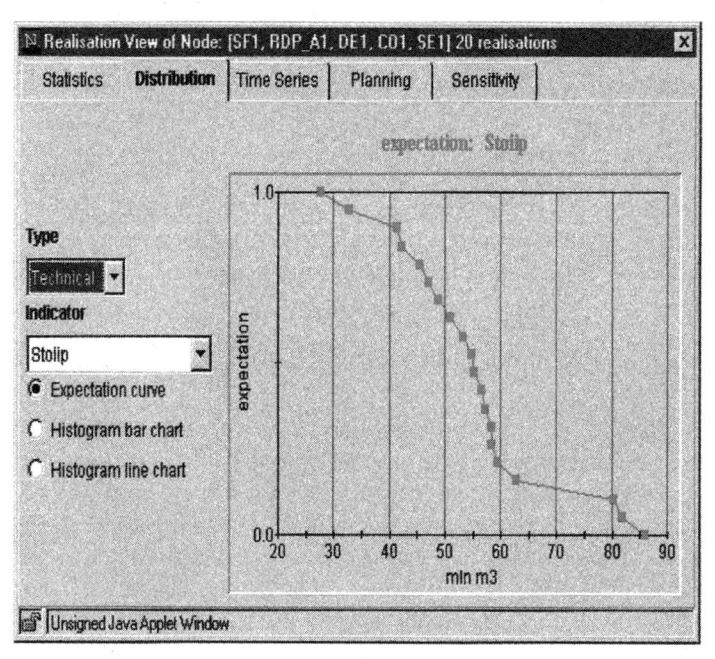

Figure 2. Monte Carlo simulated production forecasts (top) resulting in NPV expectation curve (bottom); components of an Internet based Decision Support System.

The resulting performance indicators for the various aspects are than taken into account in the decision-making process. In the following sections the economic and environmental evaluation of mine assets are discussed. The technical and operational processes and discussions are however not discussed.

3.1 SOCIO-ECONOMIC EVALUATION OF MINE ASSETS

The social economic evaluation of the natural resource asset means determining the economic and social value of the ore prospect or mine for each stakeholder involved. Normally several stakeholders have interest in the socioeconomic benefits and risks. The situation for three of them is discussed briefly.

Mining companies and their shareholders do have primarily an economic interest. In principle the economic evaluation starts with the calculation of the cash flow for each year over the full lifecycle period, as shown in Figure 2. This can be expressed for each year as:

Economic value = [{Recovered reserves * unit product price} - {Facility development (Capex) and operational (Opex) expenditures}] – [Governmental royalties and taxes]

Integrated over the whole period economic indicators can be calculated, such as net present value (NPV) or return on investment (ROI) or maximum exposure. Each mining company will have its own baseline value for these indicators where they will be compared with. The minimum threshold is that the investments made should exceed the economic profits, which could be achieved by putting the same amount of money on the bank. As the mining results contain risks and uncertainties, the economic indicators, such as net present value, are defined as expectation curves or probability density functions (see also Figure 2).

For governmental authorities it can be of strategic importance either to produce (or keep) natural resources, while it can create large royalty and tax and employment benefits. Typically the royalty and tax benefits are proportional to both the economic profits the mining company makes and the number of people employed. Therefore the authorities can use the same economic and additional social indicators, but applying different weights. The baseline figure for being acceptable economically will differ from that of mining companies, generally lower, while the government weights the social benefits more.

Also the general public and specific industrial sectors (suppliers) do have social and economic interests in mining activities, while it creates work and income. This is a growing factor in decision-making nowadays.

All stakeholders will weight their economic (such as NPV and ROI) and social (such as employment) indicators differently in each major decision throughout the asset lifecycle. Optimal decisions create a win-win situation for all of them.

3.2 ENVIRONMENTAL EVALUATION OF MINE ASSETS

The impact of mining on the surrounding environmental systems should also be evaluated. The environmental systems could be subdivided into earth (geo) and life (bio) systems. It is obvious that the second category (ecological and human systems) is of utmost importance and can be affected directly or indirectly through impacted earth systems.

Generally four major interactions between the dynamic environmental (i.e. earth systems) and the social economic systems occur in mining, which is a social economic activity in itself as well [6]:
1. geo-resources, an opportunity where earth material is extracted from the earth system (such as minerals, coal, groundwater, building materials);
2. geo-space, an opportunity where facilities are brought inside or on top of the earth systems (such as wells, mine tunnels and transport systems, waste storage, pipelines);
3. geo-hazards, where earth systems are a risk to social economic activities (such as floods, earthquakes, erosion and sedimentation);
4. geo-environment, where earth systems are disturbed and form a risk to social economic systems (such as soil and groundwater contamination, land deterioration, subsidence).

The first two interactions correspond to the opportunities for mining. The last two contain the risks for a sustainable development of mines and the surrounding environment. Besides these interactions with earth systems also interactions with the life systems occur, which can also be structured into bio-resources (use of extracted wood or fertile soil), bio-space (reclaiming nature areas). Natural bio-hazardous risks (unfavorable living conditions of mine areas, such as deserts, swaps or Polar Regions) and man-induced bio-environmental risks (extinction of species, pollution of drinking water). Basically all stakeholder groups (mining company, government and public) insist on controlling mining activity with respect to environmental impact [1]. All strive for a sustainable development throughout the asset life cycle and thereafter for various reasons.

The **mining companies** are responsible for investigating all four interactions indicated, focusing on the 2 risk groups. This is usually done in the early phases of a potential new mine in environmental impact studies resulting in an environmental control plan and in later phases by monitoring the mining activities (mining material management), the environment (earth and life systems management) and their interactions (emissions and immisssions). For old mines this information very often lacks and developments are regularly not sustainable. **Governmental authorities** responsible for planning and development of an area, also want to safeguard the sustainable development in an area, taking into account other social economic activities (habitation, agriculture, transport, etc.) and the environmental systems [7]. However, laws with a duty to provide all information to governments and standard procedures for information processing are often lacking. In a recent study of the commission of Geological Sciences for Environmental Planning of IUGS and UNESCO confirms that all hazards mentioned above really exist in a large number of European countries in various degrees of severity [2]. In that report it is recommended to involve more geoscientific expertise in environmental planning, assessments and monitoring.

The **general public or private sectors** are often organized in interest groups (habitants, environmentalists, other industrial sectors). They are more and more another important factor in the process for creating environmentally sustainable and politically acceptable mining activities.

Currently, the meaning and the responsibilities for the environmental information is not defined clearly: who is doing mining materials information management (mining company and/or government) and earth and life systems information management (government, public or all)? A clear protocol and procedures will avoid discussions based on incomplete and inconclusive information. In this protocol an explicit place must be present to handle different scenarios for the future outcomes of a mining operation, depending on the possible discrete outcomes of the size of the ore, different development facilities and external economic variables. Typically these can be organized in decision and chance trees, where each branch of the tree represents a particular scenario for the future. The final result of the assessment is therefore a tree with value assessments for each decision criterion of each branch as well as of the aggregated total, as shown in Figure 3.

3.3 INTEGRATED DECISION-MAKING AND COMMUNICATION

The social economic and environmental indicators can be integrated to score alternative mining development options of the scenario tree. All stakeholders are weighting their environmental indicators (emissions and immissions) in each major decision throughout the asset lifecycle. Other variables may play a role in the decisions, e.g. political and safety arguments (see Figure 3).

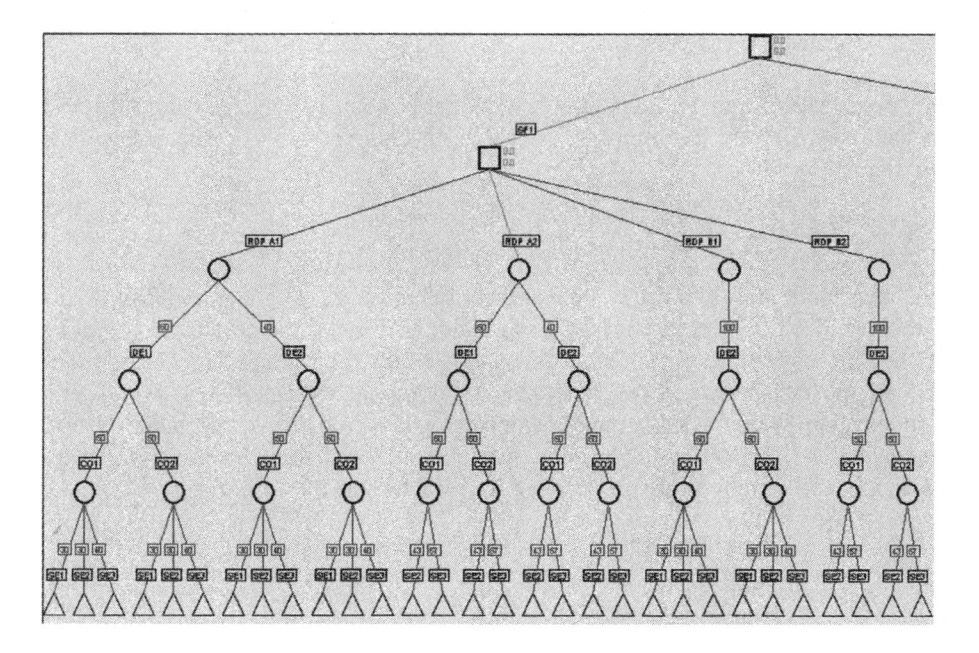

Figure 3. Scoring alternative scenarios of a decision tree (left) in a scorecard (right).

Depending on the political situation the decision process can be either an open process where all stakeholders are represented and participate or a process where decisions are taken in smaller circles. It is of utmost importance for the inevitable consensus negotiations, to agree with the acceptable stakeholders on the variables to be scored and weights to be given to each of them before the decision process is started. This requires a good platform of communication where these stakeholders participate. Each stakeholder will weight the various scores biased by its interest, classified as acceptable, negotiable or unacceptable. These issues typically need to be resolved or negotiated before an integrated and consensus type of decision can be taken.

In many situations it is possible to create win-win situations for all stakeholders by taking the right decisions.

4. Improving communication and integrated decision-making by applying ICT

ICT can contribute in various ways in improving business processes and decision-making in the mining asset Life cycle. Nowadays all technical work carried out is carried out with the help of computer programs. In this chapter I only would like to highlight three important ICT developments supporting the process of integrated and open decision-making:

- Communication
- Information transparency
- Decision transparency

In each case it can be seen how a systematic approach and ICT together can contribute significantly to the mining activities in terms of quality, time, money and environment [5].

4.1   COMMUNICATION

More and more the life cycle assessment processes are a group process. Many stakeholders need to participate at early stages not to frustrate the progress. On the one hand this means horizontal communication between the core business processes of one phase to the next phase (from development to production. etc.) including lateral communication with business partners and suppliers. On the other hand vertical communication transferring aggregated information towards higher decision levels including lateral communication with other stakeholders, as shown in Figure 4.

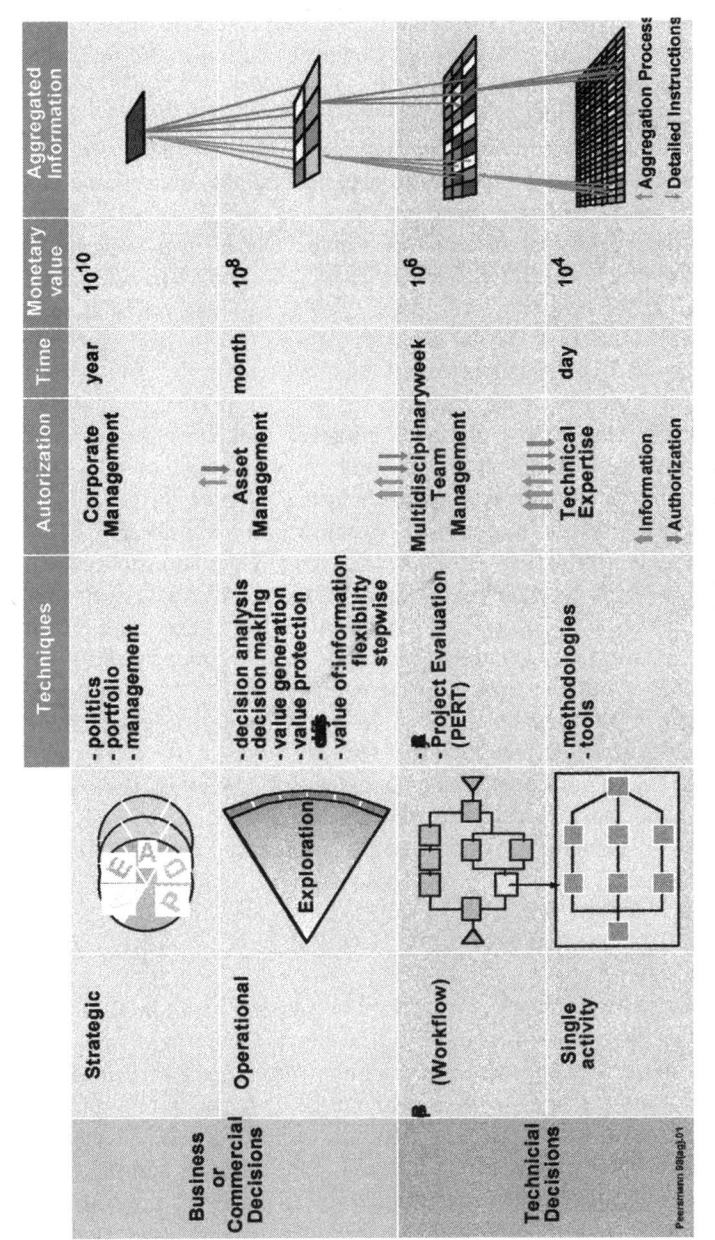

Figure 4. Vertical communication patterns.

It is very clear that using the Internet for communication is more and more attractive for societies where computers are common. The development and use of the following functionality is taking place to support this:

1. Standard Internet Browser Communication facilities, which include: email to exchange documents and data, web browsers to explore available information and programs, group ware for streamlining group processes.
2. Via the Internet already a vast of information can be found. It is however of different quality. Information Clearinghouses try to play an important role in the information exchange and access process. Typically, so-called meta-information is stored containing data set descriptions and only sample data. For the real exchange of or access to geoscience data the data providers need to be contacted. NITG-TNO is involved in both a European wide initiative of geological surveys (GEIXS. *http://geixs.brgm.fr/index.html*) and a national initiative containing any type of geo-information in the Netherlands from all kind of public institutes (NCGI, *http://www.ncgi.nl*). Fortunately both work with the same standard for describing, mainly geographical data sets (the European CEN standard for geo-information).
3. Free access to public geo-information will be norm, typically public institutes will take care of this. TNO-NITG for example provides public seismic, well and maps data received from the oil and gas industry on behalf of the Dutch government using the Internet (*http://www.nitg.tno.nl/ned/projects/6_vrijgave/index.shtml*).
4. Another example is Platform Reuse Brokerage system on the web (ACG Brokerage, *http://www.WEB-platform-brokers.com/*). This is a system under development at NITG-TNO, which matches platform facilities for sale (end of life cycle) with clients looking for facility equipment (development phase of life cycle). The system contains worldwide data and includes reservoir, facility specifications, HSE and other variables to establish matches. This system also manages the data and for matches clients are charged.

4.2 INFORMATION TRANSPARENCY

It is of utmost importance to streamline in organizations **the information flow and processes** resulting in the decision information. Typically, the socio-economic and environmental indicators of current and baseline situations, are aiming at a situation where all information on a mine asset is always ready at hand. A typical information flow contains the sequence of observations, interpretations (see Figure 7), forecasting and determination of indicator values (Figure 6). In real life the outcome after all these steps is never the same, while procedures differ from place to place. To improve this situation the information itself and the information processing functions (aggregation. integration) need to be standardized. Also the computer systems need to be integrated to support this information flow and processing.

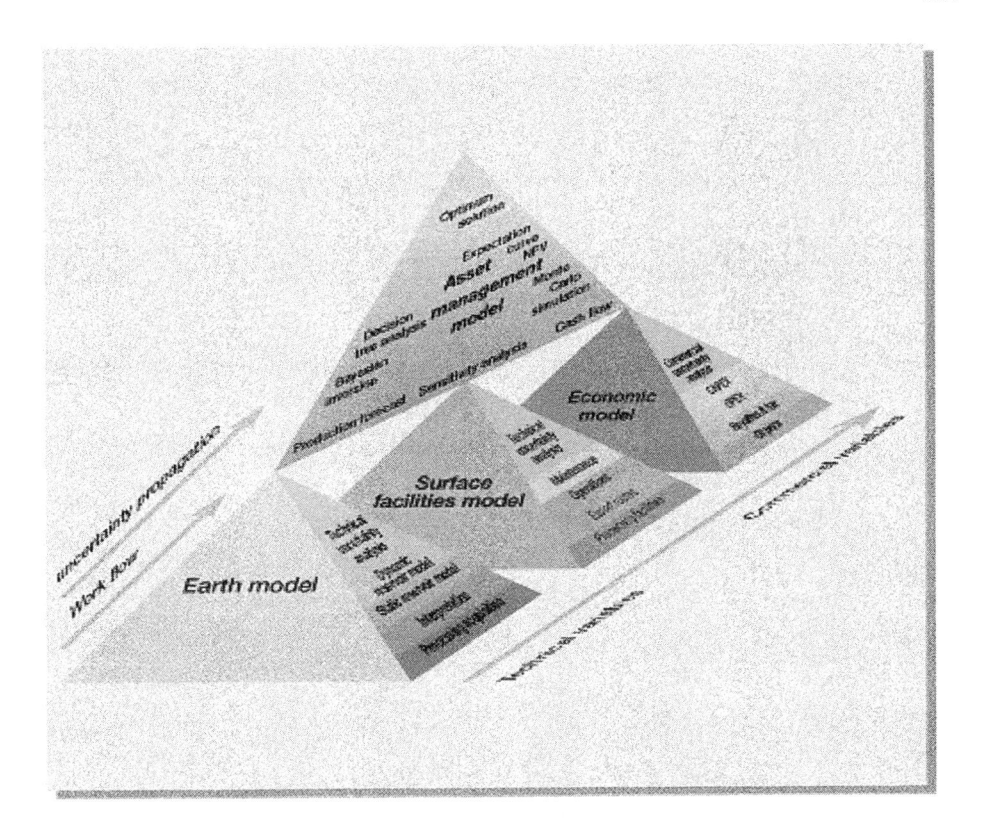

Figure 5. Scheme of the information systems involved in creating decision support indicators.

In the end all mine asset information should be easily available and accessible. In this way all stakeholders can evaluate their indicators and positions in negotiations. It should be seriously considered to improve the environmental situation in mining by reporting on the environment (nature or recreational) besides reporting on economic and infra structural consequences. The Web Atlas of the North Sea Oil and Gas Fields (owner PGS Reservoir, *http://www.nitg.tno.nl/webatlas*) is an example of such an asset-centered model, comparable to the mineral deposit model described by Wanty *et al.* (2002) [7]. 1998. Another very important use of such databases is the analysis possibilities using textual search and image processing techniques, as indicated in Figure 6.

288

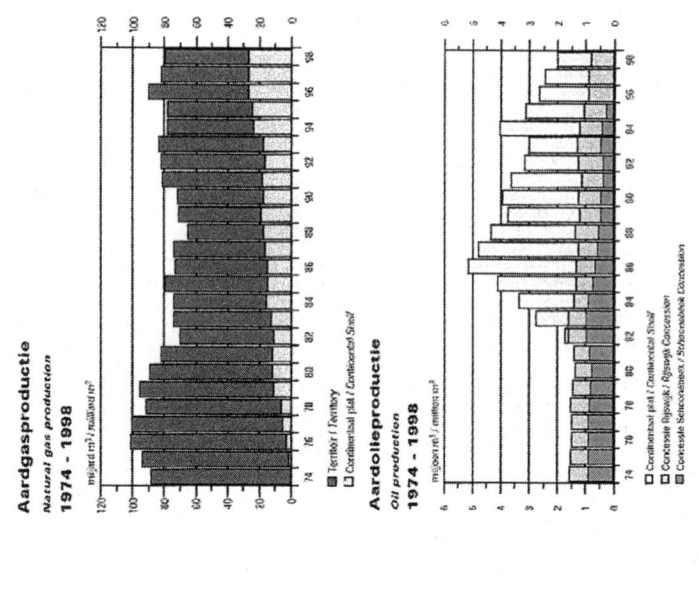

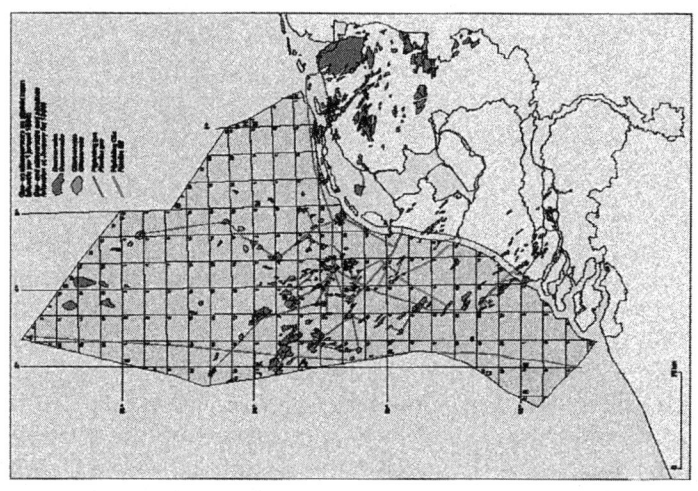

Figure 6. Map of oils an gas fields of the Netherlands (left) and production figure for gas and oil (right).

4.3 DECISION TRANSPARENCY

Currently, two major shortcomings can be identified in decision-making. First of all, uncertainties are very poorly accounted for: only the worst, mean and best cases of one or two scenarios. This typically results in too optimistic estimates of the uncertainties and therefore risks and opportunities are overlooked. In the DSS project at TNO-NITG **decision support method and tools** are developed which enable users to specify many more scenario's and evaluate these in extensive decision and chance node trees, which care evaluated multi-criteria including probabilities and uncertainty propagation.

The tools are written in such a way (pure Java) that sponsors (Shell, Statoil and Eni-Agip) can use them over the Internet or download the code for internal use. In principle the same code could with minor modifications, be used for governmental hydrocarbon asset assessments. More information is provided at the 'decision and risk management' project web site (http://www.nitg.tno.nl/eng/projects/1_drm/). Examples of the functionality were shown in Figures 2, 3 and 5).

Experiments are now set up to compare the use of such tools by different stakeholders.

A second mean is to be able to visualize the subsurface and surface in 3 dimensions over the Internet [4]. This is very important with regard to communication. TNO-NITG has developed a Java based tool for inspection of earth models (Figure 7).

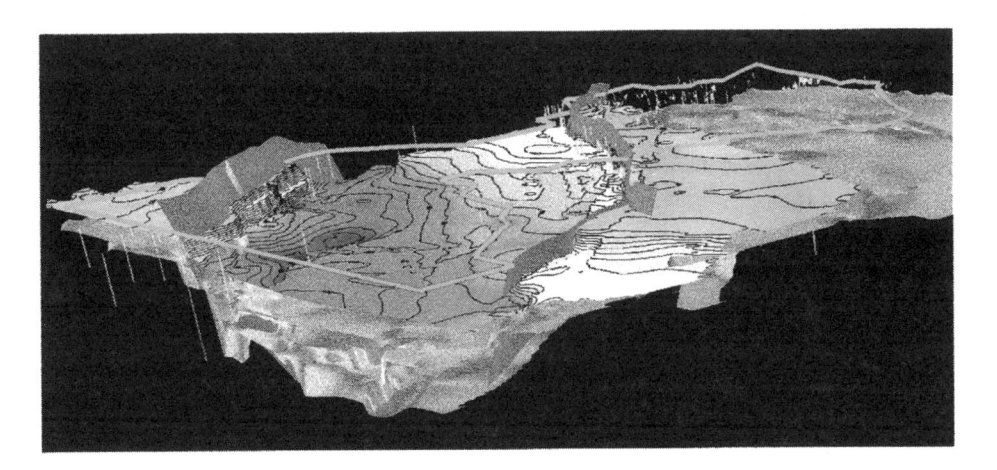

Figure 7. 3D model of the Dutch subsurface (3D Java viewer).

6. Conclusions

It is concluded that there is a need for a systematic and integrated approach in assessing social economic and environmental impacts in the asset lifecycle of ore prospects or mines. It is very important for mining companies to start co-operating with other stakeholders, such governmental ministries and surveys and regional and local authorities. To achieve this a clear and sound communication, information and decision transparency is essential.

In this paper examples were shown how Information and Communication Technology (ICT) products and systems could greatly facilitate these new business requirements in the mining industry.

References

1. Baily. R.G. (1996) *Ecosystem Geography*, Springer, New York.
2. Mattig, U., de Muldcr, E.F.J. (1998) *Report on the Geo-Environmental Inquiry Report, Central and Eastern Europe*, Environmental Geology, IUGS, pp 215-233.
3. Peersmann. M., Floris, F. (1997) *Uncertainty estimation in hydrocarbon field volumetrics for supporting decision making in the E&P Asset Life cycle Management.* J. of Petroleum Geoscience, Vol. 4, No. 1, pp. 33-40.
4. Ritsema, I.L. and Gerritsen, B.H.M. (1996) *Spatio-teinporal Modelling Geoscientijic Features*, Gocad meeting, Nancy. France, pp. 145-152.
5. Ritsema, I.L. (1997) *Geo-Informatics*, Diligentia, The Hague, The Netherlands, pp. 56-72.
6. Speelman. H., Ridder, J. (1997) Mission of the Netherlands Institute of Applied Geosciences. TNO-NITG. Delft, The Netherlands. TNO-NITG Year Report 1998, Delft, The Netherlands
7. Wanty, B.R., Berger. G.S., Plumlee, G.S. and King, T.V.V. (1998) Geo-Environmental Models, An introduction, NATO-ASI, Mátraháza, Hungary, this volume.

INDUSTRIAL ECOLOGY AND BIOREMEDIATION
Theoretical framework and technological tools for sustainable development

ENRICO FEOLI

ICS-UNIDO and Department of Biology, University of Trieste, Via Valerio 32, 34100 Trieste Italy

Abstract. The concepts of industrial ecosystem, ecological industrial carrying capacity and the concept of ecological footprint are reviewed and discussed in the context of sustainable development. It is stressed that bioremediation is offering useful technological tools for Industrial Ecology and in particular for the ecological sustainability of mining activities.

1. Introduction

The aim of this contribution is to introduce the participants of this workshop to the conceptual tools of Industrial Ecology (IE) and to illustrate why bioremediation should be considered a discipline of IE since it offers technological tools for restoring degraded sites and for decontaminating polluted sites.

Mining is a fundamental activity of the industrial production system, it is a temporary use of a relatively limited land, however during this temporary use, environmental disturbance is inevitable owing to all the activities related to taking minerals from the ground and turning them into useful products. Notwithstanding the area directly affected can be relatively limited, mining has always a relevant impact on the surrounding ecosystems both in terms of spatial displacements and/or rearrangement of their components and in terms of pollution. The typology of the impacts is very rich (its review is out of the scope of this contribution), however, irrespective the typology and magnitude of the impacts, it should be useful to analyse the mining activities in the framework of IE with the aim to find solutions to reduce the impacts. Strictly speaking IE is suggesting a theoretical framework that foresees a set of actions to avoid the accumulation and the dispersal of waste in the environment and to limit as much as possible the release of pollutants from the industrial plants. According to IE these objectives should be achieved by a combination of the improvement of technology at the plant level and the implementation of a product policy at societal level. From the literature (e.g., [1]) and from the Journal of Industrial Ecology (see http://www.yale.edu/jie/) it looks as if IE does not consider that industrial plants and mines are damaging the environment also for the reason that they occupy space that before was used by agricultural activities and/or by urban areas or was occupied by natural ecosystem. When industrial plants are built the site destruction is always unavoidable and when industrial plants are functioning contamination by pollutants may be unavoidable in several cases. Among the industrial activities mining is the one that more than others needs environmental restoration for reassessing degraded landscapes and/or for decontaminating polluted sites during and after the life of the mine. This is the reason why IE and bioremediation are here considered together. As public concern about the quality of the environment intensified in recent decades, laws and regulations have been written and enforced that prescribe the quality of land, air and water that must be maintained during mining operations and left when mining ceases. This requires that we understand what actions are needed to maintain the environmental quality both to factor the environmental costs in the economics of the mining plan and to find the suitable measures for efficient environmental restoration. When bioremediation is required in restoration plans it is necessary to know the biology and ecology of plants and micro-organisms that should be used. In this way IE would not remain only an analogy of ecology but it will fully include ecology in its applications.

2. Industrial Ecology

2.1. INDUSTRIAL ECOSYSTEM AND ECOLOGICAL FOOTPRINT

Industrial Ecology (IE) has been defined by White (1994), [45], as: "the study of the flows of materials and energy in industrial and consumer activities, of the effects of these flows on the environment, and of the influences of economic, political, regulatory, and social factors on the flow, use and transformation of resources. The objective of industrial ecology is to understand better how we can integrate environmental concerns into our economic activities". Furthermore: "IE seeks to optimise the total industrial materials cycle from virgin material to finished product to ultimate disposal of wastes" [16]. According to Frosch (1992), [15], IE is based: "upon a straightforward analogy with natural ecological systems... The system structure of a natural ecology and the structure of an industrial system, or an economic system, are extremely similar". Grubler (1994), [17], conceptualises the industrialization as a "succession of phases, characterized by the pervasive adoption of technology clusters". In this respect he rediscovers the technological space of Tewolde (1982), [37]. The cluster is defined as "a set of (interrelated) technological, organizational and institutional innovations driving industrial output and productivity growth". The analogy with the ecological succession is very strict and it is more evident in Grubler's statements: "Such a succession is, however, not a rigid temporal sequence as various clusters coexist (with changing weights) at any given time. Older technological and infrastructural combinations coexist with the dominant technology cluster, and in some cases previous clusters (compared to the dominant technology base in the leading industrialized countries) are perpetuated, as was largely the case in the post-World War II industrial policy of the former USSR".

The technology clusters are analogous to the species clusters (species associations) in ecology. In the natural succession the changes of clusters of species is gradual and, in some case of isolation, the succession does not proceed in the same way as in other places. In analogy of what happens in ecology, trajectories of industrial ecosystems can be detected and modelled in the multidimensional space defined by the variables describing the system [12]. In natural successions the trajectories end in the hyper-volume of equilibrium (climax ecosystem). This happens when the carrying capacity of the environment is reached, i.e., when the environment is not able to sustain more biomass of what it is presently sustaining.

The concept of multidimensional space is essential because IE is analysing the industrial system "within and outside" the industrial plants by using a systemic approach that "can help illuminate useful directions in which the system might be changed" [33].

According to [1], the production system should be planned in such a way that it would most closely resemble to the natural mature ecosystems (climax ecosystem). In these ecosystems all the materials enter into functional cycles maintaining the structures and functions of ecosystems in such a way that the total production is equal to the total consumption and only heat is dispersed [33]; [16]. If the concept of the industrial system would be consistent with the concept of natural ecosystem it should be self-sustainable also for what the environmental variables are concerned. These are, besides the necessary row materials on which the products are based, energy, manpower, air, water and biodiversity. [14], has an approach consistent with IE. He cites a statement of Ulanowicz (1989), [41], written in the review of the Odum's book [30] "despite all the advances in modern technology, society remains irrevocably dependent upon natural systems for life-support a condition that is unlikely to change in the foreseeable future". This brings to the concept of industrial ecosystem as the system including all the components that are ensuring the ecological sustainability of the industrial system. For this reason the boundaries between the industrial systems and the other ecosystems (urban, agricultural, natural) are very difficult to draw. All the industrial ecosystems are more or less functionally interconnected in the biosphere notwithstanding they may belong to different states, nations, and economies. The overall industrial system of the biosphere consists of all the industrial plants (including mines) that are placed in different states for various historical, political, economical and environmental reasons (resource availability). All the industrial plants have strict links with the areas from where row material, energy, water, labour, information are coming and also with the areas where the products and pollutants are "going". In particular mining, with respect to other industrial activities, has an obvious strict link with the area where the row material is available.

As in Ecology the definition of the ecosystems is based on the biological structures and functions, in IE the definition of industrial ecosystems is based on structures and functions of the industrial systems. An industrial system may be defined at the level of a single factory, at the level of a set of factories of the same typology (i.e., food factories, or refineries, etc.), or at the level of set of different factories that are located in specific industrial settlements (industrial areas, industrial parks, etc.), or sets of industrial settlements belonging to small administrative units (e.g., municipality), or to administrative units of higher level such as a province or a state. An industrial system may be defined also at the level of set of states (e.g., the industrial system of the European Union), etc. Because the concept of self-sustainability is implicit in the concept of ecosystem, the definition of any industrial ecosystem is not complete if we do not define the area (portion of the biosphere) sustaining the industrial system under all the ecological point of views. This area is considered by many Authors as the "ecological foot print" of the industrial system (see [43]). It is analogous to the concept of minimal area in ecology, i.e., the area necessary to support one ecosystem in its complete functionality. It follows that the sustainability of a given industrial system by a given area is achieved only if the rate of depletion of renewable resources of the area utilized by the industrial system is lower than or equal to the rate of their renovation. In other words a given area has the necessary carrying capacity for a given industrial system if it is larger of or at least as large as the footprint of the industrial system. In principle, similar industrial systems could give origin to different industrial ecosystems. This depends on the type of ecosystems they directly impact as a consequence of their location: a refinery in the Gulf of Trieste (North Adriatic Sea) defines a different industrial ecosystem from the one defined by a refinery in the Persian Gulf; a mine in the African desert defines a different industrial ecosystem from that of a mine in the Alps, etc. However, today is not anymore possible to associate an industrial system only with the area where it is located. In fact it is hard to say that the area, for instance of the province of Trieste (in Italy) has the necessary carrying capacity for the corresponding industrial system. More striking examples could be Singapore, Malta, etc., where the area of the land (and the sea) hosting the industrial system is evidently much smaller than the ecological foot print of the corresponding industrial systems. This leads to the consideration that the industrial system is global and has to be analysed accordingly following specific hierarchical patterns that cannot be considered here as they would bring us into socio-economic and political issues that fall outside of the scope of this contribution. However, we can conclude that if the ecological footprint of the total industrial system of the Earth is larger than the area of the biosphere, the total industrial system is not ecologically sustainable.

2.2 ECO-EFFICIENCY AND LIFE CYCLE ASSESSMENT

The industrial ecosystem is a dynamical entity changing in time according to the market demand and according to technological changes. According IE, the industrial system should develop in such a way as to minimize the sink of waste as the natural ecosystem is doing in the last phases of the ecological succession (climax). For this man has to find technological solutions for using the wastes as resources and/or to change the technology so as to minimize the energy and material consumption, i.e., by developing the so-called clean technology [10].

The message of IE is that today we have to deal with the sustainability of industrial systems by integrating the different human systems with environmental system in a very practical context. It means, according to Ehrenfeld (1995), [9], to study "how we humans can continue rearranging Earth, but in such a way as to protect our own health, the health of our natural ecosystems, and the health of future generations of plants and animals and humans". In this context Tibbs (1992), [38], proposed seven practical objectives of IE, these are:

(1) Improving the metabolic pathways of industrial processes and material use;
(2) Creating loop-closing industrial ecosystems;
(3) Dematerialising industrial output;
(4) Systematizing patterns of energy use;
(5) Balancing industrial input and output to natural ecosystem capacity;
(6) Aligning policy to conform with a long-term industrial system evolution;
(7) Creating new action-coordinating structures, communicative linkages, and information.

Ehrenfeld (1995), [9], considers Tibbs' practical framework as a convenient way to organize a complex set of notions into two categories, one strictly containing the technological elements (objectives (1) to (4) above) and the other including the organizational and strategic elements (objectives (5) to (7) above). He focuses the discussion on IE under the perspectives of developing a product policy and suggests guidelines to develop such a policy accordingly. The most striking concept that we have to take into consideration in objective (5) is the *eco-efficiency* of the industrial system

The concept of *eco-efficiency* of an industrial system is related to the concept of the ecological footprint and it may have only a relative meaning. In fact we can only say that one industrial system among a set of industrial systems considered equivalent (for technological or economical reasons) is more eco-efficient than the others if its ecological footprint is lower than those of the others.

On such a basis a parameter of the eco-efficiency of an industrial system is the area of the biosphere necessary to sustain the corresponding industrial ecosystem. It follows that the ecological efficiency of an industrial system is improving when the area for its self-sustainability is decreasing. Reducing the areas of industrial ecosystem should be the target of international and national environmental policies.

In analogy with Ecology, where the life cycle of the species and turnover of materials is fundamental to predict the trajectories of ecosystems, in IE the concept of "life cycle" (LC) has been introduced as the history of products and all the parts of the industrial system (that are also products of the same or other industrial systems) and the system itself, from their "birth" to their "death". In each stage of its life (raw material collection, manufacturing, primary use, secondary use, ... , disposal) a product has a more or less intense and direct or indirect impact on the environment owing to that it occupies space and/or consumes the environmental resources. The reason to study the LC is to get information on how to generate a product by a process design that minimizes the overall environmental impact of the industrial system. It is therefore necessary to integrate the so-called Life-Cycle Assessment (LCA) with the industrial Design For the Environment (DFE). This will lead to the development of clean technology application to industrial production processes. There are some examples of applications in [4], with the perspective to replace as much as possible the waste management by the product life-cycle management. According to [8] and [9], both LCA and DFE require normative inputs to permit weighting and aggregation of impacts along diverse environmental axes that "address" the trajectories of the industrial systems in the direction of a higher eco-efficiency.

The life-cycle assessment is one of the applications of IE that would help the industrial enterprises within an industrial system to conform their standards with ISO14000 [26]. The ISO 14000 is a set of evolving guidelines that are supposed to help the industrial systems to improve their ecological efficiency. However other perspectives have to be introduced from IE. These need education and new research directions (objectives (6) and (7) of Tibbs, 1992 [38]). [35], suggests a shift to an economy that sells the functionality of products rather than the products themselves to decelerate materials flow. Another practice that should be used in this direction is the reuse of the product as much as possible before recycling. However, recycling has great implications for energy and mineral resource conservation, waste reduction and pollution prevention. Recycling programs reduce the solid waste stream, saving landfill space and valuable energy and natural resources. It is known that the secondary recovery of aluminium uses 5 percent of the energy required to recover the same amount of aluminium from imported bauxite ores. At the same time, 63.5 percent of the 99 billion aluminium cans produced in the United States in 1996 were recycled, rather than being added to a waste disposal dump.

Notwithstanding ISO 14000, the 1992 UNCED conference, the 1997 Kyoto conference, and other international events, the application of clean technology to the industrial system, the recycling and the necessary changes in the product policy according to the life cycle of the products are still at the initial phase. As a consequence of this slowness, today we are still pressed to find solutions that are related to end-of-pipe pollution treatment and waste management. Bioremediation is one techniques of the end of pipe treatment. However, it is not only related to clean the effluents before the discharges, it has a more spatial meaning that is particular important in mining activities.

3. Bioremediation

Bioremediation deals with a set of technological tools belonging to remediation technology. To avoid terminology problems, remediation is here considered as the set of technologies that are useful to halt ecosystem degradation and to redirect a disturbed ecosystem in a trajectory ensuring the self-sustainability of the ecosystem under the perspective discussed in the context of IE. Remediation includes: restoration *sensu stricto* (when the aim is to reconstruct the original ecosystem); restoration *sensu latu* (when the aim is to halt degradation and to redirect a disturbed ecosystem in a trajectory resembling that presumed to have prevailed prior to the onset of disturbance) and rehabilitation or reclamation (when the primary goal is to restore the ecosystem productivity for the benefit of local people irrespective the type of intervention) (see [3]). Reclamation is particularly important after mining activities or even during mining in the surroundings of the mines. Reclamation in mining areas usually includes the following steps: levelling of fill soil by bulldozers; the placement of topsoil or an approved substitute on the degraded area; reseeding with native vegetation, crops and/or trees; and years of careful monitoring to assure success. Ultimately, reclaimed sites are returned to many productive uses. Although underground mines do not have as much surface disruption, they do have reclamation responsibilities for stabilizing tailings ponds during use and reclaiming the area when mining is completed. Any surface subsidence must also be accounted for and included in mining plans. As always, surface and groundwater must be protected from acid drainage and metal components higher than the ambient water levels. Mining companies should constantly seek for better methods of reclaiming mined-out lands because land reclamation is now an integral and normal part of the mining process. For example in the USA the responsible coal operators are guided by the principle that the right of coal extraction carries with it the responsibility of restoring the land. Before mining operations even begin, the coal operator's plan must offer specific details on how the land reclamation will be accomplished. Afterwards when mining is underway, the reclamation work is closely monitored by various company, state, and federal officers. Reclamation is required by both the state law and the federal Surface Mining Control and Reclamation Act (SMCRA). In the USA in 1997, the mining industry began a major program to go beyond protecting the environment in current operations by addressing the problems of the past. The Western Governors' Association and the National Mining Association launched a joint initiative to accelerate reclamation of the abandoned hard-rock mines in the West. The overall goal of the initiative was to increase public and private investment in remediation, consolidating financial resources and technological expertise to promote on-the-ground cleanup. Most mine waste is simply dirt and rocks;

it needs not, and should not, be treated in the same manner as hazardous wastes generated by other industries. Although the Bevill Amendment of 1980 (www.nreca.org/NREEA/meetings/2001Spring.ppt) directed the EPA to determine if any mining waste should be regulated as hazardous, the intent of the U.S. Congress remains clear: the avoidance of unnecessarily burdensome regulation of the mining industry.

It is obvious that reclamation could be a very expensive activity and studies are needed to apply the best technology as a function of the problem. In this respect operative strategies are suggested (e.g., [21]) so that they foresee four steps before starting with the remediation activities (e.g., in Italy D.M.471/99), namely:

1. The characterisation of the plan for reclamation in which cartographic, historical and socio-economic data are collected in order to formulate the conceptual model of remediation that is required;
2. The technical description of the site reporting the results of all the necessary hydro-geological, geological, chemical, ecological and biological analysis;
3. A feasibility study in which the applicable technologies will be compared in function of the results in phase 2 (a cost benefit analysis is here necessary);
4. The detailed remediation plan with the description of the technology to be used and with the description of the control plans after the execution of the remediation.

The technologies of remediation can be classified in three main categories that can be combined to achieve the desired results:

1. chemicals, when only chemical treatments are used to transform the pollutants in a less toxic and mobile compounds;
2. physical, when separation techniques based on absorption or thermal processes are applied to absorb (or to desorb) or to destroy the pollutants (e.g., volatilisation);
3. biological, when biological organisms are used to decompose or uptake the pollutants (this is called bioremediation).

All the three technological categories can be applied *in situ* when the soil or water is not transported to other sites, or *ex situ* when the contaminated soil and water are collected in special treatment plants.

Reclamation is viewed under the perspectives of successional process of vegetation defined first by Clements (1916), [6]. It consists in the following phases: migration, ecesis (establishment), reaction, competition and stabilization. These are topics of vegetation ecology, an ecological discipline aimed to understand all the aspects of vegetation concerning its composition (diversity), structure (spatial pattern and spatial relationships), functions (growth), dynamics (the ways and the causes of vegetation changes in space and time) and its role within the biosphere. Bioremediation is a practical activity that aims to use biological organisms for rehabilitation of degraded and degrading areas owing to different types of human activities. According to the Committee of In Situ Bioremediation (1993), [7], the current practice of bioremediation started more than 20 years ago "to clean up an oil pipeline spill in Pennsylvania, and since then bioremediation has become well developed as a means of cleaning up easily degraded petroleum products".

For some Authors (e.g., [7, 21]), bioremediation only considers the decontamination of contaminated soils and waters by microbes (mainly bacteria). However, bioremediation is including two main branches: microbial remediation and phytoremediation. The ecological theory applies for both [28, 24, 20], however, in the former case the decontamination is based on the transformation of organic contaminants, or metal precipitates by an oxidation-reduction reaction produced by respiration (aerobic and anaerobic) and/or fermentation of bacteria.

In the latter case the decontamination can be achieved by the capacity of higher plant to uptake materials from the soil [26, 27, 31]. Microbial remediation is presented under the ecological perspective of Tiedje (1993), [39]. He considers the basic principles of ecology, namely: specificity, diversity, biogeography and natural selection. According to Tiedje (1993), [39], the microbial activity is related to three key questions:

1) Is the contaminant degradable?
2) Is the environmental habitable?
3) What is the rate-limiting factor and can it be modified?

The first question concerns the biodegradability of the pollutant and the frequency of organisms of the specific biodegradative property in the microbial community of the soil or water.

The second question concerns the toxicity of the contaminant(s) that can make it difficult or impossible for microbes to live, and the availability of sufficient life-sustaining growth factors (nutrients, appropriate electron acceptors, etc.) necessary to sustain the life of the biodegrading micro-organisms.

The third question concerns the limiting factor identification once biodegradability and habitability have been established. Thus treatments to overcome a rate limitation can be implemented (oxygen supply, nutrient supply, adjustment of pH, dilution of toxicants, etc.).

Phyto-remediation is presented under an ecological perspective by [19] and [32], and under a very practical perspective by [31]. They distinguish four subsets of metal phytoremediation that were targeted for commercialisation:

" 1. Phytoextraction, in which high–biomass, metal-accumulating plants and appropriate soil amendments are used to transport and concentrate metals from the soil into the above-ground shoots, which are harvested with conventional agricultural methods. 2. Rhizofiltration, in which plant roots are grown in aerated water, precipitate and concentrate toxic metals from polluted effluents. 3. Phytostabilization, in which plants stabilize the pollutants in soils, thus rendering them harmless. 4. Phytovolatilization, in which plants extract volatile metals (e.g., mercury and selenium) from soil and volatilise them from the foliage. "

In order to correctly apply the phytoremediation techniques the following concepts of vegetation ecology have to be considered:

(1) *plant community* is a state of the vegetation system given by a combination of populations of different plant species living together in an area that is environmentally homogeneous.

(2) *species niche* is a portion of the ecological space defined by the ranges of all the factors in which the species can live. Different species may have the same niche, the

more similar the niches are, the higher the competition between the species is supposed to be.

(3) *plant association* is the vegetation type that defines the fundamental hierarchical level in the hierarchic classification of the biosphere [29]; [42]. The plant communities that are considered similar enough are grouped into the same community type called plant association. According to the Braun-Blanquet's approach [44] it can be defined by a combination of differential species (characteristic species). A characteristic species of a plant association is statistically more frequent and/or abundant in that association rather than in other associations. In the biosphere a plant association is spread within the geographic ranges of its characteristic species. The extension and the shape of the area occupied by an association depend on the extension of the environment favourable to the combination of its characteristic species in the geographical space.

(4) *community niche* is the portion of ecological space occupied by a plant association [13]. The community niches may be wide or narrow in the multidimensional ecological space irrespective the extension of the association in the geographic space. The same community niche may correspond to different plant associations in terms of species composition (flora). This depends on the geographic ranges of the species. Once defined the plant associations, vegetation can be studied on the basis of other characters (e.g., structural, functional, chemical, as in [11] and [12].

(5) *ecological succession* is the unidirectional sequence of plant associations corresponding to different states of a vegetation system toward the steady state called climax. This is the plant community in equilibrium with the climate of a certain area. During the succession, species are substituted by other species that are more suitable for the new environment being created by the previous states (reaction).

A correspondence may be found between these concepts and the principles mentioned by Tiedje (1993), [39]: specificity corresponds with the concept of niche, diversity with that of plant community, biogeography with that of plant association, natural selection with that of succession. It follows that in developing phytoremediation plans the possibility has to be carefully evaluated to address the "engineered phytoremediation" according to clear answers to the three questions of Tiedje (1993), [39]. The establishment of a favourable revegetation depends on whether the plant material (propagules) used for revegetation match the niche and whether we are able to help to establish a process of ecesis (establishment) and reaction that throughout proper competitive relationships between different species will lead to a successful stabilization process. In line with vegetation ecology, [40] provide guidelines for reclamation of disturbed lands. They stress the importance of selecting the plant material when starting an engineered rehabilitation activity. The influence of degraded and polluted environments on the genetics of organisms has to be taken into consideration. Many species, termed bioindicators, have been adapted to extreme environments [18, 2] and are very active bio-accumulators of heavy metals [25, 36, 22]. They can be used for colonizing toxic environments and also for up-taking the "toxicity". The use of genetically engineered organism is also a possibility for decontamination [5], however, it is now considered only at the academic stage and has implications of bio-safety as discussed by [23].

4. A conclusion towards integration

The industrial system is the one that is responsible for direct and indirect impacts on biosphere due to all the life cycles of its products from the phase of mining to the phase of waste disposal. Industrialization has the principal responsibility of the economic growth of human population of the last five centuries and of its demographic "explosion" of this last century. Industrialization is the product of the application of technology in all the productive systems. It is a "self-fertilizing" system owing to all the positive feedbacks it has had from all the technological applications in different human activities. In its evolution, directly or indirectly, the industrialization has changed and is changing the landscape in the biosphere. The relationship between the industrial system and the components of the biosphere is becoming more and more critical. The sustainability of the industrial system is under question. For this reason there is a growing interest in understanding the natural systems in the perspective of life-support systems. In this respect we can say that from the ecological point of view the industrial system is sustained in a very important way by the vegetation (natural and artificial). Through the photosynthesis, the vegetation captures from the solar input the energy necessary to sustain all the human activities by regulating the water and the other biogeochemical cycles. This energy storage is free of charge once the price for seeds, fertilizers, labour, machines, etc. has been paid. This very simple fact supports Ehrenfeld's (1995), [9], suggestion, that the level of economic activity ultimately should be consistent with solar input.

According to [9], IE would mark a necessary step of changing the human behaviour and economy towards what he calls the "deep ecology paradigm", or more "realistically" the "sustainable paradigm". He states: "I have drawn on the "industrial ecology" notion rather than on the "deep ecology" as the basis for a product policy strategy. The pessimistic and suspicious technological theme of deep ecology is inconsistent with modern social structures. Going directly to deep ecology from the present system would create a serious disruption. Perhaps, it is the only way of thinking and acting over the **long run,** that will maintain a sustainable balance between humans and the world, but there is little or no ability to listen to its claims today and act accordingly. Industrial ecology, on the other hand, speaks to many themes already present or slowly emerging and offers a learning, guiding framework that can move from the present base to a more sustainable world without such a trauma".

Reclamation (which includes Bioremediation) is an activity that will be always necessary to keep the industrialization sustainable because industrialisation is a dynamic process that involves both time and space. Industrial sites and mines are continuously abandoned and the sites occupied by them have to be reclaimed in order to conserve biodiversity [34] and to keep the biosphere with a suitable industrial carrying capacity.

It should be clear that the concept of ecological sustainability is different from the concept of sustainability. Many argue that mining is by definition a non-sustainable activity because the raw material is worked out soon or later. The concept of sustainable development does not consider the product itself but it considers its functionality and the possibility of the future generation of utilising such a product in a healthy environment.

References

1. Allenby, B.R. and D.J. Richards (eds.). 1994. The Greening of Industrial Ecosystems. National Academy of Engineering, National Academy Press, Washington, D.C.
2. Arianoutsou, M., P.W. Rundel and W.L. Berry. 1993. Serpentine Endemics as biological Indicators of Soil Elemental Concentrations. In: Markert, B. (ed.) "Plants as Biomonitors. Indicators for Heavy metals in the Terrestrial Environment". pp. 177-189. VCH Verlagsgesellschaft mbH, Weinheim.
3. Aronson, J. C. Floret, E. Le Floch, C. Ovalle and R. Pontanier 1993. Restoration and Rehabilitation of Degraded Ecosystems in Arid and Semi-Arid Lands. I. A View from the South. Restoration Ecology, (March): 8-17. Society for Ecological Restoration.
4. Braungart, M. 1994. Product Life-Cycle Management to Replace Waste Management. In: R. Socolow, C. Andrews, F. Berkhout and V. Thomas (eds.) "Industrial Ecology and Global Change", pp. 335-348. Cambridge University Press, Cambridge.
5. Brown, R.A., W. Mahaffey and R. Norris. 1993. In Situ Bioremediation: The State of the Practice. In: Committee on In Situ Bioremediation (eds.). 1993. In Situ Bioremediation. When does it work? pp.121-135. National Academic Press, Washington, DC.
6. Clements, F.E. 1916. Plant succession: An analysis of the development of vegetation. Carnegie Institution Publication 242. Washington, D.C.
7. Committee on In Situ Bioremediation (eds.). 1993. In Situ Bioremediation. When does it work? National Academic Press, Washington, DC.
8. Ehrenfeld, J.R. 1994. Industrial Ecology and Design for Environment: The Role of Universities. In: B.R. Allenby and D.J. Richards (eds.) " The Greening of Industrial Ecosystems", pp. 228-240. National Academy of Engineering, National Academy Press, Washington, D.C.
9. Ehrenfeld, J.R. 1995. Industrial ecology: a strategic framework for product policy and other sustainable practices. In: E. Ryden and J. Strahl (eds.) "Green Goods" . Proceedings of The Second International Conference and Workshop on Product Oriented Policy. Stockholm, September 1994. Ecocycle Delegation (Kretsloppsdelegationen), Stockholm, Sweden.
10. Ehrenfeld, J.R. 1997. Industrial Ecology: A New Paradigm for Technological Innovation. Journal of Cleaner Production 5 (1-2): 77-85.
11. Feoli, E. 1984. Some aspects of classification and ordination of vegetation data in perspective. *Studia Geobotanica* 4, pp. 7-21.
12. Feoli E. and L. Orloci (eds). 1991. Computer Assisted Vegetation Analysis. Kluwer Academic Publisher, Dordrecht.
13. Feoli, E., Ganis, P. and Zerihun Woldu. 1991. Community niche an effective concept to measure diversity of gradients and hyperspaces, in Feoli, E. and Orlóci, L. (eds) 1991 *Computer assisted vegetation analysis*, pp. 273-277. Kluwer, Dordrecht.
14. Folke C. 1992. Socio-Economic Dependence on the Life-Supporting Environment. In: C. Folke and T.Kaberger (eds.)"Linking the Natural Environment and the Economy: Essays from the Eco-Eco Group. Kluwer Academic Publisher, Dordrecht.
15. Frosch, R.A. 1992. Industrial Ecology: A philosophical introduction. Proceedings of the National Academy of Science 89 (February): 800-803.
16. Gradel, T. 1994. Industrial Ecology: Definition and Implementation. In: R. Socolow, C. Andrews, F. Berkhout and V. Thomas (eds.) "Industrial Ecology and Global Change", pp. 23-41. Cambridge University Press, Cambridge.
17. Grubler A. 1994. Industrialization as a Historical Phenomenon. In R.Socolow, C.Andrews, F. Berkhout and V. Thomas (eds.) "Industrial Ecology and Global Change", pp. 43-68. Cambridge University Press, Cambridge.
18. Jain, S. 1983. Genetic Characteristics of Populations. In: Mooney, H.A. and Godron, M. (eds) "Disturbance and Ecosystems. Components of Response". pp. 240-258. Springer Verlag, Berlin.
19. Jordan III, W.R., M.E. Gilpin and J.D. Aber. 1987 (eds.). Restoration Ecology a Synthetic Approach to Ecological Research. Cambridge University Press, Cambridge.
20. Kikkawa, J. and D.J. Anderson. 1986. Community Ecology. Pattern and Process. Blackwell Scientific Publications, Melbourne.
21. Kofi Asante-Duach, D. 1996. Management of contaminated site problems. CRC, Lewis Publishers, New York. pp.410.
22. Kovacs, M., G. Turcsanyi, K. Penksza, L. Kaszab and P. Szoke. 1993. Heavy Metal Accummulation by Ruderal and Cultivated Plants in a Heavily Polluted District of Budapest. In: Markert, B. (ed.) "Plants as Biomonitors. Indicators for Heavy metals in the Terrestrial Environment" pp. 495-505. VCH Verlagsgesellschaft mbH, Weinheim.

23. Levin, M. 1995. Safety Considerations in Biotreatment Operations. In G.T. Tzotzos (ed.) Genetically Modified Organisms. A Guide to Biosafety. UNIDO, Cab International, Wallingford, UK.
24. MacMahon, J.A. 1987. Disturbed lands and ecological theory: an essay about a mutualistic association. In: Jordan III, W.R., M.E. Gilpin and J.D. Aber. 1987 (eds.). Restoration Ecology a Synthetic Approach to Ecological Research. pp. 221-237. Cambridge University Press, Cambridge.
25. Malyuga, D.P. 1964. Biogeochemical Methods of Prospecting. Consultants Bureau, New York.
26. Marcus, P. A. and J.T. Willig (eds.) 1997. Moving Ahead with ISO14000. Improving Environmental Management and Advancing Sustainable Development. J. Wiley, New York.
26. Markert, B. (ed.) 1993. Plants as Biomonitors. Indicators for Heavy metals in the Terrestrial Environment. VCH Verlagsgesellschaft mbH, Weinheim.
27. Meyers, R.A. (ed.)1998. Encyclopaedia of Environmental Analysis and Remediation. John Wiley, New York.
28. Miller, R.M. 1987. Mycorrhizae and succession. In: Jordan III, W.R., M.E. Gilpin and J.D. Aber. 1987 (eds.). Restoration Ecology a Synthetic Approach to Ecological Research. pp. 205-219. Cambridge University Press, Cambridge.
29. Mueller-Dombois, D. and H. Ellenberg. 1974. Aims and methos of vegetation ecology. Wiley, New York.
30. Odum, E.P. 1989. Ecology and Our Endangered Life-Support Systems. Sinuaer Associates, Sunderland, Massachusetts.
31. Raskin, I. and B. D. Ensley. 2000. Phytoremediation of Toxic Metals. Using Plants to Clean Up the Environment. John Wiley & Sons, New York. pp. 304.
32. Redente E. F. and E. J. Depuit. 1988. Reclamation of drastically disturbed rangelands. In :Tueller, P.T. (ed.) Vegetation Science Application for Rangeland Analysis and Management. pp. 559-584.Kluwer Academic Publishers, Dordrecht.
33. Richards, D.J., B.R. Allenby and R. Frosch. 1994. The Greening of Industrial Ecosystems: Overview and Perspective. In: B.R. Allenby and D.J. Richards (eds.) " The Greening of Industrial Ecosystems", pp. 1-19. National Academy of Enineering, National Academy Press, Washington, D.C.
34. Schlesinger, W. 1994. The Vulnerability of Biotic Diversity. In: R. Socolow, C. Andrews, F. Berkhout and V. Thomas (eds.) "Industrial Ecology and Global Change", pp. 245-260. Cambridge University Press, Cambridge, USA.
35. Stahel, W.R. 1994. The Utilization-Focused Service Economy: Resource Efficiency and Product-life Extension. In: B.R. Allenby and D.J. Richards (eds.) " The Greening of Industrial Ecosystems", pp. 178-190. National Academy of Enineering, National Academy Press, Washington, D.C.
36. Streit, B. and W. Stumm. 1993. Chemical properties of Metals and the Process of Bioaccumulation in Terrestrial Plants. In: Markert, B. (ed.) "Plants as Biomonitors. Indicators for Heavy metals in the Terrestrial Environment". pp. 31-62. VCH Verlagsgesellschaft mbH, Weinheim
37. Tewolde Egziabher. 1982 Technology generation and the technological space. in "Project on Research and Development Systems in Rural Settings" United Nations University Press (Tokyo).
38. Tibbs, H.B.C. 1992. Industrial Ecology - An agenda for environmental management. Pollution Prevention Review, Spring: 167-180.
39. Tiedje, J.M. 1993. Bioremediation from an Ecological Perspective. In: Committee On In Situ Bioremediation (ed.) "In Situ Bioremediation. When does it work?". pp. 110-120. National Academic Press, Washington, DC.
40. Toy, T. J. and W.L. Daniels. 1998. Reclamation of Disturbed Lands. In: Meyers, R.A. (ed.) "Encyclopaedia of Environmental Analysis and Remediation". (Vol. 7) pp. 4078-4101.John Wiley, New York.
41. Ulanowicz, R.E. 1989. Book Review of Ecology and Our Endangered Life-Support Systems. Ecological Economics 1:363-365.
42. Walter, H. 1979. Vegetation of the Earth and Ecological Systems of the Geo-biosphere. 2nd ed. Springer-Verlag, New York.
43. Wacknagel, M and Yount, J.D., 2001. Footprints for sustainability: the next steps. Environment, development and sustainability, 2:21-42.
44. Westhoff V. and van der Maarel, E. 1978. The Braun Blanquet Approach. In R.H. Whittaker (ed.) "Classification of Plant Communities. Junk, The Hague, pp. 287-399.
45. White. R. 1994. Preface. In : B.R. Allenby and D.J. Richards (eds.) "The Greening of Industrial Ecosystems", pp. V-VI. National Academy of Enineering, National Academy Press, Washington, D.C.

PART 5. CASE STUDIES

Change detection using remotely sensed data permits assessment of the impact of mining activity on the landscape, the soil layer, the vegetative cover, and on the land's water resources. In addition, spatial modeling and the monitoring of mining operations can provide an indicator of the vulnerability of groundwater to pollution, the expected effects of underground mining on land subsidence, and the characterization of water quality for protection measures. Several case studies in the Czech Republic and in southern Spain offer examples of how time-dependent observations provide guidance in resource protection and the disposal of mining waste.

The paper by Fabbri and others describes the results of systematic photo-interpretation of aerial photography sets over a coal mining area near the city of Ostrava in the Czech Republic during the years from the 1950's to the 1990's. In addition, they provide examples of aquifer vulnerability models that use GIS techniques. Woldai and Fabbri's paper compares aerial photo-interpretation with the results of information extracted from satellite images to detect and inventory the effects of copper, gold, and silver mining in the pyrite belt in the region of Huelva in southwestern Spain. The paper by Rapantova and Grmela pays special attention to the effects of mine closures and subsequent mine-waste disposal on groundwater and surface water. In a second paper, they provide detailed evidence of the consequences of mining on land subsidence in the Ostrava-Karviná area of the Upper Silesian coal basin in the Czech Republic.

LAND-USE CHANGE AND VULNERABILITY AS A RESULT OF COAL MINING ACTIVITIES.

Application of Spatial Data Analysis in the Upper Silesian Region of the Czech Republic

FABBRI A.G.[1], WOLDAI T.[1], BABIKER I.S.[2], KITUTU KIMONO M.G.[3], HOMOLA V.[4]

[1] *International Institute for Geo-Information Science and Earth Observation Sciences-ITC, Enschede, The Netherlands*
[2] *Khartoum University, Khartoum, Sudan*
[3] *Geo-Mineral Environmental Consult Ltd., Kampala, Uganda*
[4] *University of Ostrava-VŠB, Ostrava-Poruba, Czech Republic*

Abstract. Because the environment is considered to be a resource in itself, it is worth characterizing and assessing in terms of such meaningful indicators as land use/cover change and aquifer vulnerability. This contribution aims to validate past and present spatially distributed data such as maps, aerial photographs, satellite images and other types of numerical or qualitative data, for the purpose of representing natural and human-induced processes that affect environmental quality. In areas of intense underground coal mining, systematic photo-interpretation of airborne and space-borne images can identify changes in land-cover through time that may reveal trends in mining development and its effects towards degradation or rehabilitation of the environment.

Two applications are discussed in which spatial data analysis contributes to the environmental characterization of an area of the Czech Republic that has been deeply affected by underground coal mining and by the industrial activities dependent on coal as a source of energy. Although the applications contain several aspects of generality for geo-environmental analyses, they also reveal a particular need for data-quality assessment within administrative and social conditions that only recently have allowed cartographic data and aerial photographs to become freely available. Hence, sensitivity analysis and reliability testing can lead to operational strategies in predictive modeling and in the necessary data capture.

1. Introduction

Land-use changes are the symptoms of the environmental condition or health status. Vulnerability is the predisposition or propensity of the land or media to be affected by damaging processes. A resource is a product of the physical environment used by man. Considering the environment as the set of conditions, circumstances and influences under which an organization or system exists and that it therefore may be affected or described by physical, chemical, and biological features, the following fundamental kinds of environmental services can be contemplated [21]:

1. General life support,
2. Supply of raw material and energy, and
3. Absorption of the waste products of economic and social activities.

It becomes obvious to assume then that an interdependence exist among human health, economy, and environment (Sadar, 1996, p. 15) [14]. Such interdependence leads to attach to the environment the characteristics of a resource. Therefore, the application of environmental economics, environmental geology, and other interdisciplinary research areas that deal with holistic or systemic approaches becomes a fundamental issue.

The exploitation of natural resources, such as coal, uranium, and raw material, is strictly linked with the environmental impact of development activities and with associated socio-economic transformations. Impacts can affect the landscape, the water table, air quality, and the access to other resources.

The quantitative characterization of natural resources has been made through the construction of indicators of resource quality distribution. Such indicators, which became tools for mineral exploration, consisted of the distribution of mineral deposits of given genetic types, mineral occurrences and mineral indicators; i.e., occurrences of mineralogical settings that have characteristics typical of the ones in which mineral deposits have been found. Typicality of settings (i.e., conditions similar to those observed in key locations) is also the basis of environmental indicators. An indicator has been defined as "a synthetic representation of a complex and equivocal reality; i.e., a characteristic or a set of characteristics that allow capture of a given environmental phenomenon [18].

One example of environmental indicators used in this contribution is the pattern of land-use changes through time that can be deduced from the study of aerial photographs, satellite images, land-use maps, and direct field observations. Another example is the cartographic quantitative representation of the vulnerability of aquifers to pollution as a result of combining several layers of thematic and point interpolation maps.

Spatial data analysis has been applied in studies of the environmental conditions in the Ostrava-Karvina area of the Czech Republic where underground black coal mining and the development of ancillary industries, such as metallurgical, chemical, or thermal plants, have inflicted deep scars in the landscape and have severely damaged the environment.

The purpose of this contribution is to demonstrate how data is transformed into information that can be validated in quality and in its significance as being representative of a present condition or of a future trend. Natural and human processes that affect the environmental quality can be comprehended and managed by an operational strategy in predictive modeling that is here exemplified by the two applications made in the study area. They take advantage of the analytical power of a geographical information system, GIS, to interpret processes, in time or in space, that relate with decision making for environmental protection. First a review of the effects of coal mining for the study area is made that is accompanied by a description of the data collected and used and of the problems encountered in spatial data acquisition. Next an application on land-cover changes is discussed that is followed by another on aquifer vulnerability assessment.

Finally, considerations are made on the importance of the analysis of such spatial data that has become easily-available in the Czech Republic.

2. Study area and databases

2.1 COAL MINING AND ITS IMPACT

The coal mining industry is the most important source of power in the Czech Republic. In the Ostrava-Karvina area of the North Moravian region, coal exploitation has increased as population has increased. The coal deposits in the Ostrava-Karvina coal field were first discovered in 1767. During 200 years of intensive activity, coal mining in the region made the Ostrava-Karvina coal belt the primer producer of hard coal. Between 1985 and 1993, approximately 10 million tons per year were produced for consumption and export in the country, of which half came from the North Moravian region.

The Ostrava-Karvina region supplies about 80% of the production of coal of good quality in the Czech Republic [20]. The geological coal reserves of the Czech part of the Upper Silesian basin (which extends over Poland) are about 12.8 billion tons, of which 4 billion tons of mainly coking coal are in the Ostrava-Karvina region, as shown in Table 1.

Table 1: Coal reserves in the mining areas of the Ostrava-Karvina coalfield (after [15]).

Reserves	Million tons	%
Geological	4005.495	100
Usable	881.272	22
Really mineable	353.910	8.8

Research activity carried out in the Ostrava-Karvina area has led to report serious environmental problems caused by underground mining. According to Schejbal (1994) [15], waste rock dumped on the surface has created anthropogenic forms of relief. Subsidence basins, partially filled with water owing to underground water, and slurry ponds are the most conspicuous landscape damages. Owing to subsidence, for instance, about 96 urban settlements have been lost in the Czech Republic, most of which are in the Ostrava-Karvina region.

308

Figure 1: Waste rock dumps forming a hill along the Ostrava-Karvina road. The dumps on the left side are Fresh, and those on the right have shrubs and grass growing on them.

Figure 2: A slurry pond from the CSA mine constructed after clearing a forest. Also, a hill of waste rock partly covered by bush can be seen in the vicinity.

In addition, the rapid growth of population and metallurgical industries has accompanied the extraction activities. The use of coal resources led to the existence of industries and residential areas in the same surroundings. The high density of settlements, estimated to be 1,000 persons per square kilometer, is attributed to mining. Such areas have the highest growth in population with the highest levels of diseases and mortality caused by malignant cancer, that have been attributed to such factors as air pollution, contaminated foods, and water [3].

All this has resulted in severe changes in landscape and land-use patterns. Coal mining in the Ostrava-Karvina area resulted in subsidence, waste rock dumps (see Figure 1), slurry ponds, and industrial plants (see Figure 2). The impacts are felt in the loss of villages, loss of agricultural land, damages to the landscape, soil degradation, air/water/soil pollution, and growth in urbanization, as shown in Figure 3.

Some of the land-use manifestations in the Ostrava-Karvina area

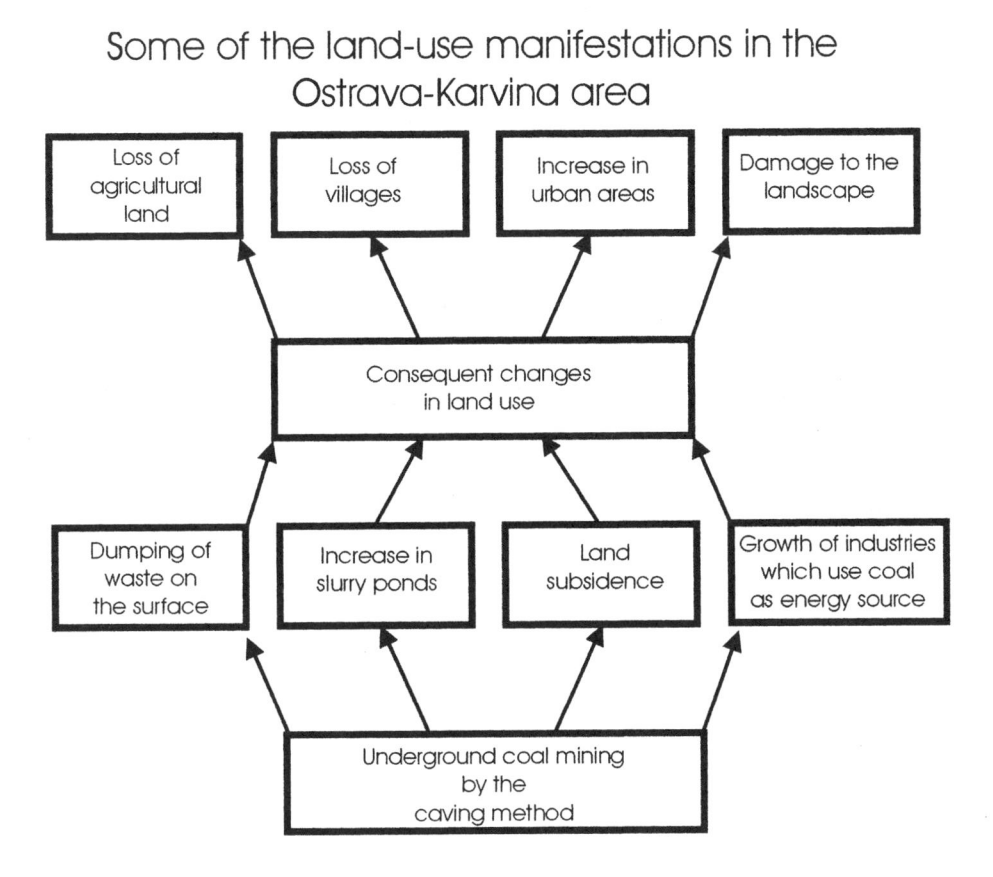

Figure 3: Problem tree. Coal mining and its effects on land use.

Waste rock dumps contain siltstones and clay-stones, usually with added mixtures of coal matter. Although the carboniferous material is considered to be practically harmless to the environment [10], it may hamper the vegetation growth for some time and alter the ecosystem because of the surface dumping.

Beginning in 1990, the amount of waste rock dumped on the surface has decreased because some of it was used for filling subsidence basins. Also, the reduction in the price of coal has resulted in the closure of some of the mines. At present, some of the areas covered by waste rock have been recultivated into forests or agricultural lands [16], but it has been slow owing to the financial constraints that have resulted from the low price of coal. The mining companies cannot easily adapt to new technologies to minimize the impacts on the environment owing to the low price of coal and the low profits. In addition, the waste waters from the processing of coal are pumped into slurry ponds with their smelly black waters, that have increased during the years of intensification of mining. Before the mining companies started taking measures, they were partly a source of pollution to surface waters. Coal-mine drainage in the study area can have a higher salt content of up to 100 g/l [19].

The Ostrava and Karvina districts have the largest undermined area in the Czech Republic, that has been estimated to be 313 Km^2. Reichman (1992) [13] described the impacts of mining on the environment in the Czech Republic. The flowchart in Figure 4 summarizes the main environmental pollution hazards.

Many conflicts between the communities and the mining companies in the area have occurred because of the loss of infrastructures and the pollution that has resulted directly or indirectly from coal mining [13]. On the map compiled by the Cesky Geologicky Ustav (1992) [4], it can be seen that the largest area of conflicts owing to mining activities is the Ostrava and the Karvina districts. The Karvina area has the largest and richest coal reserves; consequently, many conflicts have occurred and will continue to occur until the mining companies can assure the community that no further problems will result from the mining activity. As shown in Table 2, the Karvina area has the greatest impact from land subsidence because of its richer and thicker coal seams [16].

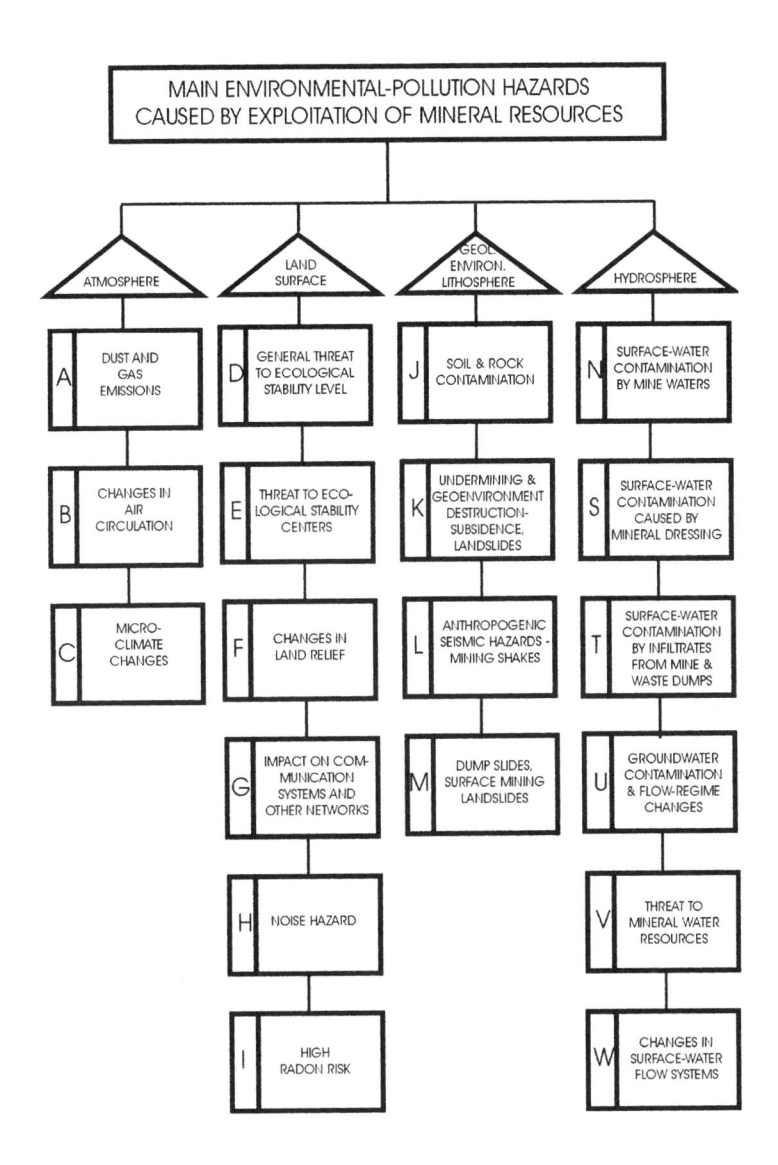

Figure 4: A summary of impacts of mining in the Czech Republic (after [13]).

Table 2: Surface subsidence from mining activities in the Karvina district (after [17]).

Period	Subsiding area (hectares)				
	< 1 meter	1-5 meters	5-10 meters	>10 meters	Total
Before 1990	1,500	9,000	15,000	2,500	28,000
1990-2000	4,000	10,000	6,000	--	20,000
2000-2010	12,000	5,000	1,000	--	15,000
Total	17,500	21,000	22,000	2,500	63,000

2.2 TWO TARGETS AND DATA COLLECTIONS

Targets were set to assess the environmental impact of mining in the Ostrava-Karvina region: (1) the study of land use changes through time, and (2) the assessment of the vulnerability of aquifers to pollution. Although target 1 was achieved by using aerial photographs and satellite images and by performing systematic photo-interpretations, target 2 required the compilation of several maps to be integrated into an index of vulnerability by a semi-quantitative model. Here, we summarize the data collection for the two targets, which was performed by direct fieldwork and by obtaining maps, tables, and reports from Czech institutions, which have a particularly strong tradition in archiving geoscience and environmental data. Among these were the Geological Survey of the Czech Republic and the Geological Documentation Information Service of the Czech Republic (GEOFOND) in Prague, and the Technical University of Ostrava (VŠB), in collaboration with which this research was undertaken.

2.3 PROBLEMS IN DATA ACQUISITION

Tables 3 and 4 show the lists of data types and sources collected for the studies of targets 1 and 2 by Kitutu Kimono (1998) [9] and Babiker (1998) [2], respectively.

Table 3: Spatially distributed data for the study of land-use change in the Ostrava-Karvina area of the Czech Republic (modified after [9]). Abbreviations: GS ACR, General Office of the Army of the Czech Republic; ITC, International Institute for Aerospace Survey and Earth Sciences, Enschede, The Netherlands; CUGK, Czech Bureau of Geodesy and Cartography; CGU, Czech Geological Survey.

Source	Year	Scale	Format	Input	Output	Reliability	Used for
GS ACR	1954	1:23,000	Aerial photos	Visual interpretation	Land-cover map	High	Land-use area
GS ACR	1979	1:26,000	Aerial photos	Visual interpretation	Land-cover map	High	Land-use area
ITC	1989	1:50,000	SPOT panchromatic	On-screen digitization	Land-cover map	High	Land use area
ITC	1995	1:50,000	SPOT panchromatic	On-screen digitization	Land-cover map	High	Land-use area
CUGK	1989	1:25,000	Civil topo. maps	Digitization & interpolation	Digital Elevation Model	Low	Elevation
GS ACR	1995	1:25,000	Military base maps	Digitization & interpo-lation	Digital Elevation Model	High	Elevation
CGU	1970	1:50,000	Geologic map of Carboniferous	Digitization	Map of thickness of coal seams	Medium	Distribution of thickness
CGU	1970	1:50,000	Geologic map of Carboniferous	Digitization	Map of quality of coal	Medium	Distribution of quality

One of the main problems in collecting spatial data in these studies is the inconsistency in the coordinate systems used to register the different data types. The civil base maps and other thematic maps in the former Czechoslovakia did not display any coordinates for 1:50,000 scale, and 1:25,000 scale maps. The more-recent military topographic base maps, now available, have two different types of coordinate systems; the Geographical coordinates (latitudes and longitudes) and the Old Soviet UTM coordinates (Universal Transverse Mercator). Borehole location data were supplied with the Old Soviet UTM coordinates, and the remotely sensed data (SPOT Panchromatic images) were not georeferenced. In addition, the corner coordinates of the civil thematic maps, such as geology, soil, and hydrogeology, could only be obtained in the Krovak coordinate system by calculations performed at VŠB and that required transformation to the conventional UTM system.

Table 4: The data used to construct a database for aquifer vulnerability assessment in the Ostrava area of the Czech Republic (modified after Babiker, 1998) [2]. Abbreviations: CUGK, Czech Bureau of Geodesy and Cartography; CGU, Czech Geological Survey; GS ACR, General Office of the Army of the Czech Republic; VŠB, Technical University of Ostrava; GEOFOND, Geological Documentation Information Service of the Czech Republic; CHU, Czech Hydro-geological Survey; ITC, International Institute for Geo-Information Science and Earth Observation, Enschede, The Netherlands.

Source	Years	Scale	Format	Data type	Used for
CUGK	1982	1:50,000	Map	Topography	DEM, depth to water, land-use maps
GS ACR	1995	1:25.000	Map	Topography	
CGU	1989	1:50,000	Map	Geology: Ostrava and Hlucin	Vadose-zone map
	1986	1:50,000	Map		
CGU	1991	1:50,000	Map	Hydrogeology: Osytrava and Hlucin	Aquifer media, ground-water level and water-quality maps
	1986	1:50,000	Map		
CGU	1995	1:50,000	Map	Soil: Ostrava and Hlucin	Soil media map
	1986	1:50,000	Map		
VŠB	1992	1:500,000	Map	Mining and the environment	Risk of ground-water contamination
GEOFOND	1959-1997		Table: 478 points Table: 351 points	Bore-hole data piezometric level Hydrogeological parameters	Ground-water level map Hydraulic- conductivity map
CHU	1901-1960 1931-1960		Tables	Annual rainfall	Net recharge map
ITC	1995	1:50,000	Digital raster image: 10 × 10 m	SPOT Panchromatic image	Land-use map

The Krovak mapping system is a Gaussian equiangular conic projection in the skew position. The ellipsoid used in this system is the Bessel [Date, 1841; length of semi-axes (m) 6377397.2 Major (a), and 6356079.0 Minor (b); and ellipticity f 1/299.15] that is projected into the plain by using the reference sphere. The Krovak mapping was established in Czechoslovakia in 1922 for cadastral maps but was used later for definitive military mapping. Since 1968, the sheets of the Basic Map System of Czechoslovakia have been built into this projection. Coordinates in this system can be transformed into UTM coordinates by using long and complicated approximate mathematical formulae owing to the secrecy still existing on the conversion procedures.

The Old Soviet UTM system is the national system and is used in Czech military topographic base maps in addition to the Geographical coordinate system. It uses the Krasovsky ellipsoid (Date: 1940; length of semi-axes (m) 6378245 Major (a), and 6356863.0 Minor (b); and ellipticity f 1/298.3), which is one of the widely used ellipsoids in the world. Although this is a different metric system, transformations to the UTM system are easily obtained.

In these studies, all coordinates were transformed from the original national coordinate systems (Krovak, Old UTM or Geographical) into the UTM. This projection has facilitated the registration of all the spatial data, including the satellite images, into a Geographical Information System (GIS) for subsequent editing and analysis.

Such a problem represents a general situation not only in the Czech Republic, but also elsewhere in the former Communist countries where, for political and military security reasons, cartographic data and aerial photographs were not available to the general public. All aerial photographs used to be secret, and accurate topographic maps could only be used by the military. This led to the growth of two distinct types of base maps: one without coordinates and with likely distortions and another for the military only. At present, such information is becoming available so that an assessment of accuracy for all thematic spatially distributed data is now feasible. The impact of inaccuracies of civil base and thematic maps is still to be evaluated, however, and becomes a priority in GIS processing and modeling. The value of analyzing records of the past, such as aerial photographs, satellite images, and old maps, is considerable in environmental studies, as will become evident in the following sections.

3. Land-cover changes through time

The Ostrava-Karvina study area, that was selected for the analysis of land- use/cover changes, is located in the northeastern part of the Czech Republic, near the border with Poland, as shown in Figure 5. It covers about 15,150 ha (151.5 Km^2) and is approximately 21 Km east-west and 7.2 Km north-south. The area has experienced a strong dynamic change in land use owing to the high concentration of mining and related industries. It is relatively flat, laying between 190 m. and 300 m. above sea level, the few visible hills being man-made features resulting from the dumping of waste rock.

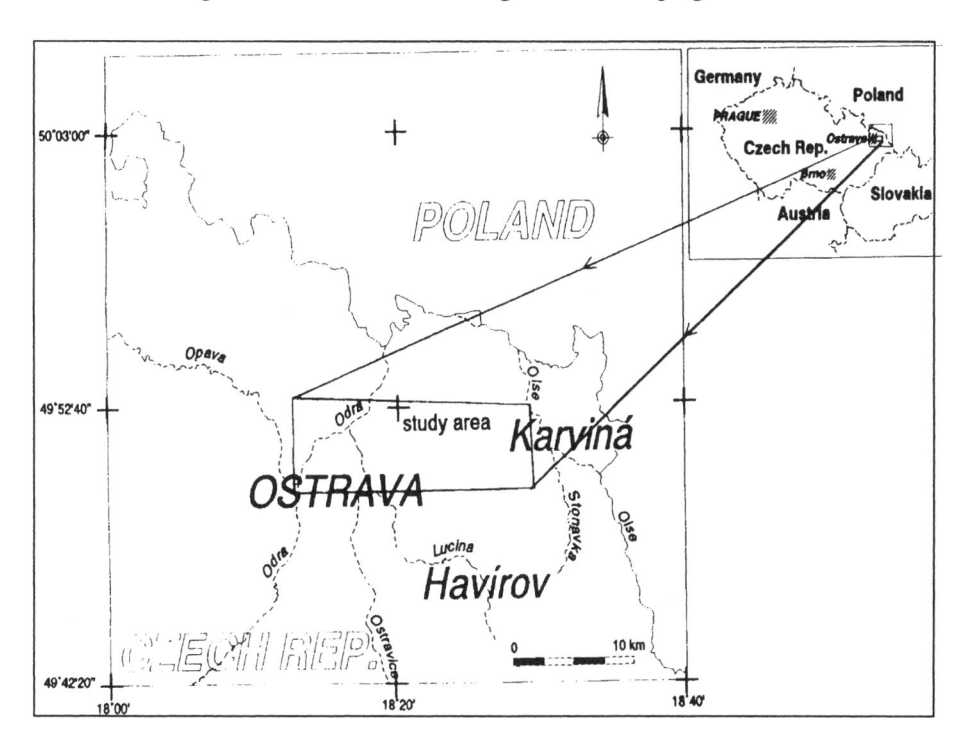

Figure 5: Location of study area for land-use/cover change in the Ostrava-Karvina region of the Czech Republic.

In the study area, a database was constructed for this analysis, as shown in Table 3. Aerial photographs for 1954 and 1979 were visually interpreted under a stereoscope. On-screen digitizing for visual interpretation was done for the SPOT panchromatic images, for 1989 and 1995. Image and photo characteristics, such as tone, texture, color, shape, and pattern, were translated into land-cover/land-use attributes. The interpretation process was guided by information from 1989 and 1995 topographic maps and also from field observations. In the absence of land-use maps to guide the classification, the land-cover types that were used were derived from direct field observations, topographic maps, and remotely sensed data by applying the land-cover/use classes shown in Table 5. The separate land-cover maps that were produced for the different years were then digitized for entry to a GIS. The subsequent analysis consisted of overlaying the raster images obtained for each map and generating the overlay maps and the corresponding tables of associations of successive land uses. Figure 6 shows the land-cover maps for 1954 and 1995. The ones for 1979 and 1989 are not shown. Figure 7 shows the land-cover changes for 1954-79, and 1989-95. The changes for 1979-89, that were also generated, are not shown. Figure 8 shows the graphical representation of the results of land-cover analysis for the years 1954, 1979, 1989, and 1995. Figure 9 shows the areal extents of the land-cover types obtained from the map patterns for the years 1954-1979, 1979-1989, and 1989-1995.

Table 5: Land-cover/ use classification used in the Ostrava-Karvina study area of the Czech Republic.

Land cover type	Land use	Expression on aerial photographs	Expression on Spot panchromatic image
Field	Agricultural land for crops, such as potatoes, corn, and vegetables	Regularly shaped areas, sometimes with linear texture showing crop alignment	Regularly shaped, with light tones and linear texture
Infrastructure	Includes built-up areas, roads, railway tracks, mines and industrial plants	Mines and industries have tall chimneys sometimes with smoke; built-up areas have buildings, roads, and railway tracks	Bright tones and smooth texture. Some of the structures were regular in shape. Roads and railways are distinct and bright linear features within the built-up areas
Forests	Coniferous forests used for timber logging and natural reserves	Dark and rough texture	Dark and rough texture
Water bodies	Lakes used for fishing and only one used as a protected lake area	Dark and very smooth texture usually near forests or grasslands	Dark and smooth texture
Shrubs	Natural grasslands with shrubs; sometimes they are abandoned areas owing to subsidence	Dark scattered tones within a slightly lighter tone	Dark scattered tones within a slightly lighter tone
Barren lands	Include slurry ponds and waste rock dumps	Slurry ponds have dark tone and smooth texture, in addition they occur near the mines. The waste rock dumps have light tones and irregular shapes	Slurry ponds have dark tone and smooth texture; in addition, they occur near the mines. The waste rock dumps have light tones and irregular shapes

To put change analysis into perspective, 1954 was selected as a base year to establish the state of the environment. The areal extent of each land-cover class for each of the four years (namely 1954, 1979, 1989, and 1995) was then compared to obtain an overview of the magnitude of the changes. For a total extension of 151.5 Km^2, the area percentages covered in 1954 by the six classes are shown in Table 6:

Table 6: Area percentages of the six land-cover classes mapped for 1954.

Class	%	Comment
Agricultural fields	70.0	highest value
Forests	9.3	
Shrubs	1.2	
Barren lands	1.0	Low value
Water bodies	1.8	
Infrastructures	16.7	Mostly residential

The histograms in Figure 9 show that agricultural fields decreased strongly from 1954 to 1979, while both the forests and the infrastructures steadily increased. Water bodies increased slightly during the same perid but then stabilized in extent. Barren lands increased noticeably up to 1989 and dropped slightly afterwards. Shrubs increased during 1954-1979 and then dropped afterwards.

In terms of "naturalness", the base year (1954) had a low proportion of area covered by forests and shrubs, which are considered to be environmentally friendly. The largest areas were occupied by agricultural fields. Barren lands occupied low areal proportions (relatively low mining activity). The infrastructures were mostly residential.

As reported in local sources, agricultural land was abandoned because of flooding from underground water and dumping of waste rock. Fields and forests are considered to be old classes on the basis of the situation in 1954. Because fields, which are used for agriculture, occupied 70.0% of the study area, it was assumed that they would be affected most. Because forests were considered to be a limited resource that occupies about 9.3% of the area, any further change would be greatly damaging to the environment.

The loss of agricultural fields and forests owing to mining can be evaluated in terms of the yearly rate of change, as follows in Table 7.

Table 7: Losses in ha/y and in %/y of agricultural fields and forests. Negative values represent gains.

Years	Agricultural fields Loss in ha/y (%)	Total ha (%)	Forests Loss in ha/y (%)	Total ha (%)
1954-1979	182.44 (1.20)	4556.1 (30.07)	-1.00 (-0.01)	-25.2 (-0.17)
1979-1989	47.69 (0.31)	476.9 (3.15)	-15.04 (-0.10)	-150.4 (-0.99)
1989-1995	51.40 (0.34)	308.9 (2.39)	48.26 (0.32)	289.6 (1.91)
Total change		**5341.9 (35.26)**		**114.0 (0.75)**

Even though not enough data from past years were available for this study, a clear change in agricultural land and forests could be seen in the land-cover change maps in Figure 7. This confirms that the change in land cover owing to coal mining activities is measurable and significant.

About 5341.9 hectares of agricultural land was lost from 1954 to1995, as shown in Table 7. The highest loss, however, seems to have occurred from 1954 to 1979 when about 4556.1 hectares were lost. From 1979 through 1989 and 1989 through 1995, lower losses were registered in agriculture (476.9 and 308.9 hectares, respectively). The high impact on agriculture during from 1954 to 1979, may give an indication that mining was intensified without proper measures to minimize on the effects that seem to have altered the land cover. Mines were initially nationalized in 1945 and since 1968 this seems to have stimulated the increase in the use of coal energy and, consequently, lead to rapid growth in industries and population. These rapid changes finally led to changes in land-use patterns, such as agricultural land turned into built-up areas, to meet the demands of the increasing population, that was seeking employment in mining companies and related industries.

In addition, agricultural land was lost owing to land subsidence. Schejbal (1994) [15] reported that about 3 Km^2 were covered by depressions filled with water. This is evident when comparing the aerial photographs for 1954 and 1979. Water bodies in depressions, some of which had regular shapes and were not existent in 1954, had emerged by 1979. The regular shapes of the water bodies in the succeeding years were comparable with those of the surrounding agricultural fields. This leads to the conclusion that agricultural land that had subsided was flooded by underground water because the water table was shallow in some places. Some of the emerging water bodies, which later formed permanent lakes, however, were along the Odra river whose locality is shown in Figure 5. This phenomenon may be due to changes in river systems owing to land subsidence caused by undermining, as reported by Reichmann (1992) [13].

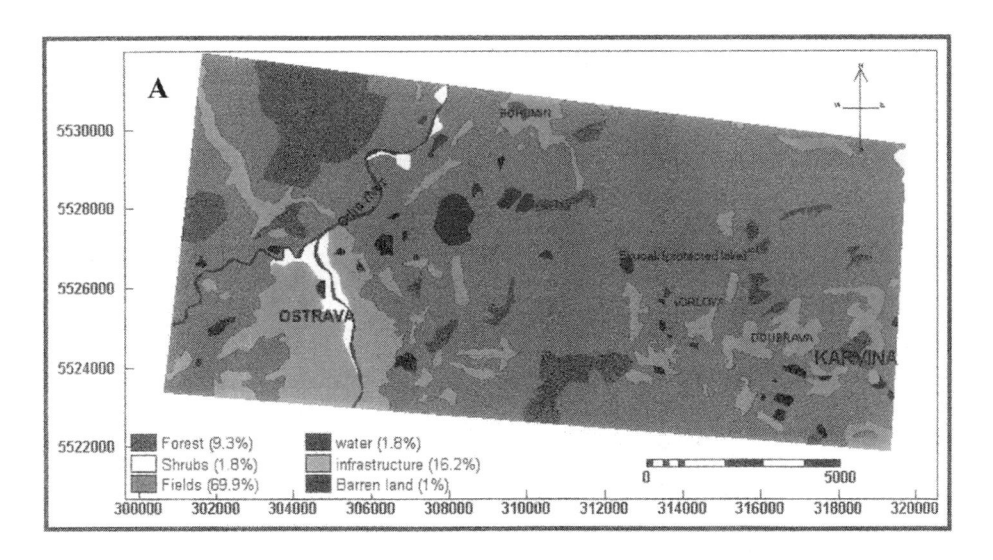

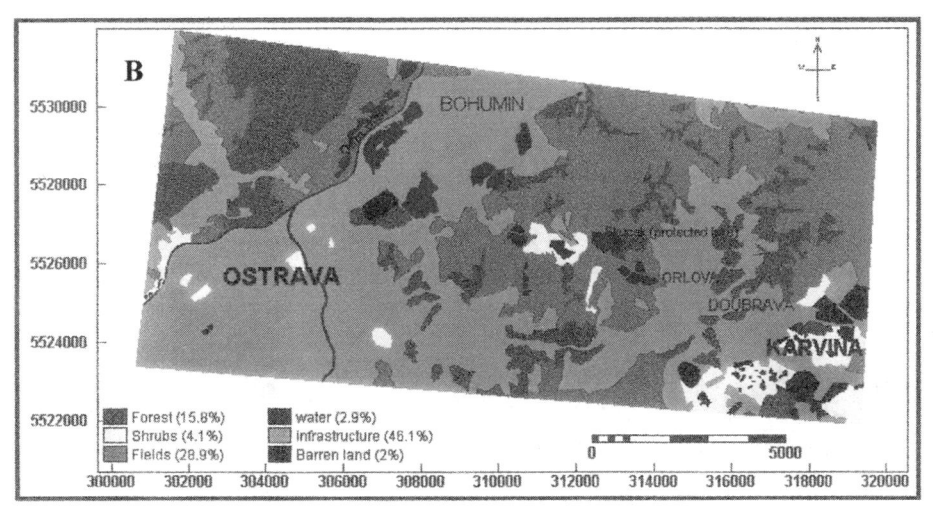

Figure 6: Land-cover map for the Ostrava-Karvina study area of the Czech Republic,. (a) for the year 1954, obtained by photo-interpretation of aerial photographs; an (b) for the year 1995, obtained by photo-interpretation of a SPOT panchromatic image.

320

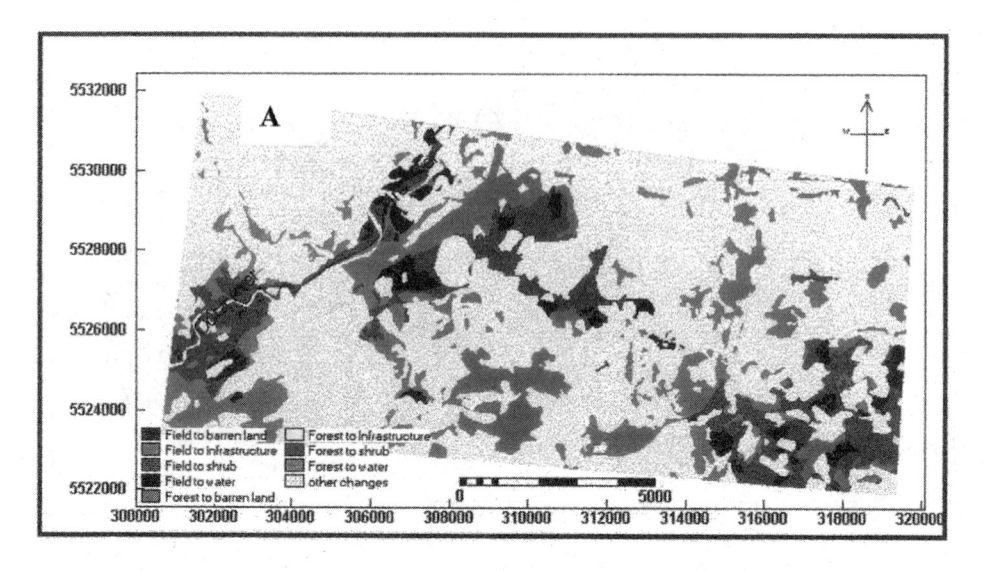

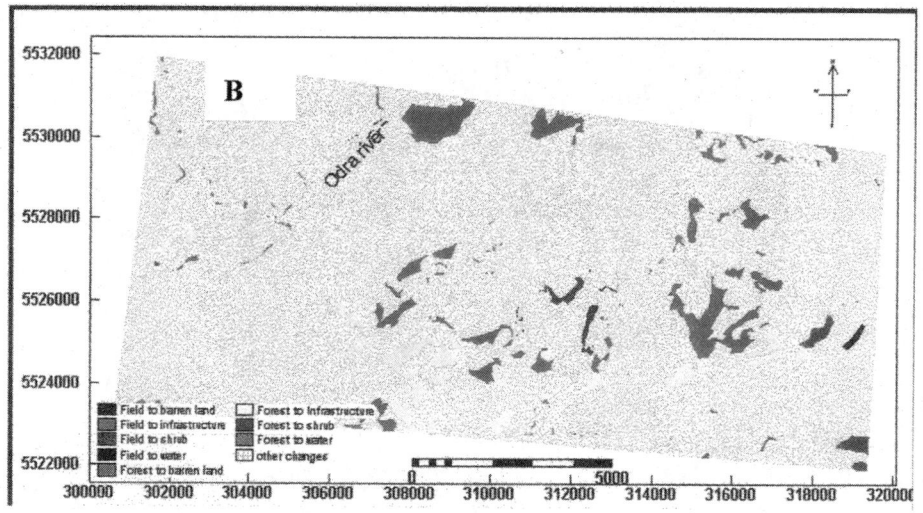

Figure 7: Land-cover change map of the Ostrava-Karvina study area. (a) from 1954 to 1979, obtained by map overlay of the land-cover maps for 1954 and 1979; and (b) from 1989 to 1995, obtained by map overlay of the land-cover maps for 1989 and 1995.

The decrease in the trend of loss in agricultural land from 1979 through 1989 and from 1989 through 1995, as shown in Figure 8, may be partly the result of rehabilitation of areas covered by waste rock dumps and slurry ponds to agricultural fields. It may, however, also be due to the reduction in mining activities after the community became aware of the damage to the environment. Hence, more pressure was put on the mining companies to curb the damages.

Forests are a natural resource that adds to the quality of the environment. In the study area this resource was relatively small in 1954. During the following years the forest areas increased, but the results, shown in Tables 8 and 9, still indicate that some of the areas with forests in 1954 were later changed to other uses as more land was needed for expansion of industries, dumping areas for waste rock, and built-up areas. The increase in areas covered by forests was due to the rehabilitation of areas covered by waste rock dumps and slurry ponds after mining expanded into forests (Table 9). This was observed directly in the field. Many techniques of agriculture and forestry have been applied to the revegetation of land disturbed by mining, but in most cases, this process has required careful planning because the waste rock does not have a natural soil medium and may take a long time for vegetation to grow.

Figure 8 shows the graphical representation of the data provided in Table 8. The change from field to infrastructure, which is the most pronounced, follows a downward trend. The curve shows a sharp decrease between the change periods of 1954 to1979 and of 1979 to 1989, which may indicate that many agriculture areas were turned into infrastructures during the latter period. Between the periods of 1979 to 1989 and of 1989 to 1995, however, the change from agriculture to infrastructure seems to decrease slightly because the curve is almost flat. Again, the changes from field to shrubs follows a similar trend, except that the areas are lower in magnitude. Changes from fields to water and fields to barren land also show a downward trend, but the magnitude of the changes is lower compared with the previous changes.

On the contrary, changes from forest to infrastructure and to water show an upward trend, which indicates that many areas covered with forests were turned into infrastructure and that others were flooded by underground water owing to subsidence. In fact, the change from forests to infrastructure reached a noticeable level from 1989 to 1995. Visualization of the land-cover change maps in Figure 7, gives an indication that the highest magnitude of all these changes was between 1954 and 1979 and that it was concentrated in the southeastern part of the study area.

The change from forests to barren land was only in the southeastern part during the three change periods. Strikingly, the change from fields to shrubs was localized in three areas; one is in the southeastern part, another is towards the north, and the third is in the western part. This may mean that this change is an indicator of areas where mining is going on below the surface. The changes from fields to infrastructure were scattered over the whole area from 1954 to 1979, but their distribution became more localized during the subsequent periods because it was concentrated mainly in the western and central regions.

Changes to water bodies appear during the three change periods and are situated in the central and northwestern parts of the study area. Another interesting observation is that the changes from field to shrubs and to water seem to occur in the same locality. By carefully looking at the change map in Figure 7a, it can be seen that in the central regions, areas with artificially regular shapes indicate change from field to water, which is a clear indication that agricultural land was flooded by water. Most of the agricultural fields, as observed in the aerial photographs and direct ground truth, have regular shapes, which is in agreement with the observations that the regular shapes of the water bodies indicate flooding of former agricultural fields.

In addition the changes from field to water was concentrated along the Odra River, whose location in the study area is indicated on the land-cover maps in the Figure 7. This was more pronounced during from 1954 to 1979, which may be an indication that the water bodies had been created as a result of the changes in the river system owing to land subsidence (as reported by [13]). The change from forest to infrastructure was higher from 1979 to 1989 and from 1989 to 1995, and is concentrated in the southern part of the study area. From 1954 to 1979, however, the change from field to infrastructure was more dominant in this same locality, which gives an indication that after the fields were exhausted, owing to the demand for built-up areas to meet the increased population, forests had to be encroached upon. The change from forests to shrubs gives an indication that the forests were being harvested for timber. Timber was used by the construction companies and as support structures in the underground mines as is usually the case in areas of underground mining

The trend identified can be explained by the decrease in mining activities. This is only one aspect of the situation in the study area, however, owing to changes that have improved mining technology that are contributing to improve the environmental quality, e.g., backfilling of mine works, greater depth of mining, etc. (Nada Rapantova, VŠB, personal communication).

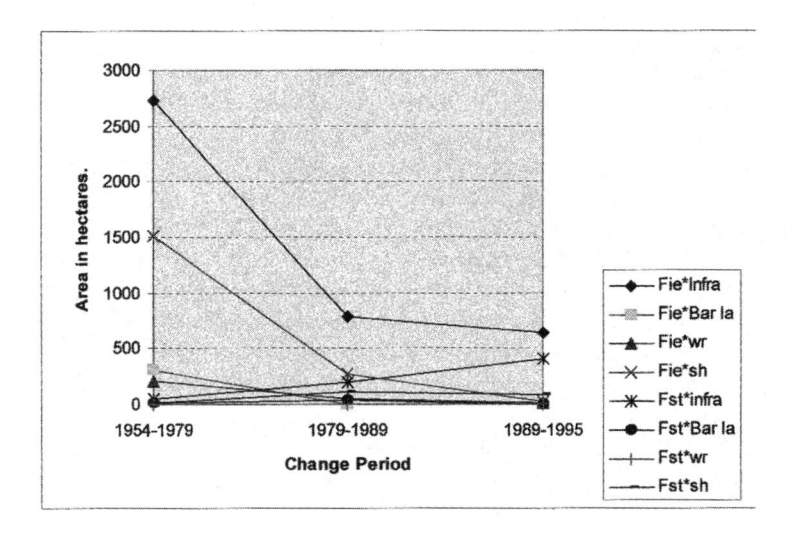

Figure 8: Graphical representation of the results from land-cover change analysis. Fie indicates agricultural fields; Fst, forests; Infra: infrastructure; Bar la: barren land; wr, water bodies; and sh, shrubs.

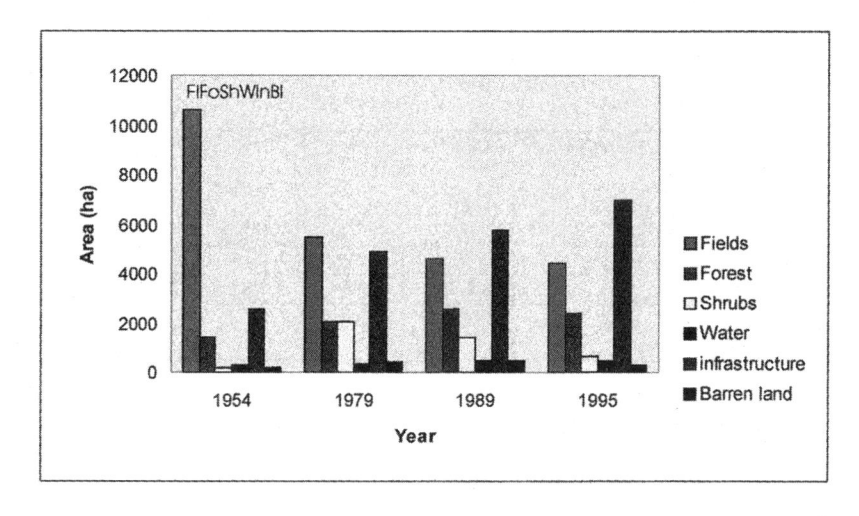

Figure 9: Graphical representation of the areal extent of land-cover types in the Ostrava-Karvina study area obtained from the land-cover maps of the years 1954 (see Figure 6a), 1979, 1989 and 1995 (see Figure 6b). The terms Fields to Barren land are abbreviated by Fi to Bl on the top left of the histogram for year 1954.

Table 8: Change table for loss of agricultural land and forest owing to mining activities. Source is the overlay of land-cover maps for 1954, 1979, 1989, and 1995 (see Figure 6 for 1954 and 1995).

Period	New land cover types	Area, in hectares and % of old land-cover types that changed to new types			
		Fields	%	Forest	%
1954-79	Infrastructure	2726.8	17.9	45.5	0.3
	Barren land	308.5	2.03	11	0.07
	Water	213.8	1.4	0	0
	Shrubs	1509.6	9.96	6.6	0.04
	Total	**4758.7**	**31.29**	**63.1**	**0.41**
1979-89.	Infrastructure	785.7	5.2	192	1.26
	Barren land	0.9	0.0005	34	0.22
	Water	52.5	0.34	3.1	0.02
	Shrubs	272.3	1.9	114.6	0.75
	Total	**1111.4**	**7.4**	**343.7**	**2.25**
1989-95.	Infrastructure	635.8	4.19	409.6	2.7
	Barren land	9.7	0.06	0.3	0.0002
	Water	6.8	0.04	3.6	0.02
	Shrubs	20.5	0.14	83.6	0.55
	Total	**672.8**	**4.43**	**497.1**	**3.27**

Table 9: Change table, in hectares, for increase in agricultural fields and forests from the rehabilitation after mining. Source is the overlay of land-cover maps for 1954, 1979, 1989, and 1995 (see Figure 7 for 1954-1979, and for 1989-1995).

Period	New land cover types	Infrastructure	Barren land	Water	Shrubs	Total	%
1954-79	Field	166.7	8.1	12.1	15.7	**202.6**	**1.3**
	Forest	77.8	7.1	3.4	0	**88.3**	**0.58**
1979-89	Field	526.7	15.7	11.3	80.5	**634.2**	**4.19**
	Forest	236.3	27.5	2.3	228.0	**494.1**	**3.26**
1989-95	Field	280.4	6.5	23.4	300.2	**610.5**	**4.02**
	Forest	83.5	54.9	3.1	66.0	**207.5**	**1.37**

Areas with land use change were further analyzed to select the changes that are direct indicators of mining impacts. An indicator is either a single parameter or a mathematical manipulation of a series of associated parameters to represent a simplified description of an environmental variable. Patterns of land cover can be indicators of the quality of the environment in relation to human activity. The change maps were reclassified to show some of the changes owing to mining. To keep the number of change classes reasonably small, a few classes that were considered to be direct or indirect indicators of mining effects were infrastructures, barren lands, water bodies, and shrubs. The infrastructure class was selected on the basis of a report that the area has had a high explosion in population and industry. In addition, barren land, which consists of waste rock dumps, and slurry ponds is directly related to mining; consequently their presence can be used to infer the effect of mining activities. In the case of water and shrub classes, their presence sometimes may be an indicator of land subsidence owing to mining.

A systematic approach, that considers environmental quality as a function of land-use/cover distribution and change, is being applied to the study area. It becomes feasible owing to the wealth of information that could be extracted by photo-interpretation and field verification described in this section.

4. Aquifer vulnerability models

4.1 PURPOSE, STUDY AREA, AND HYDROGEOLOGIC SETTING

The study area selected for aquifer vulnerability assessment is also located in the northeastern part of the Czech Republic. It covers the city of Ostrava and part of the Odra River up to the border with Poland, as shown in Figure 10. Its dimensions are 12.5 Km east-west and 18.4 Km north-south, and covers approximately 230 Km^2. Geologically it represents part of the Ostrava-Karvina coal field, which is, by far, the most significant mineral resource in the Czech part of the large triangular structure called the Upper Silesian Basin.

The hydro-geological setting of the study area is characterized by a significantly shallow aquifer located in the Quaternary sediments. The coarse or medium-grained basal Neogene (Lower Badenian) clastics, known among the miners as "Ostrava detrit", which may reach 280 m in thickness, form a reservoir for a large amount of overpressured, mineralized, and gassified groundwater. This is an unexploitable ground water that causes serious problems to the mining activities and is considered to be a dangerous zone for miners [7].

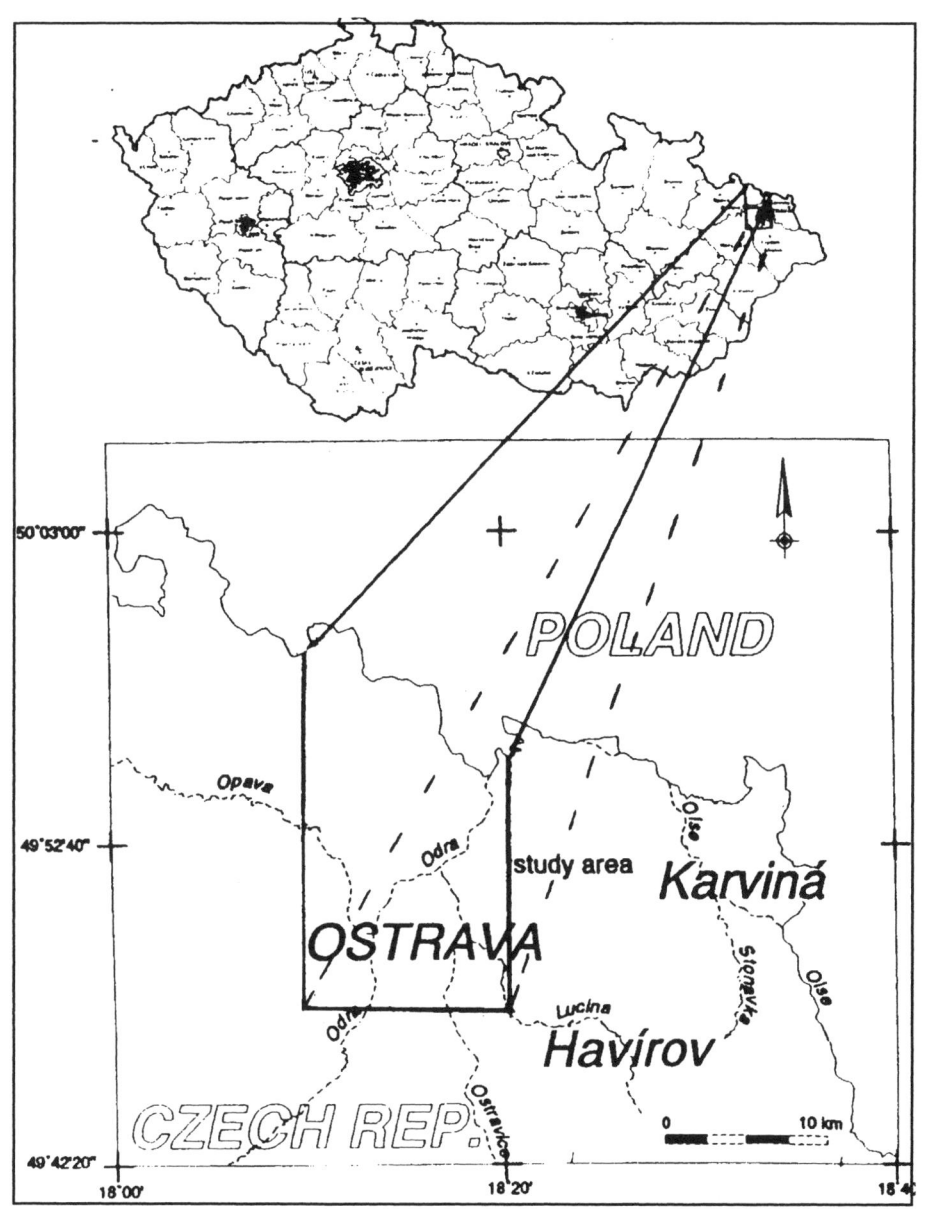

Figure 10: The location of the study area for aquifer vulnerability assessment.

The shallow aquifer represents the main exploitable portion of the hydro-geological reserves in the study area. They generally comprise unconfined porous aquifers that are gravely, sandy and loamy in composition with scattered lenses of clay and moderate to high permeability. Figure 11 shows a block diagram of the hydro-geological setting of the study area. Another minor fissure aquifer is located in the shallow sandstones and arkoses of the Lower Carboniferous units and possesses moderate permeability. It is found to the northwest of Ostrava. The river systems in the study area (Odra, Ostravice, and Opava) are draining those surrounding shallow aquifers. The average groundwater resources are $1.43 * 10^9$ m^3, and the average estimates of mineral water are $0.006 * 10^9$ m^3. Only 40% of the underground water reserves are being used mainly for residential water supplies [20].

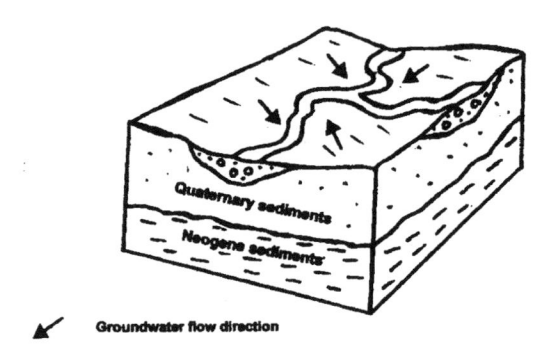

Figure 11: A block diagram illustrating the hydrogeological setting of the study area.

The Ostrava-Karvina agglomeration represents a typical, disturbed environment owing to intensive urbanization with high concentration of industry and population in the Czech Republic. In the study area, a database was constructed to generate several aquifer vulnerability maps, as shown in Table 4.

The surface- and groundwater resources are endangered by the discharged mine and industrial waters. Although "acid mine drainage" is minimal (low content of sulfur and phosphor in the mined coal), the mine waters are of sodium-chlorite type being "medium-mineralized" in terms of total dissolved matter [15]. Underground mining influences the geo-hydrodynamic systems and surface streams in the following ways [8]:

- It creates new priority paths for ground water.
- It artificially opens closed systems with tight level and frequently connects them hydraulically.
- It creates mixed ground water.
- With the aeration zone thickness, it alters hydraulic gradients of surface streams and shallow ground water.

The depressions owing to the subsidence of the overburden rocks on top of the mining workings and caves are usually free of natural drainage. They are often filled with waste rock or other waste material, used as mud pits or slurry ponds, which makes them a potential source of contamination not only to underground water, but also to surface water.

In the study area, the underground water quality was not impacted directly by the undermining, but was more liable to surface contamination from chemical industrial complexes, coking plants, dumps, and the like. Grmela (1997) [8] argued that irreplaceable ground water resources must be protected from the influences of undermining so that their use (in demanded quality and quantity) remains preserved even at the cost of the loss of a part of the mineral deposit.

Although the hydro-geological setting is relatively simple, the study area is considered to be special in terms of the environmental situation. Because of coal mining, intensive land use, urban growth, and increasing water demands, the particularity of the study area indicates the importance of aquifer vulnerability assessment. For this reason, the present study aims to assess the vulnerability of the aquifer in Ostrava and its vicinity to provide more tools for decision-makers for future land-use planning in the study area. Two empirical methods have been used in this study to test the feasibility of constructing and using local databases of readily available spatial information for predictive modeling of aquifer vulnerability. The following subsections introduce one of the models used, the data types needed for its application and the consequent verification of the robustness of the resulting cartographic expression of vulnerability.

4.2 VULNERABILITY MODELS

This application used DRASTIC, which is a method that represents one of the empirical types of approaches developed to evaluate the ground water pollution potential of any area in the United States. The method was developed by the National Water Well Association in cooperation with the U.S. Environmental Protection Agency [1]. The method was designed to assist planners and administrators in the evaluation of the relative vulnerability of the different areas to ground water contamination from various sources of pollution. Moreover, it provides priorities for protection, monitoring, or clean-up efforts, as well as a better understanding by the industry personnel for the relations between various practices and ground water pollution potential [1].

DRASTIC deals only with vertical vulnerability and was prepared by using the concept of hydro-geological setting, which is defined as "a composite description of all the major geologic and hydrologic factors which affect and control groundwater movement into, through and out of an area" [1]. The DRASTIC method is based on the following assumptions:

- The contaminant is introduced into the aquifer at the surface of the ground and not directly into it.
- The contaminant is flushed into the ground water table by direct precipitation.
- The contaminant has the mobility of the water.
- The contaminant is considered non-reactive, i.e., the attenuation processes such as dilution, dispersion, biodegradation, sorption, ion exchange, etc., are not considered.
- The parameters included in the model critically influence the ground water vulnerability.
- The parameters are rated and weighted adequately in the DRASTIC formulation procedure.

- Data needed to apply the model are commonly available or can be easily obtained and possess acceptable degree of precision, accuracy, and resolution.
- The area to be assessed (i.e., the hydro-geological setting) should be 100 acres or larger.

It is important to mention here that the DRASTIC method was designed for the assessment of relative vulnerability from a regional perspective and not to replace on-site specific investigations. DRASTIC was designed as a screening method to delineate areas for further detailed investigation.

The DRASTIC model is based on the following seven parameters where the acronym "DRASTIC" has been originated:

Depth to water.

It represents the depth from the ground surface to the water table level.

Net **R**echarge.

It represents the amount of water per unit area of land that penetrates the ground surface and reaches the water table.

Aquifer media.

It refers to the consolidated or unconsolidated rock that serve as an aquifer, i.e., the saturated zone.

Soil media.

It represents the uppermost weathered portion of the vadose zone, is characterized by significant biological activity, and averages a depth of 6 feet or less from the ground surface.

Topography (slope).

It refers to the slope and slope variability of the land surface.

Impact of the vadose zone.

It is defined for the unconfined aquifer as the unsaturated zone above the water table and for the confined aquifer as the unsaturated and the part of the saturated zone located above the confining bed.

Hydraulic **C**onductivity of the aquifer.

It indicates the ability of the aquifer to transmit water. Hence, it determines the rate at which a contaminant will flow within the ground-water system.

The choice of the above mentioned factors originated from the understanding of the vulnerability concept and was based on their direct contribution in the ground water contamination process as a result of their physical and chemical characteristics, as well as on the availability of mappable data. The seven parameters are measurable entities that can be mapped from different sources and scales. They represent part of several principal parameters proposed by authors working in various aquifer vulnerability assessment methods and are associated with the following:

- The hydro-geological framework: characteristic of the soil, unsaturated zone, aquifer materials, and depth to ground water.
- The groundwater flow system: the direction and velocity of the ground-water flow and topography.
- The climate: amount of recharge to ground water.

Each of the seven parameters is classified into ranges or significant media types that have an impact on the aquifer pollution potential. The ranges of the different DRASTIC parameters have been evaluated with respect to each other and to indicate the relative contribution of each range in the pollution potential.

The ranges for each DRASTIC factor have been assigned a subjective rating that ranges from 1 to 10 by using a value function (data versus value). Rate 1, which is the least rating value, indicates the minimum effect of the specific range with respect to the vulnerability assessment, rate 10 stands for the maximum effect. Originally, Aller *et al.* (1987) [1] used such functions to generate the ranges and ratings of the different parameters to simplify the computation processes. For the factors D, R, S, T, and D, one value per range is provided. For the factors A and I, however, the user is allowed to choose either a typical rating or a variable rating that can be adjusted to suit a specific situation. Ranges and rating values are listed in Table 10.

In addition to the ratings, the different DRASTIC factors have been evaluated by assigning weights to them to indicate their relative importance. The relative weights, which range from 5 (the most significant) to 1 (the least significant), were established by using the Delphi Technique (consensus) approach [1]. The DRASTIC system gives a special consideration to areas that have strong agricultural activity and provides another weight classification (the "Agricultural DRASTIC"; see Table 11). In this method, the ratings and the weights are constants and cannot be modified in the model proposed by Aller *et al.* (1987) [1].

DRASTIC represents a numerical ranking method to determine a value for any hydro-geological setting by using an additive model. The pollution potential is computed by using the following formula to express it in terms of an index:

$$\text{DRASTIC index} = D_r\,D_w + R_r\,R_w + A_r\,A_w + S_r\,S_w + T_r\,T_w + I_r\,I_w + C_r\,C_w, \qquad (1)$$

where **D, R, A, S, T, I,** and **C** are the seven DRASTIC parameters, *r* is the rating value, and *w* is the weight assigned to each factor.

In other words, the seven parameters are mapped into ranges or significant media types, and rated and weighted according to their relative significance, and the pollution potential is computed as the product sum, as schematically shown in Figure 12. The DRASTIC index helps to identify areas that are more likely to be susceptible to ground-water pollution, therefore, only providing a tool for relative evaluation, but not an absolute answers. The higher the DRASTIC index value, the greater the pollution potential.

Table 10 lists the ratings assigned to the seven DRASTIC parameters, in addition to the corresponding ones for SINTACS. Table 11 lists the DRASTIC weights. and the corresponding SINTACS weights.

Another empirical method, SINTACS, which was developed by Civita (1994) [6], was partially derived from DRASTIC. SINTACS has been used to evaluate the intrinsic vulnerability of a karstic aquifer in the Alpi Apuane, in northern Italy. The acronym SINTACS originates from the initials (in Italian) of the seven principal factors of DRASTIC ("Soggiacenza" as depth to water, "Infiltrazione" as net recharge, "Non Saturo" as impact of vadose zone, "Tipologia di Copertura" as soil media, "Caratteristiche dell' Acquifero" as aquifer media, "Conduciblità Idraulica" as hydraulic conductivity, "Acclivitá di Superficie Topografica" as slope) [11]). Table 11 lists also the SINTACS weights.

The DRASTIC method

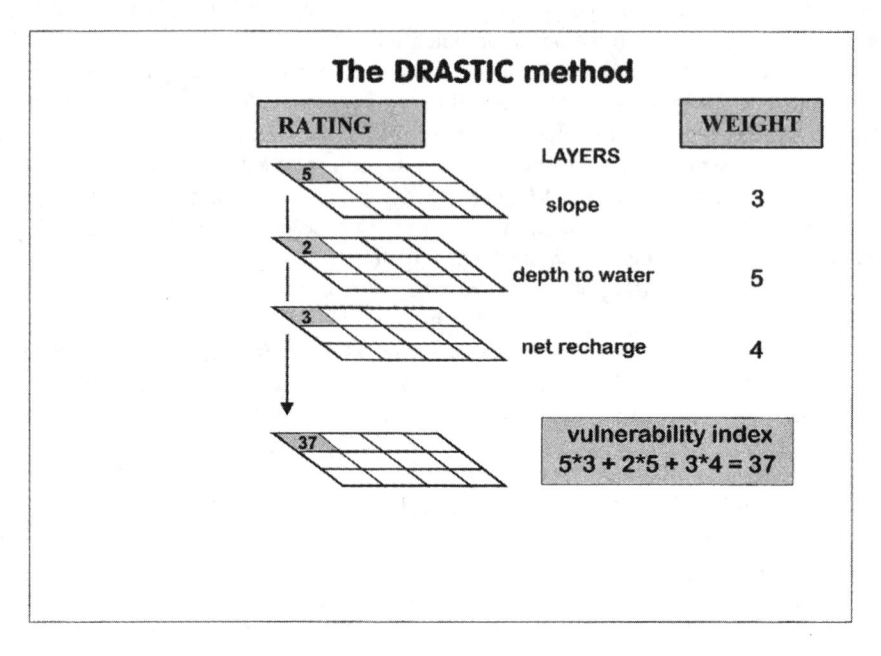

Layer	Rating	Weight
slope	5	3
depth to water	2	5
net recharge	3	4
vulnerability index	37	5*3 + 2*5 + 3*4 = 37

Figure 12: Conceptual framework of the DRASTIC method for aquifer vulnerability assessment. Only three parameters are shown (after Napolitano, 1995, p. 17, Fig. 2.4).

SINTACS differs from DRASTIC in ranges, ratings, and weights for the seven factors, as shown in Tables 10 and 11. In SINTACS also, the seven factors have been assigned weights according to the Delphi technique, where four different weight classes have been identified by several experts. The weight classes are not fixed in number (i.e., new classes can be introduced when it is necessary) and are used in parallel and not as alternatives. That means that the input data are coded to represent the real situations in the study area where it is possible to use different weight classes to meet the variation in a particular situation.

To evaluate the pollution potential the study area is discretized into a grid of finite square elements (FSE), which is 0.0625 Km^2 in area. The SINTACS index is then computed by using the following formula:

$$\text{SINTACS index} = \sum P_{(1,7)} * W_{(1,n)}, \qquad (2)$$

where $P_{(1,7)}$ is the rating of the seven parameters, and $W_{(1,n)}$ is the weight in each class which can vary from 1 to n.

The four proposed weight classes reflect the following scenarios:
- A relatively uncontaminated scenario, in which the surface drainage network is not linked to the underground water.

- Urbanized and/or industrial zones or highly agro-chemical supported agricultural areas.
- Areas with closely linked surface and ground water owing to drainage.
- Areas that exhibit karstic activity with a strong and rapid connection between ground and surface water.

Full descriptions and discussions of SINTACS have been made by Napolitano (1995) [11] and Babiker (1998) [2]. Table 11 shows the SINTACS weights for different scenarios.

Illustrations of some DRASTIC parameter maps (i.e., Depth to water, Topographic slope, and Soil media) are provided in Figure 13, as well as the corresponding DRASTIC rated maps used to compute the aquifer vulnerability map in Figure 14a. Two vulnerability maps are shown in Figure 14, that use the DRASTIC and the SINTACS methods. To help in comparing the two maps, the classes and colors in Table 12c use an arbitrary threshold of the vulnerability values. A different representation of the same values, which was proposed by Chung and Fabbri (1999) [5], is shown in Figures 14c and d. The vulnerability values were first sorted in descending order and successive decreasing groups of values, which correspond to 5% of the study area, were assigned the colors of a pseudo-color table. In this representation, which does not require the introduction of arbitrary classes, the two vulnerability maps seem to differ less than the ones in Figures 14a and b.

To assess the robustness of the results of the DRASTIC and SINTACS maps, Napolitano and Fabbri (1996) [12] and Babiker (1998) [2] performed several types of sensitivity analyses, such as: attribute, area, map removal, and single parameter sensitivities.

In particular, two versions of the Depth to water and the Topographic slope maps that had different accuracies were available. For example, the former from drill-hole data and a published hydro-geological map, and the latter from a 1:25,000-scale topographic base map and a 1:50,000-scale civil base map. Although the topographic slope had a very minor influence on the vulnerability, depth to water was found to have a greater influence, and therefore, was used for subsequent reliability representation. In addition, the Impact of the vadose zone, which caused the largest map removal variation in vulnerability, was used to characterize the reliability of the two types of vulnerability assessment.

Reliability was defined as a degree of trust in the vulnerability maps. The results of the sensitivity analysis were of assistance. The maximum impacts on the assessment were generated by substituting a less accurate Depth to water layer in the computation and by excluding the Impact of the vadose zone layer. The following assumptions were then made for the worst situation:

- The maximum impacts (in the worst case) are expected if the vulnerability computed by using the less accurate Depth to water layer *and* excluding the impact of the vadose zone parameter.
- The reliability of the output depends on the fluctuation of the intersecting polygons between the six vulnerability classes for which three scenarios were proposed:

1. If the polygon exhibited no change in the vulnerability class upon introducing the above-mentioned perturbations, then it is considered to be highly reliable.
2. If the polygon has moved one class higher or lower, then it is considered to be moderately reliable.
3. If the polygon has moved more than one class higher or lower, then it is considered to be unreliable.

The aquifer vulnerability maps were computed from the initial data layers by using an erroneous Depth to water parameter and by removing the Impact of vadose zone media. They were to be compared for DRASTIC and SINTACS on the basis of the above assumptions. The six vulnerability maps were classified, according to Table 11, and rescaled by using a unique numerical vulnerability class number that ranges from 1, which qualifies the "very low" class, to 6, which corresponds to the "extremely high" class. Hence, three reliability classes were identified and presented in Figures 14e and f for DRASTIC and SINTACS, respectively. In the illustration, the colors, red, green, and blue correspond to the highly, moderately and unreliable classes.

The highly reliable class tends to cover wider area in the SINTACS reliability map compared with the DRASTIC map. In DRASTIC, the moderately reliable class covers a larger area. This can be attributed to the better representation of the local situations in the SINTACS method by using three different weight classifications simultaneously. An obvious coincidence between DRASTIC and SINTACS is the area under the unreliable class.

Table 10a: Classification table for depth to water according to DRASTIC and SINTACS.

Rating	DRASTIC range (m)	SINTACS range (m)
10	0 - 2.05	0 - 2
9	2.05 - 4.52	2 - 3
8	4.52 - 6.70	3 - 5
7	6.70 - 8.60	5 - 7
6	8.6 - 11.00	7 - 10
5	11.00 - 14.05	10 - 13
4	14.05 - 17.62	13 - 20
3	17.62 - 22.40	20 - 30
2	22.40 - 27.10	30 - 56
1	>30.48	>56

Table 10b: Classification table for the net recharge map according to DRASTIC and SINTACS.

Ratings	DRASTIC range (mm/year)	SINTACS range (mm/year)
8	177.8 - 254	200 - 230
7	-	170 - 200
6	101.6 - 177.8	130 - 170
5	-	110 - 130
4	-	80 - 110
3	50.8 - 101.6	70 - 80
2	-	40 - 70
1	<50.8	<40

Table 10c: Scores assigned to the aquifer media according to DRASTIC and SINTACS.

| DRASTIC || SINTACS ||
Rating	Aquifer media	Rating	Aquifer media
8	Sandy gravel	8	Coarse alluvial sediments
6	Bedded sandstone and shale	8	Sand
5	Till	6	Sandstone and conglomerate
2	Massive shale	5	Thin morenic deposits
		2	Clay, silt and peat

Table 10d: Scores assigned to the soil media according to DRASTIC and SINTACS.

| DRASTIC || SINTACS ||
Rating	Soil media	Rating	Soil media
10	Gravel	10	Gravel
9	Sand	8	Sand
6	Sandy loam	6	Sandy loam
5	Loam	5	Loam
4	Silty loam	4	Silty loam
3	Clay loam	3	Clay loam
2	Muck	1	Clay
1	Clay		

Table 10e: Classification table for the Topography according to DRASTIC and SINTACS.

Rating	DRASTIC range	SINTACS range
10	0 - 2	0 - 3
9	2 - 6	3 - 5
8	-	5 - 7
7	-	7 - 10.5
6	-	10.5 - 13.5
5	6 - 12	13.5 - 16.5
4	-	16.5 - 19.5
3	12 -18	19.5 - 23
2	-	23 - 27.5
1	>18	>27.5

Table 10f: Scores assigned to the vadose zone media according to DRASTIC and SINTACS.

| DRASTIC || SINTACS ||
Rating	Vadose zone media	Rating	Vadose zone media
8	Sandy gravel	8	Coarse alluvial deposits
6	Sand, gravel and fines	6	Sandstone and conglomerate
6	Sandstone	6	Thin alluvial deposits
6	Bedded sandstone and shale	2	Marl and clay
3	Shale	2	Clay, silt, and peat
3	Silt/clay		

Table 10g: Classification table for hydraulic conductivity according to DRASTIC and SINTACS.

Ratings	DRASTIC (m/sec)	SINTACS (m/sec)
10	$>9.4*10^{-4}$	$>5*10^{-2}$
9		$1*10^{-2} - 5*10^{-2}$
8	$4.7*10^{-4} - 9.4*10^{-4}$	$5*10^{-4} - 1*10^{-3}$
7		$1*10^{-4} - 5*10^{-4}$
6	$3.3*10^{-4} - 4.7*10^{-4}$	$5*10^{-5} - 1*10^{-4}$
5		$1*10^{-5} - 5*10^{-5}$
4	$1.4*10^{-4} - 3.3*10^{-4}$	$4*10^{-6} - 1*10^{-5}$
3		$8*10^{-7} - 4*10^{-6}$
2	$4.7*10^{-5} - 1.4*10^{-4}$	$1*10^{-7} - 8*10^{-7}$
1	$<4.7*10^{-5}$	$<1*10^{-7}$

Table 11: Weight classifications for DRASTIC and SINTACS.

DRASTIC			SINTACS				
	Normal	Agricul.		Normal	Agricul.	Drainage	Karstic
D	5	5	S	5	5	4	2
R	4	4	I	4	5	4	5
A	3	3	A	3	3	5	5
S	2	5	T	2	5	2	3
T	1	3	S	1	2	2	5
I	5	4	N	5	4	4	1
C	3	2	C	3	2	5	5

Table 12a: The National color code for DRASTIC index ranges (after Aller et al., 1987).

Drastic index range	Class	Color	Printing specification color
<79	Low	Violet	Pantone purple C
80 - 99	Low	Indigo	Pantone Reflex Blue
100 - 119	Low	Blue	Pantone Process Blue C
120 - 139	Moderate	Dark green	Pantone 347 C
140 - 159	Moderate	Light green	Pantone 375 C
160 - 179	High	Yellow	Pantone yellow C
180 - 199	High	Orange	Pantone 151 C
>200	High	Red	Pantone 485 C

Table 12b: Classification of vulnerability index (Civita, 1994).

Class	Index range	Class	Index range
Very low	<40	High	153 - 163
Very low	40 - 80	High	164 - 174
Low	81 - 90	High	175 - 186
Low	91 - 105	Very high	187 - 198
Medium	106 - 116	Very high	199 - 210
Medium	117 - 128	Extremely high	211 - 260
Medium	129 - 140	Extremely high	>260
High	141 - 152		

Table 12c: Classification table for the vulnerability index based on arbitrary thresholds selected in this study.

Index range	Class name	Color assigned
<80	Very low	Violet
80 - 105	Low	Blue
106 - 140	Medium	Green
141 - 186	High	Yellow
187 - 210	Very high	Orange
>211	Extremely high	Red

It must be remarked that the depth of groundwater is very time-dependent due to the land subsidence in mining areas (up to tens of meters within a five year period), and to operational changes in dewatering of mines (mine closure) and other factors (Nada Rapantova, VŠB, personal communication). Because this parameter has the highest possible weight in DRASTIC, the necessary precision would require new geodetic and groundwater measurements. It means that the application of the method in underground mining areas is somewhat problematic due to the relatively fast changes in morphology and groundwater level.

336

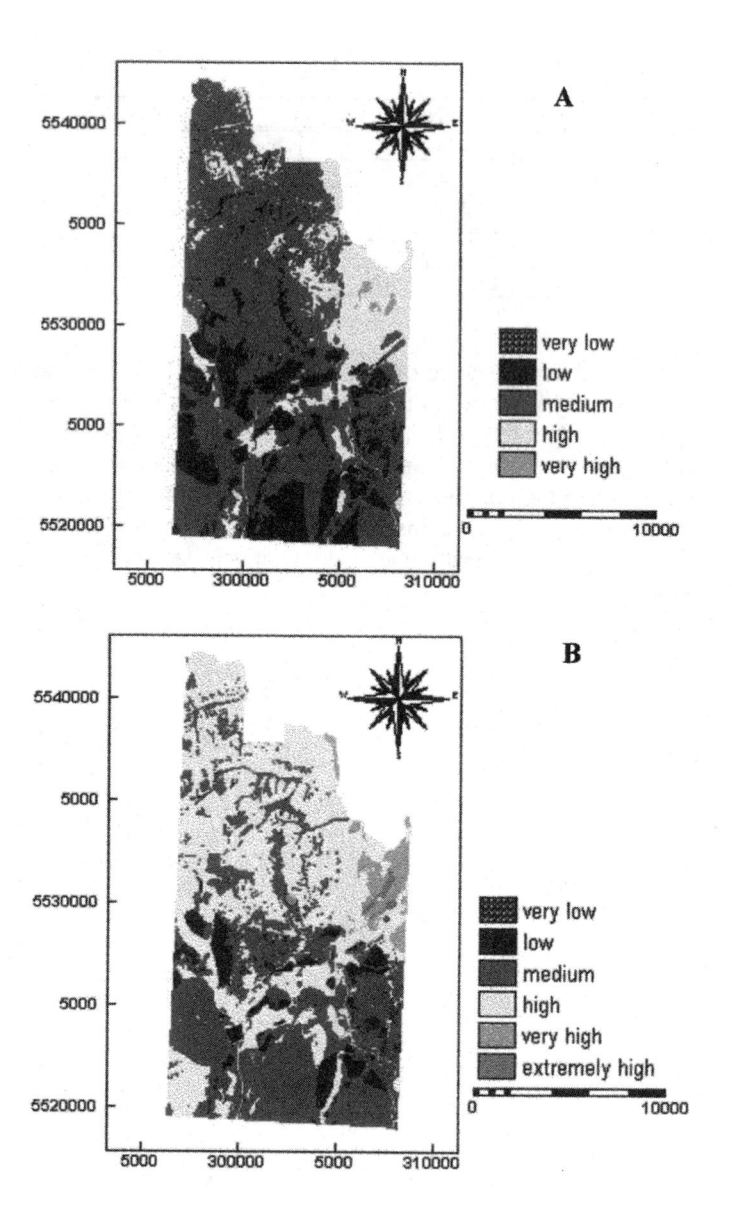

Figure 13: Three parameter maps and the corresponding DRASTIC rating maps used in this study. (a) Depth to water map; (b) Rated Depth to water map; (c) Topographic slope map; (d) Rated Topographic slope map; (e), Soil media map; and (f) Rated Soil media map.

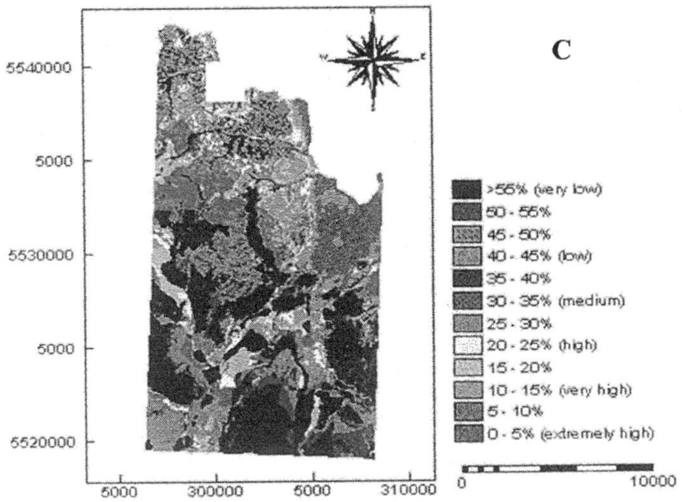
C
5540000
5000
5530000
5000
5520000
5000
300000
5000
310000
>55% (very low)
50 - 55%
45 - 50%
40 - 45% (low)
35 - 40%
30 - 35% (medium)
25 - 30%
20 - 25% (high)
15 - 20%
10 - 15% (very high)
5 - 10%
0 - 5% (extremely high)
0
10000

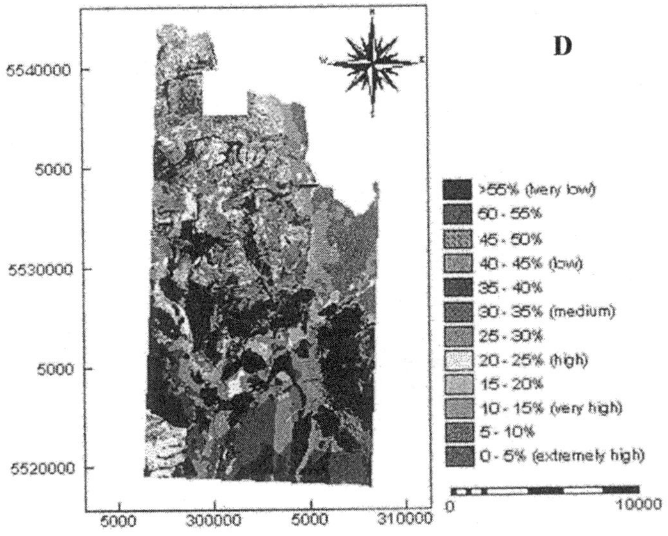
D
5540000
5000
5530000
5000
5520000
5000
300000
5000
310000
>55% (very low)
50 - 55%
45 - 50%
40 - 45% (low)
35 - 40%
30 - 35% (medium)
25 - 30%
20 - 25% (high)
15 - 20%
10 - 15% (very high)
5 - 10%
0 - 5% (extremely high)
0
10000

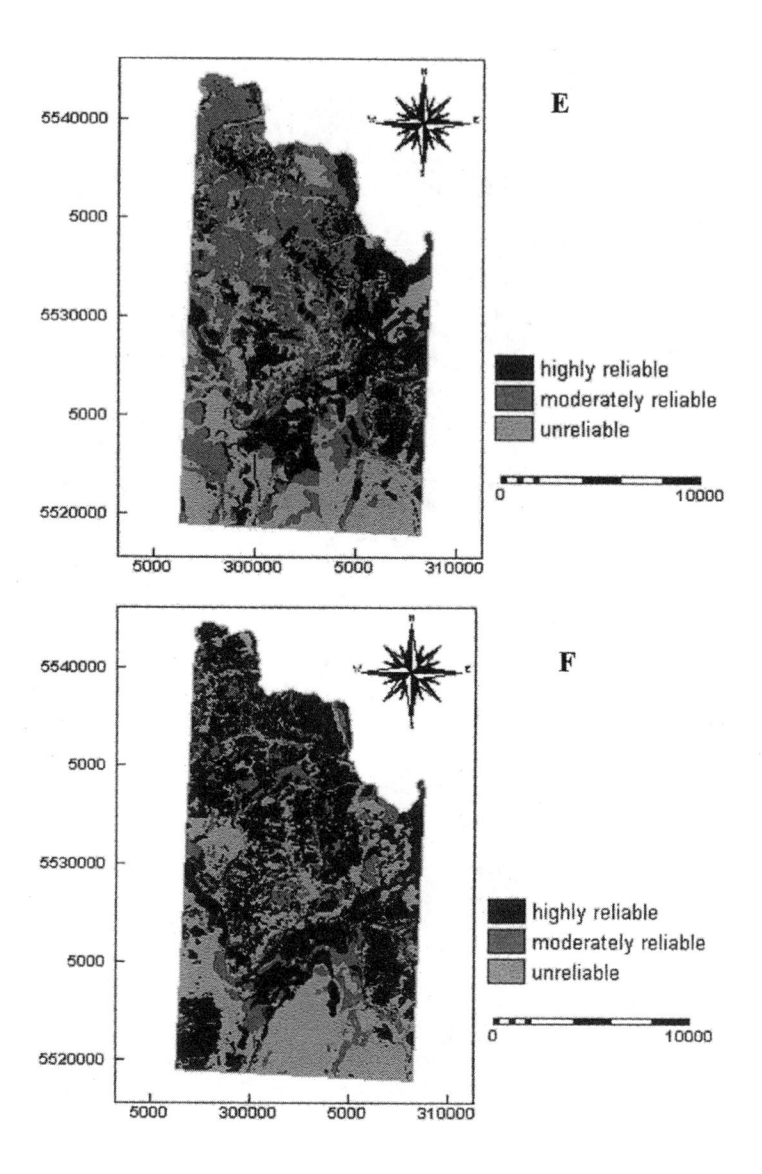

Figure 13: Continued.

339

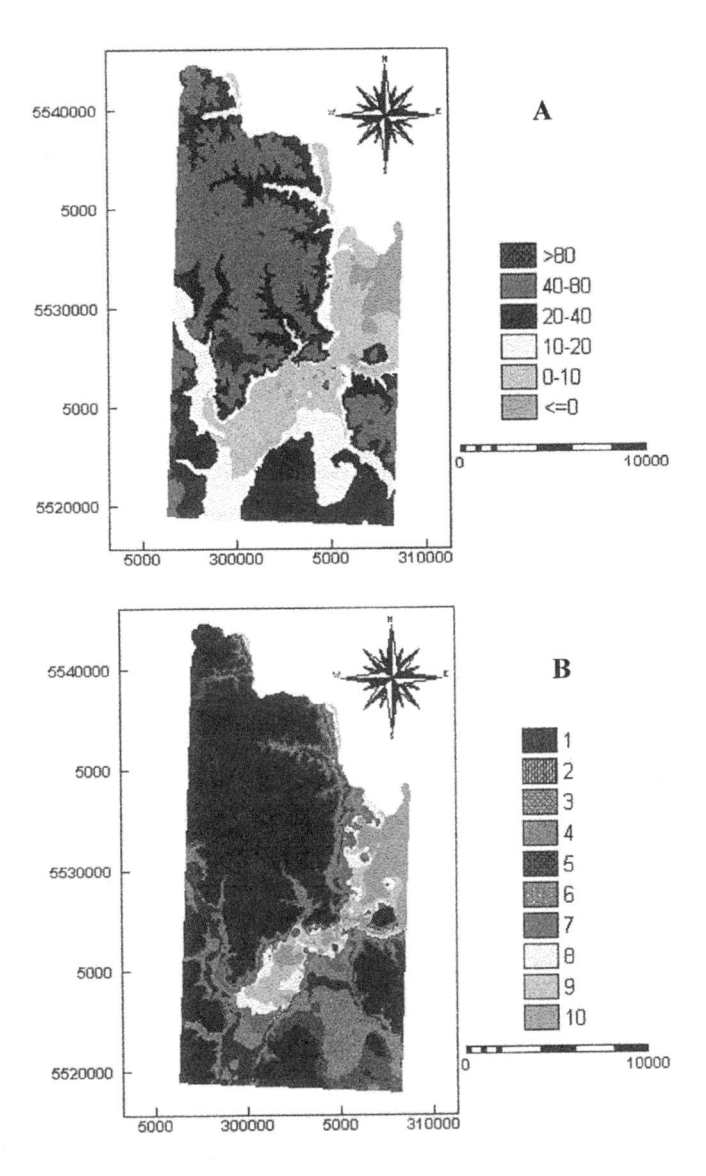

Figure 14: Two representations of DRASTIC and SINTACS aquifer vulnerability maps and the corresponding reliability maps. (a) DRASTIC vulnerability map using the classes in Table 12c; (b) SINTACS vulnerability map using the classes in Table 12c; (c) Normalized representation of the DRASTIC map in (a); (d) Normalized representation of the SINTACS map in (b); (e) Reliability of the DRASTIC map in (a); and (f) Reliability of the SINTACS map in (b).

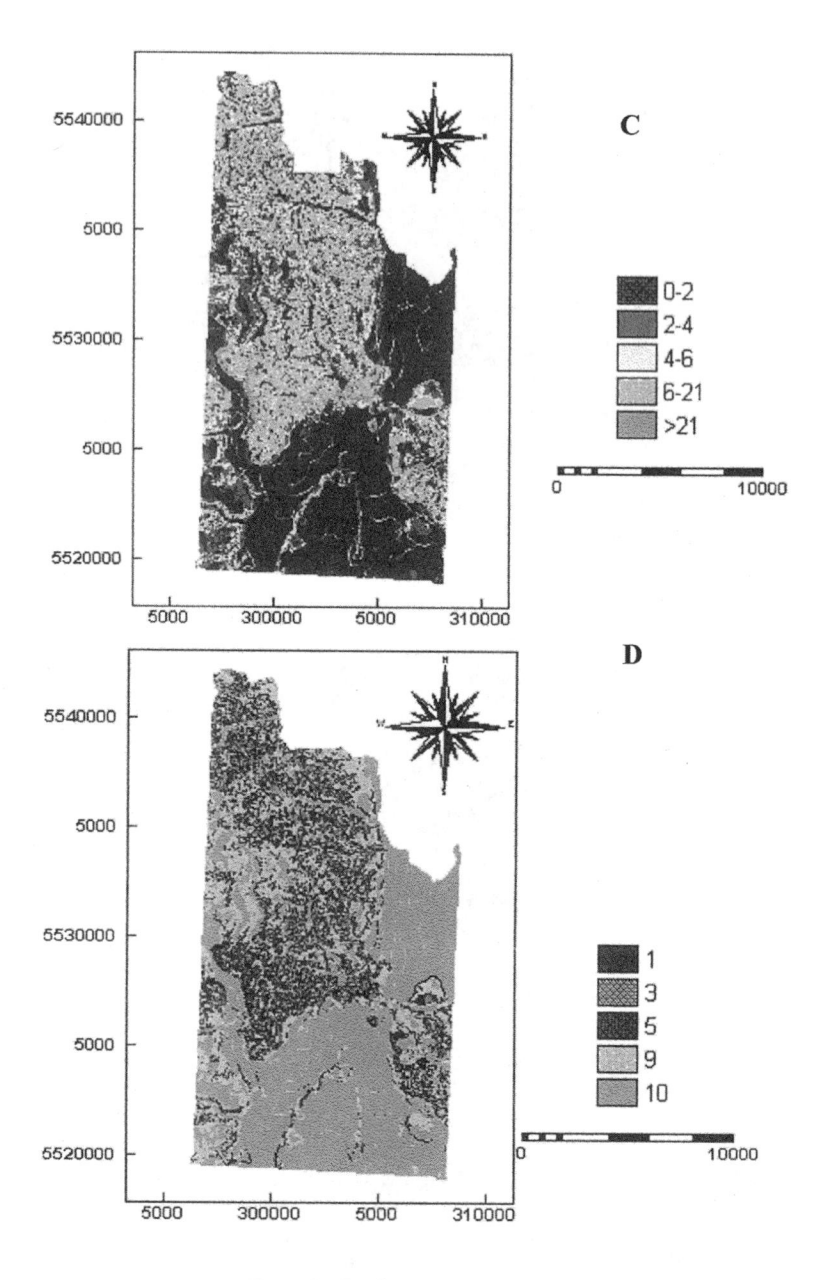

Figure 14: Continued.

341

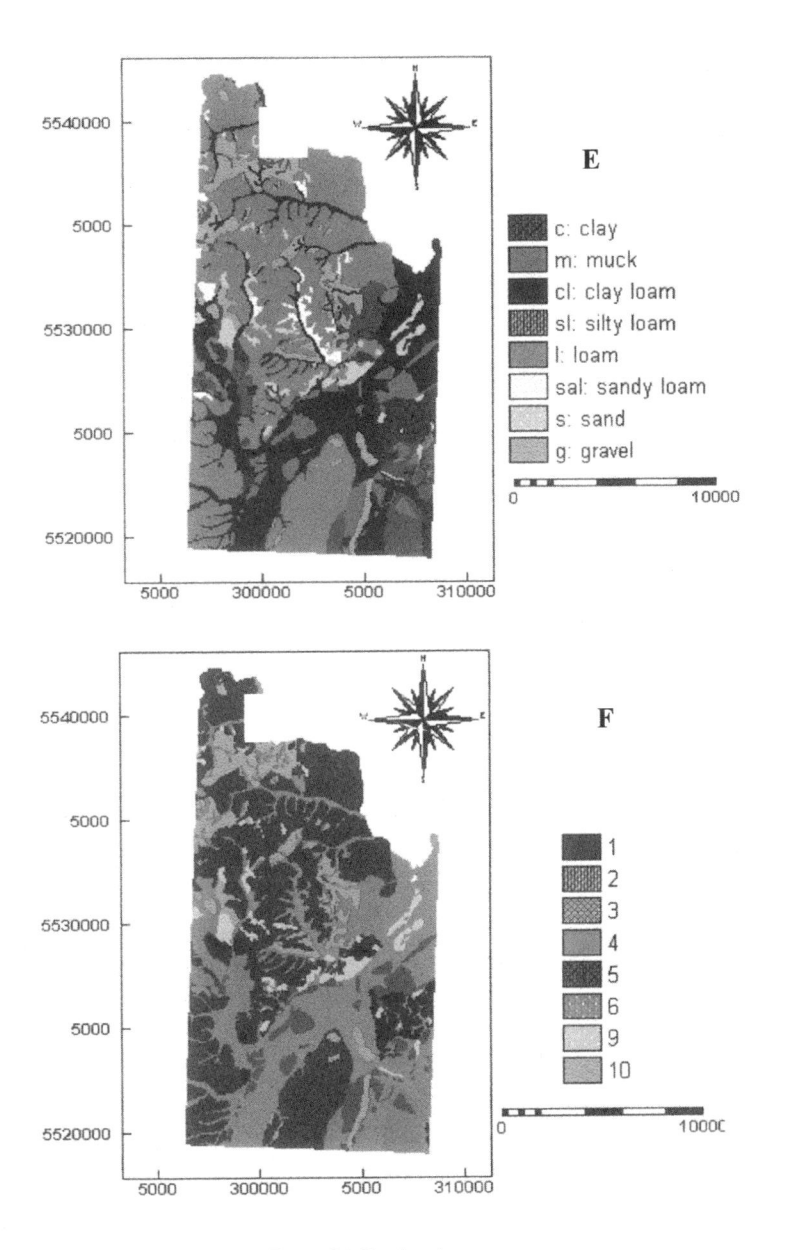

Figure 14: Continued.

5. Relevance of spatial data analysis

The two applications discussed reveal that a wealth of information is available in the study region to characterize land-use/cover changes and media vulnerability in areas affected by coal mining. This is also true in areas in general that are affected by industrial activities. In addition, it is evident that in environmental studies and in exploration applications, regional evaluations are necessary to decide on where to direct more-extensive detailed searches. In such evaluations, the spatial representation of regional conditions through time becomes critical in understanding the rates of changes and to predict in time and in space the occurrence of desirable or undesirable events.

During the past 13 years, techniques in spatial data analysis have progressed greatly, with the development of remote sensing and GIS. To put such techniques into practice, however, pilot studies are important if we are to achieve sufficient confidence or trust in either the data or the methods used and to proceed in decision-making.

Many modifications and improvements to the applications discussed here can be made with more efforts in data collection and validation: much more data is presently available with several local and regional institutions. In the Czech Republic, and in the former Czechoslovakia as well, geoscience information has been traditionally collected and systematically archived by institutions that now have the potential to expand their mandate from qualitative conventional maps of environmental conditions or of resource distribution to quantitative "added-value" maps aimed at decision-makers and other stakeholders in society.

The following points need to be addressed in a systematic manner:

1. Development of spatial communication products with predictive representations.
2. Use of techniques that maintain aspects of generality for different applications to the impact of industrial activity.
3. Assessment of data quality with respect to the prediction models being applied and development of the tools to communicate such a quality in simple and practical terms.
4. Provision of a reliability value of the available spatial data so that added- value products can be generated.
5. Development of operational strategies of predictive models in mining affected areas by using low-cost techniques first.

Although this contribution points at much applied research work that needs to be done in similar situations, the impacts of the geoscience profession on the socio-economic and the health domains should not be underestimated. For instance, the relations among geology, quality of ore, landscape setting, and economic and environmental damage should be considered in parallel in space and time if a solution is sought to real problems. This contribution has provided two basic examples of how to generate useful spatial representations of environmental conditions by using low-cost data that now is becoming broadly available.

6. Acknowledgments

This research was made possible by the enthusiastic collaboration of staff members of VŠB, the GEOFOND Institute, and Dr. Jiri Hruska, Prague. The authors are grateful for the help in the field and for the valuable advice provided in the study area. Particular gratitude is due to Prof. Jan Foldyna and Prof. Arnŏst Grmela of VŠB.

References

1. Aller L., Bennet T., Leher J.H., Petty R.J., And Hackett G., 1987, DRASTIC: A standardized system for evaluating ground water pollution potential using hydrogeological settings. EPA 600/2-87-035, 622 p,1-16.
2. Babiker I. S., 1998, GIS for Aquifer Vulnerability Assessment Applying DRASTIC and SINTACS. (A Case 2.   Study in the Ostrava-Karvina Region, Czech Repubblic).Unpublished M.Sc. Thesis, ITC, Enschede, The Netherlands, 130 p.
3. Beska F., Volf J. And Kysela T., 1992, An analysis of incidence of malignant tumors in the North Moravian Region. Abstract in: Coal, Energy and Environment, Ostrava, Malenovice, Czechoslovakia. Vysoká škola banská Ostrava, Czechoslovakia, and Southern Illinois University at Carbondale, USA.
4. Cesky Geologicky Ustav, 1992, Mapa geofaktoru Zivotniho prostredi cr Ostrava (Map of conflicting geological factors of Ostrava, Geological Survey of the Czechoslovakia) 15-343.
5. Chung C.F., And Fabbri A.G., 1999, Probabilistic prediction models for landslide hazard mapping. Photogrammetric Engineering & Remote Sensing PE&RS. v. 65, n. 12, p. 1389-1399.
6. Civita M., 1994, Le Carte della Vulnerabilita degli Aquiferi all'Inquinamento (Italian). Teoria & Pratica. Pitagora Ed.,Bolonga, 325 p, 1-325.
7. Dopita M. And Kumpera O., 1993, Geology of the Ostrava-Karvina coalfield, Upper Silesian Basin, Czech Republic, and its influence on mining. *International Journal of Coal Geology*. Elsevier Science Publishers B.V., Amsterdam, 23, 291-321.
8. Grmela A., 1997, Protection of groundwater resources quality and quantity in mining areas. VSB - Technical University, Department of Geological engineering, Ostrava, Czech Republic, 865-873.
8. Kitutu Kimono M. G., 1998, The Environmetal Impacts of Underground Coal Mining and Land Cover Change Analysis Using Mukti-Temporal Remotely Sensed Data and GIS. (A Case Study of the Ostrava-Karvina Area in the Czech Republic). Unpublished M.Sc. Thesis, ITC, Enschede, The Netherlands, 92 p.
10. Matysek D. And Bielesz M., 1992, Mineralogy of weathering process in the spoil dumps in Ostrava-Karvina coal district. Abstract in: Coal, Energy and Environment, Ostrava, Malenovice, Czechoslovakia. Vysoká škola banská Ostrava, Czechoslovakia, and Southern Illinois University at Carbondale, USA.
11. Napolitano P., 1995, GIS for aquifer vulnerability assessment in the Piana Campana, southern Italy, using the DRASTIC and SINTACS methods. Msc thesis, International Institute for aerospace survey and earth sciences (ITC), Enschede, The Netherlands, 172 p, 5-28.
12. Napolitano P., And Fabbri A.G., 1996, Single-parameter sensitivity analysis for aquifer vulnerability assessment using DRASTIC and SINTACS. Proceedings of the Vienna Conference "Application of Geographic Information System in Hydrology and Water resources. Published by IAHS, Vienna, Austria, 559-566.
13. Reichmann F., 1992, Impact of mining on the environment of the Czech Republic. A map of 1:500 000 scale. Cesky Geologicky Ustav ve spolupraci s Kartigrafii, Praha, Czech Republic , 20 p.
14. Sadar M. H., 1996, Environmental Impact Assessment. Second Edition. Ottawa, Carleton University Press Inc., 191 p.
15. Schejbal C., 1994, Coal and Gas resources in the Ostrava-Karvina district and influences of their utilization on the environment in the Ostrava agglomeration. Ostrava, Czech Republic, 11 p..
16. Schejbal C. And Ciganek J., 1995, Problems of mine closure and reviving of landscape in the mining area. VSB - Technical University of Ostrava, Czech Republic, Europe, International conference, 681-691.
17. Schejbal C., Novacek J., Vidlar J., And Flecko P., 1995, The present situation and prospects of coal separation in the Czech republic. Publication of VŠB - Technical University of Ostrava, Czech Republic, 10 p.

18. Schmidt Di Friedberg, P. And Posocco, F., (eds). 1987. L'Impatto Ambientale. Nuovi Sentieri Editore, Belluno, Italy, p. 20-22.
19. Suschkla J. And Skowronek A., 1992. An example of coal mine activities impacts on surface waters. Abstract in: Coal, Energy and Environment, Ostrava, Malenovice, Czechoslovakia. Vysoká škola banská Ostrava, Czechoslovakia, and Southern Illinois University at Carbondale, USA.
20. UNCED - United Nations Conference On Environment And Development, 1992, National report of the Czech and Slovak Federal Republic. Prague, 141 p.
21. Winpenny J., 1991, Values for the environment: A guide to Economic Appraisal, HMSO, London, 1-2.

THE IMPACT OF MINING ON THE ENVIRONMENT
A Case Study From the Tharsis-Lagunazo Mining Area, Province Huelva, SW Spain

T. WOLDAI and A. G. FABBRI
Department of Earth Systems Analysis (ESA),
International Institute for Geo-information Science and Earth
Observation (ITC),
Hengelosestraat 99, P.O. Box 6,
7500 AA, Enschede, The Netherlands

Abstract. The Tharsis-Lagunazo area, in the Province of Huelva, southwestern Spain, has a long history of mining. For over three thousand years, the contour of the land has continuously been modified and re-modified by mining activities and unplanned mine wastes to provide the present landscape. This study, making use of Landsat TM imagery from 1984 and black and white aerial photographs from 1973, was able to assess the implication and impact of mining on this area. From these datasets, it was possible to detect the number of open-pit mines, waste rock dumps, tailings, slime dams, land use/cover changes and subsurface groundwater pollution.

Preliminary results of the investigation show the indiscriminate dumping of solid mining waste to be rampant wherever land is available. All in all, no diagnostic changes with regards to the extent of the mining pits could be deciphered from the two datasets available. No new open pit sites and no extension in area coverage of the existing ones could be confirmed from the remotely sensed data and field mapping. Changes however, were evident within the bushy vegetation, known as "jara" in Spain, and the eucalyptus growing areas surrounding the mines including new dumpsites in areas identified as barren from the 1973 aerial photos. These are evident when comparing the aerial photos of 1973 with the Landsat TM of 1984. The type of landscape defacement without any concern for rehabilitation is another major environmental concern clearly vivid from both the remotely sensed datasets used.

1. Introduction

The Tharsis study area is located on the Spanish side of the Iberian Pyrite Belt proper, in the Huelva province, southwest Spain, as shown in Figure 1. The economic importance of the study area and its surroundings can be traced to proto-historic times. Numerous archaeological artefacts indicative of mining activities during the Tartessian, Phoenician and Roman domination of the Iberian peninsula, including reports outlining ore exploration rules during the Roman times, namely in the 2nd century B.C., have been found in many mines (see [1]; [7]). This early mining was directed to the gossan and gypsiferous material, an oxidized weathering product capping most of the massive pyrite bodies, containing relatively high amounts of copper, but especially gold and silver. Much of these have since been largely removed by mining operations.

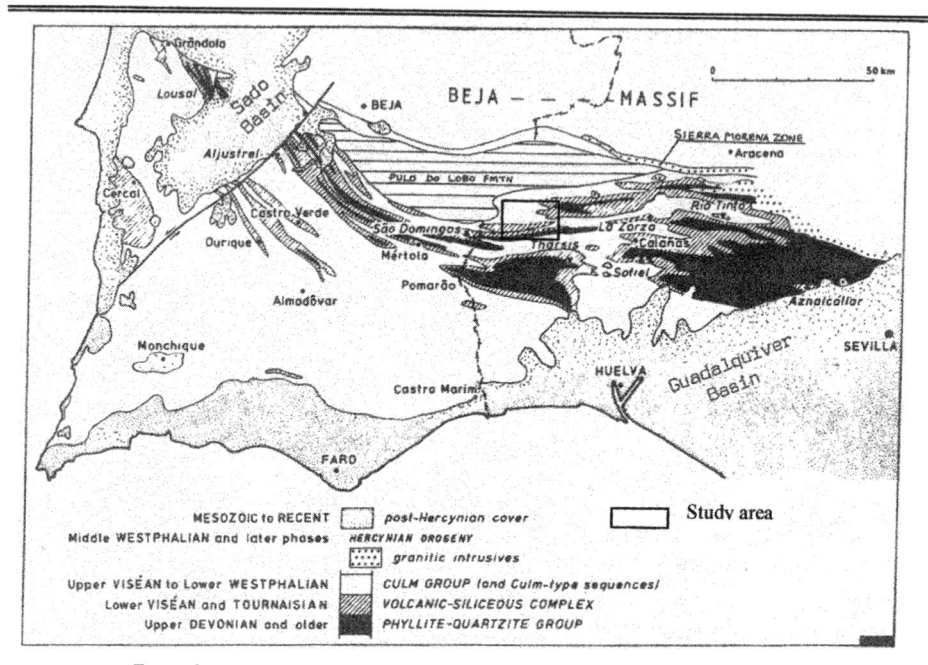

Figure 1. Location and general geological map of the Iberian Pyrite Belt (after [5]).

In the entire area, there is not one single surface mine, which was not mined already by the Romans. For over three thousand years the contour of the land has continuously been modified and re-modified by mining activities and unplanned mine wastes to provide the present landscape. The impact of such activities on the environment is clearly visible throughout the Pyrite Belt, where the land is severely degraded and barren. The ecosystem, including the morphology, soil, vegetation, and the ground and surface water regime are disrupted. Natural trees are hardly present. Instead, the toxic soil seem to favour the growth of low "sticky" bushes, the most dominant of which are <u>Cistus ladanifer</u> ("gum cistus", known as "jara" in Spanish) and <u>Cistus populifolius</u> (poplar-leaved cistus). In the last few years, intensive man-made terracing of the terrain (under a reforestation programme) has increased. Mine wastes have been used sometimes to fill the terraces making it rather difficult to define the location of past mining dumps and to understand their properties. The man-made terracing makes way for eucalyptus plantation (basically for industrial consumption more than rehabilitation).

In such an area, with a long history of mining and where the original baseline situation is unknown or missing, it is challenging to define the areas in which the problem arises. Priority in this direction, aside from locating the mining sites, will be the delineation of the exact location of the mine works; waste tips and land cover changes. This can be achieved using remotely sensed data. The Tharsis-Lagunazo mining area in Huelva is an excellent site for using such data to confirm these expectations. As such, in this work the degree is assessed to which spectral information derived from remotely sensed data can be used to extract land-cover changes and contamination from mining activities.

Landsat TM data have been processed and analyzed and where necessary compared with the 1:25,000 scale aerial photographs of 1973. Image responses to mining-related activities and land-cover changes have received special attention, because of their implication for the local environment. Throughout the area, mining wastes from the open pit mines have been dumped on the surface, resulting in toxic chemicals polluting not only the rivers but also the soil. The old contamination and those deposits outside the known mining sites are especially interesting because they are already covered by soil and vegetation and occasionally difficult to map by conventional mapping methods. This contribution assesses the impact of mining using and comparing both the Landsat images and the aerial photographs.

2. Remotely sensed data

The Landsat Thematic Mapper (TM) data used in the study area corresponds to path/row 203/034 and was acquired on 30 June 1984 with zero cloud cover. A 500 m by 500 m pixel sub-scene covering the Tharsis-Lagunazo area was extracted for enhancement. Radiometric correction was primarily applied to the six bands of Landsat TM (excluding the thermal band 6) to remove the atmospheric effects. Geometrical registration of these datasets subsequently followed using the 1:50,000 scale topographic map supplied by the Spanish Servicio Geografico del Ejercito (1982 [6]).

From the remote sensing point of view, the first interval of interest lies in the wavelength region covered from $0.4\mu m$ to $1.0\mu m$, which ultimately encompasses TM bands 1 to 4. These wavelength regions provide a means of mapping ferric iron oxide, some of which may be related to altered rocks [4] or wastes derived from the mines. The presence or absence of transition metals influences the spectral reflectance of minerals in these visible to near-infrared regions. Iron (Fe) is by far the most diagnostic transition element for terrestrial remote sensing because its abundance and variations on the Earth's surface are greater than for all other elements. It is the most common cause of the charge transfer and electronic transition absorption features seen in rock and soil reflectance spectra [3].

Vegetation in general, shows a small reflectance maximum in TM band 2 (which is why vegetation appears green) with a sharp increase in slope (high reflectance in the near-infrared region - TM band 4). The reflectance gradually decreases through the range of the TM spectral bands at longer wavelengths. The wavelength region between $1.55\mu m$ and $2.5\mu m$ provides diagnostic spectral information about the composition of minerals and rocks. In addition, these wavelength regions, corresponding to TM 5 and TM 7 data, can provide evidence for detecting and mapping hydroxyl (OH-) and water (H2O). The TM 5 provides evidence of the reflected peak of clay minerals, while band 7 covers absorption features of diverse Al-OH and Mg-OH bands in phyllosilicates.

Iron minerals and vegetation have similar reflectance in the TM 1 (0.45-0.52µm) and TM 2 (0.52-0.60µm) bands. As a result, these bands are not useful for separating both materials. TM 3 (0.63-0.69µm) covers the region where ferruginous minerals can be distinctively separated from vegetation due to the very high reflectance of iron minerals and the conspicuous absorption feature of vegetation. Vegetation, especially green vegetation (with high chlorophyll content) has a unique higher reflectance compared with other terrain covers in TM 4 (0.76-0.90µm). It also includes the iron minerals crystal field absorption feature at 0.9µm. Both features can be used to separate oxides from vegetation. TM 5 (1.55-1.75µm) and TM 7 (2.08-2.35µm) bands are very helpful for differentiating vegetation from the OH- minerals and iron oxides because they have distinct reflectance curves.

In the course of this research, several processing steps were made before the remotely sensed data could be used successfully to address the impact of mining and surface cover types. Only those processing tasks offering particular relevance in the environmental evaluation of the research area are emphasized in this paper. The primary aim was to obtain well-enhanced images that could be used in visual interpretation and analysis. Additional data available for the study area include a 1:50,000 aerial-photo coverage. This was also analyzed during the investigation.

3. Image processing of Landsat TM

3.1. NORMALIZED DIFFERENCE VEGETATION INDEX (NDVI)

In the study of mining-related activities and land-cover types and changes, the use of NDVI is by far the most convenient method to discriminate the areas covered by vegetation, bare soil/rock and waterbodies. NDVI is the most commonly used vegetation index that transforms multispectral data into a single image band representing vegetation distribution. It is calculated using two Landsat TM bands; one band containing visible or red reflectance values the other band near-infrared reflectance values. In the case of the study area, this was obtained by dividing (band3-band4) by (band3+band4). NDVI values range from -1 to 1 and compensate for changes in illumination conditions, surface slopes and aspect.

In the absence of any prior knowledge of the area, the NDVI method offered a good understanding and accurate mapping of the mine sites, the mine dumps, bare soil types and vegetation cover. The different rock/soil types in the study area have relatively similar spectral properties and in Landsat TM3 and 4 especially, they are hardly separable. As a result they have vegetation indices nearing zero, as shown in Table 1. In the NDVI image of Figure 2a, the bare soil/rock types are represented by their yellowish color. Vegetation-covered areas yield low negative values having a relatively high reflectance in the near infrared and a low reflectance in the visible range of the spectrum. On Figure 2, they are clearly seen by their dark blue to light green tints. Where the 'jara' is open however, it shows a clear association with the bare soil.

Table 1: Normalized vegetation indices for different land-cover types.

Landsat TM band 3	Landsat TM band 4	NDVI	Land use types
8	2	0.60	Water
6	25	-0.61	Vegetation
32	35	-0.04	Bare soil
24	20	0.09	Mine dump

The most significant feature within the image is that the area covered by water bodies, mine pits, slagheaps and tailings shows positive index values ranging from 0.09 to 0.6. The mining and waste dump areas have their vegetation indices around 0.09. In contrast, water has slightly higher visible reflectance than in near-infrared reflectance. Thus, this feature yields positive index values of 0.6. In the NDVI image of Figure 2a, all areas with mine dumps, bare soils (irrespective of areas previously of mine dump origin or not) and waterbodies show a distinct red colour while areas with green chlorophyll (during acquisition) show a blue colour. Subsequent processing using density slicing allowed easy separability between these targets. In Figure 2b, the water bodies obtain a blue colour that separates them from the mine waste dumps, which appear as distinct red. For over three thousand years, unplanned waste dumps from the various mines in the area have occurred in different places -- some deposits had been even dumped in areas where no historical record exists. After density slicing, it can be seen that areas that were originally thought as barren for example, have their spectral signature corresponding to the mine dumps identified in the Tharsis and Lagunazo mine sites. On Figure 2b these appear orange in colour. Field investigation and laboratory analysis carried out on samples collected from such mine dumps clearly demonstrate and verify this affinity. In the absence of any baseline data, it was difficult to ascertain their origin precisely. One possible conclusion is that they are material transported from the nearby mines. No distinction is possible between the cultivated area and the eucalyptus vegetation to the East and South of Tharsis. Nor it is possible to delineate the 'jara" vegetation dominant in the study area.

3.2. BAND DIFFERENCES

In this research, the band difference approach was preferred to ratioing. The spectral consideration noted above was made in the application of this approach. Ratioing involves the division of the digital number (DN) of a pixel in one band by the DN of that pixel in another band and is a non-linear operation. Such subtle information contained in both extremes of the DN range, however, is quite often destroyed during linear stretch. Band differences on the other hand, are used to extract spectral information in a linear way. According to Crôsta and Moore (1989 [2]), the resulting information can be linearly stretched with no loss of information.

Iron oxide in general shows higher reflectance in TM3 with an absorption band in TM1. Subtracting TM1 from TM3, as shown in Figure 3, results in brighter signatures, which may indicate higher iron content in the image. Field investigation confirms the light toned areas to be covered mostly by gossans, lateritic soil and red clays. Significantly, the image also shows brighter signatures to the south and southwest of the study area. Field investigation shows that the area is levelled by bulldozer for reforestation with most of the soil been transported from the neighbouring mine dumps.

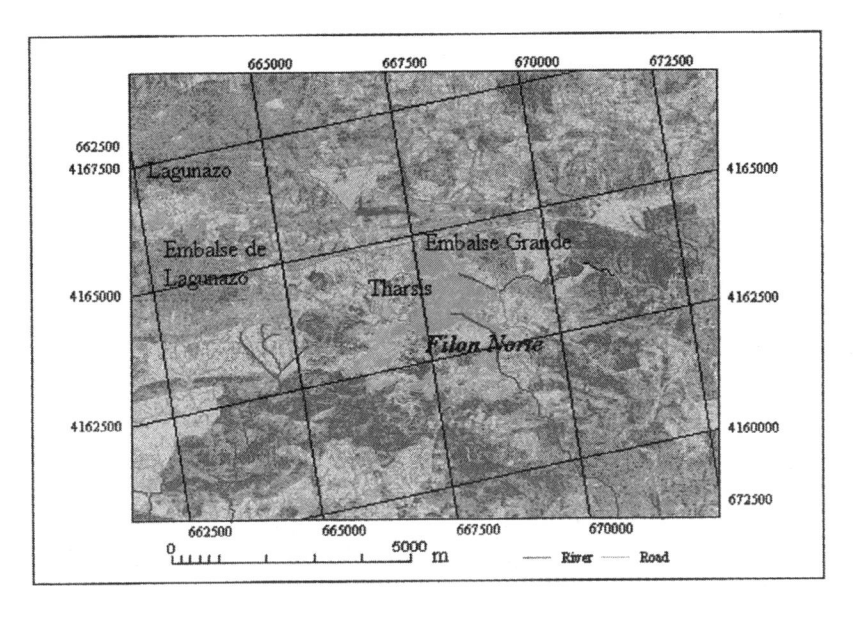

351

(a)

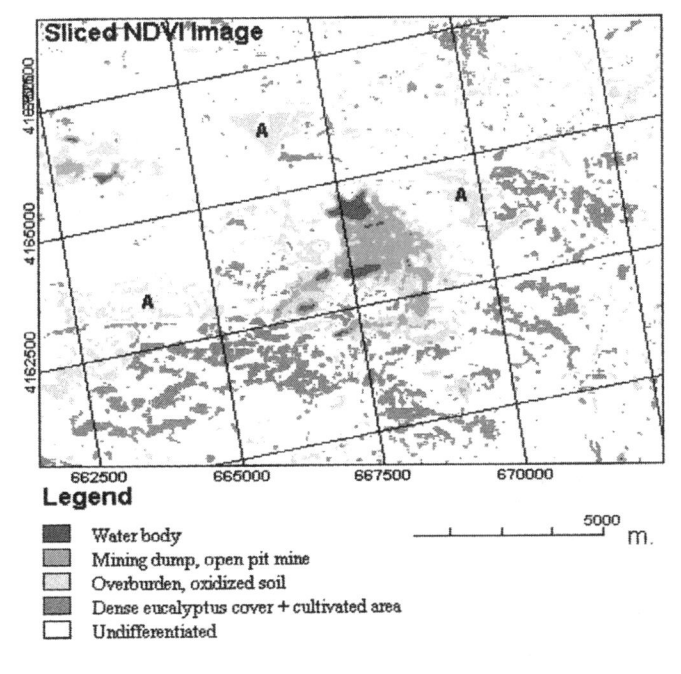

(b)

Figure 2. (a) Normalized Difference Vegetation Index (NDVI) image of the Tharsis – Lagunazo study area. Purple to red colour corresponds to mining dumps and areas associated with mining activities, dark blue to eucalyptus and cultivated areas with the light green areas related to jara and mixed vegetation and land use types. In image (b) density slicing is applied to (a) whereby further subdivisions into waterbodies, mine waste dump and old mining dump-sites (designated as A).

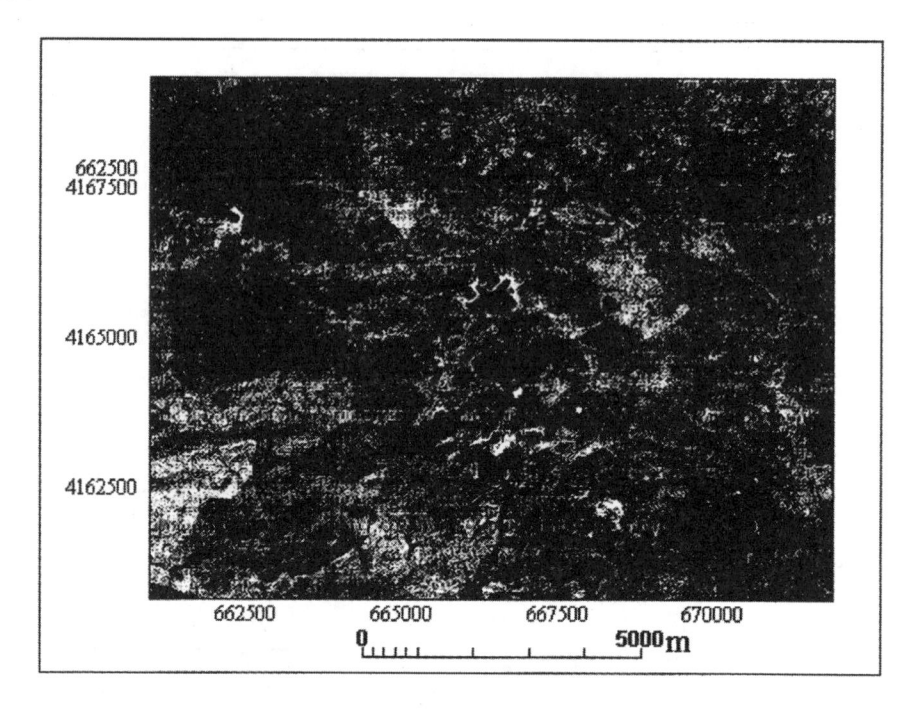

Figure 3 Landsat TM3-TM1 image of the Tharsis – Lagunazo study. The brighter signature in this image corresponds to areas with higher iron oxide concentrations.

TM4 covers the higher range of the vegetation 'red edge' reflectance feature in the near infrared region, where green vegetation has a unique high response compared with other terrain covers. It has an absorption band in TM 3. Besides, band 4 includes the iron mineral crystal field absorption feature at 0.9 microns, producing a differential spectral response of these two targets. Subtracting TM 3 from TM4 results in brighter signatures on Figure 4 (shown as B), which may relate to the cultivated portion (exhibiting high chlorophyll content) of the image followed by other vegetation types in light grey. In Figures.2a/b, it was mapped as having similar spectral signature to the eucalyptus to the southwest of the Tharsis mine. The mine dumps including the area associated with gossans, altered volcanic products, polluted mining extracts and water bodies all appear very dark in tone. The response obtained from the band difference TM 4 minus TM 1 was highly complementary to that of TM4 minus TM3. The only difference was in the bare soil/rock exposures to the southwest of Tharsis, the crushed rocks and uneconomical ore deposits to the east of Tharsis and in the Lagunazo mine area where loss of information was apparent in the TM 4 minus TM1 difference image.

353

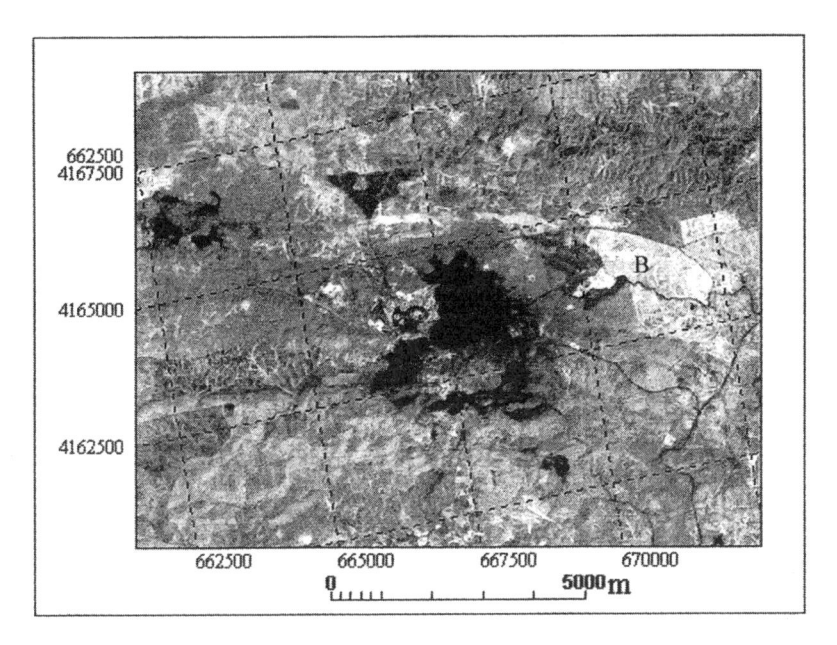

Figure 4 Landsat TM4-TM3 image of the Tharsis – Lagunazo study area. The areas occupied and affected by mining activities are seen as darker tone contrary to those occupied by green vegetation (mainly cultivation) and green grass appearing as white. All other land use types appear grey.

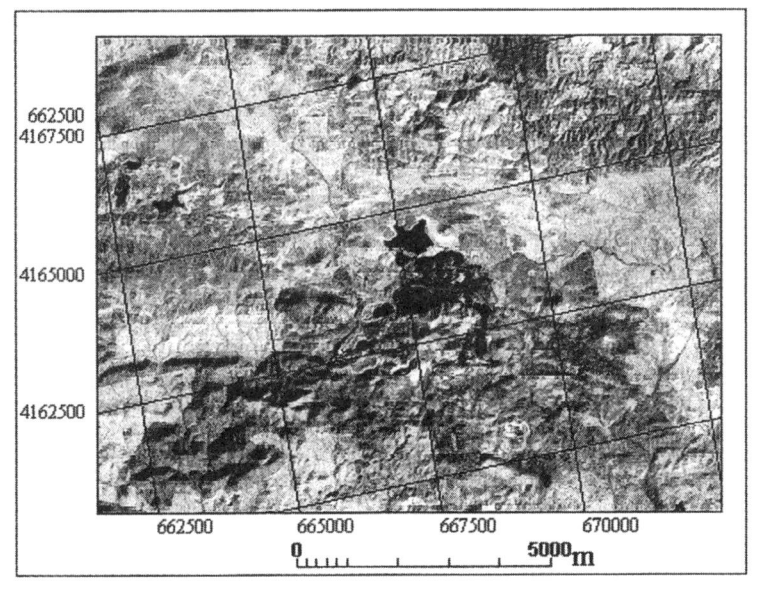

Figure 5 Landsat TM5-TM1 image of the Tharsis – Lagunazo study area showing the mine dumps, open pit mines and water bodies in dark tones relatively to the eucalyptus jara covered areas, which appear dark grey.

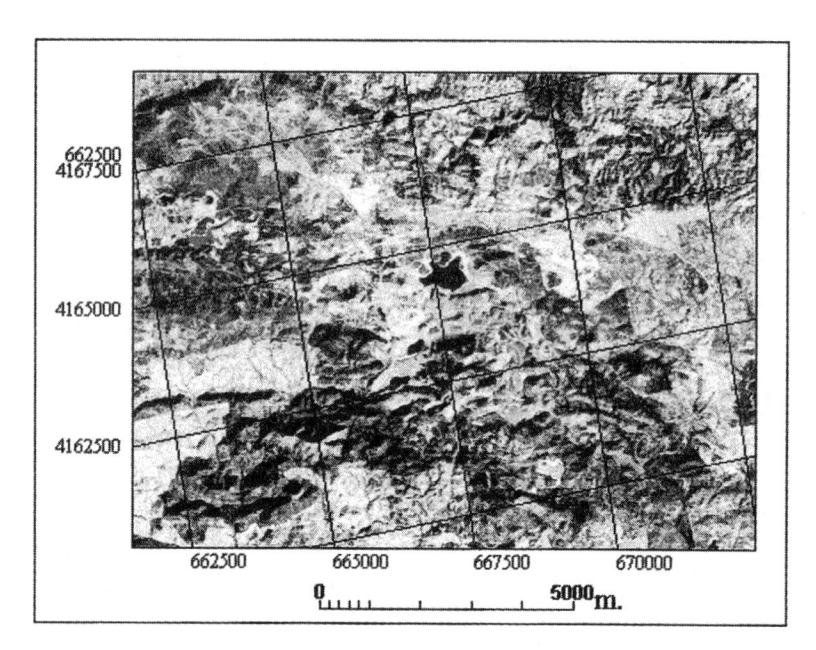

Figure 6. Landsat TM7, 3, 1 composite image (in RGB order) of the Tharsis-Lagunazo study area. Significant features in this image are the slag heaps (appearing red) seen to the southeast of the Embalse Grande and west of the Embalse de Lagunazo.

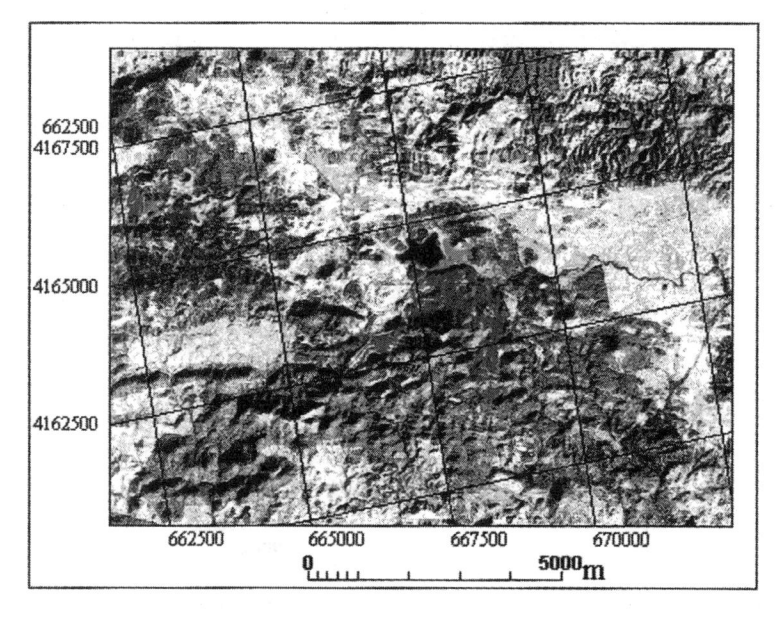

Figure 7 Saturation-enhanced Landsat TM data of the Tharsis-Lagunazo mining area. The reddish brown area represents dense eucalyptus cover while the area showing dark green corresponds to jara cover. Cultivated areas appear red, with those under grass being light yellow and the bare rock/soil area white. The mine dump in the study area is blue, while the oxidized gossan/slag remains to the east and southeast of the Embalse Grande lake and south of Lagunazo are in green.

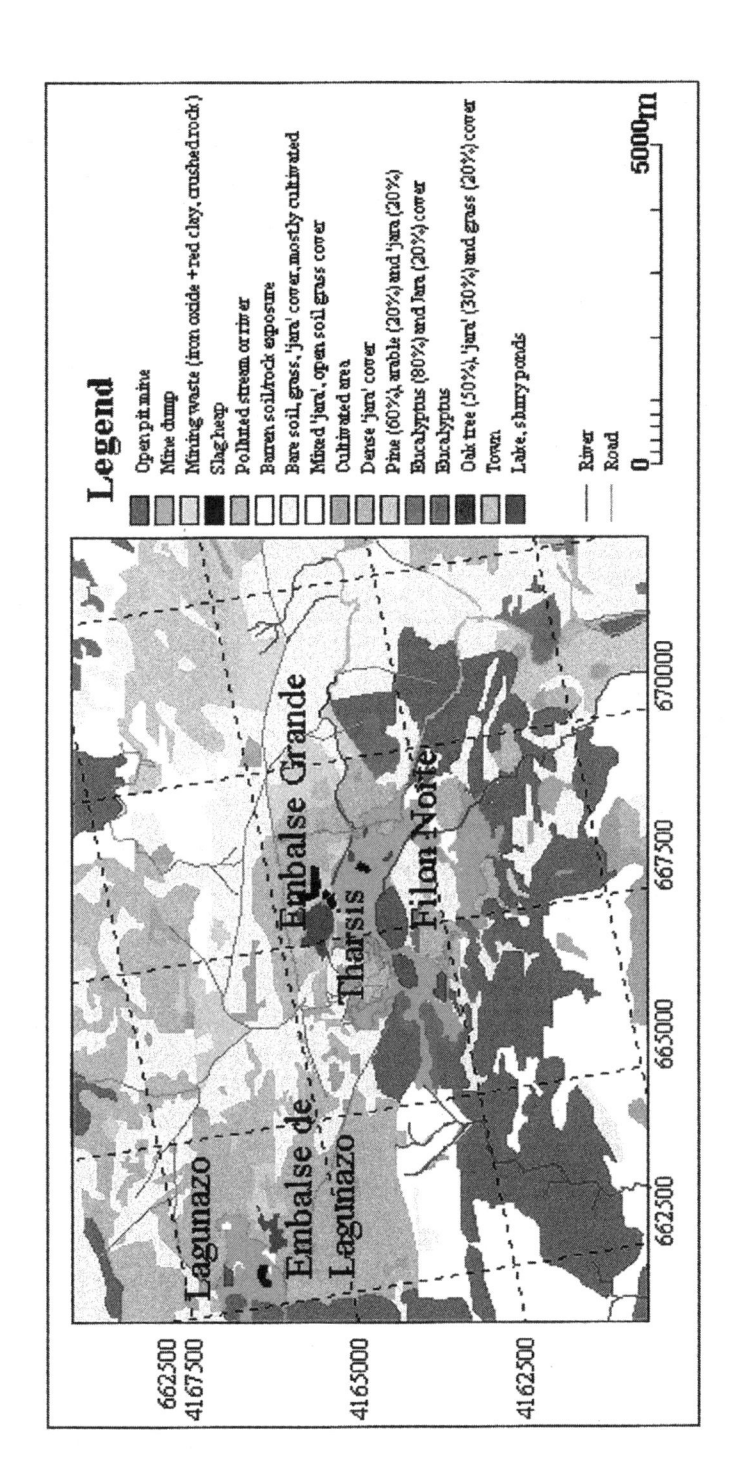

Figure 8 Land-cover types in the Tharsis-Lagunazo Mine areas.

TM 5 minus TM 1, shown in Figure 5, on the contrary, gives a distinct demarcation boundary between the mine dumps and water bodies, both appearing dark in tone and vegetation (eucalyptus and Jara) in grey. Significantly, no distinction is possible among the various other land use types. Even some of the areas associated with iron oxide, slagheaps appearing as dark in Figure 4, are not easy to delineate from this image.

3.3. FALSE COLOR COMPOSITE

Just as the data in an image can be displayed as grey tones of single bands, they can also be displayed in colours. In this work, several false colour combinations were generated. By visual inspection the False Colour Composites of Landsat TM4, 3, 1 and TM 7, 3, 1 (in RGB order) were selected. The first composite did not add more information than what was obtained by other processing methods described above. Instead the combination by TM7, 3, 1 (in the RGB order) seems aesthetically appealing and easy to interpret. The combination shown in Figure 6 is relatively adequate for delineating the various vegetation covers from those occupied by the mine dumps, barren soil/rock cover and soils enriched in iron oxide.

Of particular significance in the composite image of Figure.6, are the areas denoted as red to the Southeast of the Embalse Grande and to the west of the Embalse Lagunazo. Field investigation reveals this to consist of slag heaps and gossaniferous terraces. Unlike in Figures 2, 4 and 5, the distinction between the eucalyptus (green) and the jara (deep purple) and the water bodies (dark and blue) is possible but most grass cover and all bare soil/rock areas, irrespective of their composition, appear to vary in colour from light purple to white. Precise delineation of the mine dumps around Lagunazo and Tharsis is easier from this image than from any of the images considered above. What is not possible is to identify the old waste dumps (away from the mine centres) and the cultivated areas.

3.4. SATURATION ENHANCEMENT

The saturation enhancement [7,8] was used in an attempt to improve the spectral contrast between objects with small differences in saturation and hue (= colour). It produces colour enhancement on a three-band input image. In the study area this was done using Landsat TM bands 4 (760-900 nm near infrared), 5 (1550-1750 nm short-wave infrared) and 3 (630-690 nm red). These input data were first transformed from red, green, and blue (RGB) space to hue, saturation and intensity (HSI) space. A gaussian stretch was later computed on the saturation band so that the data fill the entire saturation range. In order to produce output bands that have more saturated colours, the inverse operation (decorrelation) was applied to the HIS to return to the RGB space.

Figure 7 represents the composite image generated from the saturation enhancement. While textural and pattern recognition form the basis of morphological interpretation from the remotely sensed data, some of the diagnostic features related to mining activities and land use types were more easily identified from this image than from the normal Landsat TM bands 4, 5, 3 combination. The saturation enhancement causes little distortion in the perceived hues, but color saturation was exaggerated, making it possible to distinguish very subtle reflectance differences. An important aspect of this technique was that interpretation was straightforward.

A clear separation of the various features related to mining activities and land cover types was possible from Figure 7, and where vegetation was not totally obstructing, differentiation was possible between the barren soil (cleared for terracing or other activities shown as white), contaminated soil (blue), mine waste dumps, polluted rivers and lakes (dark blue). The morphological expressions of the various rock/soil types within the study area were sharper and better enhanced. In an area where the terrain is rather subdued and where the continuity of the various terrain units is very difficult to establish in the field, the information that can be delineated by such a method and combination is highly significant to the research. At least, it allowed pinpointing the mine concentration with ease. The Embalse Lagunazo is filled with a highly polluted pool of acidic water and shows the same spectral signature as the mining dumps while the Embalse Grande shows relatively minor pollution and is therefore showing darker tones. The area covered by transported mining dumps, comprising iron oxide mixed with red clay, crushed rocks, fragments of adjoining rocks (mostly slates) and uneconomical ore deposits is discriminated by its distinct blue color. The oxidized gossan/slag remains to the east and southeast of the Embalse Grande Lake and south of the Lagunazo mine appear green. The open pit mines to the south of Tharsis show darker colours. The cultivated areas appear as red and where bare soils with some admixture of grass and jara exist, the composite image is marked by yellow colour contrary to open barren soil/rock, which appears as white.

4. Interpretation of the results of image processing

The techniques and types of image enhancements (especially in vegetated terrain such as, the current study area) are often critical in rock/soil discrimination and most of all in separating the mining-induced impact in the study area. To begin with, in any study of such nature, the most powerful technique to be used will be the NDVI method. The advantage of the NDVI approach (see Figure 2) is that even if one is not aware of the land cover types in a given area, a quick assessment of areas affected by mining activities and residual soil (enriched in iron oxide), polluted water bodies (all appearing red), eucalyptus and cultivated areas (with high chlorophyll content appearing blue), other vegetation and land cover types (green) and bare soil (light yellow) can be made immediately giving the researcher targets for further investigation. Differentiation within the areas shown in red is further enhanced using density slicing or using the Crôsta and Moore [2] band-difference approach. With density slicing approach, the polluted water bodies and past and recent mine waste dumps were easily separable. Using the Crôsta and Moore band-difference [2], it was possible to delineate the soils enriched in higher iron oxide (TM3-TM1, in Figure 3), the mine dumps including the areas associated with gossans, slag heaps, altered volcanic products, polluted soil and cultivated areas (in TM4-TM3 in Figure 4). Furthermore, the mine dumps and the water bodies affected by mine waste alone were easily separable by their low DN values from the other land cover types (in TM5-TM1, Figure 5).

Missing information regarding the various vegetation types and land use/cover was obtained using the false colour composites and the saturation stretch images, as shown in Figures 6 and 7. While the cultivated and eucalyptus areas can easily be discerned from Figure 6, clear demarcation of the jara-, eucalyptus- and pine-covered boundaries was easy to outline from Figure 7. Composite TM7, 3, 1, shown in Figure 6, was ideal for precisely mapping the jara cover. Significantly, the image offered additional information regarding the slag heaps in the Tharsis and Lagunazo mines, much more than the other false colour composite did, but failed to discriminate between the barren soil/rock exposures and the area occupied by higher iron oxide content. Sharpness and clearer demarcation of vegetation boundaries was obtained using the saturation stretch approach. All, the mining-induced activities and the land-cover types are represented in Figure 8.

5. Aerial photointerpretation

5.1 IDENTIFICATION OF MINES, DUMPS AND OTHER IMPACTS

Of particular significance to the study is the identification of the mines. They are marked by open pits, as shown in Figure 9, filled occasionally with a pool of acidic water. The water level in the mine pits is possibly below the groundwater level. As such, leakage into the surrounding is not envisaged. The ore itself never shows at the surface, but is covered by a "gossan" or "overburden" consisting of the adjoining "country" slate, or porphyry, see Figure 10. The overburden varies in thickness at different mines, being very rarely as thin as 20 feet but sometimes as thick as 160 feet.

In Tharsis for example, the ore was extracted by underground workings at some periods in its mining history. In spite of the difficulties due to the removal of the vast "head" of unproductive matter, it was entirely raised in as "opencast" and the mining instead was done from a number of open pits. The mine dumps, gossan and slagheap remains in Tharsis and Lagunazo, form large, steep- sided hills around the mine pits, obviously suggesting that they were dumped near the open pits for economic reasons.

Field observation reveals them to comprise crushed rocks; fragments of adjoining rocks (mostly slates), iron oxide mixed with red clay, and uneconomical ore deposits. The exposure of the waste to the effect of interaction with rainwater poses great threats to the surrounding environment. Preliminary results of the investigation show the indiscriminate dumping of solid mining waste (wherever land is available) to be rampant. The waste, in many cases is piled as high as the equipment will permit. In instances, like in Tharsis and Lagunazo, dumps were leveled to make terraces. Still, some were cleared and removed to different sites (possibly because of a new mine in site at the time of their deposition) without giving much consideration to their toxic nature and proper consideration in their new sites. Most of these, in spite of their easy recognition in most of the processed TM datasets, were not identifiable in the field; nor could they fully be delineated from the 1973 aerial photograph available of the area. The patterns in which they appear suggest a possible transported mine dump. Regardless of the type of disposal or the composition of the mining dump, all waste transported from one site to the other carries inherent pollution for water resources. All these features can easily be delineated from the various processed images used in this research and as a result are represented in the final map shown by Figure 8.

Figure 9 Filon Norte open pit mine in Tharsis.

Figure 10 Overburden consisting of adjoining 'country' slate or porphyry and some gossan in the study area.

The rivers and streams in the study area cut into v-shaped valleys, with the intermittent tributaries arranged in sub-rectangular to sub-dendritic patterns. Field observations indicate that the main underlying geological units significantly influence the morphology. The higher ridges are apparently formed from more resistant rock units, while the low-lying areas are often underlain by argillaceous or tuffaceous rock units and/ or more rapidly weathering intrusives. The morphology around the man-made hills is such that the streams flow around their bases and collect water seeping through the dumps. Rainfall percolating through dumps carries small amounts of metals such as copper or zinc into the surface waters leaving the main rivers and streams with high levels of toxic materials (shown as orange on Figure 8). Their spectral signature in Landsat TM correlates highly with the mine dumps in Tharsis and Lagunazo and their field appearance is totally red in color.

The study area also shows artificial reservoirs created by earth dams. The impounded waters, such as the Embalse Grande and the Embalse de Lagunazo, are delivered to the mines, and occasionally to the villages, for domestic and industrial use. They are polluted by toxic chemicals from the mine wastes. The lake deposits rich in Cu (e.g., in the Embalse Grande) show a spectral signature different from that of the adjacent lakes with a rich Fe content (e.g., in the Embalse Lagunazo). Other water-filled mine pits appear as dark.

The gently sloping riverbanks are covered with jara. The latter occur in dense, patchy thickets spread widely over the test area, and for reasons not clearly understood, around the mines. Equally significant in this study is the delineation of the growing eucalyptus and cultivated areas. While the NDVI is an excellent source of information regarding these land-use units, their distinct discrimination was only evident from the saturation stretch image of Figure 7. Separation into other vegetation cover such as, oak and pine trees, from the remotely sensed data alone, was difficult. By overlaying the results obtained from visual interpretation and field observations, however, closer matching of these vegetation types was possible (see Figure 8).

5.2. CHANGE DETECTION

The 1973 aerial photograph was compared with the 1984 Landsat TM data for evaluating the land cover changes. No diagnostic changes with regards to the extent of the mining pits could be deciphered from the two datasets. No new open pit sites and no extension in area coverage of the existing ones could be confirmed from either the remotely sensed data or field observations. Changes were evident, however, within the jara- and eucalyptus-growing areas surrounding the mines including new dumpsites in areas identified as barren from the 1973 aerial photos. For the Tharsis area these are evident when comparing Figures 11 and 12.a with Figure 12b including Figures 13a/b with Figure 8 for the Embalse Lagunazo area. A large portion of the area devoid of vegetation or cleared and filled by mining waste in the 1973 aerial photos (Figures 11, and Figures 12a and 13a/b) seems to be blanketed by the jara (see Figure 8 and Figure 12b). The latter, is a fast growing plant. The 1973 aerial photograph (Figure 11) for example, clearly demonstrates that this vegetation is in contrast with that shown in Figure 12b, and has covered a large portion of the area around Tharsis.

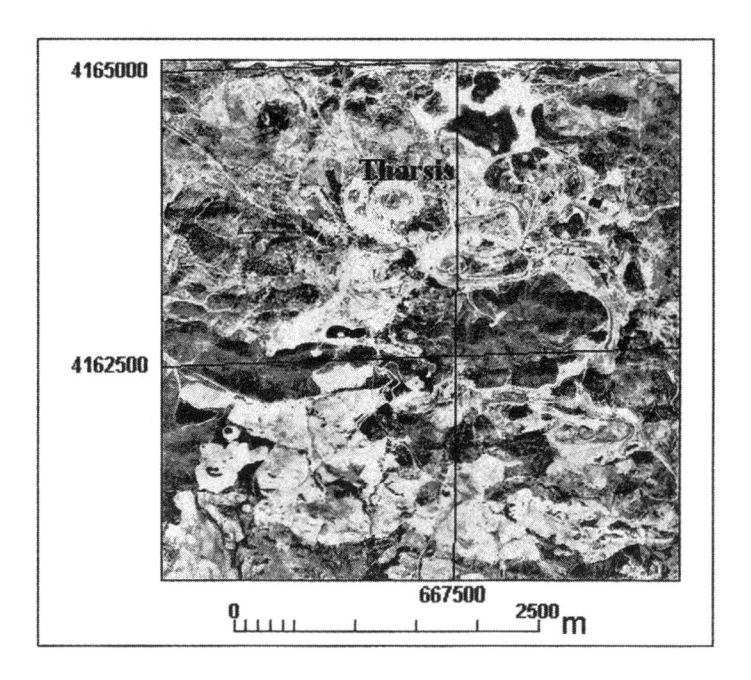

Figure 11 Aerial photo covering a window of the Tharsis Mine area.

The eucalyptus varieties, namely *Eucalyptus globulus* and *Eucalyptus camakiulensis are* intensively grown for the paper industry. As a result, they rapidly change the landscape with the construction of terraces and road networks in the area. Under the afforestation programme, most of these trees seem to have been planted in areas where a slate type of rock is present. The slaty unit of the Culm and the phyllite and slaty type rock unit of the Phyllite Quartze Group are easily cleared, modified and terraced. In the study area, as a result, changes in the natural landscape are very eminent. Thus, although the boundaries seen from the remotely sensed data are artificial, the spectral variation is very clear and at times even sharp. Eucalyptus can grow anywhere. However, its presence in most parts of the study area is significant because of the lithologic and structural importance it manifests in the image. The densely eucalyptus-covered part dominates to the south and southwest of Filon Norte and shows a red colour in Figure 7 against a green bushy vegetated terrain. The eucalyptus cover changes are easy to see from the aerial photographs and satellite imagery of the area. On the aerial photograph of 1973, a large area to the south of Tharsis mine for example, shows a patch that is barren and cleared for eucalyptus growth. Changes in land cover are evident from the Landsat TM data of Figure 7, where the eucalyptus growing area is extended much more to cover a wider area. Agricultural activity is evident only around the towns and villages.

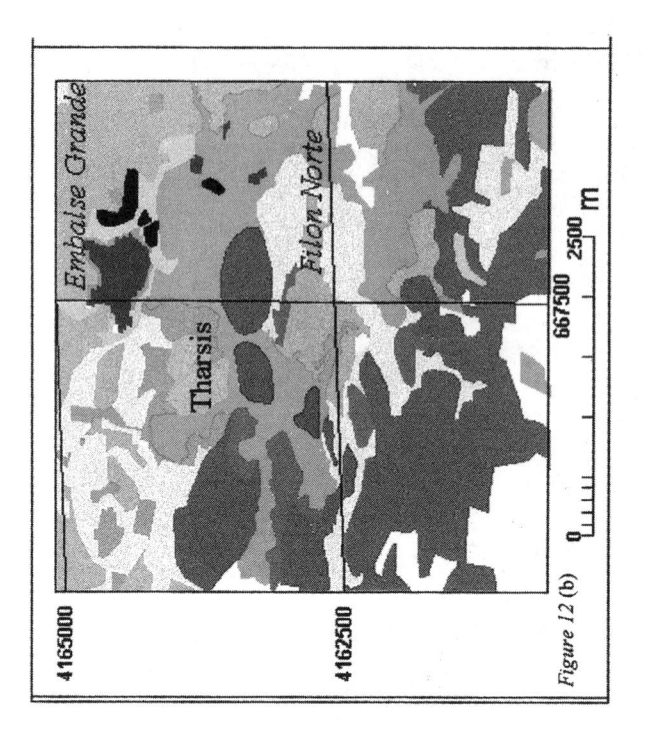

Figure 12 A window of the Tharsis Mine areas representing: (a) the land cover types as interpreted from the aerial photos, and (b) the land cover type map of the same area as deciphered from the Landsat TM data. The legend used in (a) also applies to (b).

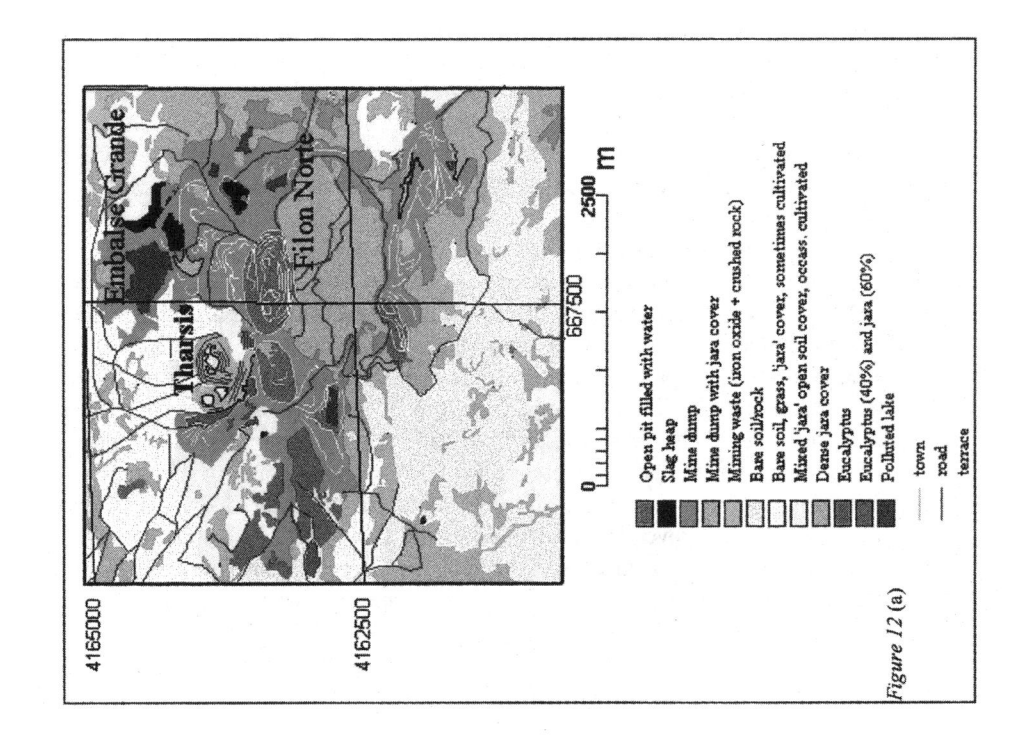

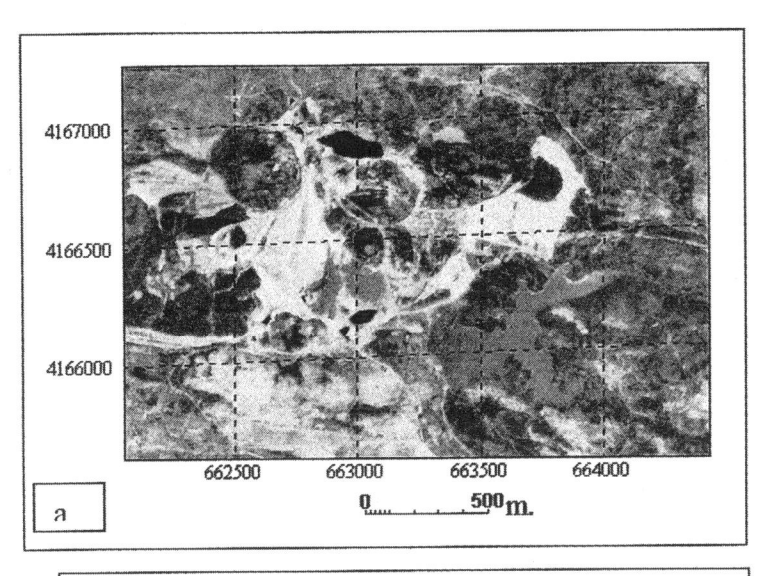

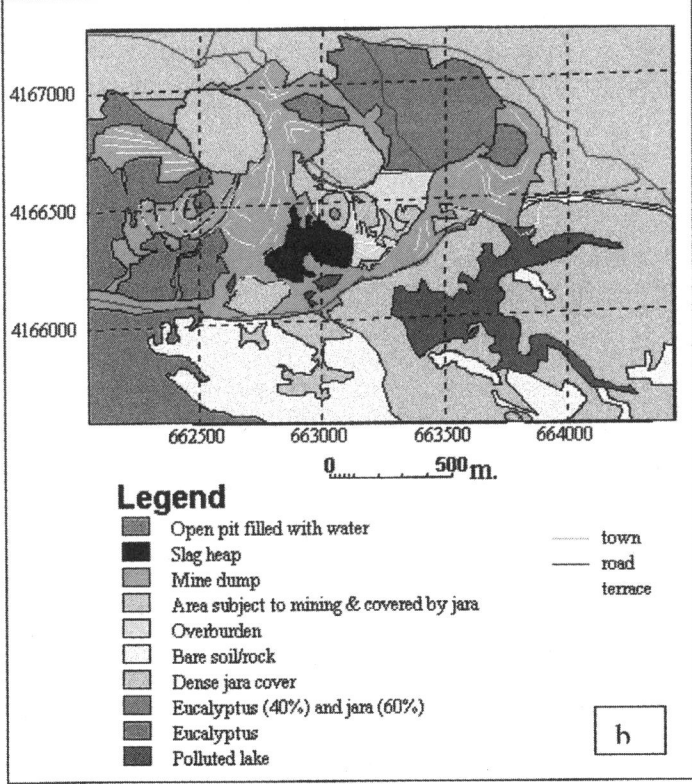

Figure 13 (a). Aerial photo coverage of the Lagunazo mine. (b) Land-cover types and mining induced impact in the Lagunazo area derived from (a) and verified in the field.

6. Concluding remarks

In many parts of the world, mining operations are carried out within a social context that bore a total disregard for the surrounding environment. This has led to the generation of a number of environmentally related problems -- the toxication of the vegetation, land degradation and contamination, groundwater (surface and subsurface) pollution, mine dump disposal and landscape defacement. The Tharsis-Lagunazo area is no exception; it has a long history of mining, which equivocally fit into this categorization.

The use of various Landsat TM bands and the complementary nature of the information sources obtained after processing allow the recognition of various landcover types and mining-induced impacts. They were not discernable in one TM band alone. The advantages of these data lie not only in the acquisition and interpretation of monoscopic images in optical form, but also in the merging of the various remotely sensed bands. They are also due to the multi-spectral power to discriminate surfaces of different composition, and under favourable circumstances to the production of high-resolution image documents on which the environmental mapping of the area can be based. The interpretation of the aerial photographs was essential to support and confirm the findings from the processed TM bands.

Obviously, we are tempted to say that, with the detection capabilities of remotely sensed data it becomes increasingly difficult to hide waste and pollution "under the carpet."

References

1. Carvalho, D., Goinhas, J.A.C. and Schermerhorn, L.J.G. (1971) Livro-guia da excurcao no.4 – Principais jazigos minerais do Sul de Portugal, Director-General de Minas e Servixos Geolægicos Lisboa, *Congr. Hisp-Luso-Amer. Geol. Econ-lo Madrid, Lisboa.*
2. Crôsta, A.P. and Moore, J.M. (1989) Enhancement of Thematic Mapper imagery for residual soil mapping in SW Minas Gerais State, Brazil: A prospecting case history in Greenstone Belt terrain. *Proceedings Of The Seventh Thematic Conference On Remote Sensing For Exploration Geology*, Calgary, Alberta, Canada, 2-6 Oct., 1989, 877-889.
3. Goetz, A.F.H., Rock, B.N. and Rowan, L.C. (1983) Remote sensing for exploration; an overview, *Economic Geology* **78**, 573-590.
4. Knepper, D.H. Jr. (1989) Mapping hydrothermal alteration with Landsat Thematic Mapper data. In: Lee, K. (Ed.): Remote sensing in exploration geology, field trip guidebook T182, *American Geophysical Union*, Washington D.C., USA, 13-31.
5. Schermerhorn, L.J.G. (1975) Spilites, regional metamorphism and subduction in the Iberian Pyrite Belt: some comments. *Geol. Mijnb.* **54** (1), 23-35.
6. Spanish Servicio Geografico del Ejercito (1982) *Topographic sheet of Cala● as*, Sheet no.959, scale 1:50,000, Madrid, Spain.
7. Woldai, T. (1995) Lithological and structural mapping in a vegetated low-relief terrain using multiple-source remotely sensed data: a case study of the Calanas area in southwest Spain, *ITC Journal* **2**, 95-114.
8. Mulder, N.J. (1981) Spectral correlation filters and natural color coding. *ITC Journal* **3**, 237-252.

ENVIRONMENTAL IMPACT OF MINE LIQUIDATION ON GROUNDWATER AND SURFACE WATER

N. RAPANTOVA and A. GRMELA
Institute of Geological Engineering
Faculty of Mining and Geology
VSB - Technical University of Ostrava
17. listopadu Str., 708 33 Ostrava -Poruba, Czech Republic.

Abstract. The possibility of disposing of selected kinds of industrial waste into abandoned mines, or into the exploited parts of active mines, belongs at present to topical problems in the Czech Republic. In the Czech part of the Lower Silesian Basin, the mining activities have stopped in 1995 and the mines are now under a process of liquidation. Mine workings in both the Odolov and Kateřína Mines were simply flooded and their shafts were filled by sorted rock material. The first negative impacts on groundwater quality were documented for this method of liquidation. The Mine Jan Šverma Mine was liquidated instead by filling of the mine workings with a floated self-solidifying ash mixture. Finally the mine workings were sealed with the fill that had the hydraulic properties of impermeable rocks. All the mines considered represent independent geo-hydrodynamic systems without connection with other hydro-geological structures in the study area. The hydro-geological problems of the two liquidation methods are assessed in this contribution.

1. Causes and present state of the problem in the Czech Republic

Mining activity and its environmental impact are not exclusively connected with exploitation, but they play a role also a long time after the exploitation has stopped. The impact is displayed, for instance, by the influence of undermining on landscape morphology and the damage or destruction of both surface and underground constructions. This depression period may last for at least five years after the termination of exploitation in an underground mine, or after the formation of subsequent cavings in old mine workings after their abandonment. Documented examples of such an extreme phenomenon are the collapse and caving of the old Salma II shaft in the Matusek's park in Ostrava (the shaft was abandoned in 1926, and an accident occurred in 1973), the caving in the Jaklovec adit, and the subsequent creation of a cavity in the football stadium in Ostrava in October 1997. The prevention of such accidents with disastrous surface impacts can be effectively achieved by a suitable selection of a proper liquidation method of the old mine workings. One of methods currently used is the so-called "wet conservation", during which the water column in flooded old mine workings gives rise to hydrostatic counter-pressure. However, applications of this method need not result in the "permanent" stability of massifs fractured by mining, but frequently they only postpone the process of uncontrolled caving.

During the last ten years the exploitation in the Czech ore mines has in fact been totally stopped: only one uranium mine has remained active and the exploitation of hard coal has stopped in entire districts, e.g., in the Rosice-Oslavany basin, the Lower Silesian basin, the Ostrava part of the Upper Silesian basin. This suddenly has released a big capacity of mining areas assigned for liquidation. This capacity exceeds the present economic possibilities for systematic and reliable liquidations of all accessible mine workings by their standard backfilling. The liquidation of mines, along with ever-increasing problems of environmental protection, opens the possibility to utilize mine areas cleared in this way for **underground liquidation of waste**. Waste would enter the technological process of mine liquidation as a "**secondary raw material**", which would have to comply with special environmental-hygienic requirements. This kind of "disposal" would fill in the mine areas and eventually play the same role as the standard backfilling of mine workings.

The disposing of hazardous industrial waste to appropriate mine workings of abandoned deep mines is possible only on the assumption that those mines or their parts **will not be developed in the near future**. It is necessary to be aware of the basic principle of this type of repository, i.e., that after closure and sealing of the repository, **its monitoring is difficult or in fact impossible**. Deposited materials cannot be exploited again from mine workings for future contingent industrial processing from geomechanical, technical, and especially economical reasons. Under special technical conditions of repository sealing it is possible to consider such closed deep mines or other similar underground cavities in deeper parts of the Earth's crust to be an object permanently **out of the sphere of the environment.**

Underground mines under closure or other suitable underground areas can be used for disposing of secondary raw materials, which their producer might consider hazardous industrial waste [10]. This kind of utilization of secondary materials in mine workings of underground mines presumes the fulfilment of one basic requirement: **the prevention of leakage of harmful substances from deposited waste through other mine workings into the environment with intensity** "*exceeding the level specified in special regulations*" [9].

2. Liquidation of mining activity in the Lower Silesian Coal Basin

Here we deal with the problem of the influence of mining activity in its final phase (mine liquidation) on geohydrodynamic systems and surface water. Individual cases of influence on water quality and quantity in various locations of the Czech part of the Lower Silesian hard-coal basin are also presented. Figure 1 shows the location of the basin and Figure 2 shows the location of the three mines that will be discussed..

The exploitation in the Czech part of the Lower Silesian coal basin was stopped between 1992 and 1994. Individual mines were liquidated by two methods as follows:

- Recovery of selected mine workings, filling of shafts, flooding of a mine. Two examples are the Kateřina and Odolov mines.

> Kateřina Mine: Uncertain outflow locations pumping 20-35 $l.s^{-1}$ of sulphates over 1900 $mg.l^{-1}$ necessary construction of new treatment plant

Odolov Mine: March 1995 outflow from adit Ida 160 $l.s^{-1}$ content of Fe up to 310 $mg.l^{-1}$ emergency situation in surface water contamination.

- Systematic filling of mine workings by self-solidifying ash mixture.

Jan Šverma Mine: Discharged mine water corresponds to given limits of Government Regulations No.171/92 Sb.

Figure 1. Location of the Czech part of the Lower Silesian coal basin.

Monitoring of quality of pumped and discharged mine water provided diametrically different results. The first method is accompanied by accidental states in consequence of excessive content of iron and/or sulphates. Monitoring of quality in the second method - in the area of the Jan Šverma mine - confirms that the changes of quality of pumped and discharged mine water are not so dramatic to pose a risk for the environment of the area.

3. Liquidation of Jan Šverma mine in Žacléř

A project of mine liquidation by the method of its flooding with self-solidifying ash mixtures with the admixture of secondary materials has been approved for the Jan Šverma mine in Žacléř. The secondary materials, if evaluated in accordance with present legislation concerning waste management, would be characterized as hazardous waste. They consist mainly of industrial waste from asbestos-cement production, neutralizing processes in chemical industry, surface treatment by phosphatisation, desulphurisation of flue gases, metal treatment and galvanizing, blasting, surface treatment of metallic and non-metallic constructions, etc. However, any kind of admixture to be used must be endorsed in advance and extracts from prepared flooding mixtures must be tested (potential transition of pollutants from the mixture into water environment of the mine must be quantified).

According to the character of used secondary material and its extract, resulting backfilling mixtures can be applied in selected mine workings and in respective hydrogeological sections of the mine, regarding the approved zonation of the mine. From the hydrogeological point of view, this zonation (see Figure 3) takes into account mainly the natural vertical zonality of the intensity of water exchange with earth surface (hydraulic properties of rock environment influenced by mining activity).

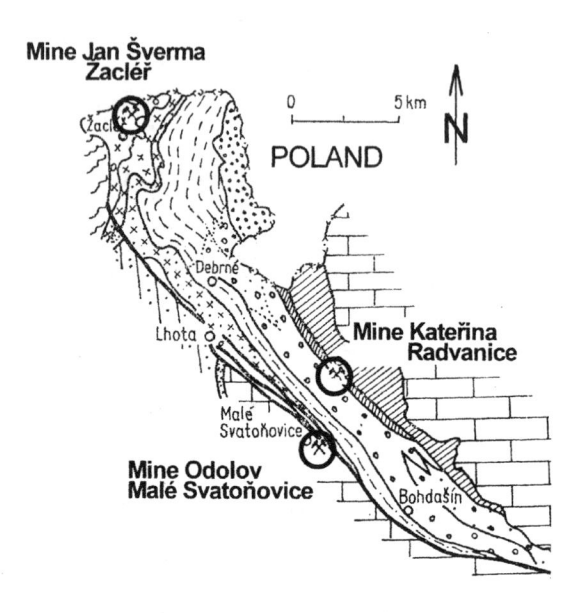

Figure 2. Location of the three mines: Kateřina, Odolov and Jan Šverma.

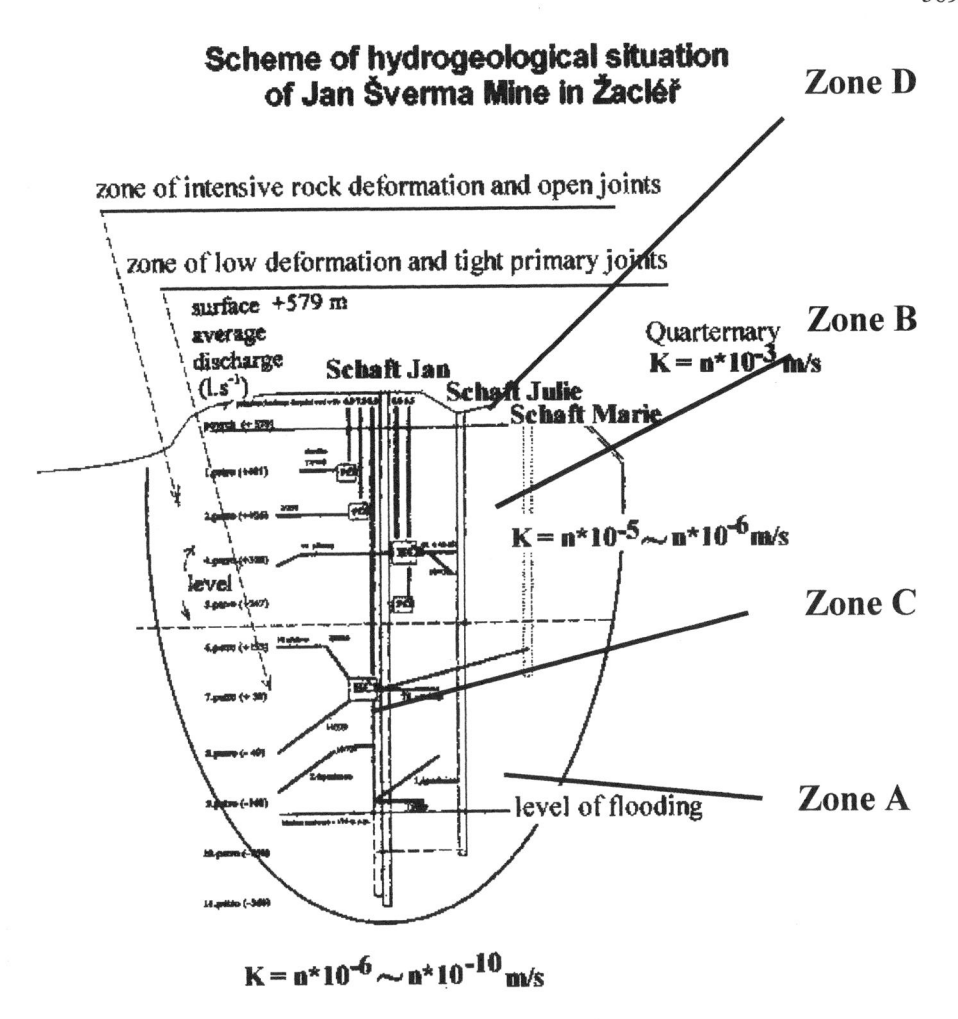

Figure 3. Zonation of the Jan Šverna Mine. **Zone D:** rock massif section situated between the surface and the first mine level (approx. 90m below the surface). **Zone C:** inclined and vertical mine workings connecting particular levels between the zones of shallow and deep groundwater circulation. **Zone B:** zone of shallow ground-water circulation – approximately from the first mine level (90 m below the surface) down to the depth of 400 to 500 m. **Zone A:** zone of deep-water circulation with limited water exchange with water surface - deeper than 400 to 450 m below surface.

It should be emphasized that the Czech legislation does not view this as waste disposal in mine, but as utilization of *"secondary material"* in the technology of the liquidation process. Secondary material is looked at quite differently in this particular case. An example is provided in Figure 4.

```
                            ASH
              ┌──────────────┴──────────────┐
       raw material                      waste
```

raw material
Filling of excavated spaces, shafts, mine working reconstruction, building activities.

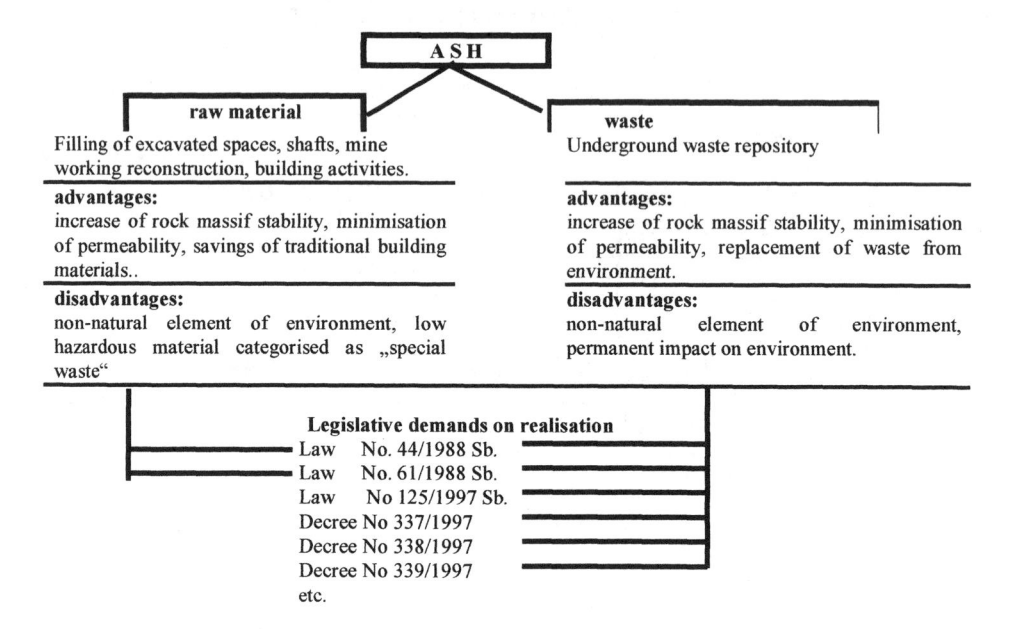

advantages:
increase of rock massif stability, minimisation of permeability, savings of traditional building materials..

disadvantages:
non-natural element of environment, low hazardous material categorised as „special waste"

waste
Underground waste repository

advantages:
increase of rock massif stability, minimisation of permeability, replacement of waste from environment.

disadvantages:
non-natural element of environment, permanent impact on environment.

Legislative demands on realisation
- Law No. 44/1988 Sb.
- Law No. 61/1988 Sb.
- Law No 125/1997 Sb.
- Decree No 337/1997
- Decree No 338/1997
- Decree No 339/1997
- etc.

Figure 4. Considerations on legislation concerning secondary material in the Czech Republic.

Secondary raw material is material produced from waste by sorting or processing, which is suitable for further utilisation [6]. If secondary raw material enters the technological process of exploitation plant for the activities supervised accordingly to the Mining Law, it is not considered to be a waste further on. It is not subdued to the laws and regulations valid for waste [6] and other linked legislation on waste. Utilization of such secondary raw materials (filling mixtures) represents a technological problem, which is solved entirely, concretely, and unequivocally by the decision of the Regional Mining Office.

The liquidation of the Jan Šverma Mine by the method of flooding of all accessible mine workings with self-solidifying ash mixture, and possibly with admixture of hazardous waste, was proposed under the following conditions:

⇨ Minimization of negative environmental impact on the hydrosphere;
⇨ Discharged waters from mine workings to surface stream must correspond to given limits;
⇨ Systematic flooding with maximum filling of all mine workings from bottom up (from the lowermost levels of the mine upward).

The management of the Jan Šverma Mine and the GEMEC Ltd. Co. Ostrava used an approach to the method of liquidation that was in accordance with the laws and regulations valid for waste (an agreement with the state administration, and institutions of environmental protection and hygiene). Considerations on the admixture of materials, ranked as hazardous waste, and as basic ash mixture are that:

- ⇨ It must satisfy the conditions of the decrees of the Ministry of Environment [4], [5], and [7];
- ⇨ It must be approved by hygienic and water management institutions;
- ⇨ It is assessed according to the water extract of the resulting filling mixtures;
- ⇨ It can be used only in the mine workings selected according to the hydro-geological zonation of the mine.

The liquidation of the Jan Šverma Mine was in stages, each with several steps and sub-steps as follows:

First stage. Pre-liquidation activities:
- Definition of natural hydrogeochemical background;
- Assessment of quality and quantity of pumped mine water.

Second stage: Liquidation activities:
- Hydro-geological zonation of the mine;
- Set up of the conditions for environmental protection,
  - ➢ conditions of suitability of mine workings for application of secondary raw materials,
  - ➢ system of impervious barriers,
  - ➢ criteria for suitability of used secondary raw materials,
- monitoring of quality and quantity of pumped mine water,
  - ➢ monitoring of recycled mine water quality,
  - ➢ monitoring of quality and quantity of mine water discharged to surface streams.

The basic results of the monitoring of quality and basic principles for the evaluation of mine workings from the viewpoint of suitability of flooding mixture selection are presented here. The results prove that the actual liquidation method is acceptable from the viewpoint of environmental influences. If all conditions of the liquidation method (selected and approved secondary raw materials, their acceptable proportion in the mixture, control and monitoring of water extract quality, and water treatment prior to its discharge to the stream system) are satisfied, **the external hydraulic system will not be influenced above given limits of pollutants** by water discharged to the stream system, and the current state of environment will not be changed.

With respect to favourable hydrogeological conditions, mining activity on the hard-coal deposit of the Jan Šverma Mine did not require a detailed and systematic hydrogeochemical monitoring of the mine water quality. Therefore, no relevant data - from the viewpoint of mathematical-statistical relevance of the database for representative processing - were collected at the time of exploitation. Moreover, sporadic analyses of mine water are of very different quality and the spectrum of the parameters monitored does not provide a comprehensive picture of the trends of alterations caused by the mining activity.

Permission and realization of backfilling of mine workings using secondary material has necessitated a thorough monitoring of quality and quantity of pumped mine water, recycled mine water and mine water discharged into surface streams in the course of liquidation. Therefore, a representative data set has gradually been gathered in this way, which serves as a base for the analysis of the extent of the environmental influence. We can now assess former mining activity, but also the impact of the liquidation method on revitalization of environment in the area.

3.1. CONDITIONS AND CRITERIA OF SUITABILITY OF MINE WORKINGS FOR APPLICATION OF SECONDARY MATERIALS FOR THE JAN ŠVERMA MINE

The following are the conditions of suitability of mine workings for application of secondary materials concerning the location:

- is not situated in spots with uncontrolled priority hydraulic ways interconnecting other hydraulic systems :
  natural : pervious tectonic zones;
  artificial : boreholes, mine workings, pervious areas fractured by cut fissures,
- is situated in sufficient depth under the surface, where groundwater flow through the rock mass is minimized or stagnant (sufficient barrier between the aquifer with intensive water exchange with earth surface and the system with quasi-stagnant groundwater),
- is not situated in a zone of extreme geomechanical phenomena - natural (earthquakes) or artificial (rock bursts). Actual properties of rock environment will not change in future as a result of mining activity in the deposit or other artificial interference in the newly established natural regimen.

The following are the conditions concerning the system of impervious barriers:

- The mine locality can be hydraulically insulated from aquifers that are being used for the purpose of water management (or that are planned to be used for that purpose),
- The mine workings must be hydraulically insulatable so that the watering of their surroundings does not allow future transport of contaminants into adjacent areas through-flowing hydraulic systems.
  - every mine working filled with mixture containing secondary material is closed by insulating objects (barriers) able to resist the pressure of 10 MPa or by a mine working section filled with mixture without any content of secondary material with the characteristics of hazardous waste,
  - the mine will be divided into zones with limited application of secondary material in each of them (see zones A through D below).

The following is an example after [1]:

Zone A: The storage of secondary raw materials is allowed here providing that the resulting water extract, after the ripening, satisfies the legal limit [7] I/I with exceptions.

Zone B & C: The storage of secondary raw materials is allowed here providing that the resulting water extract, after the ripening, satisfies the legal limit [7] I/II with exceptions.

Zone D: This zone must be filled only with filling materials, whose extract satisfies the limit I/II - without the admixture of secondary raw materials.

The criteria for suitability of used secondary raw materials are:

- The backfilling mixture and its extracts must be tested according to the actual legislation [7] and liquidated mine water must permanently satisfy the requirements of environmental protection of the locality, according to [8], for instance;
- The secondary material and the whole mixture must not be self-ignitable, inflammable and must not produce toxic, inflammable or explosive substances under the present mine conditions and future natural conditions;
- The backfilling mixture and its extracts must not increase mine water contamination (chemical, radiological nor biological) above the present state (or a limit state determined in a different way) in water discharged into streams or other surface water systems [8], etc.

3.2. RESULTS OF MONITORING OF MINE WATER QUALITY

Concerning the results of the monitoring of mine water quality we have to characterize the mine waters. At present, mine water is a mixture of two water types of a different kind and origin: natural and anthropogeneously influenced waters.

Natural waters:

>>> **water of atmospheric precipitation origin** – it infiltrates into the mine:

> ➤ through fissure systems in the area of the loosening of the sub-surface Carboniferous zone (ca. 80 m deep),
> ➤ in the zone of intense fracturing and secondary fissures opened as a consequence of mining activity (ca. 400 m deep),
> ➤ along hydraulically active faults,
> ➤ along priority ways represented by mine workings (old shafts, adits, boreholes).

>>> **water originated in porous and fissured Carboniferous aquifers** - it infiltrates into the mine:
> - through fissure systems and/or hydraulically active faults. Its residence time in rock environment is substantially longer than that of the aforementioned type of surface water. It acquires the character of altered petrogeneous water.

Anthropogeneously influenced waters:

>>> **water hydrogeochemically altered by residence in mine workings** – it is enriched by the products of chemical weathering of rocks, and anthropogeneously contaminated. The occurrence of this water is relatively rare at the Jan Šverma Mine locality in Zacleř; it shows increased value of total mineralization (up to 10 - 13 $g \cdot l^{-1}$), increased content of SO_4^{2-}, etc.

>>> **water extracted from filling mixtures** – it is hydro-geo-chemically altered by the residence in mine workings and enriched by the extracts from waste materials. This water is the main object of interest in evaluating the possible danger to the environment. It represents the highest proportion of mine water in liquidated parts. Its total mineralization is approx. 2 - 2,5 $g \cdot l^{-1}$, it has an increased content of SO_4^{2-} (ca. 1,5 $g \cdot l^{-1}$), an increased content of non-dissolved substances (turbidity) and, unsystematically, increased contents of heavy metals (Zn, Cd, Pb, Hg) as well.

Dewatering of a mine (see Figure 5) is performed by pumping of water to the surface. The main pumping stations are at 4th level (+ 338 m a.s.l.) and at 7th level (+50m a.s.l.). At 1st, 2nd and 5th levels there are additional pumping stations.

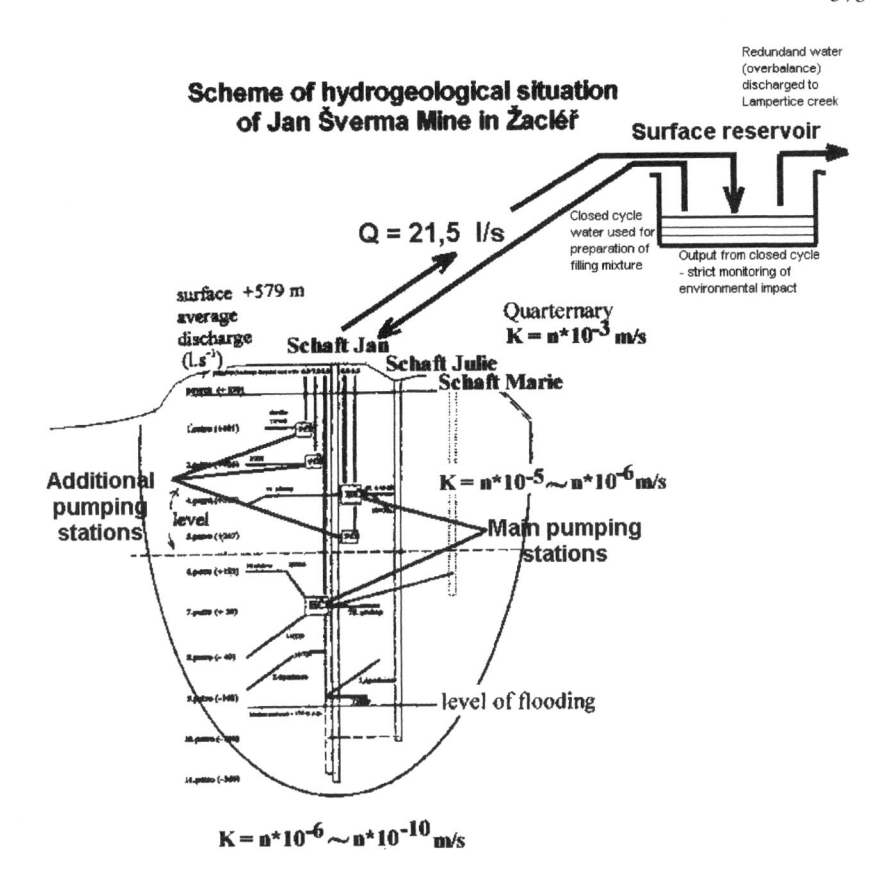

Scheme of hydrogeological situation of Jan Šverma Mine in Žacléř

Figure 5. Scheme of mine dewatering

The efficiency of liquidation and its contamination impact on environment of the region of exploitation field are evaluated on the basis of quality of mine water discharged into the local water stream (i.e., the Lampertice Creek). The quality of mine water pumped to the surface is also monitored. Water is, however, in the maximal possible volume recycled and used for preparation of filling mixtures. Their surplus is then, after simple treatment, discharged into water streams. Monitoring of mine-water quality is performed at the mine-water purge outlet into the Lampertice Creek, in the surface reservoir of pumped mine water and also at individual levels of the mine (4th, 6th, 7th and 8th level). Water samples from mine levels are analysed in case that substantial limit excess (see limits in Table 1) is found in the surface reservoir of pumped mine water. The monitoring scheme timetable is presented in Table 2. Circled numbers represent the sampling phases. Phase 1 is carried out routinely, subsequent phases only in case of exceeded limits in the samples. The aim is to define location of the contamination source.

Table 1 The table of limits for monitoring of mine water quality during liquidation of the Jan Šverma Mine

parameter :		513/92 Sb. limit for water extracts Ic (canceled)	338/97 Sb I / II	171/92 Sb. surface water	Mine Šverma Žacléř internal limit (levels and extracts) ① ②		**limit outlet of mine water→ stream**
pH		6,5-8,0	6,5-8	6,0-9,0	6-10	5,5-13	6,0-10,0
conductivity	mS.m^{-1}	40	40			300	300
non-diss. mat	mg.l^{-1}					100	100
dissolved mat	mg.l^{-1}			1000	2400	3000	2400
SO$_4^{2-}$	mg.l^{-1}	250	250	300	900	1500	900
Cl$^-$	mg.l^{-1}	100	100	350		100	350
Ca^{2+}	mg.l^{-1}	200		300		300	300
Mg^{2+}	mg.l^{-1}	120		200		400	400
NH$_4^+$	mg.l^{-1}	0,1	0,5	2,5	1,5	5,0	2,5
BOD$_5$	mg.l^{-1}			8			8
CHOD (Cr)	mg.l^{-1}	8	8	50		40	50
Al	mg.l^{-1}	0,2	0,2			0,2	0,2
As	mg.l^{-1}	0,05	0,05	0,1		0,05	0,1
Ba	mg.l^{-1}	1,0	1,0	2,0		1,0	2,0
Cd	mg.l^{-1}	0,005	0,005	0,015		0,05	0,05
Co	mg.l^{-1}	0,05	0,05	0,1		0,05	0,1
Cr	mg.l^{-1}	0,05	0,05	0,3		0,05	0,3
Cu	mg.l^{-1}	0,1	0,1	0,1		0,1	0,1
F	mg.l^{-1}				1,5	10,0	1,5
Fe	mg.l^{-1}	0,1	0,1	2,0		2,0	2,0
Hg	mg.l^{-1}	0,001	0,001	0,001		0,006	0,006
Ni	mg.l^{-1}	0,1	0,1	0,15		0,1	0,15
Pb	mg.l^{-1}	0,05	0,05	0,1		0,05	0,1
Zn	mg.l^{-1}	3,0	5,0	0,2	1,0	5,0	1,0

Table 2a. Monitoring scheme - time table [3]

Sampling location /performed activity/		Reference sampling (each 6 month)			Periodical sampling (month)		
			Set up contamination			Set up contamination	
Surface	outlet	❶ C			❶ B		
	reservoir	❶ C			❶ B		
Mine	4.level	❶ B			①		❶ P
	6.level		❷ O	③ P		②	❷ P
/liquidation/	7.level	❶ B			①	❶ O	
/liquidation/	8.level		❷ O	③ P		②	❷ P
/liquidated/	9.level		❷ O	③ P			❷ P
		Selected accredited laboratory			Laboratory of J.Sverma Mine		

C - complex analysis   B - basic analysis   O - orientation analysis   P - purposive analysis

Table 2b. Monitoring scheme - kind of analysis [3]

Kind of analysis	Analysed parameters						
	physical	sensorial	major ions	minor ions	organic matter	radioactivity	Special analyses
C complex	pH, conductivity non-dissolved and dissolved matter	smell, turbidity colour	Na,K,Ca, Mg, Mn, NH$_4$, Fe, Cl, SO$_4$ HCO$_3$,CO$_3$	Hg,Cd,Pb, CuCo,Cr, Ni,Zn,V, Al, Ba + selected	BOD CHOD$_{Cr}$ Non-polar matter, PAH, PCB	Total volume activity, activity α and β	According to the character of raw material
B basic O orientation P purposive	pH, conduct.		selected	selected selected selected	selected	selected	

It is unambiguously obvious from the monitoring of mine water quality on particular mine levels and in the surface reservoir, and of water discharged to the stream system (after its treatment and refinement), see Figure 6, that the **mine liquidation influences the chemism of mine water only on levels that are active from the viewpoint of the liquidation,** and that upper mine levels are not affected with alterations. At present, the 8th and 9th levels have been already liquidated and the liquidation activity proceeds, above all, on the 7th level. As a matter of principle, the liquidation is carried out **from the bottom up, systematically, with the maximum filling of all mine workings,** and with no unfilled mine workings left.

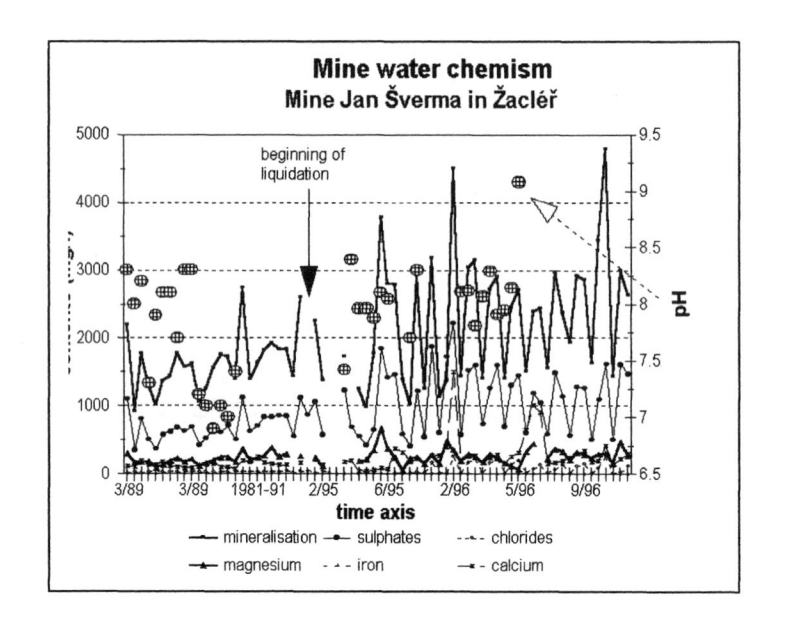

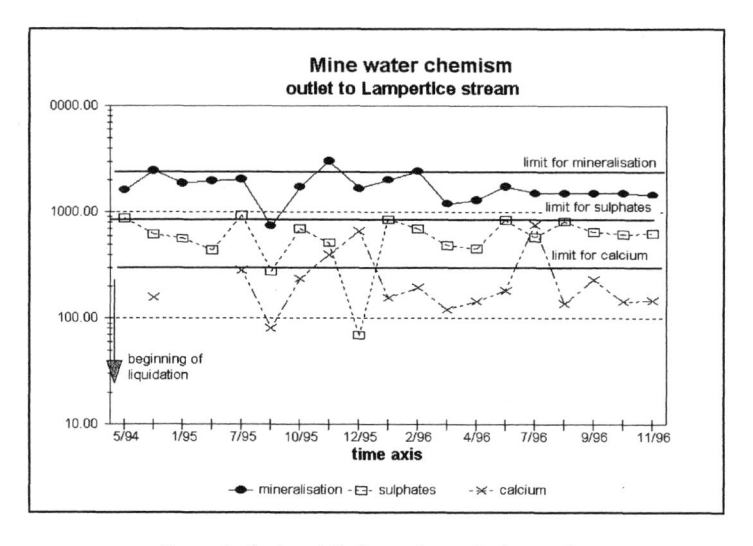

Figure 6a (top) and 6b (bottom): monitoring results.

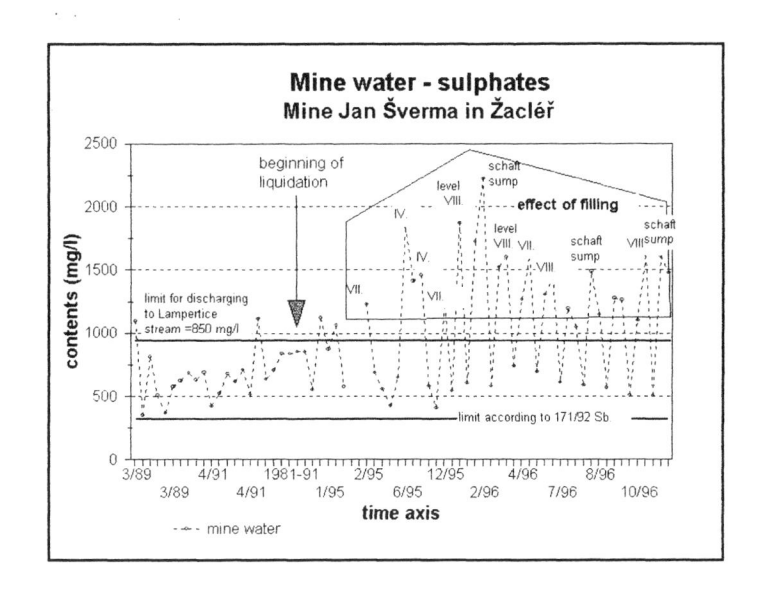

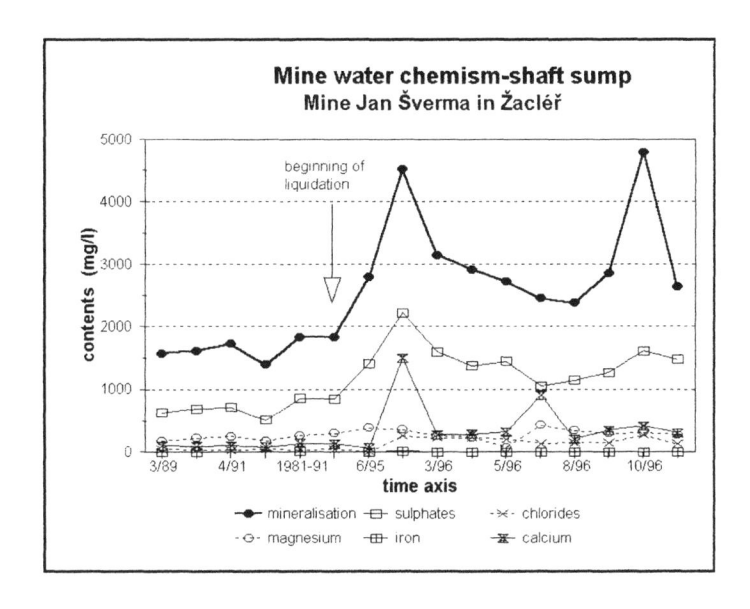

Figure 6c (top) and 6d (bottom): monitoring results.

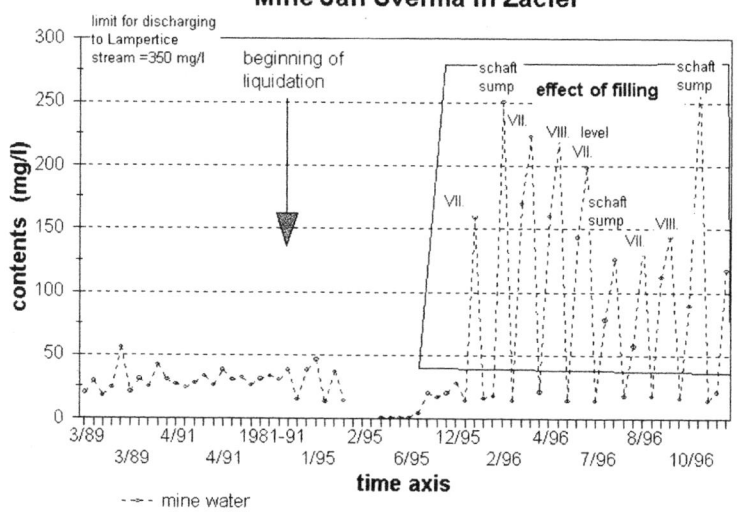

Figure 6e (top) and 6f (bottom): monitoring results.

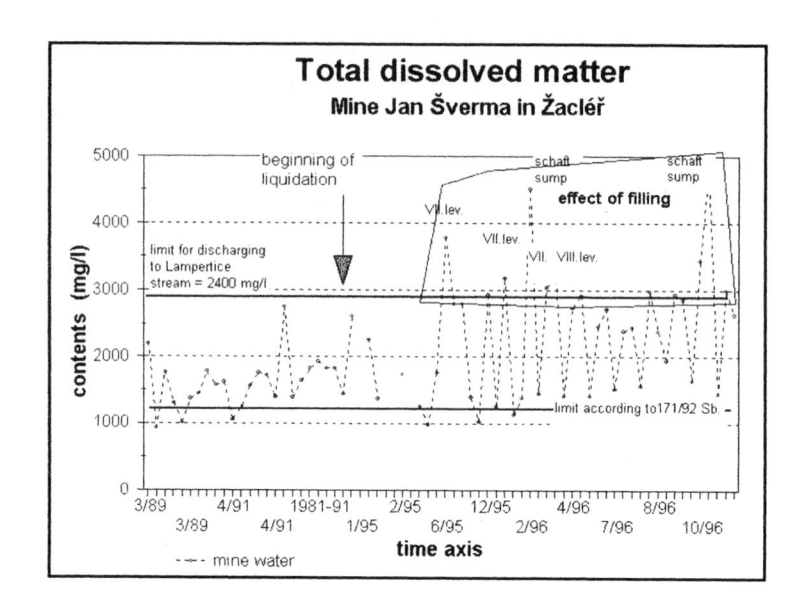

The results of sampling of mine water on particular mine levels, in the surface reservoir and in places where waste water mouths into the stream system of Lampertice stream document **inexpressive alterations of quality of both pumped and discharged mine water.**

However, this cannot be considered as a permanent negative influence on the hydrosphere. The fact that some monitored parameters sporadically exceed deter-mined limiting values may be accidental and unsystematic, as shown in Table 3. Systematically exceeded limits (e.g., the pH factor) are not exclusively a consequence of activities connected with mine liquidation (its method, type, of standard and secondary raw materials) but also a consequence of natural background and natural quality of mine water infiltrated into mine workings.

Table 3. Results of mine water quality monitoring [1]

	Total number of analyses	Total number of limits exceeding			
		internal limit	%	Limit for discharging	%
outlet	317			6	1,9
reservoir	317	7	2,2	24	7,6
IV. level	266	6	2,3		
VII.level	266	8	3,0		
VIII.level	266	36	13,5		

	Number of limits exceeding				
	outlet	reservoir	IV.l	VII.l.	VIII.l.
amonium	1	2	4	3	6
chlorides				3	7
conductivity				2	4
diss. matter		3			3
turbidity					8
sulphates	2	14			3
Fe					2
CHOD$_{Cr}$	1		1		1
Ni					1
Hg		2	1		1
Al	2	2			
V		1			
totally	6	24	6	8	36

4. Liquidation of the Odolov Mine in Male Svatoňovice and the Kateřina Mine in Radvanice

Table 4 shows the activity during 1950-1997. The mining activity in the Odolov mine in Male Svatoňovice was terminated and the water pumping stopped in 1991 [2]. Liquidation was carried out using a method consisting of recovery of material from selected mine workings, simple flooding and filling the shafts with sorted rock material. Mine-water filled exhausted areas until 1995. In March 1995 there was an overflow of mine water from the flooded mine into the mouth of the Ida adit (drifted in 1864, length 1.800 m), i.e. into the lowest level of the Odolov mine. Water yield of out-flowing mine water amounted to 90 to 160 litres per second and this water contained disastrous content of iron (up to 310 mg·l^{-1}) and sulphates (1.770 mg·l^{-1}). This outflow of mine water caused a grave environmental accident in surface streams.

The mining activity in the Kateřina mine in Radvanice was terminated and water pumping stopped in 1994. Mine water filled exhausted areas until 1998. It entered the surface through unconcentrated hidden outflows (20 to 35 l·s^{-1}). Pumping of mine water from the shaft (approx. 27 l·s^{-1}) had to be started, mine water level had to be artificially lowered and water had to be treated prior to its discharge due to very high concentrations of sulphates (up to 1.900 mg·l^{-1}).

Table 4. The mining activity during 1950-1997 for the three mines analysed.

Year	Locality	Yield / ∅ year - l.s^{-1} /	Salinity / mg.l^{-1} /	Fe / mg.l^{-1} /	SO$_4$ / mg.l^{-1} /
1950-1990	Odolov Mine	50-80	Situation in time of mining activity		
1950-1990	Kateřina Mine	60-70			
1950-1990	Šverma Mine	30-40	Situation after liquidation of mine		
1995	Ida Adit	68	2948	145	1771
1996	Ida Adit	77	2562	137	1684
1997	Ida Adit	77	2052	92	1377
1997	Kateřina Mine	27	3143	14,5	1878

The resulting impact on water quality of water-bearing systems and the environment as a consequence of mine water contamination from rock and mine environment will now require substantially higher decontamination costs and additional treatment of discharged water, which would not be necessary in case of controlled, systematic and conceptually correct liquidation of mine workings.

5. Concluding remarks

The results of mine-water sampling at particular mine levels, in the surface reservoir, and at the outlet of waste water into the Lampertice Creek prove that there are no unexpected changes in the quality of pumped and discharged mine-water. The detected changes cannot be classified as a permanent negative environmental influence on the hydrosphere. Exceeding given limit values for some of the parameters monitored appears mostly or only accidental and unsystematic. The systematic overrunning of some limits, e.g., of the pH value, is not a consequence of activities connected with the liquidation of a mine (its method, the basic and secondary raw materials utilized) but it results from the natural background and consequently it is due to the natural mine water quality.

On the basis of the knowledge acquired, it was decided that mine workings below the level of shallow groundwater circulation (i.e., approximately 400 to 450 m below the surface) are suitable for the utilization of secondary raw material so that they can be filled by selected and approved secondary raw materials, which are categorised as "hazardous industrial waste".

An existing project of sealing the whole repository system can be considered as suitable and efficient in the protection of the external hydraulic systems. Each mine working, in which the hazardous waste is dposited, is sealed by a section of 20 to 50 m in length, and then flooded by a self-solidifying mixture without waste. At the same time the main crosscuts at particular shafts and mine levels, and the inclined adits between mine levels, are liquidated without admixture of waste.

Owing to the fact that all rules and monitoring systems of mine-water quality, according the decision of the Trutnov Mining Office, are kept, we can say that the negative impact on the external hydraulic system was not observed. The quality of mine water discharged into the stream has not exceed the given maximum allowable limits of pollutants.

The system of mine liquidation described and proposed here is more expensive and time consuming but acceptable from the point of view of environmental protection (especially because of the low impact on the hydro-chemical regime of hydrodynamic systems and surface water streams). It represents a procedure that not only removes certain types of hazardous wastes out of reach, but it also minimizes the impact on land subsidence and that on geo-mechanical events related to the non-controlled sinking of mine spaces. The possibility of sudden expulsions of methane to the surface is also minimized.

The mine workings took place in the rock massif with suitable depth below the surface. The hydro-geological, structural-tectonic and geotechnical conditions are suitable for the construction of permanently sealed repositories of selected hazardous industrial wastes. It certainly is not farsighted to liquidate the mines by employing only non-hazardous wastes that also represent potential raw material and can be utilized in the building industry or for land reclamation. The case studies presented in this contribution are in support of this conclusion.

References

1. Grmela, A. (1997): *Hydrogeochemical alterations of mine water owing to selected waste disposing into abandoned mines-case study*. Proceedings of Hydrogeochemia'97,Comenius Univ.,Bratislava.Pg.20-29,1997-05-28, Slovakia.
2. Koros, I. (1998): *Problems of mine water after the termination of coal exploitation in Lower Silesian basin* (in Czech). 10th Conf. on Mining and Geology. Proceedings, pg. 52-56. Straz pod Ralskem, Czech Republic.
3. Rapantova, N. (1997): *Results of Contaminant Transport Monitoring in Aquatic Environment Performed during Liquidation of the Mine J. Sverma in Zacler*. VIII. Symp."„Wspólczesne Problemy Hydrogeologii", Univ.of A. Mickiewicz, Poznań-Kierkz, Poland.
4. Decree No. 339/1997 of the Ministry of Environment (ME) from December 11, 1997, evaluating dangerous properties of waste.

5. Decree No. 337/1997 of the ME from December 11, 1997, submitting the Waste catalog and specifying other lists of waste
6. Law No. 125/1997 from May 13, 1997, concerning waste - basic regulation for waste management
7. Decree No. 338/1997 of the ME from December 11, 1997, specifying details of waste management
8. Instruction No. 171/1992 of the Czech government, determining indicators of acceptable contamination of waste-water
9. Law No. 17/1992 concerning the environment
10. Decree No. 104/88 of the Czech Mining Authority, concerning special interventions into earth crust

PROTECTION OF GROUNDWATER RESOURCES QUALITY AND QUANTITY IN MINING AREAS

A. GRMELA and N. RAPANTOVÁ
Institute of Geological Engineering
Faculty of Mining and Geology,
VŠB - Technical University Ostrava,
17. listopadu Str., 708 33 Ostrava -Poruba, Czech Republic.

Abstract. This contribution provides an overview of the temporary and constant impacts of underground coal-ining activity on the environment with a special emphasis on groundwater quality and quantity protection in the Czech Republic. Recommendations are made for the monitoring of mining influences on water resources in the Ostrava-Karviná coal-mning district.

1. Causes and present state of the problem in the Czech Republic

Mining activity is always of a temporary or constant impact on the environment, results in new geomorphologic formations, and represents a strong anthropogenic intervention, especially on the hydrologic and hydrogeologic conditions of the areas affected. Not in all cases, however, it is a completely negative intervention, or an intervention that could not be later eliminated by means of recultivation carried out with sufficient sensibility [2]. On the contrary, a variety of exploited localities are being utilized as showplaces of nature **beauty** today (e.g., Soos, the Amalino udoli valley in the Kasperske Hory mountains, the Hermanovice quarry, the Panska skala rock in Kamenicky Senov), or as localities of extraordinary importance (e.g., the radioactive springs of the Svornost shaft in Jachymov).

Since long ago, the mining and exploitative activity of mankind has been an integral part of its existence, a condition of its development, and must be, as such, accepted and assessed by all of us. Mining activity was a condition for the rise of big and rich cities, provided with material for their construction, has given rise to the technique of metal and non-ore treatment, and still is the cause of the creation of grand cultural works which are the source of international boast today.

The burden to the environment caused by mining activity is the heaviest in the course of active exploitation, when auxiliary underground operations, dumps, mud pits, preparation and sorting plants, and power and transport equipment do not represent an organic complex of the original surroundings. Also, the opposition of the part of the society against such an activity is strongest during that time. Luckily, this period is, from the historical viewpoint, relatively short and vanishes. Similarly, the influence of mining activity on the surface and groundwater systems is usually temporary [2].

From the viewpoint of topical problems of groundwater and surface water protection it is necessary to distinguish two basic exploitation methods:

- Open-pit exploitation of deposits
- Underground exploitation of deposits.

The open-pit exploitation of deposits induces changes in geohydrodynamic systems apparently and directly (e.g., alterations of surface run off, changes of hydraulic gradient of shallow aquifers). Results are in groundwater draw-downs, in the surroundings of open pits, increase of flow rates, and changes of flow directions, and the destruction of natural spring areas. In these exploitation methods no undermining affects can be observed.

Underground exploitation of deposits influences geohydrodynamic systems and surface streams in the following way:

⇒ It creates new priority paths for groundwater;
⇒ It artificially opens closed confined hydrodynamic systems (stagnating water bodies or connate (fossil) waters in overburden, deposit, and to a degree, underlying rock complexes – it frequently connects those systems hydraulically;
⇒ It produces mixed mine water usually pumped to mine surface and discharged to the streams;
⇒ It results in changes of hydraulic gradients of water streams and shallow groundwater owing to undermining, including the alterations of aeration zone thickness.

2. Environmental impact of mining activity on groundwater and surface water

Mining activity is always of a temporary or constant impact on the environment, generating:

➢ new geomorphologic formations, and
➢ strong anthropogenous impact on environment especially on hydrological and hydrogeological conditions

Mining and exploitative activity of mankind has been an integral part of its existence. It represents a condition of its development and must be, as such, accepted and assessed because:

➢ It is negative, especially in the period of active exploitation;
➢ It is temporary: the period from the historical viewpoint is relatively short and the activity eventually vanishes;
➢ It is modifiable because the influence of mining activity on the surface and groundwater systems is usually temporary.

2.1. INFLUENCES OF UNDERMINING

The underground exploitation of deposits gives rise to worked-out areas that are, after they reach a critical size, caved by overburden rocks. Under certain circumstances, such movement of rocks is transferred up to the surface, accompanied by the creation of a so-called depression basin with characteristic movements and deformations. Movements and deformations, rising in a depression basin, hinge on a variety of factors. These factors stem especially from geological composition of the territory in question, and from space, time and technological-operational conditions. The resulting movement of a surface point is a complicated space-time movement, which is characterized by its horizontal and vertical components, horizontal shift v and depression s, respectively. The depression can be expressed by a general equation [1]:

$$s = f(m, a, e, z), \qquad (1)$$

that is usually presented in the following form,

$$s = m \cdot a \cdot e \cdot z, \qquad (2)$$

where $z = 1 - e^{-(1,2\, t/t_u)^2}$ is the time factor,
m is the seam thickness,
a is the coefficient of exploitation:
  caving = 0.7 ÷ 0.9; pressed-air filling = 0.4 ÷ 0.8;
  mechanical filling = 0.2 ÷ 0.4, and hydraulic filling = 0.1 ÷ 0.2;
e. is the coefficient of efficiency = quotient between worked-out area and the so-called effective area in a measured surface point ($0 \le e \le 1$).

The effective area radius is:

$$r = h \cdot \cotg \mu, \qquad (3)$$

where h is the deposition depth, and μ is the angle of draw ($55° \le \mu \le 65°$)
The horizontal shift is represented by the following relationship [1]:

$$v = \frac{s_{max}}{\sqrt{2\pi}} \cdot e^{-\pi \cdot x^2 / r^2} \qquad (4)$$

where x is the distance of the point from the edge of the working face.

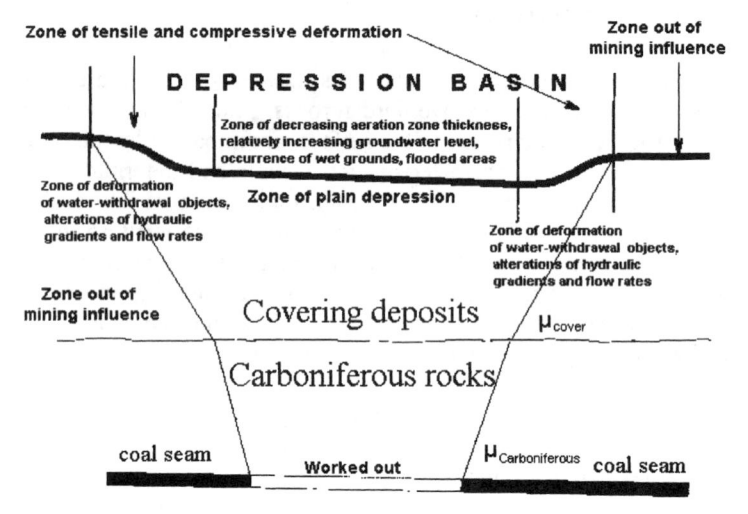

Figure 1. Development of a depression basin.

From the above mentioned results we can see that the influence of undermining is limited in time approximately to five years after finishing the exploitation in the effective area and its intensity is strongly affected by [1]:

- natural factors: thickness and geomechanical character of covering deposits,
  geomechanical character of deposited rocks,
  deposition depth ($\Rightarrow$ exploitation depth),
  thickness of the deposit exploited (e.g., coal seam), and
- technical-operational factors:
  exploitation method (long-walling, rooming, etc.)
  caving or backfilling of worked-out space.

The technical-operational factors can be influenced by man, and two forms of deformations take place in the area of depression basin:

1. Tensile and compressive deformation on the peripheries of depression basin,
2. Plain depression (without alterations of surface level).

These two types of deformation also the cause two types of negative influence that can destroy the groundwater-withdrawal areas and the technical equipment of the waterworks, as shown in Figures 2 to 10.

2.2 ALTERATION OF HYDROGEOLOGICAL CONDITIONS IN DEPRESSION BASINS

We will focus on the question of the influence of undermining on geohydrodynamic systems and surface water, especially in relation to possible effects on quality and quantity of the water retained for mass or individual supply, and in relation to the measures concerning the delimitation of protection zones of water resources. Two situations can be envisaged.

No alterations:
- ⇨ sources of contamination,
- ⇨ types of pollutants,
- ⇨ transport medium – water.

Altered or influenced:
- ⇨ conditions for transfer of pollutants from the source to transport path,
- ⇨ hydrogeochemical character of groundwater (impact of flow field changes, changes of saturation zone thickness),
- ⇨ Intensity of self-cleaning processes in the rock and soil environment (changes of residence time of groundwater, changes of aeration zone thickness).

Two basic types of deformation can be distinguish in the depression basins:

- tensile and compressive deformation in the area of depression basin boundary,
- plain depression (minimal alteration in levelling of surface).

These deformations are the causes of:

- **The stability failure of water-withdrawal objects** (wells, surface objects), damage to their support, tightness etc.
- **The decrease of groundwater level** that can lead to total dewatering of aquifer as a results of hydraulically active draw-cut fissures
- In case of unconfined aquifers, **the decrease of the covering protection layer thickness** - in flooded areas total destruction of water-withdrawal area (endorheic depression)
- **The occurrence of the drainage endorheic depression,** that in turns means:
  - ♦ Flooded terrain depressions - occurring above mine workings and worked-out stopes.
  - ♦ The frequent filling with waste rock or other waste material, utilized as mud pits (a potential source of contamination to groundwater, but also surface water).
  - ♦ The water is drained off the mud pits by the system of artificially built channels mouthing into surface streams. The channels are usually insulated against natural aquifer systems, but in the course of their operation and as a result of other mining influences such insulation can fail.

The following illustrations, in Figure 2 to 10, document the influences of undermining in the Ostrava-Karviná coal mining district.

Figure 2. Destruction of church in Karviná as a result of very rapid subsidence (the main phase of mining influences took place within one week – 1964

Figure 3. Depression basin – groundwater level arose on the ground surface (Orlová –1956).

Figure 4. Depression basin – groundwater level arose on the ground surface (Orlová–1956). Apparent denivelation of individual objects – total subsidence over 20 m.

Figure 5. Flooded areas in the central parts of undermined area (Karviná – about 1960)

Figure 6. The last attempt to save the house in the central part of depression basin (Karviná –1960)

Figure 7. Church of St. Peter from Alcantra (between Karviná a Orlová) - year 1935. The church with original tower situated on the hill.

Church of St. Peter from Alcantra
Karviná,
Czech Republic

subsidence / year

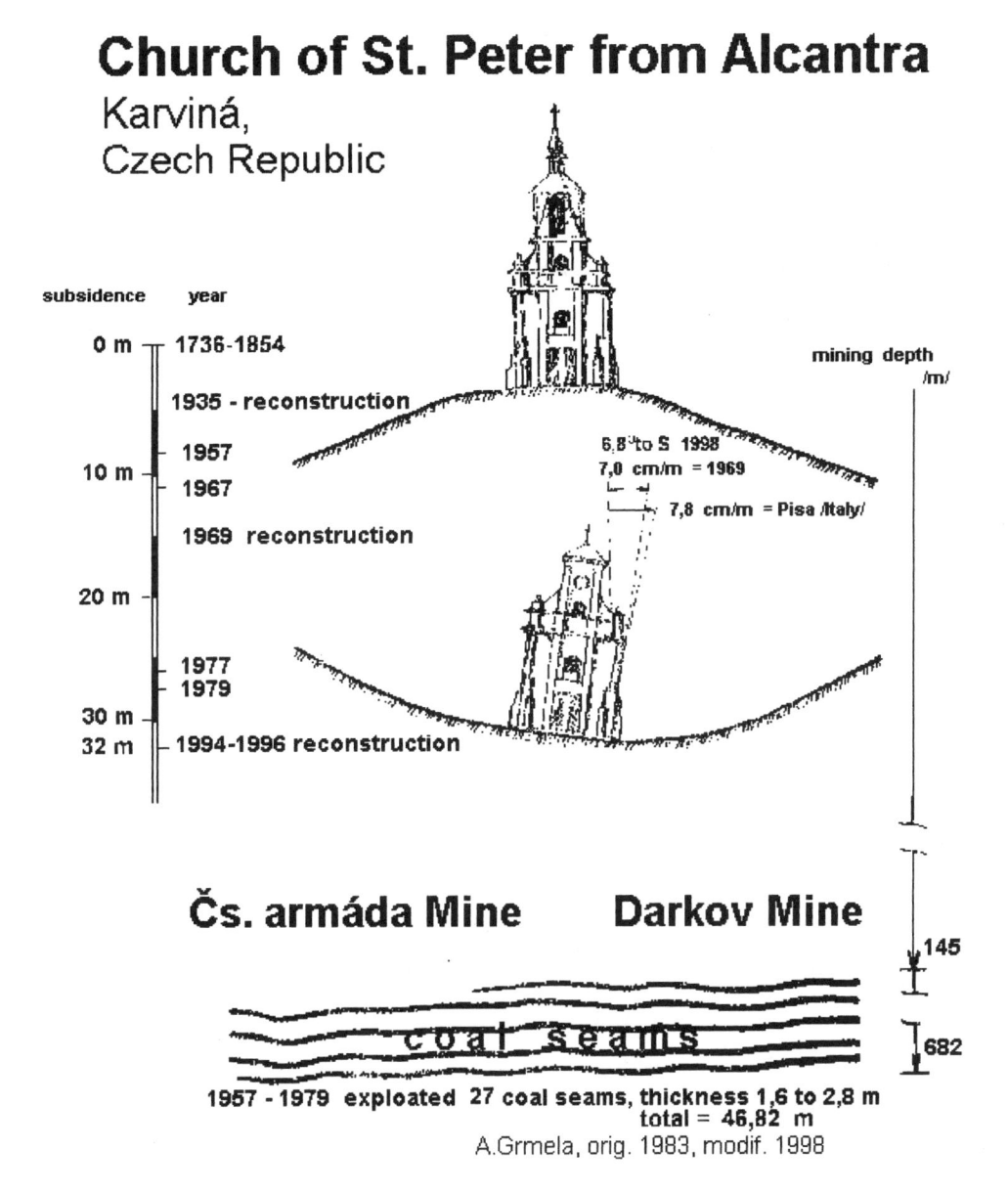

- 0 m — 1736-1854
- 1935 - reconstruction
- 1957
- 10 m — 1967
- 1969 reconstruction
- 20 m
- 1977
- 1979
- 30 m
- 32 m — 1994-1996 reconstruction

mining depth /m/

6,8 to S 1998
7,0 cm/m = 1969
7,8 cm/m = Pisa /Italy/

Čs. armáda Mine   **Darkov Mine**

145

coal seams

682

1957 - 1979 exploated 27 coal seams, thickness 1,6 to 2,8 m
total = 46,82 m

A.Grmela, orig. 1983, modif. 1998

Figure 8. The scheme of subsidence of the church in time

Figure 9. Church of St. Peter from Alcantra -1956

Figure 10. Church of St. Peter from Alcantra -1979

2.3 PASSIVE WATER RESOURCES PROTECTION MEASURES IN MINING AREAS

Mining activity may represent a negative influence on groundwater quality and quantity in water-withdrawal areas. The following effects may be expected:

- Effects of permanent influence or destruction of water resource
  - ❖ undermining reduces the protective layer of the aeration zone; water level may rise above the surface in flooded areas,
  - ❖ alterations of hydraulic gradients that cause changes of flow direction and velocity,
  - ❖ damage or destruction of water-withdrawal equipment (both surface and underground objects),
- Temporary effects
  - ❖ turbidity of groundwater due to rock-burst effects.

The effectiveness of groundwater resource protection can be influenced by:

- The selection of exploitation method (including the method of liquidation of worked -out space); and
- The plan of development, preparation and exploitation of a deposit

A key role of natural conditions in a successful resource protection was recognized in two sets of factors:

- The geomechanical properties of interstrata rocks; and
- The thickness and geological structure of inter-strata.

To plan water resource protection measures it is necessary to consider the priorities of the mineral deposit - water resource relationship in the light of the needs of both the state and the region.

2.4. MONITORING OF WATER RESOURCE IN MINING AREAS

The intensity of the effects of mining influences varies significantly in time. This should be taken into account in measures taken in the zones of groundwater protection. The measures should be assessed and determined by experts (e.g., by hydrogeologists-mining experts), on the basis of the evaluation of the results of monitoring the influence on a water resource. When determining a protection zone, the measures for the time-space protection of water resources should not be fixed to their expected service life. They must be flexible towards the changes of mining-technical conditions and mining methods hitherto results of the monitoring of mining influences, and so on. Their determination and revision should be carried out as a part of every mine plan (or after each change of it, which could involve the water resource, or alter hydraulic conditions of the aquifer system).

The monitoring of mining influences on a water resource must be carried out in two forms:

⇒ **geodetical monitoring** -observation of absolute depressions of network points in a water-withdrawal area
observation of permanent alterations of groundwater levels,
Geodetic monitoring must be in 6 to 12 month intervals and data must include objects with groundwater level monitoring.

⇒ **hydrogeological monitoring**

a) hydrogeochemical -observation of alterations of selected parameters of water quality.
Hydrogeochemical monitoring must be in 6-12 month intervals.
b) hydrological -observation of alterations of permanent changes of ground-water level and their evaluation,
-it is necessary to take corrections on natural regime measurements.
Hydrological monitoring must be in weeks or months intervals.

The assessment of hydraulic changes as a result of terrain subsidence can be done by the means of mathematical modelling. We apply the method in the following steps:

1. hydraulic model 2D - steady state, confined aquifer with leakage,
2. complementary hydrogeologic exploration - drilling of monitoring boreholes, hydrodynamic tests, geophysical measurements - tracer tests verifying of boundary conditions,
3. 2D hydraulic model – transient and 2D transport model,
4. changes of model geometry according to expected subsidence to predict changes in groundwater flow field and predict flooded areas. No calibration is possible due to use of predicted geometry.

The problems faced during modelling works are:

- Uncertainty in alterations of terrain
    - depends on prognoses and plans of exploitation
    - changeable under various economic-political situation
- Precision of aerial photogrammetry - considerable differences from geodetic levelling of monitoring wells.
- Problems with modelling present and expected future massif geometry.

3. Conclusion

All the **abovementioned problems, connected with the protection of both groundwater and surface water in the zones of influence from exploitation activity,** must follow from the solution of the basic question of priority [1]:

water source ↔ raw material

From the viewpoint of regional or national needs, mineral raw materials (ores, coal) can usually be rated as **irretrievable** - if not at once systematically extracted, they are **permanently lost for utilization by man.**

In some regions the category "**irretrievable**" also includes the **used** or **potential sources of groundwater** (drinking, mineral or thermal).

Retrievable sources of groundwater used for water management purposes can be influenced by mining activity. However, there are a number of possibilities to reduce the influence so that water intake remains feasible for as long as possible (by influencing the time horizon of eventual termination of utilization of the source).

References

1. Grmela A. (1997) : *Protection of Groundwater Resources Quality and Quantity in Mining Areas.* 6. mezinárodní sympózium „Mine Planning and Equipment Selection" VŠB-TU Ostrava, sborník ref. ss.865-872 ISBN 9054109157,. Ostrava 3.-9. září. 1997
2. Grmela, A., Tylčer, J. (1998) : *Problems of Groundwater and Environmental Protection in connection with Abandoning of Mining Activities in Czech Republic.* VII. International Mine Water Association Congress Johannesburg, Southern Africa.

PART 6. OTHER CONTRIBUTED PAPERS

The papers in Part 6 of this volume were selected among the many contributions of the participants to the four scientific sessions held during the NATO ASI.

The papers include: a discussion by Drew, in his after-dinner speech, of human attitudes to hiding waste; an overview by Karimov and Gainutdinova of the radioactive contamination in the Kyrgyz Republic; an operational PC-based coal fire hazard monitoring system using RS and GIS by Vekerdy; the use of geochemical methods in monitoring the impact of mining in streams and soils in Sweden by Szücs and others; the environmental impact due to mining in Slovakia andCroatia by Trtíková and by Čović and others; and finally, an assessment of land degradation due to mining in the Zambian copper belt using RS and GIS by Limpitlaw.

IRRESISTIBLE HOLES IN THE GROUND

Text of Afterdinner Speech Given at Mátraháza, Hungary, on September 10, 1998

L.J. DREW
U.S. Geological Survey
Mail Stop 954
National Center
Reston, Virginia 20192

I would like to continue my technical lecture on how the continental crust fractures under far-field tectonic stress, but my stress-strain ellipsoid transparencies that are so necessary to this discussion are missing from my lecture notes. I suspect that they have been stolen by someone in this room.

I judge from your laughter that you do not believe me. Actually, the topic of my talk is the philosophy of environmentalism with an emphasis on mining. During the past 7 years, I have written 25 columns for *Nonrenewable Resources* on the topic of environmentalism and the production of nonrenewable resources. First, I will discuss this subject in the context of these articles and then focus on the one that forms the basis of this talk.

There are two end-member positions in environmentalism today. One is ecocentrism, which is growing rapidly, and the other is anthropocentrism, which is older and now static. Ecocentrists are the new neomalthusians—they are pessimistic about man and his motives. But they have added a significant twist—they perceive man as being out of control. They now generally conclude that the consequences of mining outweigh its benefits. Similarly, they believe that the consequences of building dams outweigh their benefits and that, in general, the consequences of economic growth outweigh its benefits.

As an image, I offer you one of the central ideas of the ecocentrists—man has rejected his position in the ecosystem. He looked at his thumb and opposing fingers and examined his brain and openly embraced economic growth and trusted technology for his salvation. The ecocentrists recoil, saying that man has made an incorrect assumption. In summary, it can be said that the motives of individuals always function to produce environmental catastrophes.

Listen to some of their words and make your own summation. For example,
"We must live more lightly on the land."
"The ecosystem cannot be whole if men are allowed to own land."
"Rivers must be wild and free."
"Man has never really been a steward."
"Multiple land use is a myth of the business community."
"All the waste sinks are clogged."
"Man is irreversibly disturbing ecosystems.

"He must be removed from the equation of nature."

"The human population must be reduced by half and soon."

What, then, do the ecocentrists want society to do? To dematerialize itself—to substitute walking for riding in automobiles and music making for steel making. And their most aggressive idea—to depopulate the world.

Perhaps we can say that a pessimism has descended over the landscape of the human mind.

The anthropocentrists advocate a more-traditional view. They put man above other living things and recognize personal liberty—human liberty is paramount. And, therefore, they argue that man can manage nature to his and its advantage and use it as a source of human wealth. The anthropocentrists understand the importance of the creation of wealth. To that end, they recognize that the investment multiplier in the production of minerals is high; that is, a unit of investment in the mining sector produces a large effect on a country's gross national product (GNP).

Technically, metals are very important to the world economy. They make it possible to do the things that we want to do in efficient and durable ways. Durability is a key factor in our demand for metals. As materials, they have replaced most of, if not all, the materials that preceded them solely because of their durability. Similarly, we can speak of the efficiency of the fuels as we have progressed from using wood to coal to petroleum.

But, and this is a big "but," there is a problem. Many metal producers have historically avoided paying the full cost of their products. They have not dealt well with the externalities that they have created in the production process. Pollution is an externality, a waste that producers have visited upon someone else and often someone in the future. Once you make pollution, it never goes away—it just gets pushed around.

Who should pay those costs? Why, the users, of course. And who are the users? You and I. We should pay for the utility we gain by using metals. We have to insist that the producers take part of the price of their products to clean up past damage and to prevent, by means of engineering designs, the creation of pollution as part of the process of producing metals.

And the solution for the developed countries is not to make a mining colony (economic) of a lesser developed country or to allow such a country to do it to itself in the name of expanding its GNP.

Let me then say this, to the ecocentrists, the business of protecting the environment is not about using nature to produce wealth and money. Instead, it is about the business of ethics. And these ethics have come out of the religion of ecocentrism. Yes, I see ecocentrism as a religion that worships Nature and the biologic order. And in fact, I am now working on a piece that will be entitled something like "Mysticism— The Religious Basis for Ecocentrism."

To the anthropocentrists, the business of protecting the environment is about paying the full cost of using the Earth and leaving nature in good "managed" working order. For example, the anthropocentrists would let us have a rock quarry if we created no pollution and presented a design that demonstrated how the full cost was going to be paid. We ought to be able to show how the pit could be converted to some further use (for example, water reservoir) after quarry operations cease. The ecocentrists, however, would rather we not quarry the rock because we will to use it to build a road, the consequences of which are more automobiles on the carrying capacity of the environment.

In summary, we have an innate sense of the anthropocentrist position—a clean environment (no externalities), abundant wildlife (diversity), and respectful management of the land (nature gets a fair portion). As soon as I say this, however, I want to add that I sense a falling away from the rational economic ethics of yesterday to the advancing religious ethics of ecocentrism.

The foregoing discussion has been taken from the columns that I have written for *Nonrenewable Resources* during the past 7 years. At the request of the Directors of this NATO ASI, the rest of my talk this evening has been taken from a single column. Those of you who are writers have learned to deal with the praise or the silence that follows you after your work is published. One of the Directors told me that he liked this column so much that he has copied and circulated it widely; he even gave his mother-in-law a copy.

Sometimes, such praise, I think, is overdone and perhaps a carrot to gain a speaker for the risky business of speaking after dinner. I prefer to speak before lunch, to be the first or second speaker in the morning session of a professional meeting. But, then, let us accept this praise at face value and analyze why it might have been given.

For this purpose, we can pretend that we are students in a literature class. Our professor has handed us this column, "Irresistible Holes in the Ground," and asked us to summarize its content and structure as if it were a scene from a play. He guides our analysis by asking a question, "What is the writer trying to do with the form and substance of his words?"

Our first observation is that the writer claims to be an eyewitness. Stories told in this way are easy for readers to identify with; thus, this is an often-used literary form.

This story has the classic motif of light versus dark with its parallel in good versus evil. The light-versus-dark motif is a time-honored device in storytelling and has formed the basis for a genre of film-*film noir*.

The first character that we are introduced to is the heavy. He is Doug, an official from an environmental regulatory agency. We know his name is Doug because when I asked him his name, he said, "Doug. Just call me 'Doug.'" Next, there are two innocents, the writer, who is an honest man, and the miner, who is a good man. The fourth character is the villain, a shadowy man with something to hide.

Why have you become so quiet? This is only an 1100-word story about a conversation between Doug and the writer. Well, I will continue to the action—an accidental conversation on an airplane to pass the time. Here, again, the reader encounters a literary device—the chance encounter that sets the stage for intrigue. Alfred Hitchcock used this, the McGuffin, in most of his films. Hitchcock films are noted for the use of this device—the most famous being *Strangers on Train*. Or perhaps you remember the innocently switched suitcase in Polanski's *Frantic*? Once the mistake has been was made, the main character descends into a netherworld.

This may be the point at which I should tell the story. It went like this. A large man like myself, Doug squeezed his way into the middle seat on a plane bound for Washington's Dulles Airport. I had been assigned the window seat. He was nervous and could not find a niche for his elbow. His smile was weak, and he was obviously tired. I asked him if he was from the Washington area and what his line of work might be. He answered that he worked for Federal environmental regulatory agency. He had had a bad day because he was employed by the enforcement division of his agency. He said that people do not like his agency or him very much. Doug was not sure that he wanted to talk at first, but changed his mind after he heard that I work for the U.S. Geological Survey.

"You know about holes in the ground," he asked. I said that I did, thinking about open-pit mines. He next stated that holes in the ground are irresistible, a phrase that he would repeat several times.

Curious, I inquired what he might be talking about. Doug then explained about men, shadowy men, who look at holes in the ground excavated by mining men very differently than he suspected that I might look at them. He said that there are men who have things to hide, and when nobody is looking, they acquire a hole in the ground into which they put bad things, often barrels of unused paint, solvents, and other harmful chemicals, as well as tree trunks and associated organic matter. Again, he repeated that holes in the ground are irresistible. I tried to add that mining causes pollution because piles of mine waste and tailings are left. Doug thought this a minor problem compared with the bad stuff that went back into the holes and then, in time, into the ground water, rivers, lakes, and ocean.

That was Doug's story, and the writer cast it as a conflict between the regulator and the regulated. When he goes to work, Doug takes the Clean Water Act with him; sometimes, he takes sworn officers—"they carry firearms," Doug said.

The climax of our story comes when the innocent miner is taken advantage of by the shadowy man with something to hide. The ethical point is made that the shadowy man is clever in his dishonesty. The literary point is that the villain uses a small problem to cover up a large problem. In other words, the shadowy man uses a hole dug by a miner (the small problem) to hide harmful or potentially harmful "things" (the large problem).

As with any good story, there is a lesson to be learned, and this story is no exception. The next time you see a hole in the ground, look around for a villain lurking in the shadows.

ENVIRONMENTAL SECURITY AND RADIOACTIVE CONTAMINATION

K.A. KARIMOV and R.D. GAINUTDINOVA
Institute of Physics National Academy of Sciences
265-A Chui Prosp., Bishkek 720071
The Kyrgyz Republic

Abstract. The problem of environmental security is connected to changes in the natural environment as a result of increased stress and the incidence and consequences of extreme events. One of the major problems of environmental security in The Kyrgyz Republic is radiation security connected with conditions of safe storage and destruction of uranium tailings.

The preliminary results of a radiation survey are presented from an investigation of the public health status in mountain regions of The Kyrgyz Republic, near densely populated areas and ground water reservoirs.

1. Introduction

The problem of environmental security is connected to changes in the natural environment as a result of increased stress and the incidence and consequences of extreme events. Short- and long-term environmental changes may affect security—national, community, and human. Adaptation mechanisms of the community to environmental changes and human security include water security, food security, and energy security [1].

One of the major problems of environmental security in The Kyrgyz Republic is radiation security connected with conditions of safe storage and destruction of uranium tailings. Uranium tailings and radioactive atmospheric fallout are the main sources of radioactive contamination of the environment. Radiation factors in the environment of are caused, in general, by natural (rock structures and radon emanation) and technogenic sources connected with output and processing of uranium and atmospheric fallout of radionuclides.

2. A View at the Radiation Situation in The Kyrgyz Republic

2.1. SAFETY CONSERVATION OF URANIUM TAILINGS AND TRANSBOUNDARY IMPACT

2.1.1. *General Situation*
The Kyrgyz Republic is situated in Central Asia in the Tian Shan mountain system, which has a high level of seismic activity. The natural radioactive background in the republic ranges from 9 to 25 mcR/h, about the same as in other countries. In the former USSR, the average was 25.5

mcR/h, in the United States, variations of gamma background range from 8 to18 mcR/h [2]. In general, radiation background in The Kyrgyz Republic is within the limits of ecological safety except for local areas near uranium tailings, natural anomalies, and areas exposed to contamination by artificial radionuclides in atmospheric fallout. Observations of radon emanation are practically absent; in places where it has been found to collect, however, it is a real threat to human health.

For many years, mining and metallurgy were very important to the economics of the republic. The environment was negatively impacted in spite of adhering to standards responsible for conservation and safety. About 3,700 ha was exposed to the harmful impact of mining, and 43 Mm^3 of mining wastes was created [3], which may be the source of contamination of air, water, and soil.

At present, some tailings have been assessed by experts as a direct threat to human health and safety. It is impossible now to estimate the exact scale of the problem connected with radioactive contamination near uranium tailings.

2.1.2. *Ecologically Unfavorable Regions*

The most dangerous of these past mining sites have been identified in the National Environmental Action Plan of The Kyrgyz Republic [3]:

- The Kara-Balta group of enterprises—penetration of toxins into the ground water, radioactive dust, feasible accident probability and potential worsening of human health, and impact on the ecosystem;
- The former Min-Kush uranium mine—presence of radioactive wastes without control in a dangerous seismic area that could flood. Because the mine is situated in the basin of the river and storage lake, there is a potential impact on the health of the population;
- The tailings of the former Tujuk-Suu uranium mine—a dump of radioactive wastes, including of harmful chemicals and reagents; danger of wastes penetration into the river by soil erosion; and
- The former Kaji-Sai uranium mine—located a short distance from Issyk Kul lake. There is a danger of contamination because of the absence of necessary controls on people using radioactive metal. This situation can present a dangerous influence on the health of children and the ecosystem.

In The Kyrgyz Republic, there are 44 tailings sites and 28 dumps, of which 68 percent are uranium tailings. The volume of accumulated waste, which consists of radioactive materials, salts of heavy metals, and toxic material, is 56 Mm^3. Among them, solid waste makes up 34 Mt with total activity of more than 88,000 Cu.

These tailings and dumps have a harmful impact on the environment. Many of them are situated in the depressions and hollows in areas with unfavorable climatic and seismic conditions [4]. Many of these tailings are situated a short distance from populated areas and present a threat in emergency situations (earthquakes, floods, landslides, etc.). Some of them are within populated areas, and others are along open reservoirs.

Geological and seismotectonic processes have caused large landslides, and unfavorable meteorological conditions have led to the destruction of tailings and dumps and the spreading of radioactive and toxic material outside of depositories. For example, in 1958 at Mailuu-Suu, there were deaths and the destruction of dwelling and civil objects. In 1964 at Aktuyz and in 1994 at Sumsar, there was the radioactive and toxic contamination of agricultural fields and the contamination of surface and ground water.

2.1.3. *Trans-boundary Transfer*

Tailings at Mailuu-Suu, Shekaftar, and Aktuyz are the most dangerous trans-boundary objects, containing radioactive and other material harmful to human health and the environment. People in densely populated towns and villages of Uzbekistan and Kazakhstan are under the threat of trans-boundary migration from these dangerous sources. Landslides led to emergency situations, the destruction of radioactive tailings, and trans-boundary contamination.

In Mailuu-Suu, radioactive wastes contained in 23 tailings sites have a total volume of 2 Mm^3 with a total mass of more than 4 Mt. According to some estimates the total activity of these tailings is about 50,000 Cu.

In the town of Mailuu-Suu, the background level of gamma-radiation is about 23 mcR/h, and on the surface of the tailings, it is 50 mcR/h. On parts of the dumps, the same level is from 360 to 440 mcR/h, and some of the tailings are in an unstable condition and could easily contaminate the environment. Unfavorable hydro-meteorological conditions could contaminate water in Mailuu-Suu River.

The catastrophic accident that took place in 1958 on the uranium tailings, which had a volume of 1.2 Mm^3, was caused by an intensive spring downpour. About 600,000 m^3 of radioactive wastes drained into the Mailuu-Suu River and reached densely populated regions of the Fergansky Valley. The results of the accident were many human victims and destruction of dwellings and industrial and civil buildings and contamination of a vast territory.

2.1.4. *Population Morbidity*

The health status of the population has not been investigated in the ecologically unfavorable regions of The Kyrgyz Republic where there are uranium tailings. Territories exposed to radioactive contamination are not controlled.

In 1995, the specialists of The Kyrgyz Institute of Oncology and Radiology examined 5,500 inhabitants of the town of Mailuu-Suu. The general results showed that more than two-thirds of adults are ill. Among this number, there were 36% with pre-tumor diseases, which is more than four times the same index in other regions of The Kyrgyz Republic. Population morbidity by different forms of cancer in the town exceeds the average republic index by more than a factor of three [5].

2.2. TRANSBOUNDARY ATMOSPHERIC TRANSFER OF RADIONUCLIDES

The long-living radionuclides Sr-90 and Cs-137 formed as a result of atmospheric fallout from a long series of nuclear tests that made a considerable contribution to the radiation contamination of The Kyrgyz Republic.

The Chinese nuclear tests carried out until 1996 exposed the territory of The Kyrgyz Republic to contamination by radionuclides as a result of tropospheric and local fallout [6,7].

In some periods, the excess of specific beta-activity level of atmospheric fallout was more than 100 times [2].

On the basis of radiometric measurements and investigations of atmospheric circulation, the atmospheric mechanisms of transfer of radioactive material to The Kyrgyz Republic after underground nuclear explosions at the Lopnor test site were determined [6,7]. Analysis of the dates of nuclear explosions showed that they were conducted during periods when eastern jet streams existed in the lower atmosphere. Those jet streams transferred explosion products to The Kyrgyz Republic and other Central Asian countries. The level of general radiation background in the eastern Kyrgyz Republic grew up by 40% after the explosion conducted on June 10, 1994 [7]. The ground gamma-spectrometric measurements in five energy channels in air samples showed the presence of isotopes of strontium, yttrium, and cesium. Analysis of soil samples determined the considerable concentrations of technogenic radioisotopes. Soil samples taken in the eastern Kyrgyz Republic before and after the explosion showed a threefold to fivefold increase in Cs-137 radioactivity, from 10 to 46 Bq/kg [6].

Prior to that, the Chernobyl accident made a contribution to the radioactive contamination of the environment in The Kyrgyz Republic. In May 1986, the sharp increase of the total month's beta activity of tropospheric fallout was more than 80 times that of May 1994 [2].

3. Conclusions

The problem of radiation security in The Kyrgyz Republic is connected with radioactive contamination of the environment by internal and external sources—uranium tailings and atmospheric fallout. Considerable contamination of the environment has been observed after catastrophic accidents involving uranium tailings, nuclear tests conducted in China, and the Chernobyl accident.

4. References

1. Karimov, K.A. and Gainutdinova, R.D. (1999) Environmental changes within Kyrgyzstan, Proceedings of the NATO ARW on Environmental Change, Adaptation and Security. Dordrecht, The Netherlands: Kluwer Academic Publishers.
2. Radiation and Mountains (1995) IPPNW-Kyrgyzstan, Bishkek.
3. National Environmental Action Plan of Kyrgyz Republic (1995). Bishkek, Print Company ((Kyrgyzstan-Turkey)).
4. Moldobekov, B.D. (1998) Activization of catastrophes in the territory of Kyrgyzstan and its impact on regional security. Human Rights and Environment, 12, 7-8.
5. Kamarli, Z.P. (1996) General and oncological morbidity in ecologically unfavorable regions of Kyrgyzstan. Proceedings of the Bishkek Seminar on Environmental Problems, Ankara, Turkey, EPF, 55-57.
6. Karimov, K.A. (1998) Programmes on radiation safety of the population and the environment, liquidation of the consequences of the radiation impact, Nuclear Tests: Proceedings of the NATO ARW on Long-Term Consequences ofNuclear Tests for the Environment and Population Health (Semipalatinskl Altai Case Study). Springer-Verlag, Berlin, Heidelberg, NATO ASI Series 2: Environment - Vol. 36, 2-5.
7. Karimov, K.A. and Gainutdinova, R.D. (1997) Some results of environmental monitoring in Kyrgyzstan: Atmospheric transfer of contaminants, Proceedings of the NA TO AR W on Integrated Approach to Environmental Data Management Systems. Dordrecht, The Netherlands, Kluwer Academic Publishers, NATO ASI Series 2: Environment - Vol. 31, 465-472.

A PC-BASED INFORMATION SYSTEM FOR THE MANAGEMENT AND MODELLING OF SUBSURFACE COAL FIRES IN MINING AREAS (COALMAN)

Z. VEKERDY
International Institute for Geo-Information Science and Earth Observation (ITC)
Hengelosestraat 99, P.O. Box 6
7500 AA Enschede, The Netherlands

Abstract. In coal mining areas subsurface coal fires waste the coal reserves, make mining difficult or impossible, endanger human lives and the exhaust gases damage the environment. Their location, development and size largely depend on the geological setting, which also determines the possible prevention and fighting methods. Information technology combined with environmental modelling can help the fight against the existing coal fires and support the prevention of fires in the:

0. Detection and mapping of coal fires.
1. Definition of areas at coal fire risk.
2. Setting priorities in coal fire fighting.
3. Definition of optimal fighting and prevention methods.

COALMAN is a coal-fire monitoring and management software, which was programmed using Visual Basic for the user interface and the database management procedures and using ILWIS (the Integrated Land and Water Information System, developed at ITC in The Netherlands: http://www.itc.nl/ILWIS) for the GIS and remote sensing functions. It comprises of the following main modules:

0. A database.
1. Standard database management functions.
2. GIS and image processing functions.
3. Special procedures and models.
4. Quantification of the parameters of coal fires from satellite images.
5. Mapping of coal fire risk and hazard.
6. Monitoring the development of coal fires and the results of fire fighting measures.

The primary users of COALMAN will be the fire fighting and prevention team of the Ningxia Province, in China. The development is carried out as a part of a project financed by the Dutch and the Chinese governments.

1. Introduction

In coal mining areas surface and subsurface coal fires waste the coal reserves, make mining difficult or impossible, endanger human lives and damage the environment with exhaust gases, landslides, etc. China is one of the richest countries in coal but coal fires seriously endanger her reserves in a belt from Northwest- to Northeast-China stretching 5000 km east-west and 750 km north-south [5]. The coal fire problem has recently been listed in the Chinese "21st Century Agenda" as one of the five most serious geological hazards [6]. Fires cause the loss of 100-200 million tons of coal in China annually. The CO_2 emission from these fires accounts for the 2-3% of the total global production [12]. Thus, the fight against the coal fires is of economical and environmental interest in the whole country.

In the following, a practical example is given of integrating data and models into an information system, "the Coal fire monitoring and management information system" (CoalMan), which is being developed for one of the most important coal mining areas of China in the Ningxia Hui Autonomous Region.

1.1 THE PHENOMENON: COAL FIRE

Coal fires are of two major origins: naturally ignited (generally by spontaneous combustion) and man-induced (generally due to improper mining techniques, negligent acts or technical mishaps like electric sparks). Spontaneous combustion occurs due to the accumulation of heat as a result of interaction between the oxygen in air with the coal. In the beginning of the process the interaction is slow and heat accumulates only if an external heat source is available too (e.g., solar heating). Beyond a certain elevated temperature, the so-called threshold temperature (between 50 and 100 degrees centigrade), the oxidation is accelerated due to a sustained exothermic reaction and ultimately produces flaming combustion or ignition [14]. The actual threshold temperature depends on the physical and chemical characteristics of the coal.

The location, development and size of the coal fires largely depend on the geological setting. Fires occur both on the surface and in the subsurface, but always close enough to the surface where oxygen is available. It is a general experience in the test area of the project here described that coal fires never extend further from the outcrop than 200 m along the coal seam, or in other words, the subsurface fires are never deeper than some tens of metres [3]. Thus, the discussion in the following has been kept limited to the shallow subsurface fires.

Fires usually start at or close to the surface: in the outcrop, as shown in Figure 1A, or in the wall of an open cast mine, shown in Figure 2, or in the shafts of the mines close to the surface. Fires spread along both directions of strike and dip of the coal seam (see Figures 1B and 1C). As coal burns to ash in an underground coal fire, the overlying rock at first cracks then collapses resulting in a land subsidence (see Figure 1C). The cracks provide ventilation for the fire (see Figure 3), so it can progress deeper. The progress might be limited by the groundwater, by lack of oxygen or geological features.

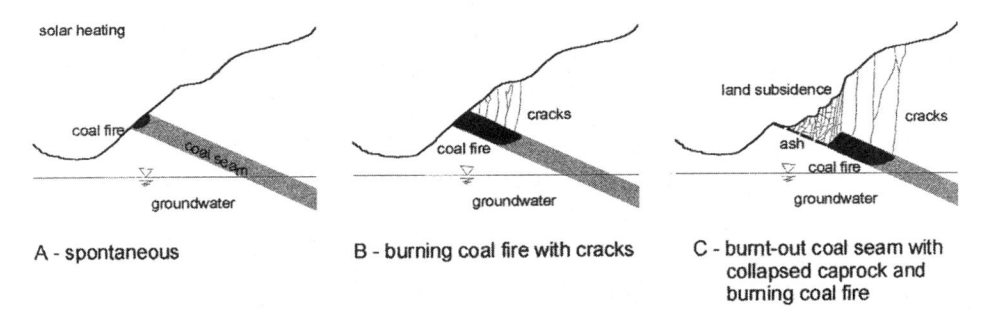

Figure 1 Schematic representation of the phases of coal fires.

Coal fires cause positive temperature anomalies on the ground surface. The closer the fire is located to the surface the less the heat diffusion is in the overburden and the higher the temperature is on the ground surface. The surface temperature anomalies are the basis for detecting the fires on remotely sensed images taken from air or space platforms. It is an intellectual and practical challenge to estimate the parameters of a fire (e.g., depth, amount of coal burning) from the distribution of the surface temperature by physical modelling. A good overview of the history of the efforts in these fields, i.e., the remote sensing based modelling of coal fires is given by [2].

It is not enough to know only the location of the fires. Fire fighting needs information about several other factors, which can help the selection of the proper prevention measures too. Perhaps the most important is the geological setting, which determines the optimal prevention and fighting methods. The available infrastructure and resources pose further constraints on the selection of the possible measures. Information about natural risk factors and mining activities is also needed in the delineation of the areas at coal fire related risk.

1.2 THE PROJECT

The Chinese and the Dutch governments launched a collaborative project to develop and implement a coal fire monitoring and fighting system in the Ningxia Hui Autonomous Region in China. Three Dutch contributors, the Environmental Analysis and Remote Sensing (EARS) Bv., the Netherlands Institute for Applied Earth Sciences (TNO–NITG) and the International Institute for Aerospace Survey and Earth Sciences (ITC) as well as a Chinese partner, the Beijing Remote Sensing Corporation (BRSC) work on the project. The project will be completed in the end of 1999.

Figure 2 Burning coal seam in an open cast mine

Figure 3 Steam breaks through the cracks above a coal fire at Rujigou in Ningxia, China. One of the frequently used fire fighting methods is the flooding of the fire with sewage from the settlements.

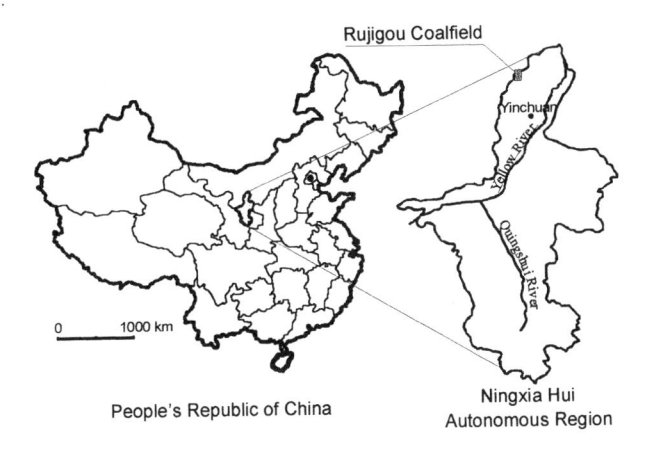

Figure 4 Location of the Rujigou Coalfield

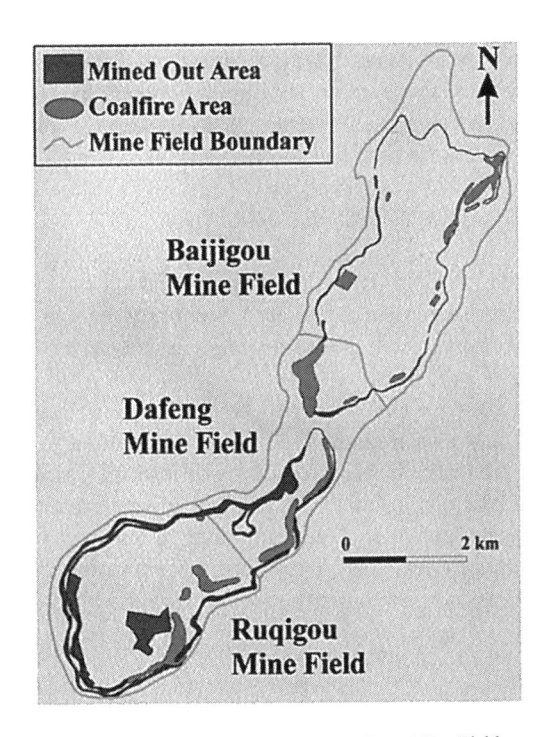

Figure 5. Coal fires in the Rujigou Mine Field

The project has two major objectives:

- To develop methods for coal fire fighting and prevention in the Rujigou Coalfield.
- To develop a GIS-based tool for the monitoring and management of the fight against coal fires.

In the following, the discussion concentrates on how the second objective of the project is met. The examples are given from the Rujigou coalfield, which is located in Northwest China, as shown in Figure 4, in the semi-arid Helan Mountains.

In the Rujigou coalfield the first coal fires started approximately a century ago and in 1997 seventeen coal fires were reported (see Figure 5): 7 in the northern part (Baiqigou Area), 5 in the central part (Dafeng Area) and 5 in the southern part (Rujigou Area).

2. The management tool: CoalMan

A personal computer based monitoring and management information system (CoalMan) is being developed for coal fires. The user interface of it is written in Visual Basic. The system combines GIS and remote sensing functions (using the software 'Integrated Land and Water Information System', ILWIS: http://www.itc.nl/ILWIS), a tabular database (in MS Access format) and some special coal fire tools (written in Delphi).

2.1 OBJECTIVES AND USERS OF COALMAN

The main objective of CoalMan is to integrate data from various sources (geological information, remote sensing images, etc.) and based on the data to allow the user to generate answers to various questions, e.g., about hazard, extent, mitigation and fighting priorities of coal fires.

The primary users of CoalMan will be the Fire Fighting and Prevention Team of Ningxia Hui Autonomous Region. The system is planned to be set up in 1999 in the central office of the team. It will provide basic information about the state of the coal fires and tools for planning of activities for the decision-makers.

The 'actual operators' of CoalMan have to have some background about fire fighting and computer operations. They need some training in using the system, e.g., to carry out data processing and to use the special analysis tools.

2.2 MAIN MODULES OF COALMAN

The main modules of CoalMan are the *database*, the *general tools* and the *coal fire analysis and evaluation tools*. All the modules are integrated in an easy-to-use graphical user interface.

CoalMan handles data from data entry to the presentation of analysis results. The data flow in CoalMan is shown in Figure 6.

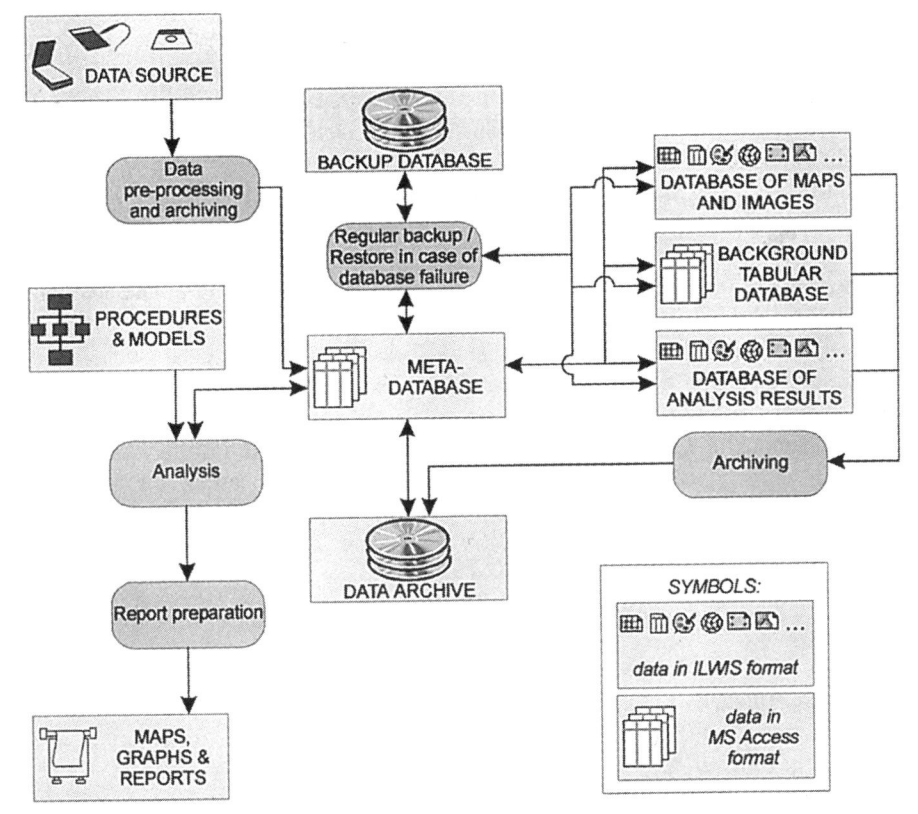

Figure 6. Data flow in CoalMan

2.2.1 *The database*
The database is subdivided into a map and image database, a background tabular database and a database of analysis results (see Figure 6).

The map and image database contains the following data types:

- maps (geology, geomorphology, topography, etc.),
- digital elevation models,
- satellite images (Landsat TM including night-time thermal images, SPOT, IRS-1C, ERS),
- airborne thermal infrared images, aerial photographs,
- thermal inframetric images, and
- field photographs.

The background tabular database stores the attribute data for the maps and other ancillary information like data of field measurements and observations (surface and subsurface temperature, mining data, coal samples, laboratory results, etc.).

Practical reasons make it necessary to handle the processing results separately. This database contains coal fire maps with their attribute tables, which are partly results of standardised analyses and partly results of special processing.

All the data mentioned above are stored on the computer's hard disk to be accessible any time, whilst the data archive is linked to the database off-line on CD-ROMs. Besides the original raw satellite images, parts of the archive are those data which are not used frequently so that it is not necessary to store them on the computer's hard disk.

A regular backup facility is available in CoalMan to avoid data loss in case of hardware and software problems. The backup is also stored on CD-ROMs.

All the data are registered in the meta-database, which plays a central role in the data management of CoalMan. It stores information about the location of each data object and some quality parameters. Location information is important because the directory structure of the database is complex, and its complexity is just growing as new data are added. Quality parameters help the assessment of the reliability of the analysis results.

2.2.2 *General tools*

Tools are available in CoalMan for general data management, for the maintenance of the database integrity and for general GIS analysis and remote sensing image processing.

- Standard database management functions.
  The user is able to enter, manage and analyse tabular data.
- Database integrity management functions.
  These functions help the user to maintain the integrity of the tabular and the map database and to register all the data objects in the meta-database. Only the data registered in the meta-database are available for the coal fire analysis tools.
- GIS and image processing functions.
  COALMAN provides functions for input, display and analysis of maps and remote sensing images with their attribute data. Although the full functionality of ILWIS is available, the proper use of them needs special knowledge about GIS and image processing. Therefore, besides the access to the full functionality of ILWIS, some tailored special procedures and models are available too for the less trained users.
- Presentation and output functions.
  These functions are used for generating formatted hard copies of the maps and reports. In fact a part of them can be considered as GIS functions (i.e., the annotation of maps), and another part of them are standard functions of MS Windows.

2.2.3 *Coal fire analysis and evaluation tools*

These tools are developed to be used in the day-to-day activities of the fire fighting team. All of them are tailored for the needs of the fire fighters and do not require special GIS or numerical modelling knowledge:

- Quantification of the parameters of coal fires from satellite images.
  The most important parameters are the size and depth of coal fires and the amount of coal burnt. The determination of them is based on pre-processing a selected part of the satellite image (thermal band of Landsat TM) and then applying numerical models. The user interface of the coal fire detection tool developed by EARS Bv. is shown in Figure 7.

- Mapping of coal fire risk and hazard.
  Map algebra is used for mapping the risk of spontaneous combustion. The system also shows some examples on how numerical modelling can be used for mapping the spread of air pollution caused by the coal fires. This module under development in the time of the writing of the present paper.

- Monitoring the development of coal fires and the results of fire fighting measures.
  Time series analysis tools support the monitoring. The basic input is the results of the analyses of single remote sensing images. The development and the movement of the fires can be monitored as well as the success of the fire fighting measures.

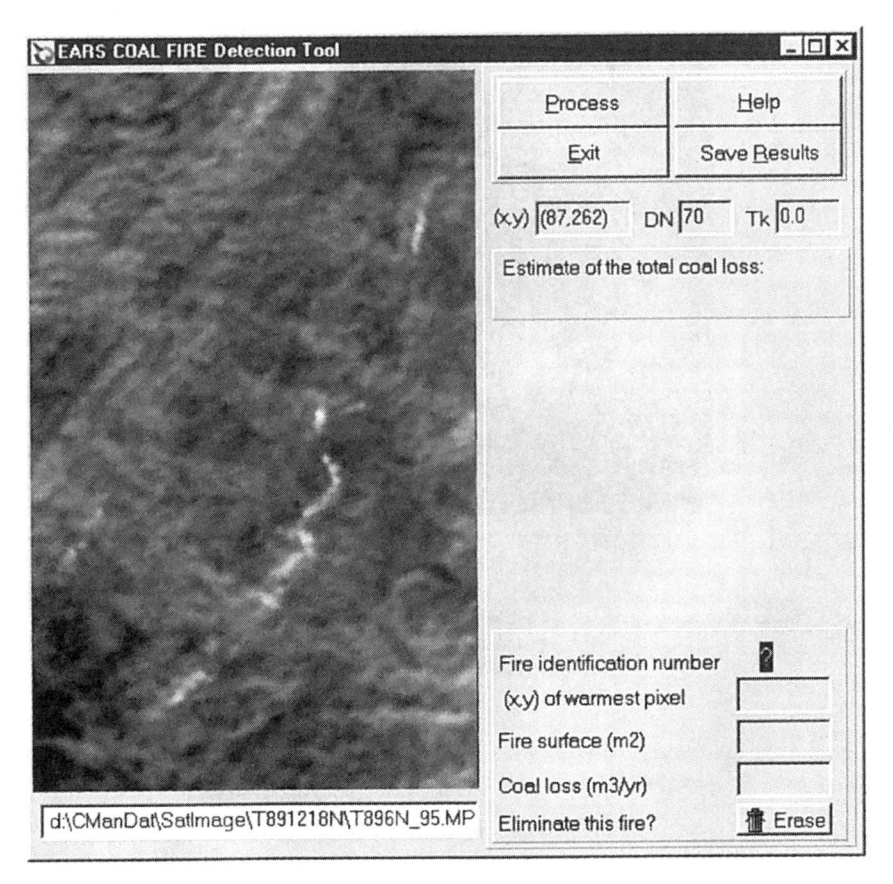

Figure 7. The user interface of the Coal Fire Detection Tool of CoalMan.

418

Special functions are needed to handle and display 3D data in CoalMan. The basic data structure is two-dimensional (since CoalMan uses ILWIS as the GIS software), but the information about the depth of the coal fires is important for the fire fighters. Therefore, special procedures are worked out to handle a series of two-dimensional maps with elevation attributes of the geological layers to represent the three-dimensional geological set-up. Cross-sections can be generated about the coal seams, as exemplified in Figures 8 and 9. The accuracy of the representation is limited to the accuracy of the individual maps, which are compiled from bore log data, using the interpolation functions of ILWIS.

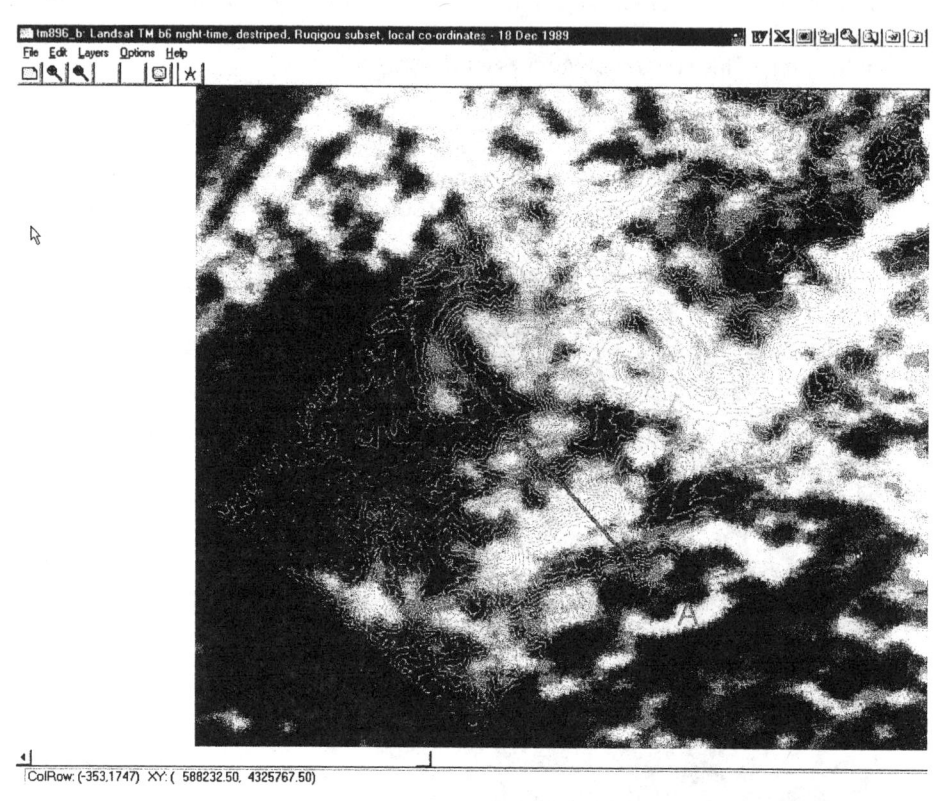

Figure 8. Selection of a geological section of coal seams in CoalMan. The user can define any – not necessarily straight – line with a thermal satellite image and the contour lines in the background. The section along the line defined in this figure is shown in Figure 9.

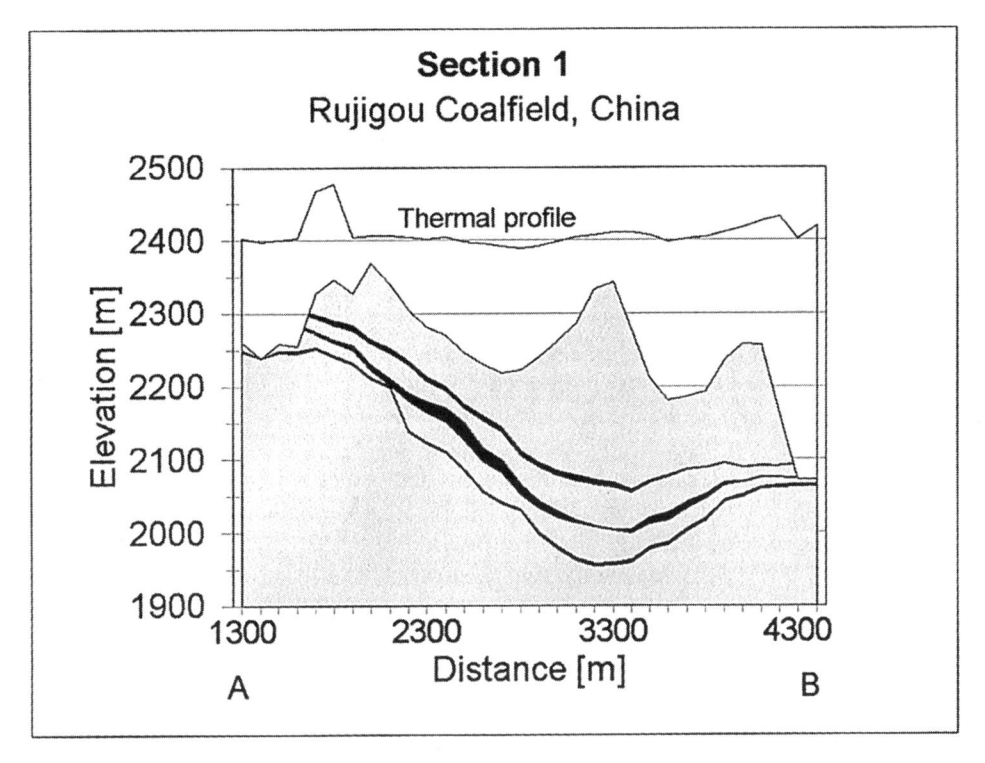

Figure 9 Geological section generated in CoalMan along a user-selected line A-B (see Figure 8)

3. Discussion

Geoinformation technology combined with environmental modelling can help the fight against the existing coal fires and support the prevention of fires in the following fields:

- Detection and mapping of coal fires.
- Definition of areas at coal fire related risks.
- Setting priorities in coal fire fighting.
- Definition of optimal fighting and prevention methods.
  These fields are discussed in detail in the following subsections.

3.1 DETECTION AND MAPPING OF COAL FIRES

Subsurface coal fires cause thermal anomalies on the surface, so thermal remote sensing is a good tool for the detection and mapping of the spatial distribution of the coal fires. Cassells [1998] [2] reports that airborne techniques have been first applied by [13], [7] and [4] in the US. Satellite techniques were widely used in the eighties and nineties – among others – in Indian coalfields by [1] and [11].

The size of a developed coal fire varies from some tens of metres to several kilometres. Successful application of NOAA images with 1 km resolution was not widely reported (e.g., [8]), but Landsat TM band 6 images with the 120 m resolution can provide a good synoptic overview.

Landsat TM images form the backbone of the monitoring of the coal fires using CoalMan. It is planned that the images are purchased annually to map the coal fire situation. Comparison of the analysis results from different dates gives an insight into the dynamics of the fires [10], as illustrated in Figure 10. Unfortunately, these images are not adequate for detecting the new, just emerging fires with small size or relatively low temperature. Limitations are also present in the monitoring of the coal fire dynamics if higher accuracy is needed than a few tens of metres. Therefore, CoalMan provides possibilities to include future high resolution satellite data as well as airborne thermal scanner data with higher resolution (in the range of a few metres) for the monitoring of the coal fires.

If simple coal fire related information is needed (e.g., the location of the coal fire), simple methods can be used for the analysis (e.g., density slicing), but for deducing more complex information (e.g., depth of the fire) physically based numerical models are needed. These are provided in the coal fire analysis tools of CoalMan.

3.2 DEFINITION OF AREAS AT COAL FIRE RELATED RISKS

The most important risk types are the following:

- risk of the occurrence of the coal fire,
- environmental risks caused by the coal fires,
- risk of damages in infrastructure and human lives.

Based on the maps of factors related to the different risks geoinformation technology provides tools (overlaying of maps, cartographic modelling, digital elevation modelling) for the risk analysis. CoalMan combines maps of coal fires, mining activity, coal seams, infrastructure and other factors for determining the coal fire related risks. The result is a qualitative risk map, although several input data can be quantitative. The conversion between the quantitative and the qualitative data is done with classification, where the user's knowledge or the in-built knowledge base of the system play an important role in defining the classes.

CoalMan includes a demonstration of the potentials of air pollution modelling for the definition of environmental risks caused by the coal fires. Such models need more detailed information about meteorological conditions than is currently available about the Rujigou Coalfield.

Radar interferometry is an emerging field of remote sensing for the detection of changes in the elevation of land surface. An experiment was carried out for the Rujigou Coalfield to map the land subsidences in the area [9]. Operational application of this method is still not fully possible, but the results proved its potentials. This promising technology could be used in later versions of CoalMan for the monitoring of land subsidences threatening the infrastructure.

421

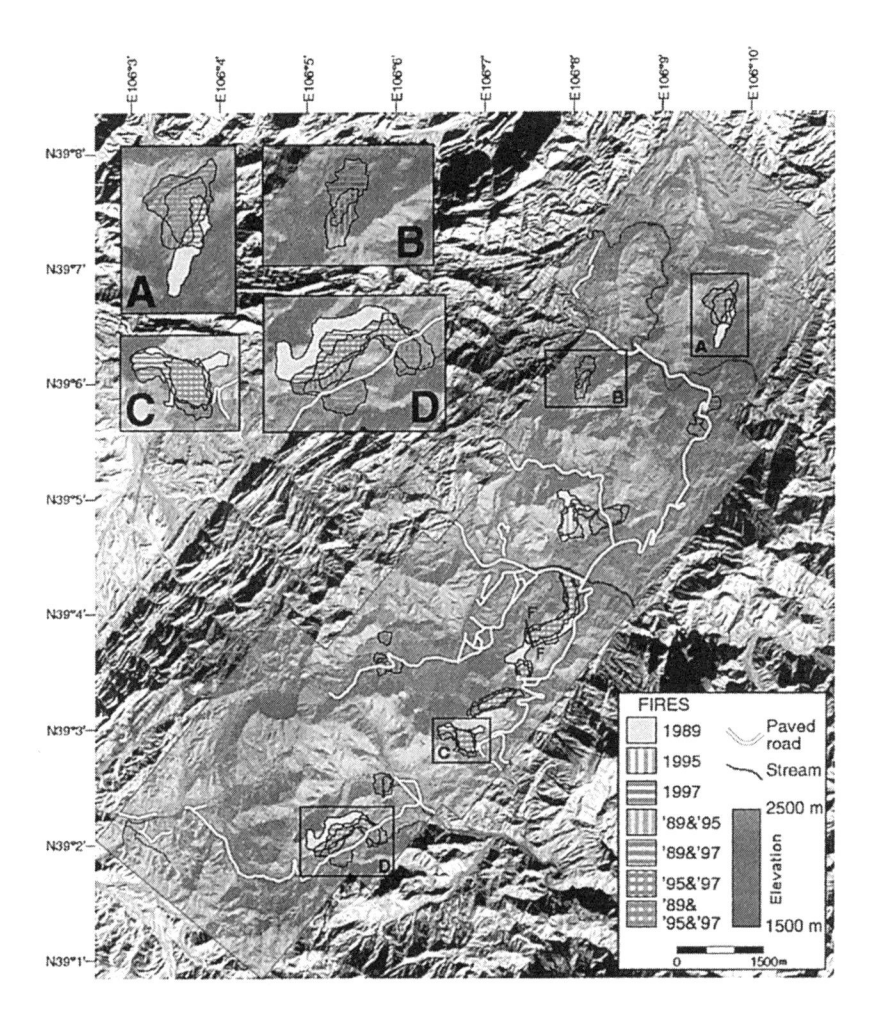

Figure 10. Representation of coal fire dynamics. The background is an IRS panchromatic image, which is overlain with the colour coded digital elevation model and the outlines of fires interpreted from Landsat TM thermal images. The small boxes (A-D) enhance different cases of fire movements. [10]

3.3 SETTING PRIORITIES IN COAL FIRE FIGHTING

Coal fire fighting is a complex and expensive measure. Optimal use of the usually limited resources requires setting priorities, depending on the tasks, risks and resources involved. Thus, interactivity in the priority setting is important: the user has to be able to analyse different scenarios. The result of the analysis is only a support for the decision-maker – it is always risky to fully rely on these results and to neglect non-computerised local experience. The fire fighting priority determination models of CoalMan are under development. The algorithms will be finalised in co-operation with the Fire Fighting and Prevention Department of the Ningxia Hui Autonomous Region.

3.4 DEFINITION OF OPTIMAL FIGHTING AND PREVENTION METHODS

The optimal fighting and prevention methods depend on the parameters of the fire, the geological and topographic circumstances, as well as the available infrastructure and resources. Geoinformation technology can provide background information for selecting the optimal method, but – as was shown in the case of setting the fire fighting priorities – the decision-maker has to evaluate whether all the important site-specific factors have been accounted for by the computer. The implementation of the decision support for the definition of the optimal fighting and prevention methods in CoalMan is planned for the second half of 1999.

4. Conclusions

Geographic information technology provides tools for efficient data management and handling in the monitoring of coal fires. Anyhow, tools without data are useless. It was experienced in the project described above that approximately 70–80 percent of the efforts had to be invested in data preparation and database set-up. Although this is a well-known fact among database experts, the importance of data preparation is usually underestimated in several applications.

The quality of the data determines the possible quality of the modelling results. Data quality information is stored in the meta-database. It is not possible to know the quality requirements of all the applications of monitoring systems in advance, therefore, the system has to provide interactive possibilities to set up different scenarios using data of different quality or reliability. Meta-information has to accompany the scenarios in reporting.

It is argued that besides the technical implications of the geographical information systems the organisational aspects are of the same – if not greater – importance. Replacing the classical information flow with a computer-based system needs adaptation of the organisation to stronger rigor in data handling. On the other hand the GIS has to fit the existing data flow as much as possible. This needs careful tailoring of the system in co-operation with the users.

5. Acknowledgements

The development of CoalMan is carried out as a part of a project financed by the Dutch and the Chinese governments. ITC provided additional funds for the coal fire research, supporting – among others – a comprehensive fieldwork.

When writing about CoalMan it is the author's privilege to acknowledge the other members of the ITC team: Professor Dr John van Genderen is the leader of all the coal fire research in ITC. Dr. Anupma Prakash was responsible for remote sensing. Dr Zhang Jianmin was delegated by BRSC to be trained on CoalMan, but did much more than a usual trainee. Associate Professor Wang Feng is the main programmer and Guan Fang was the digital elevation mapping expert. Dr. Rüdiger Gens contributed with data fusion and interferometric research.

The co-operating partners are to be acknowledged too: the EARS Bv. worked out the coal fire modelling, TNO–NITG worked on the geological aspect and BRSC provided the data and the technical and logistical support during the field work. Special thanks are due to Professor Dr Guan Haiyan, the "Coal Fire Man of China", for his leadership and support.

References

1. Bhattacharya, A., C.S.S. Reddy and T. Mukherjee, 1991: Multi-tier remote sensing data analysis for coal fire mapping in Jharia coalfield of Bihar, India. *Proceedings of the Twelfth Asian Conference on Remote Sensing*, Singapore, 30th October–5th November 1991 (Singapore: National University of Singapore), pp. 22-1–22-6.
2. Cassells, C.J.S., 1998: Thermal modelling of underground coal fires in Northern China. *PhD Thesis*, University of Dundee, Scotland.
3. Cui B.L., 1999: Head of the Fire Fighting Department of the Ningxia Hui Autonomous Region, personal communication.
4. Greene, G.W., R.M. Moxham and A.H. Harvey, 1969: Aerial infrared surveys and borehole temperature measurements of coal mine fires in Pennsylvania. *Proceedings of the Sixth International Symposium on Remote Sensing of the Environment*, 13th–16th October 1969 (Ann Arbor, Michigan: University of Michigan), pp. 517–525.
5. Guan, H.Y, 1989: Applications of remote sensing techniques in coal geology, *Acta Geologica Sinica*, Vol.2, No.3, pp. 254:269.
6. Guan, H.Y, J.L. van Genderen and H.J.W.G. Schalke, 1996: Study and survey on the geological hazards of coal fire in North China. *Abstracts 30th International Geological Congress*, Vol. 1, p. 458.
7. Knuth, W.M., 1968: Using an airborne infrared imaging system to locate subsurface coal fires in culm banks. *Proceedings of the Pennsylvania Academy of Science*, 42 p.
8. Mansor, S.B., A.P. Cracknell, B.V. Shilin, and V.I. Gornyi, 1994: Monitoring of underground coal fires using thermal infrared data. *International Journal of Remote Sensing*, Vol. 15, pp. 1675–1685.
9. Prakash, A., E.J. Fielding, R. Gens, J.L. van Genderen and D.L. Evans, 1999a: Data fusion for investigating land subsidence and coalfire hazards in a coal mining area. *International Journal of Remote Sensing* (submitted).
10. Prakash, A., R. Gens and Z. Vekerdy, 1999b: Monitoring coal fires using multi-temporal night-time thermal images in a coalfield in North-west China. *International Journal of Remote Sensing* (accepted for publication).
11. Prakash A., R.P. Gupta, A.K. Saraf, 1997: A Landsat TM based comparative study of surface and subsurface fires in the Jharia coalfield, India. *International Journal of Remote Sensing* Vol. 18, No. 11, pp. 2463-2469
12. Rosema, A., Genderen, J. L. van, Schalke, H. J. W. G. and Beijing Remote Sensing Corporation (BRSC), 1995, *Environmental Monitoring of Coal Fires in North China. Project Identification Mission Report October 1993*. Beleidscommissie Remote Sensing (BCRS) report 93-29 (Delft: BCRS), 25 p.

13. Slavecki, R.J., 1964: Detection and location of subsurface coal fires, *Proceedings of the Third Symposium on Remote Sensing of the Environment*, 14th–16th October 1964 (Ann Arbor, Michigan: University of Michigan), pp. 537–547.
14. Zhang, X.M, 1998: *Coal fires in Northwest China, detection, monitoring and prediction using remote sensing data*. PhD thesis, Technical University Delft, the Netherlands. Also published as ITC Publication No. 58, ITC, Enschede, the Netherlands, 136 p.

INTEGRATED MODELLING OF ACID MINE DRAINAGE IMPACT ON A WETLAND STREAM USING LANDSCAPE GEOCHEMISTRY, GIS TECHNOLOGY AND STATISTICAL METHODS

A. SZÜCS[1,2], G. JORDÁN[1,2,3] and U. QVARFORT[2]
[1] *Geological Institute of Hungary, Hungarian Geological Survey, 14 Stefánia, 1143 Budapest, Hungary*
[2] *Department of Earth Sciences, Uppsala University, Norbyvägen 18B, S-752 36 Uppsala, Sweden*
[3] *Environment Institute, Joint Research Centre of the European Commission, TP 280, 21020 Ispra, Italy*

Abstract. The attenuation of Cu, Fe, Mn, Ni, Pb and Zn originating from acidic ore mine leachate is studied in a natural wetland stream environment in central Sweden. A sequential chemical extraction procedure is used to investigate fractions that are expected to act as potential sinks of the six metals studied in the stream sediment. Geochemical abundances, geochemical gradients and geochemical flow patterns are analysed and modelled and the stream sediments are interpreted as an oxidising landscape geochemical barrier. Sampling locations and geochemical barriers are identified using landscape geochemical methods and GIS techniques. For data modelling robust statistical methods of Exploratory Data Analysis are used to treat small sample sizes with multi-modal character and outlying values. The spatial variability of metal retention in the stream sediments is studied by multivariate data analysis methods.

Results of data analysis show that stream sediments act as a complex oxidising-adsorption barrier and the heterogeneity of the geochemical barrier is controlled by redox gradients in the sediments, which can be sufficiently characterised by the distribution of Fe fractions. Data analysis suggests that adsorption and co-precipitation with Fe oxy-hydroxide are major processes beside the adsorption on organic matter. Mn is probably specifically adsorbed on Fe oxy-hydroxides, and beside Zn, it is least retained in the sediment. Pb, Cu and Ni are found in considerable quantity in the reducible fraction and are suggested to occur occluded in Fe oxy-hydroxides. On the other hand, organic matter provides important adsorption sites for Cu and Pb and controls exchangeable metals, too. Based on enrichment factor calculation and correlation analysis in the pore water and the oxide-bound fractions Ni, Cu and Zn are thought to represent the effects of ore mining.

1. Introduction

Oxidation of ore mine wastes produces a great amount of sulphuric acid and releases heavy metals into the hydrological system resulting in acidification of surface and subsurface waters and deterioration of ecosystems. The geochemistry of acid mine drainage (AMD) has been studied extensively particularly in mining areas where the acid-neutralising capacity of the bedrock is limited [9, 22, 52].

Pyrite, the most common source of SO$_4$ in mine waste, is stable under wide pressure, temperature and pH conditions. However, upon exposure to air by ore mining, iron is liberated and controlled predominantly by redox processes. The oxidation of pyrite is initiated by aerial oxygen according to equations (1) and (2), [39]. Under very acidic conditions (pH<3.5), however, Fe^{3+} becomes the dominant oxidant, and the oxidation of pyrite is catalysed by obligate aerobic chemolithotrophic bacteria, such as the *Thiobacillus ferrooxidans* [49], as described in equation (3).

$$FeS_2 + 7/2O_2 + H_2O \rightarrow Fe^{2+}_{(aq)} + 2SO_4^{2-}_{(aq)} + 2H^+ \tag{1}$$

$$Fe^{2+}_{(aq)} + 1/2O_2 + 2H^+ \rightarrow Fe^{3+}_{(aq)} + H_2O \tag{2}$$

$$FeS_2 + 14Fe^{3+}_{(aq)} + 8H_2O \rightarrow 15Fe^{2+}_{(aq)} + 2SO_4^{2-}_{(aq)} + 16H^+ \tag{3}$$

As a result of these reactions soluble hydrated sulphates are formed first, and then, due to continued oxidation and the rapid hydrolysis of Fe^{3+}, less soluble iron oxyhydroxide minerals [2, 47]. Ochreous colloidal precipitates described generally as Fe(OH)$_3$ are often observed on top of sediments in streams and lakes receiving mine effluents. These are formed by rapid precipitation from Fe-rich waters usually at pH>5 and are poorly crystalline [7]. During the chemical evolution of AMD, as it is indicated in Figure 1, iron oxides and hydroxides have great importance in the retention of trace metals and other compounds on their surfaces by adsorption and co-precipitation [28, 50]. Although iron oxides can exhibit ion exchange properties, they are positively charged in most natural systems (e.g., pH$_{pzc}$=8.5 for amorphous Fe(OH)$_3$; [51], and bound trace metals most frequently by specific adsorption.

Among the various sampling media, stream sediments have been widely used in applied geochemistry to investigate element distributions in surface environments (e. g., [12]; [14]; [36]). Stream sediments are a dynamic sampling medium, which in our case represents the geochemistry of nearby landscapes in terms of both the natural geochemical heterogeneity, and the pollution caused by mining. The abundant clays, Al, Mn, and Fe oxides, and humic material in stream sediments collectively provide a large reactive surface area for the effective control of dissolved metal content of aquatic systems, therefore they can be regarded as a geochemical barrier. In this study the attenuation of six metals Pb, Ni, Cu, Zn, Fe and Mn originating from abandoned mines and waste rock dumps is investigated in the stream sediments at Slättberg, in central Sweden, where acid mine leachate has been discharging for over 70 years into the wetland stream studied in this research. Stream sediments, as a potential geochemical barrier, are analysed using a sequential chemical extraction in order to investigate the control mechanisms of metal retention in the bottom sediments of the stream affected.

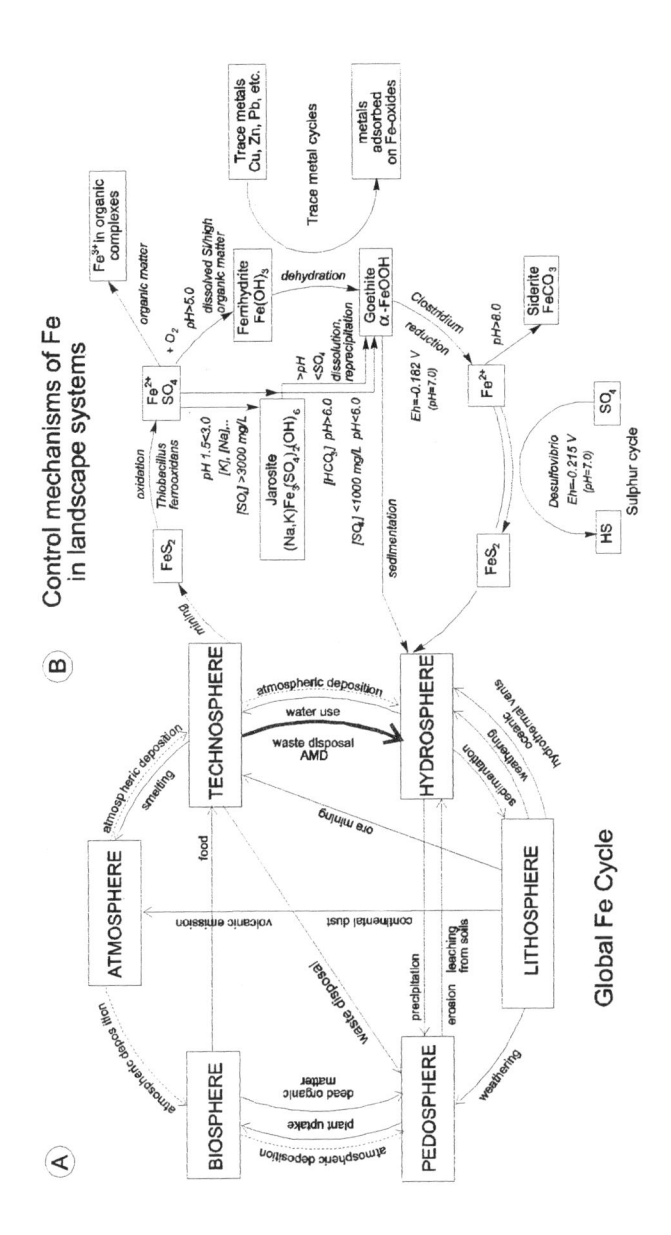

Figure 1. Global cycle of iron and matter flow control processes. A. Qualitative description of global iron cycle. B. Control mechanisms of iron in landscape systems with emphasis on the fate of iron in AMD, indicated by bold arrow in A. (Compiled after [8] and [47]).

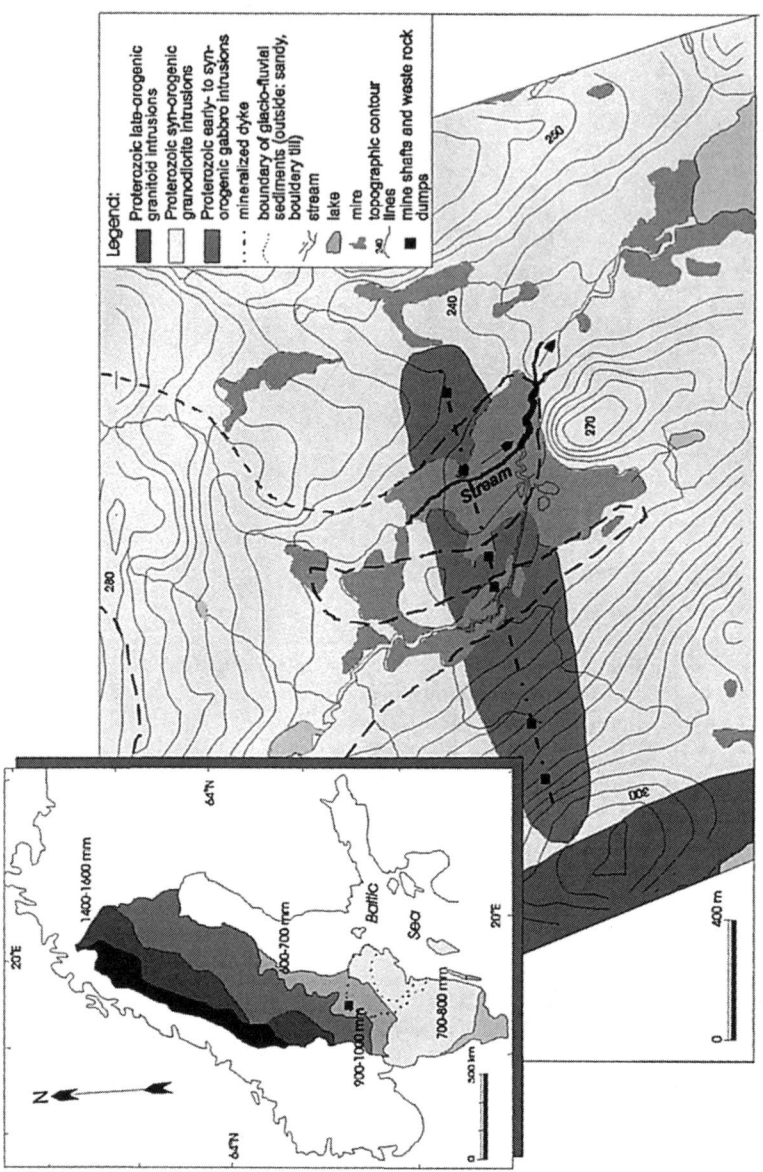

Figure 2. Location of the study area at Slättberg. *Inset*: Solid box indicates the location of the area in Sweden (60° 48'N, 15° 15'E). Shading shows physiographic zones. The study area is situated in the southern boreal zone of Scandinavia (medium grey shading) (figures show average annual precipitation). The study area falls into the region of the Svecofennian island arc lithologies of Early Proterozoic age (dotted line). *Large map*: Map is the result of GIS overlay operations of topographical, hydrological, geological and quaternary geological layers. The investigated stream segment is shown in bold.

2. Location of the study area and site description

The location of the study area in central Sweden is shown in Figure 2. The geology of the area is characterised by island-arc igneous rocks of the Early Proterozoic Svecofennian orogeny [21]. The main geological feature of the area is an approximately 1.6 km long and six metre wide mineralised amphibolite dyke in a WSW-ENE orientation, which is intruded into a hornblende bearing granitic gneiss host rock [54]. The ore association is characterised by pyrrhotite ($Fe_{1-x}S$), pyrite (FeS_2), pentlandite (($Fe,Ni)_9S_2$) with bravoite (($Fe,Ni)S_2$), millerite (NiS) and linneite (($Co,Ni)_3S_4$) and chalcopyrite ($CuFeS_2$), [35]; [60]). As a result of metasomatic processes the dyke has been strongly altered and feldspar is replaced to a great extent by sericite, chlorite and epidote. In the stream valley, the bedrock is covered by Quaternary gravely-sandy glaciofluvial deposits of 6 to 15 metres in thickness and the surrounding area is dominated by bouldery-sandy moraine. The study area is situated in the southern boreal physiographic zone of Scandinavia (see Figure 2). Vegetation of the area consists of pine and spruce forest with mosses and heather.

Six abandoned waste rock dumps and mine shafts are situated along the dyke upstream to the stream segment studied, as shown in Figure 2. Wastes derived from the excavation of the tunnels and shafts have been dumped directly adjacent to the mines, covering a surface area of about 1300 m^2 with a thickness between one to two meters each. Mining of copper, iron and nickel ores commenced in 1805 and ceased during the 1940s. However, the bulk of the waste was produced in the course of several years from 1917. Metals are leached from waste rock dumps into groundwater in high concentrations (Fe, 1110 mg/L; Cu, 40 mg/L; Zn, 18.5 mg/L; Ni, 32.1 mg/L; Pb, 0.29 mg/L; Mn, 4.4 mg/L; and SO_4, 3750 mg/L; [25], and are discharged into excavated ditches and the stream studied in this research. Down gradient to the mire the stream widens into a fen, a shallow stagnant water. Both in the ditches and in the fen ochreous precipitates are locally observable on the sediment surface.

3. Methods

In order to describe the speciation of trace metals in the solid phase, a sequential extraction procedure based on the method of [55] was applied. Four fractions were analysed to investigate the partitioning of trace metals in the stream sediments: *1.* exchangeable metals and metals bound to carbonates; *2.* metals bound to Fe/Mn hydroxides; *3.* metals bound to organic matter and amorphous sulphide; and *4.* residual metals. In this study, geochemical processes were studied as stochastic processes and statistical tools were used for data analysis. Univariate exploratory data analysis (EDA) techniques [26, 58], followed by multivariate statistical analyses, such as cluster and Q-mode principal component analysis, were used to investigate element distribution patterns, geochemical abundances and gradients in the sediment and identify sample populations. Correlation analysis and canonical correlations were applied to investigate significant relationships among the variables measured. All correlations were visually controlled and robustness of regression analysis was assured by interactive outlier rejection. Throughout geochemical modelling the use of robust and non-parametric techniques were preferred to classical statistical methods because geochemical datasets are often characterised by small sample sizes, multi-modal populations, outliers and non-normality [32].

The investigation followed the geochemical modelling procedure shown in Figure 3. According to this scheme methods of landscape geochemistry and the spatial modelling tools of GIS were integrated to enhance sampling design. Fortescue (1984; 1992), [18,19], gives a detailed review of landscape geochemistry and its methods. In our methodology first areas of homogenous matter flow patterns, i.e., elementary landscapes, are identified and classified to characterise the basic geochemical processes of the landscape investigated and its relationships with neighbouring landscapes. Three fundamental elementary landscape types can be distinguished based on the relative position of the topographic surface and the water table. Where the water table is always below the topographic surface, an eluvial landscape characterised by downward salt movement in well-drained soils is formed. If the water table and the topographic surface coincide a super-aqual landscape (e.g., bog or marsh) appears, where the dominant pattern of salt movement is upward and horizontal. A lake or stream environment, where the water table is above the land surface, is termed an "aqual" landscape characterised by circular salt movement [24]. This information can be derived simply from the hydrology layer or by the use of groundwater level and topographic data. Landscape classes can be extended with the "trans-" prefix emphasising that the given landscape is characterised by through-flow. For example, on the basis of transport intensities (and hence differences in oxidation conditions) in surface waters active stream segments are classified as trans-aqual landscapes while lakes and stagnant waters are considered as aqual landscapes. Similarly, eluvial landscapes are subdivided as eluvial, trans-eluvial and eluvial-accumulative landscapes on hill tops where erosion and downward movement of elements dominates, on hill slopes where down-gradient through-flow dominates and in valleys which are characterised by deposition [24]. The delineation and classification of elementary landscapes were obtained with a GIS by the overlay of the hydrology layer and the slope break model derived from the digital elevation model (DEM). Then, relationships between landscapes are analysed by matter transport models, such as run-off and watershed models derived also from the DEM.

Elementary landscapes are used to define geochemical control elements, such as landscape geochemical barriers. These are classified based on matter transport directions at elementary landscape boundaries, and the geochemical character of landscapes determined by their acid neutralising capacity and redox conditions on either side of the barrier [42]. For example, in eluvial landscapes, characterised by podzol soils underlain by granitic glacial till, groundwater is likely to be oxidising, and in super-aqual mire landscapes it is reducing, while active stream segments are also oxidising environments. This information is added to the attributes of elementary landscape polygons. Then, the elementary landscape is combined with the run-off model using GIS technology in order to define flow direction at landscape boundaries. For instance, if the oxidising active stream segment receives weakly acidic (pH 3-6.5) reducing gley water from the neighbouring mire an A6-type oxidising barrier is likely to form in the receiving stream sediments characterised by the element association of Fe, Mn and Co [42]. If the water flow direction is just the reverse, where weakly acidic oxidising water from glacial eluvial landscapes discharges into the reducing gley mire environment, for example, a reducing gley barrier forms (C2-type) at which Cu, U and Mo are likely to accumulate [42]. In addition, adsorption is a major process for metal retardation in organic-rich peat sediments, hence complex barriers form. The landscape geochemical base map and cross-sections obtained in this way, [29], are used in the selection of sampling sites.

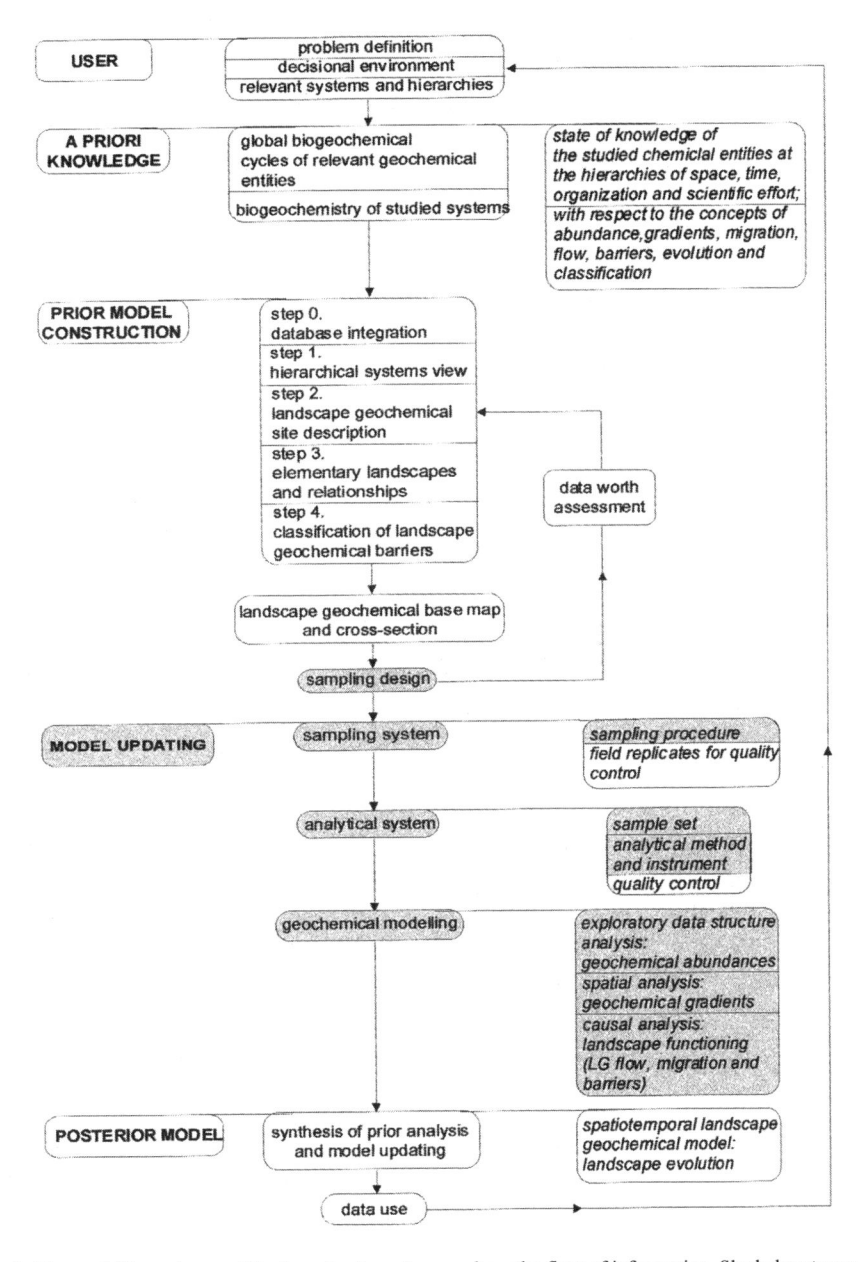

Figure 3. The modelling scheme of the investigation. Arrows show the flow of information. Shaded parts are discussed in detail in the text.

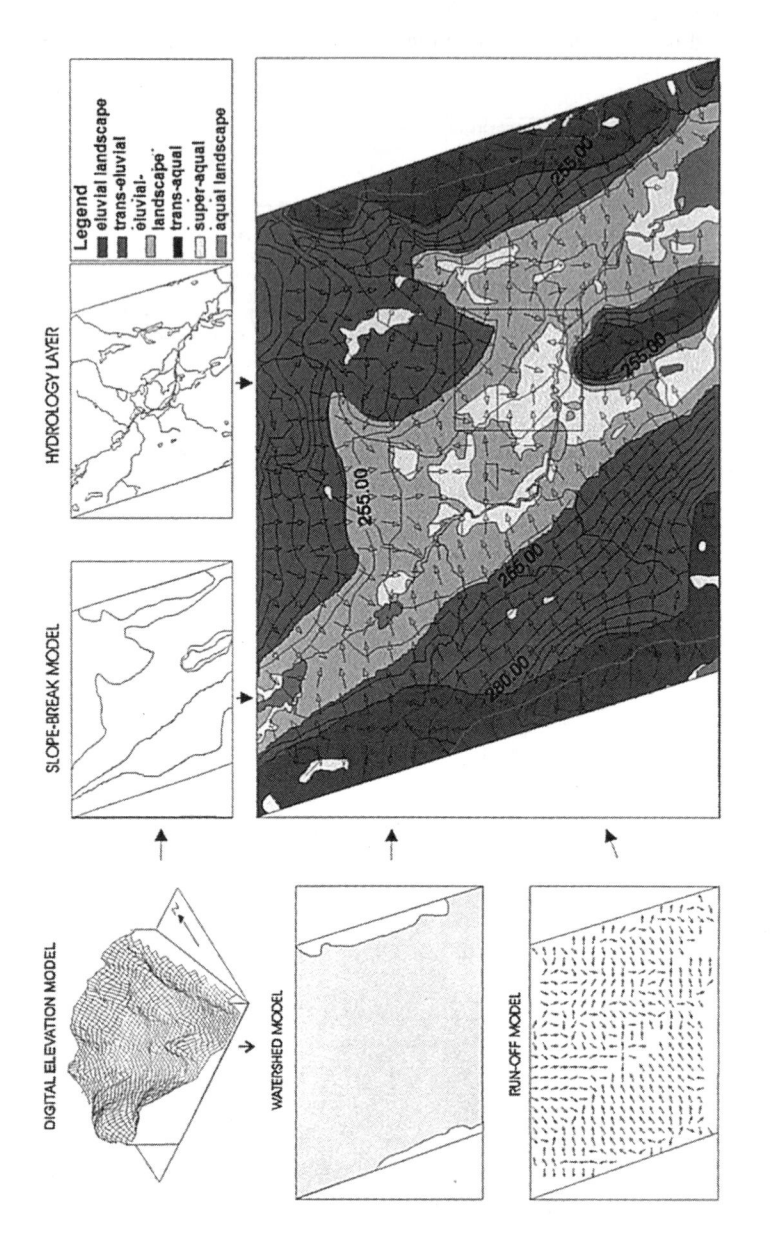

Figure 4. Derivation of landscape geochemical map using GIS techniques. Solid box indicates the area shown in Figs 5 and 7. See text for details.

4. Sampling design

Due to the importance of stream sediments in controlling trace metal attenuation in aquatic systems, the objective of our sampling design was to collect samples representative of processes at the oxidising barrier in the stream sediments, and to investigate their heterogeneity. Based on the landscape geochemical methods described above, the active stream channel was classified as a trans-aqual landscape, the stagnant fen water as an aqual landscape, and the surrounding water-saturated mire was assigned a super-aqual landscape. Using the DEM constructed by triangulated irregular network interpolation the watershed, slope- and slope-break, and the run-off models were derived, as shown in Figure 4. By the overlay of the classified elementary landscape map geochemical relationships between adjacent landscapes could be analysed and geochemical barriers were classified as shown in the landscape geochemical base map and cross section in Figures 5.A and 5.B. The wetland stream sediment was classified as a complex oxidising-adsorption barrier (A6-G6) according to Perel'man's classification scheme [42]. This forms when weakly acidic (pH 3-6.5) reducing gley water encounters relatively oxidising geochemical conditions, and is characterised by the retention of trace metals by adsorption on, or co-precipitation with, oxides and/or by organic matter chelation. However, considering the vertical redox profile likely to build up in the stream sediment column, the landscape geochemical model of trace metal retention in the upper surface sediments can be refined, as shown in Figure 5.C.

If we consider the vertical profile of the organic-rich stream sediment that receives AMD input from the overlying surface water and the discharging groundwater, depending on the redox profile developing in the sediment, three cases can be conceptualised.

The average thickness of peat sediment is 2-2.5 m in central Sweden, [20], and its hydraulic conductivity is in the range of 10^{-8}-10^{-10} m/s [46]. This is much lower compared to the average hydraulic conductivity of 10^{-6} m/s in glaciofluvial deposit [34], thus we can assume that the major flux of groundwater is supplied by the underlying glaciofluvial deposit. Hence, the groundwater remains relatively oxidising along the short flow-path of approximately 300 m from the waste rock dumps to its discharge. A vertical geochemical stratification is expected in the sediment according to the thermodynamic sequence of inorganic substance reduction [6]. In the minerogenic basal stream sediment dissolved oxygen is likely to be consumed quickly due to aerobic decomposition. On the other hand, NO_3^- is not a major constituent of the discharging mine water and is unlikely to act as a terminal electron acceptor. Because of their abundance, Fe and Mn carried by the groundwater will act as terminal electron acceptors [33], as dissolved oxygen is consumed, and the resulting dissolved reduced species will migrate upward by diffusion along concentration gradients. Provided that the redox potential gets sufficiently low (Eh <-250 mV) in the sediment column and decomposable organic matter is still presents, SO_4^{2-} is expected to be reduced by obligate anaerobic bacteria producing either dissolved sulphide or H_2S gas. Trace metals carried by groundwater are retained by either adsorption on the abundant organic matter or by metal sulphide formation in the zone of SO_4^{2-} reduction. Farther upwards in the sediment column, redox processes are more likely to be influenced by the redox conditions of the surface water. If the surface water is relatively oxidising, redox levels will increase progressively in the sediment column in a reversed succession, as it has been described above. Dissolution of reduced iron and manganese followed by upward migration will result in the precipitation of oxides and hydroxides at the sediment surface - water interface upon encountering the relatively oxidising surface conditions [53]. The term "double oxygen gradient" is used to refer to this first case, as shown in Figure 5.C.

In a second case, however, redox conditions in the sediment column might not always be appropriate for the conditions of SO_4^{2-} reduction and the zone of SO_4^{2-} reduction may be absent. Hence, instead of the double oxygen gradient a less pronounced single or low redox gradient develops (see Figure 5.C). This situation might occur at stream confluence, for example, where the stream is more erosive and sediment thickness is reduced. Also, when subsurface reducing gley water reaches the sediments from the neighbouring super-aqual mire landscape, metals propagate along the single oxygen gradient and precipitate or co-precipitate at the sediment - stream water interface.

In the third case the above model is slightly modified. The major difference is that surface water in the lowermost part of the water column can be relatively depleted in oxygen. In such cases a redox boundary layer is more likely to develop in the surface water column, rather than in the sediments, with the circulation of reduced and oxidised species between the boundary redox layer and the top sediment strata [27], (see Figure 5.C). This model could adequately describe processes in the stagnant open water of the marginal fen adjacent to the stream studied, as shown in Figure 5.A. It is noted, however, that the presence of microenvironments in the sediment can further complicate the geochemical setting.

Based on the objectives of sampling and prior landscape geochemical model, sediment-sampling locations were identified in the active stream channel downstream to the waste rock dumps, in the effluent ditches and in the fen.

5. Sampling and analysis

Twenty-three sediment samples were collected with a Plexiglas sediment-corer of 5 cm diameter in the spring (see Figure 5.A). Each sample represented a composite of the loose, upper 10-15 cm of the sediment column. At locations where the sediment was more compact and SO_4^{2-} reduction was evidenced by H_2S odour, the core was dissected into 5 cm sections and the uppermost and the lowermost sections (between 25 and 30 cm) were reserved for separate analysis. Sediment samples were extruded into polyethylene Ziploc® bags in the field and transported to the laboratory. Samples were stored at +4°C until laboratory analysis. Sediment samples for chemical analysis were sieved with a stainless steel sieve of 2 mm mesh size. The samples were thoroughly homogenised and 6 g test portions of each sample were weighed into centrifuge tubes. The selective chemical extraction procedure based on the method of Tessier *et al.* (1979), [55], was applied as described in Table 1.

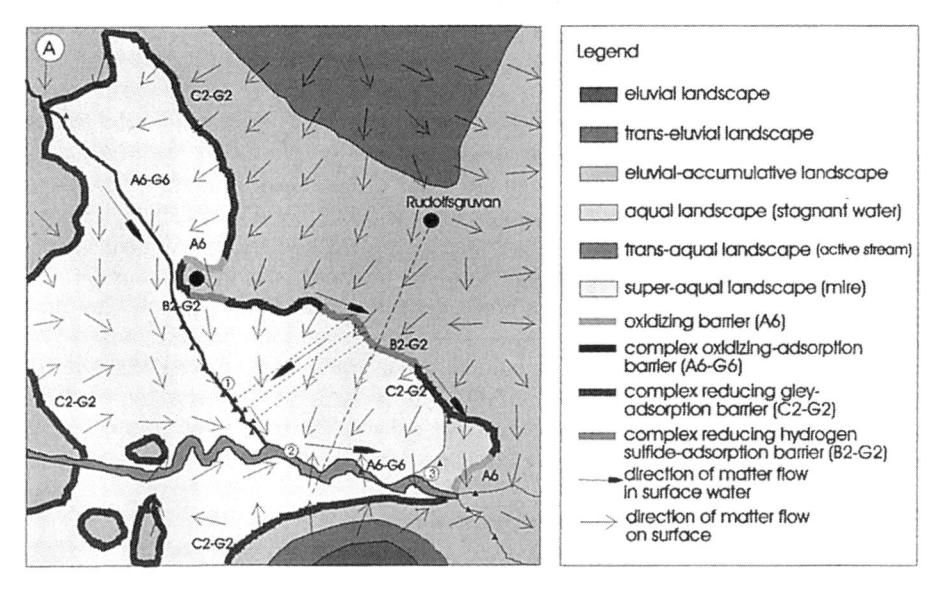

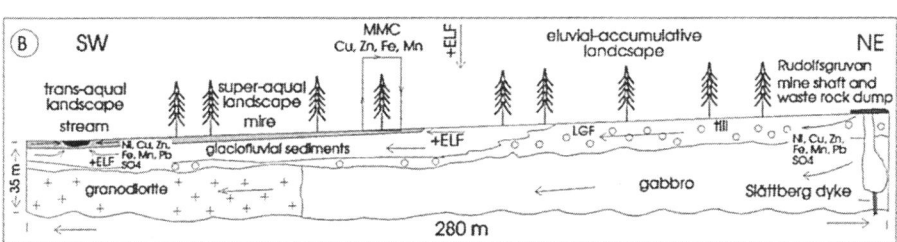

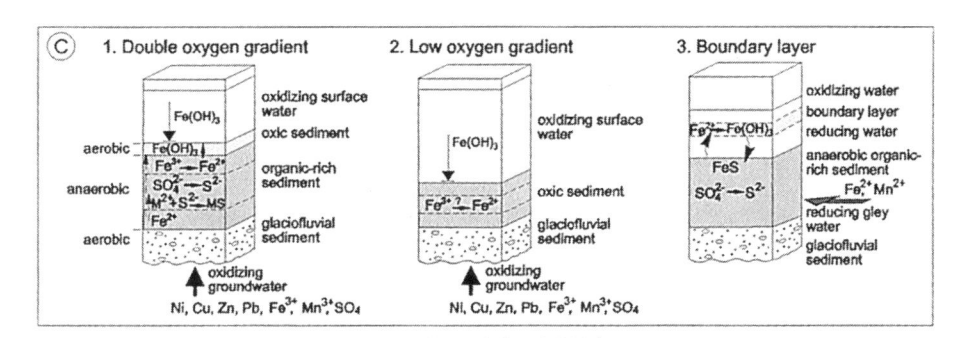

Figure 5. Sampling design with landscape geochemical methods and GIS processing. *A.* Landscape geochemical base map showing elementary landscapes of the study area, surface matter flow directions and classified landscape geochemical barriers. Locations of mine shafts and waste rock dumps are shown by the "●" symbol and sample locations by the "▲" symbol. Drainage ditches in the mire are indicated by dotted lines. *B.* Landscape geochemical cross-section indicating the expected transport patterns of the studied metals within and between landscapes. The location of the cross-section is shown by a dashed line in *A*. (MMC, main migration cycle; ELF, extra landscape flow; LGF, landscape geochemical flow; see Fortescue, 1984 [18]). *C.* Conceptual model of landscape geochemical barriers in the aqual landscape. Locations of the landscape prisms are shown on the landscape geochemical base map: prisms 1 and 2 are located in the active stream channel and prism 3 in the stagnant mire surface water (fen).

All extractions were made with chemicals of analytical grade or better. Between each extracting steps the separation of solution and particulate matter was obtained by centrifugation at 3000 rpm for 15 minutes. The supernatant was pipetted into glass tubes and acidified to pH<2 with concentrated HNO_3. The residue was rinsed with a small amount of deionised water (10 ml) and centrifuged for another 15 minutes. This second supernatant was discarded. Analysis of test solutions for Pb, Ni, Cu, Zn, Fe and Mn was conducted by flame atomic absorption spectrophotometry (Varian 1275). The pH was measured in the field with a Beckman Φ21 pH meter. Loss-on-ignition (LOI) was determined on dried samples (105°C for 48 hours) at 550°C for 6 hours. Complementary to chemical extractions, XRD analysis was also performed on selected sediment samples.

Partitioning of metals between the various fractions should be considered as operationally defined by the extraction procedure rather than a reflection of scavenging by a specific mineral phase [31, 38, 45]. Also, regarding the selectivity of the extractants, [10, 30, 55], much care should be taken in the interpretation of the results. Low selectivity has been reported for NH_4OAc in fraction 1, [23], and in our extraction no effort was made to extract the carbonate-bound fraction separately. It is noted also that only 80-90% of residual metals are extracted by hot aqua regia [1].

Table 1. Description of the applied sequential extraction method.

Pore water	Sediment pore water was obtained by centrifugation of sediment samples at 3000 rpm for 15 minutes.
Fraction 1	*exchangeable metals and metals bound to carbonates* The solid residue of the pore water extraction was leached with 1 M ammonium-acetate (NH_4OAc) adjusted to pH 5 with acetic-acid (HOAc) at room temperature for 6 hours under continuous agitation.
Fraction 2	*metals bound to Fe and Mn hydroxides* The residue from (1) was leached with 0.3 M sodium-dithionite ($Na_2S_2O_6$) added to 0.175 M sodium-citrate ($NaC_6H_5O_7$) at pH adjusted to 4.8 with HOAc. Continuous agitation was maintained for 6 hours at room temperature.
Fraction 3	*metals bound to organic matter and amorphous sulphides* The residue from (2) was extracted with 0.02 M nitric-acid (HNO_3) mixed with 30% hydrogen-peroxide (H_2O_2) in 3:5 proportion at pH 2 and heated gradually to 85°C with occasional agitation for 5 hours. After cooling 3.2 M NH_4OAc was added in 20% HNO_3 and double volume deionised water, and then kept under continuous agitation for 30 minutes in order to prevent adsorption of extracted metals on oxidised sediments.
Fraction 4	*residual metals* The residue from (3) was digested with hot aqua regia ($HCl:HNO_3$ 3:1) at boiling temperature for 1 hour, then the samples were left to cool overnight.

6. Results

6.1 EXPLORATORY DATA STRUCTURE ANALYSIS

Robust summary statistics of metal concentrations in each fraction are listed in Table 2. According to the enrichment factors, calculated as ratios of metal concentration in each fraction to the sums of fractions 1 to 4 in the sediment, metals are mostly abundant in the oxide-bound fraction (Fe 70-80%, Ni 50-60%, Pb ~40%, Zn ~30%), except for Cu and Mn. This indicates the importance of the reducible fraction in the sediment studied. The higher proportion of Mn in fraction 1 (between 50-60%) and the variability of Fe in the oxide-bound phase suggest that redox conditions are heterogeneous and Mn is probably in its reduced form. Organic matter, beside oxides, is a major sink for trace metals in the wetland stream, which is reflected by the high relative concentrations of metals in the oxidisable phase, particularly for Cu (50-60%) and Pb (30%); whereas, the residual fraction accounts in all cases for less than 20% of the total metal content.

The robust IQR/M (inter-quartile range/median) ratio indicates the variability of the measured parameters and is the lowest for Pb in all phases. This suggests that Pb is fairly uniformly distributed within the respective fractions in the sediment. On the other hand, the highest variability of Cu in the oxidisable phase indicates heterogeneity in the adsorbing surfaces of that phase. The relatively high variability of Fe in all phases is interpreted as the effect of redox changes in the sediment, which has fundamental influence on the speciation of Fe. Also, compared to the other metals, the high variability of Fe (2.42) and Mn (3.4) in the residual phase might possibly indicate that Fe and Mn originate from different mineral species. Based on the uniform enrichment of Zn in the extracted sediment fractions, as well as its high IQR/M ratio (4.09) in the pore water fraction, Zn is regarded as conservative in the studied system.

Stem-and-leaf displays, frequency histograms and box-plots of the data sets showed poly-modal behaviour with positively skewed populations and were characterised by the presence of less than 20% outlying values, as defined by the inner-fence criteria [58], as shown in Figures 6A and 6.B. Distribution shapes were similar for Fe and Mn in all fractions. For both elements, high frequency of values occurred at the lower limit of the range, representing sediment samples from the stagnant surface water, and the distributions approximated the shape of a lognormal distribution. In order to stabilise variability, the original variables were re-expressed by log-transformation for further analysis [59]. Cluster analysis (CA) and Q-mode principal component analysis (PCA) of the log-transformed data helped to further explore data structures in the multidimensional space. Using the average linkage method and the Euclidean-distance as a measure of similarity in the cluster analysis, Fe and Mn samples separated into two groups representing sediments in the active stream channel and sediments in the stagnant water areas. This suggests that Fe and Mn can selectively describe the redox conditions of the system. In the PCA analysis of Fe fractions, shown in Figure 6.C, the same groups were apparent in the plane of the first two principal components while no similar groups could be separated for the other metals.

Table 2. 5-number summaries of the measured variables. Metal concentrations in the fractions are in ppm; loss-on-ignition (LOI) is in % of dry (105 °C) sample; sample size: 23.

	5-number summaries	Fraction 1	Fraction 2	Fraction 3	Fraction 4	Pore water	LOI
Pb	Minimum	0.03	2.66	0.73	1.73	0	8.61
	Lower quartile	0.63	3.66	1.93	1.96	0	44.88
	Median	1.47	5	3.3	2.16	0	74.5
	Upper quartile	3.13	6.99	4.06	2.53	0	88.98
	Maximum	3.73	13.32	7.99	3.93	0.07	97.1
Ni	Minimum	0.43	2.96	0.23	0.83	0	
	Lower quartile	2.03	9.99	1.83	1	0.025	
	Median	3.83	13.65	3.56	2	0.045	
	Upper quartile	5.66	20.65	5.43	3.93	0.01	
	Maximum	17.55	38.63	92.91	6.63	0.58	
Cu	Minimum	0.1	0.67	1.33	0.63	0	
	Lower quartile	0.7	1.43	4.66	0.93	0.02	
	Median	1.17	5	11.32	1.53	0.045	
	Upper quartile	2.2	10.32	30.3	1.9	0.1	
	Maximum	219.78	93.24	266.4	30.3	0.45	
Zn	Minimum	3	4	8.66	5.33	0	
	Lower quartile	10.99	20.65	21.98	10.99	0.08	
	Median	23.98	43.29	32.97	17.98	0.11	
	Upper quartile	54.61	106.56	45.29	34.3	0.53	
	Maximum	5661	899.1	702.63	40.29	2.9	
Fe	Minimum	1.33	85.25	3.33	13.99	0.15	
	Lower quartile	4.93	166.83	43.29	39.63	0.26	
	Median	104.23	1132.2	131.54	92.24	0.52	
	Upper quartile	278.06	2763.9	322.68	263.07	1.22	
	Maximum	1798.2	12187.8	648.35	1964.7	14	
Mn	Minimum	1.37	0.37	0.17	0.1	0.01	
	Lower quartile	4.6	1.03	0.37	0.53	0.28	
	Median	18.65	5.59	1.07	1.9	0.47	
	Upper quartile	41.29	24.98	1.93	6.99	0.95	
	Maximum	356.31	76.92	4.36	22.98	1.55	

The spatial heterogeneity of the barrier studied could be analysed using star symbol plots [13] of the Fe fractions. Star symbol plots for samples in the active stream channel were similar in shape indicating similar geochemical conditions, as shown in Figure 7. The relative concentrations were higher in the stream which reflect the importance of Fe in this environment indicated in the plots by the dominance of the oxide-bound, organic-bound and residual fractions. Deviations from this pattern indicate heterogeneity of the stream sediments geochemical barrier. Separation of samples representing sediments in the active stream channel and sediments in the stagnant water areas are also apparent in Fig. 7.

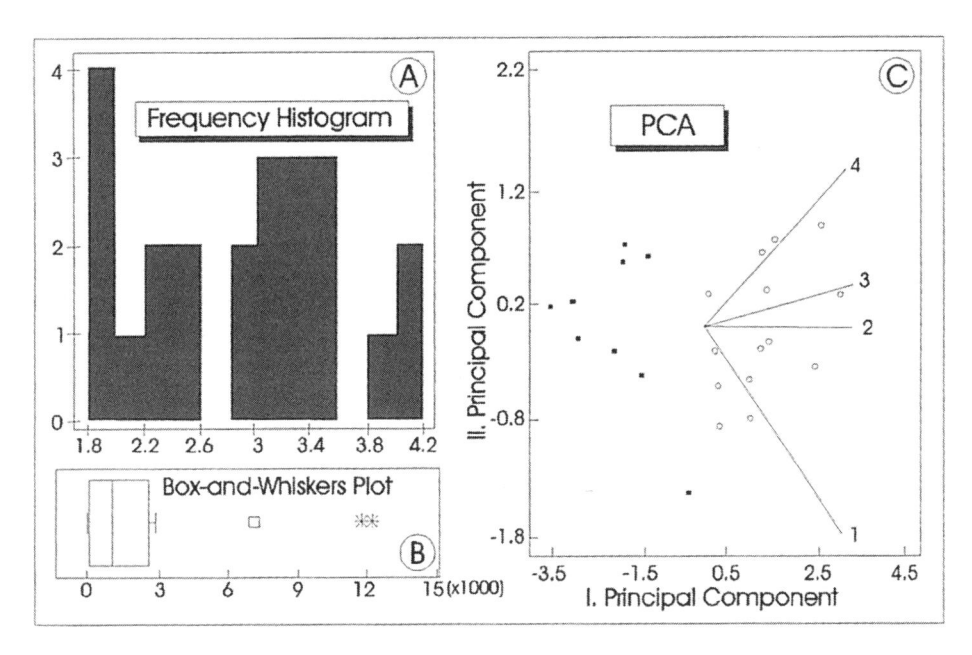

Figure 6. Univariate and multivariate exploratory data structure analysis of iron. *A.* Frequency histogram of log-transformed iron data in the reducible phase. Note the separation of populations. *B.* Box-and-whiskers plot of original iron data in the reducible phase. Outlier and far-outliers are indicated by the symbols "□" and "∗", respectively. *C.* Principal components analysis of log-transformed iron data. Mire- and stream surface water sediment samples are shown as solid boxes and empty circles, respectively, and numbers from 1 to 4 indicate the four metal fractions (see Table 1).

6.2 CAUSAL ANALYSIS

Associations among metals were studied by means of the Pearson's correlation analysis of log-transformed data. Where it was geochemically justified, correlations were subjected to partial correlation analysis in order to correct for possible induced correlations caused by a common variable. All correlations were checked for significance at the 0.05 significance level and were graphically examined to avoid spurious correlations. Results of the correlation analysis are described in Figure 8 for each fraction with correlation coefficients given in brackets.

Correlation analysis for metals in the pore water of the stream sediment revealed strong association between Ni and Zn (0.93), and Cu and Zn (0.85). These elements are associated with the mineralization and hence this fraction could most probably reflect contamination caused by mining.

In fraction 1 the correlation coefficients between Fe-Mn (0.91) and in the surface sediments also for Zn-Ni (0.82) suggest common pools for these metals. No correlation with pH is apparent in this fraction, but the strong correlation between Fe and Mn and their respective strong correlation with reducible Fe (0.92 and 0.94) suggest that they are specifically adsorbed on Fe oxy-hydroxides. The lack of any correlation of Cu and Pb, as well as their low enrichment in this fraction, suggests that these metals are preferably more strongly bound in the sediment.

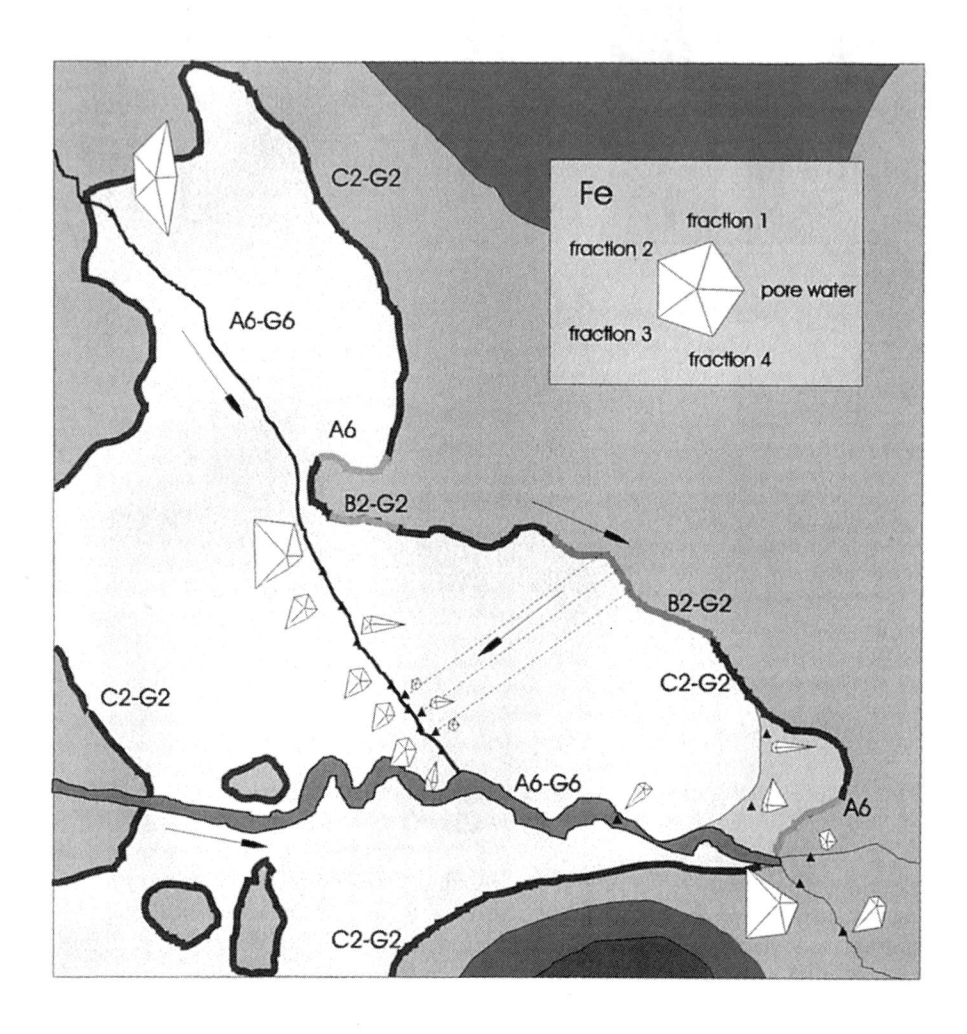

Figure 7. Spatial analysis. Star symbol plots of Fe superimposed on the landscape geochemical base map showing the heterogeneity of the studied geochemical barrier. For legend see Figure 5.

The strong correlations of Fe with Pb (0.81), Cu (0.84), Mn (0.89) and a somewhat weaker correlation with Ni (0.59) in fraction 2 are indicative of Fe oxy-hydroxide scavenging. Reducible Fe and associated metal concentrations are invariably lower in the more reducing mire surface water sediments than in the stream sediments. Although, clusters of mire and stream sediment samples can be recognised, they fall on the same regression line for Pb (see Figure 8) and Mn. In the cases of Cu (also shown in Figure 8) and Ni, however, mire surface water sediment samples are displaced away from the stream population regression line and do not show correlation with Fe. A similar pattern was observed for Zn as for Ni and Cu, but was blurred by the higher scatter of data points, thus did not lead to significant correlation.

A significant correlation was obtained between Pb and Cu (0.72) in fraction 3 in most samples (see Figure 8). This suggests that oxidisable Pb and Cu are extracted from a common pool. Regarding their high affinity for complexing with organic ligands [1], these metals are probably retained by organic matter chelation in this fraction which is in accordance with the high stability constants of Cu and Pb organic complexes [10, 51]. Based on this, one would expect strong correlation for Pb and Cu with LOI that, however, was not found in our case. On the other hand, Zn had the second largest concentration in this fraction and showed positive correlation with LOI (0.72). One explanation of these results could be that in the investigated size fraction LOI mainly represents the chemically inert, unprocessed organic debris [56]. Taking into account that Zn has the highest biological accumulation coefficient (BAC) among the metals studied (Ni(0.03), Cu(0.13), Zn(0.9), Fe(0.012), Mn(0.4); [11]; [41]) it is plausible that Zn in this fraction is extracted from the unprocessed organic debris and, therefore, represents metal uptake by plants. This would also explain the lack of correlation between Pb, Cu and LOI if the amount of decomposed organic matter is assumed to be subtle compared to the unprocessed organic remains.

In the residual fraction (fraction 4) Fe and Pb (0.83) (see Figure 8) and Fe and Mn (0.89) were strongly related to each other. Regarding these element associations and correlations of Fe and Mn with the ash content (0.97 and 0.94, respectively) it is supposed that the residual fraction represents the natural geochemical background. Mainly minerals, such as feldspar, quartz and hornblende, as shown by the XRD analysis, and perhaps amorphous oxides and clay minerals characterise the composition of the residual fraction. Moreover, the lack of association of Ni, Cu and Zn in the residual fraction might further support our earlier hypothesis that these metals originate from contamination by ore mining in the geochemical landscapes studied in this research.

7. Discussion

Enrichment factors of metals, as well as results of the correlation analysis confirm the importance of the reducible and the oxidisable phases in controlling metal speciation. Since Mn does not show correlation with any of the metals except for Fe, it is most probably found occluded in Fe oxy-hydroxides and does not form a separate phase. Correlations indicate that Pb and Mn are controlled by co-precipitation in Fe oxy-hydroxides both in the stream and mire water sediments. Cu and Ni distribution, on the other hand, can be related to Fe oxy-hydroxides only in the stream sediment samples. As these metals are primary constituents of the ore mineralization and are also characterised by associations in the pore water phase, in this fraction they are regarded as recent contamination originating from ore mine waste rocks. Therefore, the organic rich wetland stream sediment can be regarded as an oxidising-adsorption barrier with respect to these metals, as it was hypothesised in our prior model. More supportive evidence on metal co-precipitation could, however, be provided by the specific sequential extraction of the reducible fraction. Considering the Cu and Ni distribution in fraction 2 for the mire sediment samples, Al hydroxides might be more important in their regulation as they are stable under more reducing conditions [30].

Specific adsorption of trace metals on iron oxides is dependent on pH. Cu, Pb and Zn, for example, are adsorbed most strongly on iron oxides at pH≥5 [15]. Metals most able to form hydroxy complexes are specifically adsorbed most strongly on hydrous oxides and follow the decreasing order Pb(7.7)>Cu(7.7)>>Zn(9.0)>Ni(9.9) where numbers in parentheses are hydrolysis constants (pK), [1]. Thus the sequence of enrichment factors for metals in fraction 2 should follow the order of their binding strength, i.e., the stability of metal-hydroxo-complexes. However, in our samples no general similarity was found with this sequence, which indicates the complexity of the system studied. Adsorption studies in well-defined media have shown that adsorption is strongly dependent on such factors as pH, temperature, equilibration time, surface properties of adsorbent and presence of organic and inorganic ligands. At low pH ligands may enhance the adsorption of certain metals on oxides by the formation of ternary surface complexes, while at higher pH they can compete with iron oxides for the binding of trace metals (e.g., Davis and Leckie, 1978b [16]; Benjamin and Leckie, 1981a and 1981b [3,4]). In complex natural systems, the role of dissolved inorganic and organic ligands is particularly important in the influence of adsorption processes on iron oxides. Thus, in the case of the organic sediment studied, for example, it is difficult to apply directly the results of well-controlled laboratory experiments.

Taking into account these constraints and also that rigorous equilibrium conditions required for thermodynamic calculations of metal adsorption could not be assumed we applied the adsorption equations proposed by Benjamin and Leckie (1981b,c [4,5]) to further investigate the sediment samples. Adsorption density (Γ) for iron oxy-hydroxides and apparent equilibrium constants (K_a) were calculated and compared to data on trace metal adsorption on iron oxy-hydroxides compiled by Tessier et al. (1985 [57]). The calculations for Cu, shown in Figure 8, suggest also that Cu in the reducible fraction in the stream sediment could only be controlled by adsorption on Fe oxy-hydroxides. This is coherent with the results of the correlation analysis. Calculations provided similar results for Zn but not for Ni.

Regarding organic matter in the sediments, its chemical activity can be significantly different within the investigated size fraction due to the presence of both the decomposed, chemically reactive humic substances and the chemically inert, unprocessed organic debris. However, fraction 3 could also represent amorphous sulphide compounds. The formation of metal sulphides is dependent on the availability of reducible sulphur and reactive metals, low redox potential, pH influencing the solubility of sulphides, and microbial activity of sulphate reducers for which decomposable organic matter and circum-neutral pH are required [44]. It has been found that sulphate reduction is usually restricted to microsites in sediments where microbes can easily create living conditions favourable for their activity [40, 44, 48]. In a tidal marsh, for example, it was found that sedimentary oxides and sulphate reduction with subsequent formation of pyrite coexisted in the same macro-environment [43]. Thus the existence of oxidised forms extracted in fraction 2 does not exclude the possibility of sulphate reduction. In the sediments studied, metals and sulphate could be available from the discharging AMD and pH (pH is around 6 on average in the stream water) and decomposable organic matter in the sediments would not limit metal sulphide formation.

Although, the characteristic H_2S smell and black colour around organic debris could be observed in some of the sediment samples, in the field these evidence that sulphate reduction does occur locally but probably does not dominate organic matter decomposition in the sediment.

Recent metal contamination caused by mining would appear in the labile, surface-bound positions of the sediments, and its spatial distribution could be studied by element associations in the pore water and exchangeable phases. In the pore water fraction samples representing the stagnant surface water sediments fell on the same regression line as those of the stream sediment samples, metal concentrations being systematically higher in the mire pore water fraction. This indicates that metal input reaches the stream from the neighbouring landscapes and the spatial distribution pattern of metals in the sediments does not result from a single upstream source. Potential pools for the retention of metals in fraction 1 could be outer-sphere positions on minerals, such as clays, as well as organic matter surfaces and structural positions in carbonates. It is known that most of the outer-sphere positions in the sediments are occupied by alkaline and alkaline earth metals, H and Al, which exist in concentrations greater than trace metals in the stream water [51]. Carbonate minerals could be formed at the pH and redox conditions prevailing in the stream environment studied if saturation is reached in the solution. On the other hand, the formation of carbonates is limited in the acidic mire surface water sediments (pH=3.5). Assuming the presence of $CaCO_3$, for instance, metals could co-precipitate in the mineral structure [17], and present at 10^{-2} to 10^{-5} % levels which does not exclude that metals in this fraction can originate from the dissolution of carbonates. However, both stream and mire sediments form a common population falling on the same regression line for Ni, Zn, Fe and Mn. Thus co-precipitation with carbonates is not expected to be dominating in the stream and a process other than carbonate formation and co-precipitation is responsible for the regulation of metal concentrations in this fraction. XRD analysis did not indicate the presence of carbonates, either.

Canonical correlation analysis was used to analyse the interaction of several mutually related factors to model more realistically adsorption processes in the sediments studied. High canonical correlation (0.91) was found between metals in fraction 1 and fraction 3 (see Figure 8). The canonical correlation coefficient is higher than any individual correlations justifying the use of the canonical variables. Assuming that metals in fraction 3 are adsorbed on organic matter, this indicates that probably organic matter controls metals in the exchangeable phase.

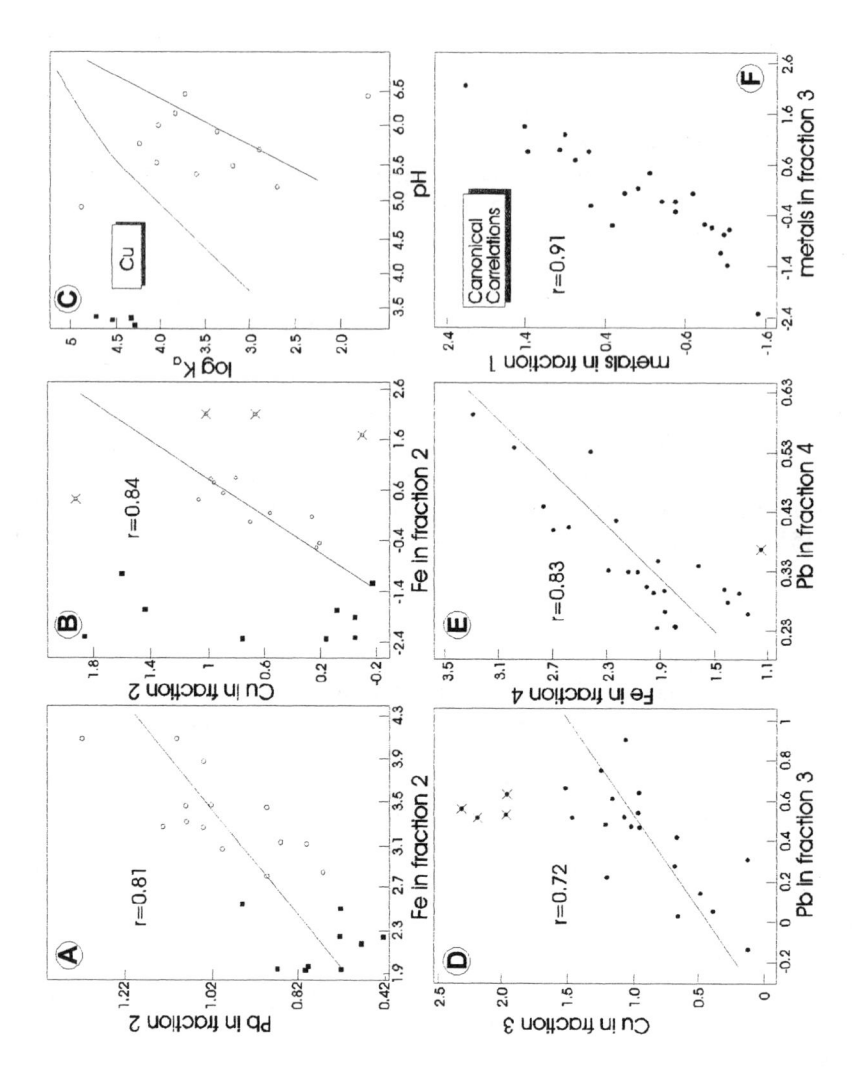

Figure 8. Correlation analysis of log-transformed data (*A, B, D* and *E*). Mire- and stream surface water sediment samples are indicated by solid boxes and empty circles, respectively, where relevant (see text). Outliers rejected from the regression are marked with crosses. (*C*) shows $\log K_a$ vs. pH for the Fe hydrous oxide system where solid lines refer to the limits compiled by [57]. Symbols refer to data from the present study. (*F*) shows the canonical correlation between metals in fractions 1 and 3.

8. Conclusions

The careful application of a sequential chemical extraction procedure can enable the determination of different metal sinks in surface water sediments and could be used to distinguish between natural and anthropogenic components of the studied metals. Modelling of the measured parameters suggests that adsorption and co-precipitation with Fe oxy-hydroxide are major processes regulating trace metal concentration in the investigated wetland stream sediment. Mn is probably specifically adsorbed on Fe oxy-hydroxides and, beside Zn, is least retained in the sediment. Pb, Cu and Ni are found in considerable quantity in the reducible fraction and are suggested to occur occluded in Fe oxy-hydroxides. On the other hand, organic matter provides important adsorption sites for Cu and Pb and controls exchangeable metals, too. Ni, Cu and Zn are suggested to represent the mineralization based on enrichment factor calculation and correlation analysis in the pore water and the oxide-bound fractions. However, differentiation between the natural geochemical background and contamination by mining would require a more specific investigation of the reducible fraction.

Results of data analysis imply that heterogeneity of the geochemical barrier is controlled by redox gradients in the sediment. In our prior model we hypothesised that the uppermost stream sediments acted as a complex oxidising-adsorption barrier for metals originating from the acid mine drainage. It is suggested by the analysis that the landscape geochemical barrier studied can be sufficiently characterised by the distribution of Fe fractions, but more data is needed for the verification of the refined hypothetical models concerning the occurrence and extent of sulphate reduction in the sediments. Despite the high overall amount of organic matter in the stream sediments, Fe oxy-hydroxides are abundant and play major role in the retention of trace metals by adsorption on their surfaces. Regarding the long-term stability of metals in the stream sediments, changes in pH could lead to the dissolution of oxide phases and the release of adsorbed trace metals. However, the drop of pH to 4 or below that value in the stream is unlikely if the AMD input remains constant. Assuming changes in landscape conditions by either a decrease in the stream flow, or considering peat growth, by the diversion of stream flow followed by spreading of the reducing mire sediments with time [37], conditions could become more reducing and cause the dissolution of oxides. Iron, as a result of the reductive dissolution of oxy-hydroxides, and consequently trace metals would be mobilised and retained probably by the subsequent precipitation of bog iron in the marginal fen [48] and by the abundant organic matter. The analysis of geochemical barriers performed by GIS-based landscape geochemical modelling in addition to a detailed geochemical modelling can contribute to the analysis of environmental geochemical problems by providing means for the systematic analysis of the relationships between related landscapes.

9. Acknowledgements

The authors would like to thank Inger Påhlsson for the laboratory analysis of samples at the Uppsala University. We are grateful to Istvan Horvath, Gyorgy Toth and Emoke Edelenyi of the Hungarian Geological Survey for making the preparation of this paper possible. Thanks are extended to Balázs Székely of the Geophysical Research Group of the Hungarian Academy of Sciences for his indispensable assistance with the GIS applications. A scholarship provided by the Swedish Institute and supplementary grants from The Soros Foundation and from the Szádeczky-Kardos Foundation are gratefully acknowledged.

References

1. ALLOWAY B. J. (ed.) (1990) *Heavy Metals in Soils*. Blackie, London.
2. ALPERS C. N., BLOWES D. W., NORDSTROM D. K. AND JAMBOR J. L. (1994) Secondary minerals and acid mine water chemistry. In *The Environmental Chemistry of Sulfide Mine Wastes*. (eds. D. W. Blowes and J. L. Jambor), pp. 247-270. Mineralogical Association of Canada, Waterloo, Ontario.
3. BENJAMIN M. M. AND LECKIE J. O. (1981a) Conceptual model for metal-ligand-surface interactions during adsorption. *Environ. Sci. Technol*. **15**, 1050-1057.
4. BENJAMIN M. M. AND LECKIE J. O. (1981b) Competitive adsorption of Cd, Cu, Zn and Pb on amorphous iron oxyhydroxide. *J. Colloid Interface Sci*. **79**, 209-221.
5. BENJAMIN M. M. AND LECKIE J. O. (1981c) Multiple-site adsorption of Cd, Cu, Zn and Pb on amorphous iron oxyhydroxide. *J. Colloid Interface Sci*. **79**, 209-221.
6. BERNER A. (1980) *Early Diagenesis: A Theoretical Approach*. Princeton University Press.
7. BIGHAM J. M. (1994) Mineralogy of ochre deposits formed by sulfide oxidation. In *The Environmental Chemistry of Sulfide Mine Wastes*. (eds. D. W. Blowes and J. L. Jambor), pp. 103-132. Mineralogical Association of Canada, Waterloo, Ontario.
8. BIGHAM J. M., SCHWERTMANN U. AND CARLSON L. (1992) In *Biomineralization of Iron and Manganese: Modern and Ancient Environments*. (eds. H. C. W. Skinner and W. Fitzpatrick), pp. 219-232. Catena Supplement 21, Catena Verlag, Cremlingen-Destedt.
9. BLOWES D. W. AND JAMBOR J. L. (eds.) (1994) *The Environmental Chemistry of Sulfide Mine Wastes*. Mineralogical Association of Canada, Waterloo, Ontario.
10. BODEK I., LYMAN W. J., REEHL W. F. AND ROSENBLATT D. H. (eds.) (1988) *Environmental Inorganic Chemistry*. Pergamon Press Inc., New York.
11. BROOKS R. R. (1973) Biogeochemical parameters and their significance for mineral exploration. *J. Applied Ecology* **10**, 825-836.
12. CARPENTER R. H., POPE T. A. AND SMITH R. L. (1975) Fe-Mn oxide coatings in stream sediment geochemical surveys. *J. Geochem. Explor*. **4**, 349-363.
13. CHAMBERS, J. M., CLEVELAND W. S., KLEINER B. AND TUKEY P. A. (1983) *Graphical Methods for Data Analysis*. Wadsworth International Group, Belmont, Calif., Duxbury Press, Boston, MA.
14. DAVENPORT P. H. (1990) A comparison of regional geochemical data from lakes and streams in northern Labrador; implications for mixed-media geochemical mapping. *J. Geochem. Explor*. **39**, 1-13.
15. DAVIS J. A. AND LECKIE J. O. (1978a) Surface ionisation and complexation at the oxide/water interface. II. Surface properties of amorphous iron oxyhydroxide and adsorption of metal ions. *J. Colloid Interface Sci*. **67**, 90-107.
16. DAVIS J. A. AND LECKIE J. O. (1978b) Effect of adsorbed complexing ligands on trace metal uptake by hydrous oxides. *Environ. Sci. Technol*. **12**, 1309-1315.
17. DEURER R., FORSTNER U. AND SCHMOLL G. (1978) Selective chemical extraction of carbonate-associated metals from recent lacustrine sediments. *Geochim. Cosmochim. Acta* **42**, 425-427.
18. FORTESCUE J. A. C. (1984) *Environmental Geochemistry*. Springer-Verlag, New York.
19. FORTESCUE J. A. C. (1992) Landscape Geochemistry: retrospect and prospect - 1990. *Appl. Geochem*. **7**, 1-53
20. FRANZEN L. G. (1985) Peat in Sweden - an important energy resource or just a paranthesis?. In *Proceedings of the Peat and Environment '85 International Peat Society Symposium*. The Swedish National Peat Commitee, Stockholm.
21. GAÁL G. (1990) Tectonic styles of early proterozoic ore deposition in the Fennoscandian Shield. *Precambr. Res*. **46**, 83-114.
22. GADSBY J. W., MALICK J. A. AND DAY S. J. (eds.) (1990) *Acid Mine Drainage: Designing for Closure*. BiTech Publishers Ltd., Vancouver.
23. GIBBS R. J. (1977) Transport phases of transition metals in the Amazon and Yukon rivers. *Geol. Soc. Am. Bull*. **88**, 829-843.
24. GLAZOVSKAYA M. A. (1963) On geochemical principles of the classification of natural landscapes. *Int.*

Geol. Rev. **5**, 1403-1430.
25. HERBERT R. (1994) Metal transport in ground water contaminated by acid mine drainage. *Nordic Hydrology* **25**, 193-212.
26. HOAGLIN D. C., MOSTELLER F. AND TUKEY J. W. (1983) *Undertanding Robust and Exploratory Data Analysis*. John Wiley and Sons Inc., New York.
27. HORNE R. A. (1978) *The Chemistry of our Environment.* John Wiley and Sons, New York.
28. JAMBOR J. L. (1994) Mineralogy of sulfide rich tailings and their oxidation products. In *The Environmental Chemistry of Sulfide Mine Wastes.* (eds. D. W. Blowes and J. L. Jambor), pp. 59-102. Mineralogical Association of Canada, Waterloo, Ontario.
29. JORDAN G., SZUCS A., QVARFORT U. AND SZEKELY B. (1997) Evaluation of metal retention in a wetland receiving acid mine drainage. *Proc. of the 30th IGC,* **19**, 189-206.
30. KARLSSON S., ALLARD B. AND HÅKANSSON K. (1988) Chemical characterization of stream-bed sediments receiving high loadings of acid mine effluents. *Chem. Geol.* **67**, 1-15.
31. KHEBOIAN C. AND BAUER C. F. (1987) Accuracy of selective extraction procedures for metal speciation in model aquatic sediments. *Anal. Chem.* **59**, 1417-1423.
32. KURZL H. (1988) Exploratory data analysis: recent advances for the interpretation of geochemical data. *J. Geochem. Explor.* **30**, 309-322.
33. LOVLEY D. R. AND PHILLIPS E.J. P. (1987) Competitive mechanisms for inhibition of sulfate reduction and methane production in the zone of ferric iron reduction in sediments. *Appl. Env. Microbiol.* **52**, 751-757.
34. LUNDIN L. (1982) Mark- och grundwatten: moränmark och mark typens betydelse för avrinningen. *UNGI Report* **56**, University of Uppsala, (in Swedish).
35. MAGNUSSON N. H. (1953) *Malm Geology.* Jernkontoret, (in Swedish).
36. MANTEI E. J., GUTIERREZ M. AND ZHOU Y. (1996) Use of normalized metal concentrations in the Mn oxides/hydrous oxides extraction phase of stream sediments to enhance the difference between a landfill emission plume and background. *Appl. Geochem.* **11**, 803-810.
37. MOORE P. D. AND BELLAMY D. J. (1974) *Peatlands.* Springer-Verlag, New York.
38. NIREL M. V. AND MOREL F. M. M. (1990) Pitfalls of sequential extractions. *Water Resources* **24**, 1055-1056.
39. NORDSTROM D.K. (1982) Aqueous pyrite oxidation and the consequent formation of secondary iron minerals. In Acid Sulfate Weathering (eds. J. A. Kittrick, D. S. Fanning and L. R. Hossner), pp. 37-56, SSSA Spec. Publ. 10. SSSA, Madison, WI.
40. PAKARINEN P., TOLONEN K. AND SOVERI J. (1980) Distribution of trace metals and sulfur in the surface peat of Finnish raised bogs. *Proc. 6th Int. Peat Congr.*, Duluth, U.S.A.
41. PEREL'MAN A. I. (1975) *Geochimia Landshafta.* Vysokaya Skola, Moscow, (in Russian).
42. PEREL'MAN A. I. (1986) Geochemical barriers: theory and practical applications. *Appl. Geochem.* **1**, 669-680.
43. RABENHORST M. C. AND HAERING K. C. (1989) Soil micromorphology of a Chesapeake Bay tidal marsh: implications for sulfur accumulation. *Soil Sci.* **147**, 339-347.
44. RABENHORST M. C. AND JAMES B. R. (1992) Iron sulfidization in tidal marsh soils. In *Biomineralization of Iron and Manganese: Modern and Ancient Environments.* (eds. H. C. W. Skinner and W. Fitzpatrick), pp. 203-217. Catena Supplement 21, Catena Verlag, Cremlingen-Destedt.
45. RAURET G., RUBIO R. AND LÓPEZ-SÁNCHES J. F. (1989) Optimization of Tessier's procedure for metal solid speciation in river sediments. *Int. J. Anal. Chem.* **36**, 69-83.
46. REYNOLDS W. D., BROWN D. A., MATHUR S. P. AND OVEREND R. P. (1992) Effects of in-situ gas accumulation on the hydraulic conductivity of peat. *Soil Sci.* **153**, 397-408.
47. SCHWERTMANN U. AND FITZPATRICK R. W. (1992) Iron minerals in surface environments. In *Biomineralization of Iron and Manganese: Modern and Ancient Environments.* (eds. H. C. W. Skinner and W. Fitzpatrick), pp. 7-30. Catena Supplement 21, Catena Verlag, Cremlingen-Destedt.
48. SHOTYK W. (1988) Review of the inorganic geochemistry of peats and peatland waters. *Earth Sci. Rev.* **25**, 95-176
49. SINGER P. AND STUMM W. (1970) Acid mine drainage: The rate controlling step. *Science* **167**, 1121-1123.
50. SINGH S. K. AND SUBRAMANIAN V. (1984) Hydrous Fe and Mn oxides - Scavengers of heavy metals in the aquatic environment. *CRC Crit. Rev. Environ. Control.* **14**, 33-90.
51. STUMM W. AND MORGAN J. J. (1981) *Aquatic Chemistry.* (2nd ed.) John Wiley and Sons, New York.
52. SULLIVAN P., YELTON J. AND REDDY K. (1988) Iron sulfide oxidation and the chemistry of acid generation. *Environ. Geol. Water Sci.* **11**, 289-295.
53. TARUTIS W. J. JR. AND UNZ R. F. (1992) Behaviour of sedimentary Fe and Mn in a natural wetland receiving acidic mine drainage, Pennsylvania, U.S.A. *Appl. Geochem.* **7**, 77-85.

54. TEGENGREN F. (1924) Sveriges ädlare malmer och bergverk. Sveriges Geologiska Undersökning, Series Ca No. 17, (in Swedish).
55. TESSIER A., CAMPBELL P. G. C. AND BISSON M. (1979) Sequential extraction procedure for the speciation of particulate trace metals. *Anal. Chem.* **51**, 844-851.
56. TESSIER A., CAMPBELL P. G. C. AND BISSON M. (1980) Trace metal speciation in the Yamaska and Saint Francois Rivers (Quebec). *Can. J. Earth Sci.* **17**, 90-105.
57. TESSIER A., RAPIN F. AND CARIGNAN R. (1985) Trace metals in oxic lake sediments: possible adsorption onto iron oxyhydroxides. *Geochim. Cosmochim. Acta* **49**, 183-194.
58. TUKEY J. W. (1977) *Exploratory Data Analysis.* Addison-Wesley, Reading, MA.
59. VELLEMAN P. F. AND HOAGLIN D. C. (1981) *Applications, Basics and Computing of Exploratory Data Analysis.* Duxbury Press, Boston.
60. WICKMAN F. E. (1994) The Siljan Ring impact structure: possible connections with minor ores in its neighbourhood. *GFF* **116**, 145-146.

ASSESSMENT OF MINING INDUCED ENVIRONMENTAL DEGRADATION USING SATELLITE DATA AND PREDICTIVE MODELS

D. LIMPITLAW
Department of Mining Engineering,
University of the Witwatersrand
Johannesburg, South Africa

Abstract. Kitwe is the largest city on the Zambian Copperbelt. Copper has been exploited around Kitwe for more than seventy years with severe environmental consequences. Tailings impoundments and a large metallurgical facility located near Kitwe are suspected of causing surface water and groundwater pollution. Several installations are located in sensitive headwaters. One such tailings impoundment is Dump 15A, runoff from which flows into the Mwambashi River, where water is drained for agriculture and for domestic consumption in Kitwe. The Mwambashi is a tributary of the Kafue River on which more than 40% of the Zambian population depend for water. The quality of this water is therefore critical to the general health of the Zambian populace. Fatal poisoning of livestock has been reported along the banks of the Mwambashi River.

To assess the impact on the water environment, a study of the spatial distribution of environmental aspects has been undertaken using a GIS. The results of the spatial data analysis are combined with geochemical modelling to determine the characteristics of surface waters and how these change over time and space.

Data acquired by Landsat TM, land cover data derived from maps, water quality data and the results of water-quality modelling can be combined to present a comprehensive picture of water quality alteration due to mine installations and their impact on the Mwambashi and Kafue rivers. Difficulties associated with this approach include the absence of historical surface water data and the low density of sampling points in the area of interest.

1. Introduction

1.1 MINING IN THE KITWE AREA OF THE ZAMBIAN COPPERBELT

Mining has been conducted near Kitwe for over seventy years. The long history of mining and the presence of other sources of pollution complicate assessment of environmental impact in Kitwe and it is thus desirable to select an appropriate indicator of environmental change. During this period, vast amounts of mine residue have been produced and disposed of on the surface. These residues, consisting of broken rock, fine particles and slag, provide an indication of the extent of mining and are therefore indicators of environmental impact. Assessing changes in these residue deposits over time provides insight into the changing stresses on the environment.

Both the slag and fine particles (tailings) contain metals in various concentrations. The tailings impoundments are particularly susceptible to both fluvial and aeolian erosion. This natural degradation of tailings impoundments represents a prominent source of metals in the environment. Furthermore, pollutants derived from tailings are persistent and remain in the system to become available again with the onset of favourable conditions [1].

The largest tailings impoundment in the Nkana Division of Zambia Consolidated Copper Mines (ZCCM), Dump 15A, was commissioned in 1971 near the Mindola Dam, outside Kitwe. This impoundment, sited over a former wetland, contained 70 000 000 tonnes of tailings by 1996. Between 1988 and 1996, 20,158,800 t of material were emplaced at Dump 15A through various pipelines. The average emplacement rate during this period was 255,175 t of tailings per month at an average concentration of 0.1% copper and 0.03% cobalt [11]. The impact of this impoundment on the surrounding environment is poorly understood.

1.2 WATER AND MINERAL POLLUTION

Flowing water is the dominant agent in landscape evolution. Sediment transportation by rivers is effected by a complex combination of processes, including suspension, deposition, erosion and redeposition of the sediment as alluvium. The alluvial matrix is continually enriched by deposition, adsorption and precipitation processes, explaining both the enhanced fertility of river basins relative to uplands and their increased susceptibility to contamination [3].

Many tailings impoundments are a significant source of contamination in a mining landscape by virtue of their large volume and fine particle composition. Through natural grading of the sediment load in rivers fine particles are transported the farthest. Such particles usually contain higher amounts of sorbed chemicals and thus chemical enrichment in eroded sediments is an important contaminant transport medium in mining areas [5].

Baudo (1987) [1] states that the heavy metals most frequently reported as pollutants are cadmium, copper, mercury, lead and zinc, the fate of which is largely dependent on the heterogeneous equilibrium established between dissolved and suspended compounds. Notwithstanding this observation, sediments act as a sink for most pollutants [1]. The only reported occurrence of chronic copper poisoning in the Kitwe area occurred through the ingestion of contaminated sediments deposited on the banks of the Mwambashi River. In 1981 chronic copper poisoning of livestock, resulting in the death of cattle, sheep and goats occurred on farms adjacent to the Mwambashi [10]. Copper was present in the river sediments and in riverside vegetation with concentrations of the metal ranging from 3,000 µg/g to 20,000 µg/g in the river sediments (Brasser and Charman, 1982 in [10]). It is therefore important to identify downstream areas on a river's floodplain that are susceptible to sediment deposition.

2. Environmental impact

2.1 SAMPLING

Chemical data used in this investigation are derived from two sources: a field trip made to the area of interest (AOI) by the author and an environmental impact scan of the mining lease conducted on behalf of the mine by consulting engineers. The number of sample sites is sub-optimal and gaps are present in the data, especially in groundwater samples. Despite this, interpolation of sample values provides insight into the spatial variation of water quality within the AOI.

Land use classes identified in Figure 1 are derived from classification of 1998 Landsat Thematic Mapper (TM) data using aerial photographs to verify classifications. Dumps and water bodies were delineated using a normalised difference vegetation index. This shows hard anthropogenic surfaces and water bodies, both of which are free of vegetation, as dark patches in the image. Areas with high vegetation density are bright.

2.2 THE KAFUE RIVER

This river receives mine effluent from most of the Zambian Copperbelt. All of Nkana Division's surface runoff finally ends up in the Kafue. Two sites provide insight into the changes in the river as it passes Kitwe.

In Figure 2, water in the Kafue above Kitwe plots near the upper apex of the Piper diamond. This implies water rich in both $Cl^- + SO_4^{2+}$ and $Ca^{2+} + Mg^{2+}$. The water has permanent hardness. After its passage downstream though the Nkana lease area, the water is characterised by temporary hardness, rich in $Ca^{2+} + Mg^{2+}$ and HCO_3^-. The total dissolved solids (TDS) concentration in the Kafue water increases as the river flows past Kitwe. The changes shown in the lower triangles of the Piper plot indicate calcite solution [7]. TDS in river water over this stretch increase from 152 mg/l to 200 mg/l. This is to be expected as several installations in the Nkana lease area show high surface water calcium contamination. Of these Dumps 15A and 33 as well as the plant area show prominent calcium enrichment of surface water. A groundwater sample collected near Dump 33 shows calcium enrichment which is absent from the Nkana Plant area groundwater sample. No analysis of groundwater is available for the Dump 15A area and this has influenced the interpolation of sample values, probably resulting in an underestimation of contamination at 15A. Calcium enrichment is, however, clearly associated with mine installations in the AOI.

2.3 THE MWAMBASHI RIVER

This river drains the Chambishi-Nkana Basin – the geological structure that hosts the copper deposits within the AOI. The Mwambashi receives runoff from Chambishi Mine north of Kitwe and from the large tailings facilities of Nchanga Division. Water from Nchanga reaches the Mwambashi via the Muntimpa Stream. Other investigators, [2, 6] have concentrated on the impact of the Nchanga on the Mwambashi and a detailed assessment of the river will not be presented here. Runoff from Dump 15A via a tributary of the Mwambashi is considered in detail.

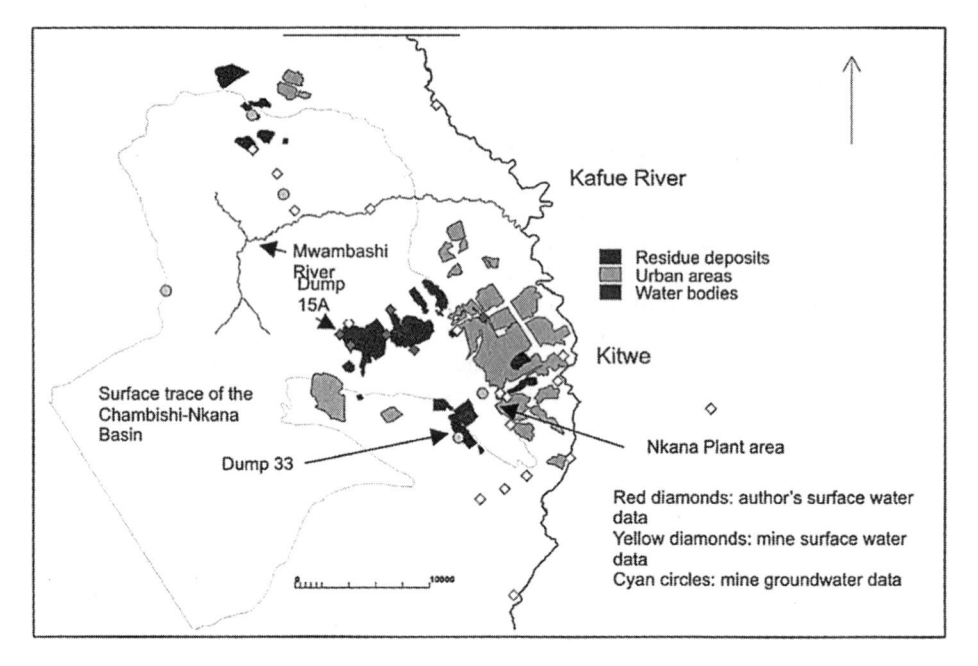

Figure 1. The area of interest showing the city of Kitwe, the Kafue River, mine waste disposal facilities and sample sites. Geology digitised from *Geological Map of Chambishi Basin and Katembula, Chisanga and Chiryongoli Domes*, RCM Geological Services, 1: 125 000, 1981.

3. Modelling

3.1 THE DIGITAL ELEVATION MODEL

The modelling of pollutants in overland flow is impossible without accurate representation of runoff. Consequently, representing the hydrologic behaviour of a catchment is important. The process of sediment loss provides the major transport mechanism for all sorbed pollutants and the interaction of pollutants with sediment loss and runoff results in the overall transport of these non-point source pollutants [5].

Topography has a significant influence on the response of a catchment to rainfall. It also has a significant effect on the ecological dynamics of the landscape [9]. In a GIS environment, the topographic attributes of a landscape may be quantified using digital elevation models (DEMs), which determine:

- the path followed by contaminated surface water flowing away from a tailings installation, and
- the area potentially affected by the pollution plume associated with this water.

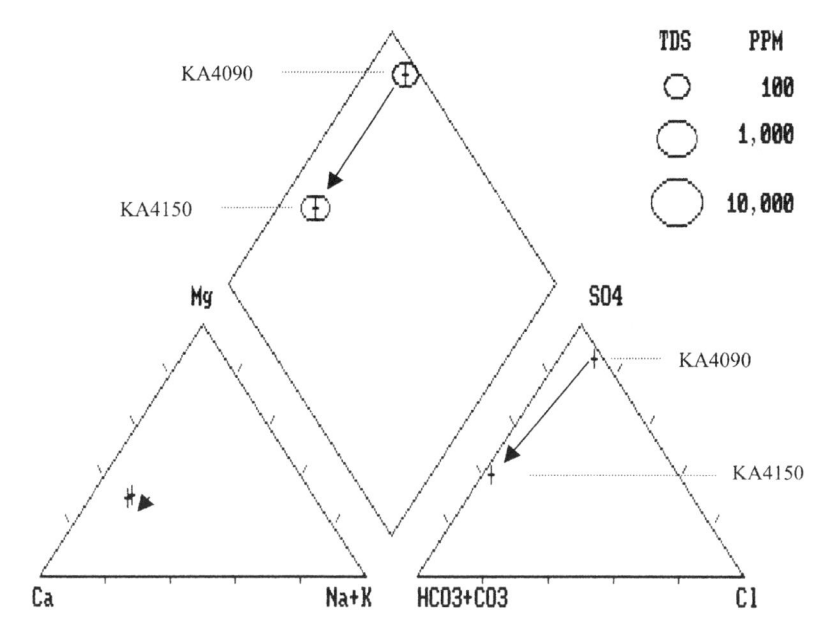

Figure 2. Piper diagram showing changes in the chemistry of Kafue waters as the river passes through the AOI. KA4090 is a sample site approximately 18 km north of Kitwe and KA4150 lies 10 km to the south of the city. . Piper diagrams facilitate the characterization of water bodies by plotting the concentrations of the major ions. Comparing the plotting positions of analyses from different sampling sites as shown above can identify changes in the chemical composition of a water body with distance. Plot produced in WATEVAL [7].

In this investigation, a 9 x 9 moving window has been used to produce streamlines with reference to a DEM which has been pre-processed to remove pits. The DEM was generated by interpolation of 50 foot (15.2 m) contour lines present on 1:50,000 government topographical maps of the area: 1228C1 Mufulira West, 1228C2 Mufulira East, 1228C3 Kitwe and 1228C4 Mabote. By the Peuker criterion, a DEM grid size has to be 4.3 times the contour interval of the source map (Sasowsky *et al.*, 1992, in [4]). Accordingly, the grid size used here is 65 m. DEMs generated with 25 m grids contained a large number of artefacts and were discarded. The accuracy of the source maps is questionable. Difficulty has been experienced in registering satellite images to these maps and a large number of pits and flat areas were created in the DEM during interpolation. As is often the case, no alternative datasets were available and therefore the study proceeded with a DEM generated from these maps.

An area for detailed examination was defined as a sub-map of the 50 km x 50 km area represented by the four topographic sheets. This sub-map is centred on Kitwe, is approximately 30 km wide, extending from the Mwambashi River in the north to about 10 km south of the city centre.

The logic statement used to remove pits in the DEM was run in the GIS software by ITC in the Netherlands, ILWIS® 2.2 (http://www.itc.nl/ilwis) [12].

iff(mindflat,mindem,iff(minsub1=1,iff(nbminp(mindem#)=5,nbsum(mindem#)/9,mindem),mindem))

where: **mindflat** is a map showing flat areas on the DEM;
minsub is a map defining the boundaries of the DEM (to avoid artefacts);
mindem is the original DEM;
nbminp and **nbsum** are neighbourhood operators; **nbminp** returns the neighbour with the lowest value and **nbsum** returns the sum of the neighbouring pixel values for each pixel;
# indicates the entire set of 3 x 3 neighbours.

Mindflat was generated using the following expression:

nbcnt8(mindem#=mindem)=8

where: **nbcnt** is a neighbourhood operator which tests the number of neighbours which satisfy a specified condition, in this case **8** neighbours which have the same elevation as the central pixel.

The modified DEM was used to propagate streamlines from waste sites to the nearest major river. These sites were identified using Landsat TM colour composite maps. The propagation was stopped at the rivers as the Kafue and its tributary, the Mwambashi, both transport pollution from other mining lease areas. This makes chemical contouring of water from the AOI difficult to interpret.

The streamline propagation expression was constructed as follows:

iff(mintg15,mintg15,nbmax8(mintg15#,mindir#=nbpos))

where: **mintg** is a raster map containing target pixels from which the propagation starts;
mindir is a direction map derived from the DEM;
nbmax8 is a neighbourhood operator that returns the highest neighbouring value for each pixel.

This expression was run through an iterative sub-routine until no further changes were made to the map.

Scanned, georeferenced aerial photographs of the AOI were used to locate decant structures and drainage channels associated with the tailings impoundments. Two seepage/failure zones where also identified in the south of the AOI and flagged as starting pixels for stream propagation.

Once the streamline propagation was complete, pollution plumes associated with each streamline were generated to identify areas liable to contamination by waterborne pollution from each source, as follows:

iff(mintg15,mintg15,nbmax(mintg15#,mindem#>mindem))

In the streamline propagation statement, only the lowest pixel is recoded while in the plume propagation statement, all pixels that occur at an elevation lower than a flagged pixel are recoded. This results in a relief-controlled spreading of the propagation with distance from the target.

The DEM modelling was undertaken to provide a spatial framework for geochemical assessment of water quality in the AOI. Several water quality data sets were collected in AOI by the mine and by the author. These were assessed for completeness and then used to characterise the chemical environment of the AOI.

3.2 SURFACE WATER

Samples collected across the AOI were loaded into the ILWIS® GIS after checking anion/cation balances in WATEVAL, a water quality evaluation program written by A.W. Hounslow and K.D. Goff (q.v. Hounslow, 1995 [7]). Most analyses were found to have a balance of around 10%, which while high, is adequate for the purposes of this study.

This Piper plot in Figure 3 shows a natural amelioration of process water-derived contaminants with distance downstream. TDS decrease, calcium concentration decreases slightly and sulfate concentration decreases dramatically.

In ILWIS, planar moving surface interpolation was performed on point maps with X-Y positions defined by sample sites and values defined by analyses for the determinant in question. The resulting maps showed a continuum of water quality values across the AOI. This is an unrealistic representation of surface water. The interpolated value maps were digitally overlaid with the plume maps to produce surface water values constrained by the spatial extent of the plumes.

From the sulphate map in Figure 4, it is clear that areas with high sulphate concentrations overlap with urban land uses in several instances. The flatter topography in the north of the map results in runoff from Dump 15A affecting a larger area that that from the older dumps draining towards the east and south. The highest sulphate concentrations in the AOI occur in the waters draining the Nkana Plant area.

The use of a DEM has shown that surface water from the Mindola dam is unlikely to flow to the Ichimpe Stream as had previously been reported [8]. This stream is an important hydrological conduit as may connect a pollution source with a drinking water abstraction station. The streamlines predicted by the DEM employed here have to be verified through the generation of another model from different data.

High calcium concentrations, shown in Figure 5, are associated with operational tailings disposal facilities. This is expected, as liming is undertaken for pH control. Copper concentrations in surface water peak in the drainage from the plant as with sulphate. Iron exhibits an interesting distribution as the highest values (360 mg/l) are found east of the Kafue River over granite rocks with no mining or mineralisation present.

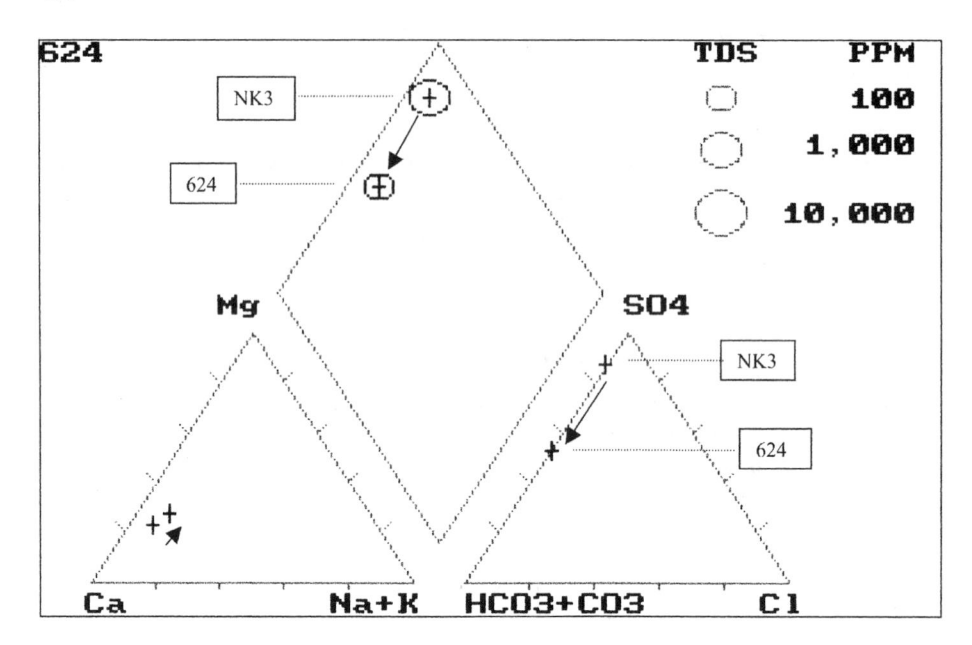

Figure 3. Changes in water chemistry in the surface stream from the dyke failure at tailings dam 33 (site NK3) downstream to sample site 624, before the confluence with the Kafue. Plot produced in WATEVAL [7].

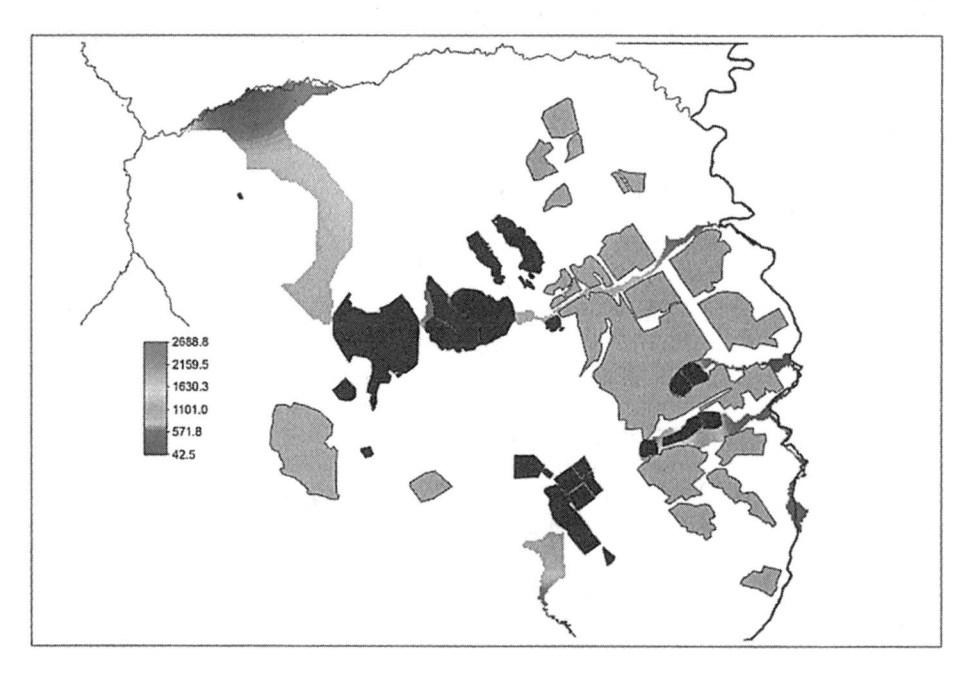

Figure 4. Spatial variation of SO_4 concentration (mg/l) in identified surface water pollution plumes.

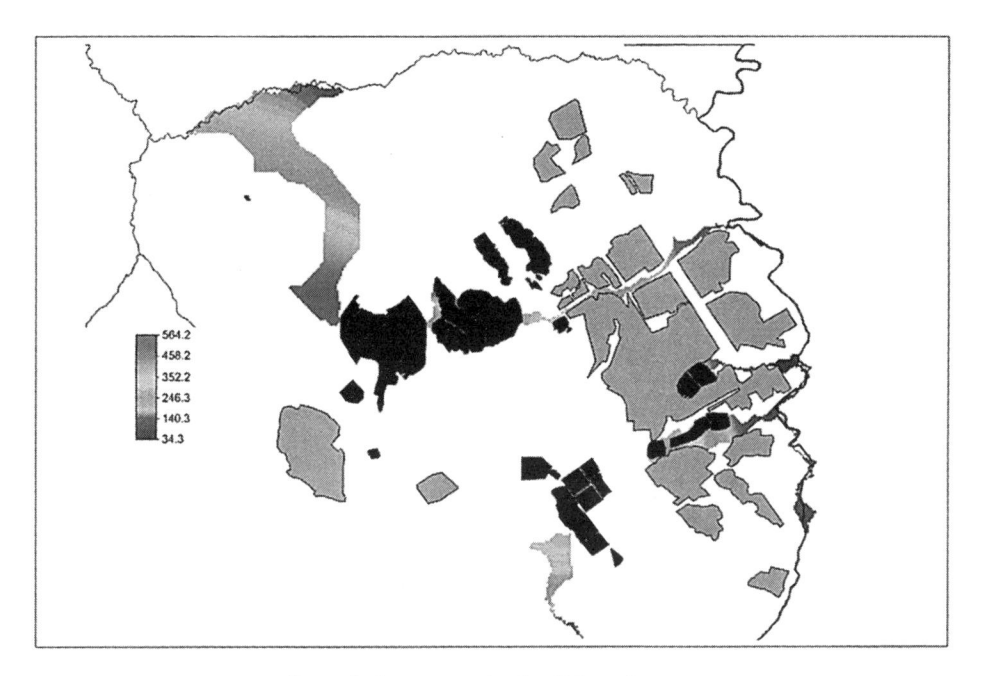

Figure 5. Ca concentration (mg/l) in surface waters.

3.3 GROUNDWATER

Groundwater samples were collected by the mine, and represent samples of water taken from below the soil horizon. The mineralised strata in the AOI occur within the sediments of the Katanga System. These rocks lie unconformably on crystalline basement granites and gneisses. Within the AOI, the Katanga sediments occur within the Chambishi-Nkana Basin, the boundary of which is shown in Figure 1. Some strata within this basin, such as the dolomites of the Upper Roan, are known aquifers and therefore treating the area as a homogenous groundwater surface is questionable. Attempts at interpolating samples within the basin separately from those occurring outside of the basin yielded poor results due to insufficient sampling density. For this reason, maps are presented which depict free flow of groundwater across the basin boundary.

High calcium concentrations in groundwater, shown in Figure 6, are associated with tailings facilities as is the case with surface water. The lack of a sample point near Dump 15A severely reduces the value of this dataset as contamination of the groundwater near this facility is most likely.

Groundwater sulphate concentrations within the AOI, shown in Figure 7, are generally low compared to sites located in other lease areas. There is a general west-east decrease in sulphate concentrations. Groundwater alkalinity (as mg/l $CaCO_3$) exhibit the inverse trend as is to be expected. Metals such as iron and copper have the highest concentrations in the north of the AOI near Chambishi Mine and exhibit a west-east trend similar to sulphate over the rest of the AOI.

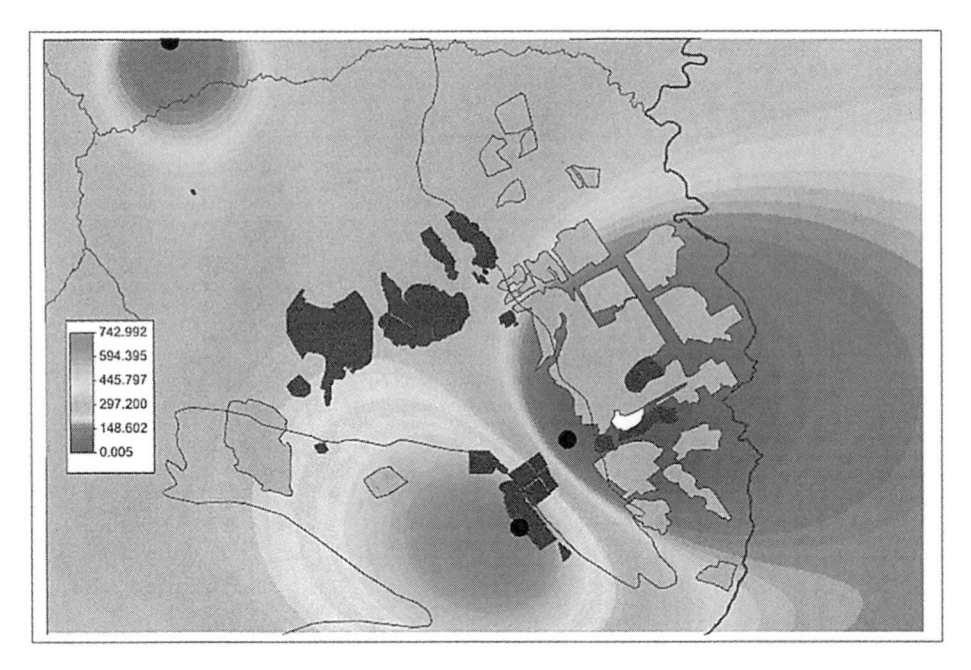

Figure 6. Groundwater Ca concentrations interpolated assuming homogeneous geology. The white area adjacent to the old tailings impoundment is of undefined Ca concentration value due to insufficient sampling points.

4. Conclusions

Elevation modelling indicates that surface flow from the Mindola Dam north overflow to the Ichimpe stream is unlikely due to intervening relief. If Mindola water is reaching the Ichimpe, then the ground water flow must play an important role. This is a potentially important pathway for present and future contamination of Kitwe's water supply and must be investigated further.

Pollution arising from old, ring-dyke tailings dams south of Kitwe is significant and is as important as that arising from Dump 15A.

This study has provided a basis for future investigations. An important aspect not quantified in this investigation are surface runoff volumes. Historical rainfall data is available and can be used in a hydrological model to estimate mass loadings of pollutants in each of the catchment areas identified here.

Together with runoff, sediment transport through the AOI landscape must be investigated as this is likely to be a significant vector for pollution transport.

The concurrent use of GIS and chemical models allows visualisation of data, highlighting data gaps and spatial trends. These can be interpreted with the aid of chemical trends in the data.

459

Figure 7. Groundwater SO$_4$ concentrations interpolated assuming a homogeneous geology. Black circles represent sampling sites; white the area has undefined SO$_4$ concentration due to insufficient sampling sites.

References

1. Baudo, R. (1987) Heavy metal pollution and ecosystem recovery, *Ecological Assessment of Environmental Degradation, Pollution and Recovery*, Lectures of a course held at the Joint Research Centre, ed. O.Ravera, Ispra, Italy, 12-16 October, 1987, Elsevier Science, Amsterdam, pp. 325-352.
2. Berglin, T. (1997) *Environmental Impacts of Mine Waste - a Study of the Possible Sources to the Elevated Metal Concentrations in the Kafue River, Zambia*, Master's Thesis, Department of Environmental Planning and Design, Luleå University of Technology, Luleå, 55 pp.
3. Davies, B.E. (1987) Mining effects on ecosystems and their recovery, *Ecological Assessment of Environmental Degradation, Pollution and Recovery*, Lectures of a course held at the Joint Research Centre, ed. O.Ravera, Ispra, Italy, 12-16 October, 1987, Elsevier Science, Amsterdam, pp. 313-323.
4. Florinsky, I.V. (1998) Accuracy of topographic variables derived from digital elevation models, *Int. J. Geographical Information Science*, Vol.12 No.1, Taylor & Francis, pp.47-61.
5. Ghadiri, H. & C.W. Rose (1992) An introduction to non-point source pollution modeling, *Modeling Chemical Transport in Soils*, CRC Press, pp. 1 – 14.
6. Hedström, A. & L. Osterman (1996), *A Study of the Suspended Solids in the Mwambashi River, a Tributary of the Kafue River, Zambia*, Masters Thesis, Luleå University of Technology, Luleå, 82 pp.
7. Hounslow, A.W. (1995), *Water Quality Data - Analysis and Interpretation*, CRC Press, Boca, Ranton, Florida.
8. Limpitlaw, D. (1998) Environmental impact assessment of mining by integration of remotely sensed data with historical data, *Proceedings of the Second International Symposium on Mine Environmental Engineering*, Brunel University, West London United Kingdom, 28-31 July.
9. Moore, I.D., A. Lewis & J.C. Gallant (1993) Terrain attributes: estimation methods and scale effects, *Modelling Change in Environmental Systems*, eds. A.J. Jakeman, M.B. Beck & M.J. McAleer, John Wiley & Sons Ltd., pp. 189 – 214.
10. Mwale, A.H. (1996) *An Overview of the Environmental Impact of the Copper Industry on the Quality of the Water and Sediment of the Kafue River, Zambia*, a paper presented at YES '96, EIZ, Luanshya District, 1996, 22 pp.

11. **ZCCM** (1996) *MONTHLY REPORTS "Nkana Concentrator – Monthly Reports on Operations"*, ZCCM Library, Kalulushi, 1988 – 1996.
12. ITC (1998) LWIS 2.2 for Windows, the Integrated Land and Water Information System. Installation and New Functionality. Enschede, The Netherlands, International Institute for Aerospace Surveys and Earth Sciences, 362 p.

ENVIRONMENTAL IMPACT OF EXPLORATION OF PYRITE AND STIBNITE IN THE MALÉ KARPATY MOUNTAINS, SLOVAKIA

S. TRÍKOVÁ
*Faculty of Natural Sciences, Comenius University,
Department of Mineralogy and Petrology,
Mlynská Dolina G, 842 15 Bratislava, Slovakia*

Abstract. Exploitation of pyrite ores and hydrothermal Sb-deposits in the Malé Karpaty Mts. culminated mostly at the break of 19th and 20th century. The host rocks of the mineralizations are black shales in the actinolitic schists and amphibolites. Streams, soils and the waters in the neighborhood of the abandoned deposits are contaminated by toxic metals, such as As, Sb, Zn, Zn, Al, and local acidification arises as well. Chemolithotrophic bacteria were identified in the mine-waters with pH 2.5 – 3. As solid secondary phases, gypsum and jarosite occur in weathered black shales and weakly crystallized young Fe ochreous precipitates mostly in streams and outflows of mine drainage. We expected the presence of ferrihydrite, schwertmannite, goethite and poorly crystalline ferric arsenate-sulphate precipitates (?) in ochres. According to the chemical composition, several types were recognized: depending on the primary mineralization As-, Si-, Al-, and SO_4-rich ochres can be present. Sb, Zn, Ni, Pb, Cd are present in lower concentrations. Accumulation of Ti, P, Ca, Na, Na and K indicate the decomposition of the rock-forming minerals.

1. Introduction

Slovakia is famous due to the abundance of ore deposits which were exploited mostly in the past centuries. Beside the largest ore-mining districts situated in the central Slovakia, the ore deposits in the Malé Karpaty Mts. played an important role. Their location and geology are shown in Figure 1.

The Malé Karpaty Mts., situated on the western rim of the West Carpathians are a mountain ridge with a complex horst/nappe structure. Economically important ore mineralizations have evolved in the lower- to middle Devonian [13] volcano-sedimentary formation, later metamorphosed in the amphibolite facies. According to Chovan *et al.* (1992) [9] two mineralization types occur:

I. Metamorphosed, primarily exhalation-sedimentary pyrite mineralization.
II. Hydrothermal mineralization, which is subdivided into: 1- molybdenum in granitoides; 2- copper-base metal with silver: 2a) Cu-Pb, Ag, (Ni); 2b) Pb-Zn; and 2c) Pb-Ag; and 3- antimony-gold: 3a) gold-sulphidic; 3b) gold-quartz; and 3c) stibnite.

Two of these subtypes were subject of the exploitation. The pyrite-pyrrhotite mineralization is a product of exhalation-sedimentary processes of submarine volcanism. Pyrite exploitation dates to the late 18th century, culminating between 1850 and 1896, [6]. The hydrothermal Sb-As-Au mineralization cuts through the pyrite-pyrrhotite one. Dominant ore minerals are arsenopyrite, pyrite, stibnite and gudmundite, and intense carbonatization took place as well. Two Sb deposits bound on this mineralization were exploited: the Pernek deposit (1790–1922), [10], where several abandoned dumps remained, and the Kolársky Vrch deposit (1790 – early 1990s) (see Figure 1). The latter one was equipped with flotation processing since 1906 [6]. Waste is deposited on three tailing impoudments.

Both of these mineralizations are bound to the lenses of black shales in the actinolitic shists and amphibolites, as shown in Figure 1.

Weathering of open deposits, dumps and tailing impoudments causes:

- pollution of water, soil, and alluvia in the surrounding area, main contaminants being As and Sb [12, 20];
- local acidification [18, 19];
- precipitation of supergeneproducts – wide variety of secondary minerals has been reported: alophane, azurite, cervantite, gypsum, halloysite, hyalite, jarosite, kaolinite, kermesite, malachite, Mn oxides and hydroxides, senarmontite, siderite, Sb ochres, schafarzikite, valentinite [1, 6, 10], chapmanite [15], scorodite [21];
- local decrease of phytocensoses, especially in close neighbourhood of dumps with black shales [18].

The most distinctively pronounced hypergenous process is the precipitation and deposition of Fe ochreous precipitates. However, there was no adequate attention paid to these minerals up to now except for the mention of "limonite" occurrence [6] as a weathering product of pyrite.

The aim of this study is a field survey of the old workings and a detailed investigation of the mine drainage and the young ferric ochreous precipitates.

2. Techniques and methods

Mapping was performed mostly on the old deposits where the information on placement of the dumps is usually lacking. Places of acid mine drainage outflows were to be recorded as well as areas where Fe ochres precipitate and settle. The pH of waters was measured and samples of water and solid secondary phases were taken for further analysis.

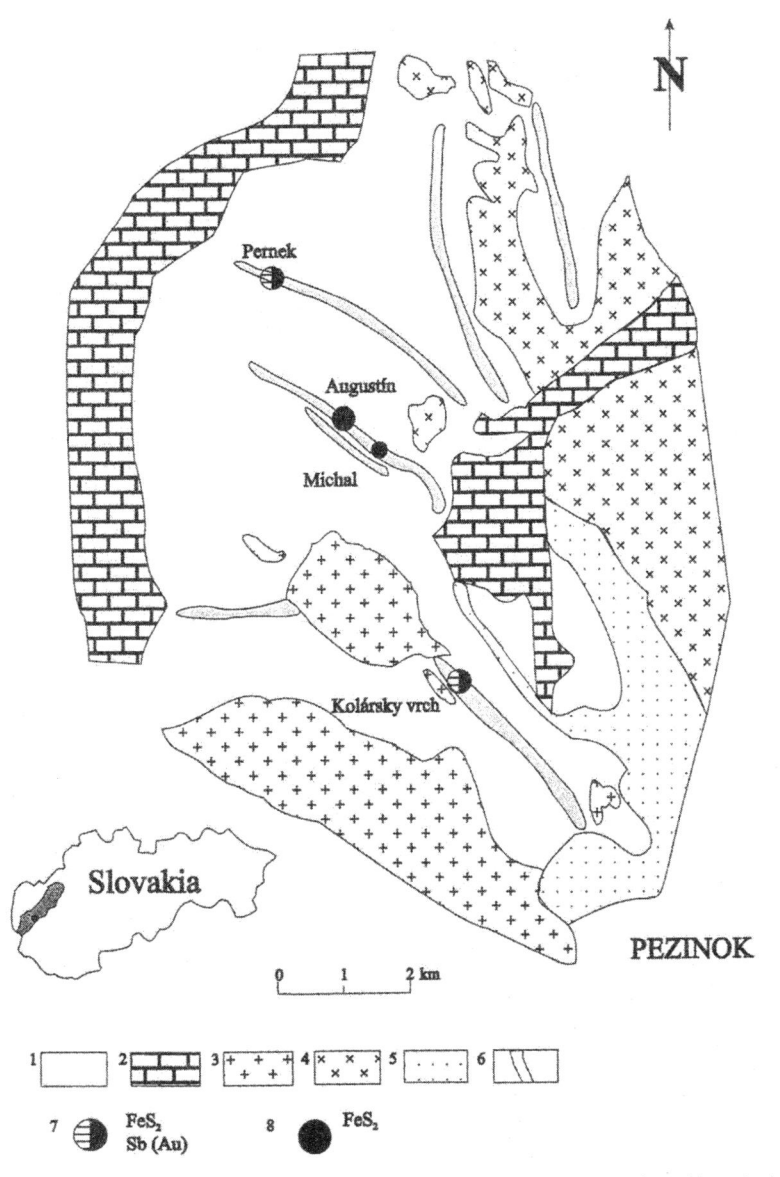

Figure 1. Schematic geological map of the pezinok-Pernek deposit crystalline complex with marked deposits and occurrences of raw materials (modified after [14]; and [6]). *Legend*: 1 – metamorphic rocks of the central deposit area; 2 – Mesozoic; 3- granites of the Bratislava Massif; 4 - granodiorites of the Modra Massif; 5 – Quaternary; 6 – productive zones with black shales and stibnite-gold and pyrite mineralization; 7 – deposits of Sb, Au and pyrite ores; 8 – pyrite deposits.

The sampling methods chosen follow those described in Bigham et al. (1996b) [5]. Sample material was poured into the plastic bottles after a wash in the local water. Organic matter (mostly remains of plants) and inorganic detritus were removed by sieving and sedimentation. The solid samples were then air-dried and properly packed to prevent the oxidation. The contents of the dissolved elements in the mine waters was estimated by atomic absorption spectrometry (AAS). In the XRD, CoKα radiation with a Philips PW1710 goniometer was used and the samples were scanned from 4 to 80° 2 Θ with increment of 0.02° 2 Θ and 0.5-0.8s per step. The amount of total oxidic Fe (Fe_{tot}) in the precipitates was determined using dissolution in concentrated HCl. Oxalate soluble Fe (Fe_o) was determined after dissolution in ammonium oxalate [16]. The samples were analyzed for Al (atomic emission spectroscopy with indicatively coupled plasma), As, Sb (AAS with hydride generation), Fe (AAS with atomization in air-acetylene flame), SO_4 (gravimetry), Cd, Cu, Mn, Pb, Si, Zn, Na and K (AAS) (Laboratories of Analytic Methods of Faculty of Natural Sciences, Comenius University, in Bratislava and at the Slovak Academy of Science, in Banská Bystrica). TEM micrographs were used for studying the morphology of precipitates (JEOL).

3. Results and discussion

3.1 THE AUGUSTÍN PYRITE DEPOSIT (POINTS 1 AND 2)

Most of the exploration was concentrated in one valley. The largest deposit was situated in the upper part of the valley (points 1 and 2, in Figure 2). Here, on the dumps and outcrops in the pyrite-bearing black shales the most intensive oxidation, dissolution and acidification take place. The high concentration of bacterial cells confirms that intensive biooxidation of the sulphides takes place. In the pyrite deposits, lithotrophic bacteria are present in the acid mine drainages. Secondary phases precipitate directly in the dump débris from acidic solutions, predominantly gypsum and jarosite. Stream and pit water is characterized by pH (2.5-3). SO_4 and Fe contents in the stream water (where ochreous precipitates occur) are relatively low as well as those of other metals, as shown in Table 1. Fe oxyhdroxide precipitates form in lesser amount, usually in the streams close to the source right below the dumps. Sulphate component and silica low concentrations in the ochres, as shown in Table 2. The other elements show low or trace contents too (Al, Ti, P, Zn, Cu, Pb etc.). These phases are poorly crystallised, nevertheless, they exhibit features typical for schwertmannite. Although the conditions are favourable for the precipitation of schwertmannite [2, 4, 5], and major part of the material is oxalate-soluble (see Table 4) it does not contain sufficient amount of SO_4 (see Tables 3 and 4). According to Bigham et al. (1994) [2], the Fe_{tot}/S_{tot} mole ratio for schwertmannite can vary from 5 to 8. The occurrence of schwertmannite is then unlikely unless we would not consider the substitution of some other anion to the sulphatic one.

3.2 THE MICHAL PYRITE DEPOSIT (POINT 3)

Deposits 1, 2 and 3 are interconnected by old adits, which are inundated by waters containing dissolved metals. Around point 3 and a few kilometers below (see Figure 2) the mine drainage flows out. In these places the most massive sedimentation of ochreous precipitates takes place, forming a red-coloured wetland of the area of about 500m². The chemical composition of the water is similar to that on Locality 1, however its pH ranges between 6.3-6.6 (see Table 4). Ochres absorb considerable amount of Al abd Si (6 wt%) but are poor in SO_4 (Table 2). They consist mainly of oxalate-soluble phase – ferrihydrite (Fe_o/Fe_{tot}=0.95). This mineral however shows poor cristallinity (2-line ferrihydrite, as shown in Figure 3). Goethite probably occurs as a minor component.

3.3 THE KOLÁRSKY VRCH DEPOSIT OF SB (POINTS 4 AND 5)

The Sb deposit with its tailing impoudment is situated at the mouth of the valley (see Figure 2). Increased contents of metals in the waters of this area refer to intensive oxidation [12]. The pH of this water ranges between 7.5-8 since the acidity here is buffered with high contents of Ca and alkalies (see Table 3). As, Sb, and Mn concentrations (see Table 1) are significantly higher here than on the other deposits. Ochreous precipitates are characterized by high contents of As and SO_4^{2-} (see Table 2). Fe_{tot}/S_{tot} mole ratio is 2.83 (see Table 4), Fe/As is 1.6. The material that consists of colloid-size particles is poorly crystalline (probably a mixture of ferric arsenate with sulphate or arsenic sulphate). Precipitates of similar composition were described by Leblanc et al. (1996) [11]. According to Carlson and Bigham (1992) [7], AsO_4^{2-} displaces SO_4^{2-} from the surface of the oxyhydroxysulphates, however, these were synthetised under *acidic* conditions. This displacement may occur either during or after precipitates of the mineral phase.

3.4 THE PERNEK SB DEPOSIT (POINTS 6 AND 7)

There is another Sb deposit situated on the western side of the mountain ridge of Malé Karpaty Mts. (see Figure 1). In the surface waters no distinct acidification was recorded, however, dumps are covered with ruderal vegetation. Waters flowing through the dumps exhibit higher Ca content. Inundated groves where pH varies between 5.5-7 "produce" a huge amount of Fe ochres in Si and Al (see Table 2). These very poorly crystallized precipitates (see Figure 3) consist predominantly of ferrihydrite or goethite and a minor amount of schwertmannite: Fe_{tot}/S_{tot} is 14-24. Higher contents of Si indicate presence of ferrihydrites. Natural Si-containing ferrihydrite were described by Carlson and Schwertmann (1981) [8]. The process of lattice substitution of Fe with Al in goethite is well known [17]. Hence the presence of also this mineral cannot be ruled out.

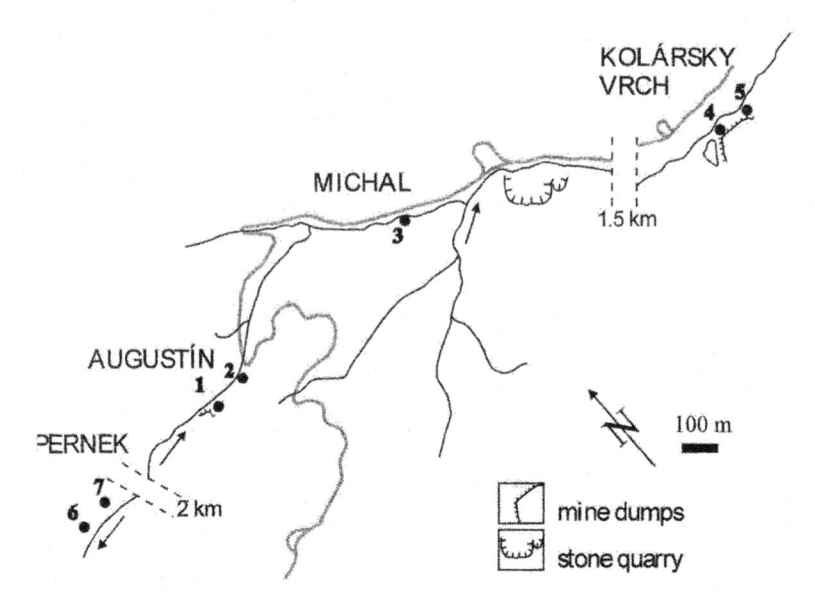

Figure 2. Schematic map of the mining area in the Malé Karpaty Mts. with location of the main sampling points.

4. Conclusions

The environmental impact of the acidification is not very pronounced, which might be due to the mountainous relief of study area. The acidity is there buffered by carbonates of hydrothermal origin in the Sb deposits. Meteoric waters acidify as percolating through the dumps and gather in the old mines; commonly they inundate them completely. Chemolithotrophic bacteria were identified in the acid mine drainages with pH = 2.5-3. Often the ochres precipitate already here and after being washed out they settle in streams and marshes. Because their volume is not negligible, they represent a constant supply reservoir of toxic elements. The chemical composition of the particular Fe ochreous precipitates is dependent on the type of deposit. Considerable sorption of Al and Si in the ochres was registered in old pyrite and stibnite deposits. The ochres, precipitating at the tailing impoundment of the Sb-deposit, are highly enriched in As and SO_4. The Sb contents in the mine drainage precipitates never exceed 530 ppm, even not in the stibnite deposits. Relatively homogeneous is the concentration of Zn and Pb – hundreds and tens of ppm, respectively. All the samples exhibit a certain sorption of Ti and P (600-3200 ppm), and in some samples also alkalies, Ca Mn, and Cu were recorded (Mn does not accumulate in the Fe ochre under acidic conditions). As a consequence of the different conditions, the mineral composition varies widely as well. The presence of ferrihydrite, schwertmannite, goethite and poorly crystalline ferric arsenate-sulphate precipitates (?) in ochres is likely. For more accurate identification of the mineral phases present, further analytical techniques have to be used such as color measurement, DTA, IR spectroscopy, Mössbauer spectroscopy as described in [2, 3, 4, 5, 7, 8].

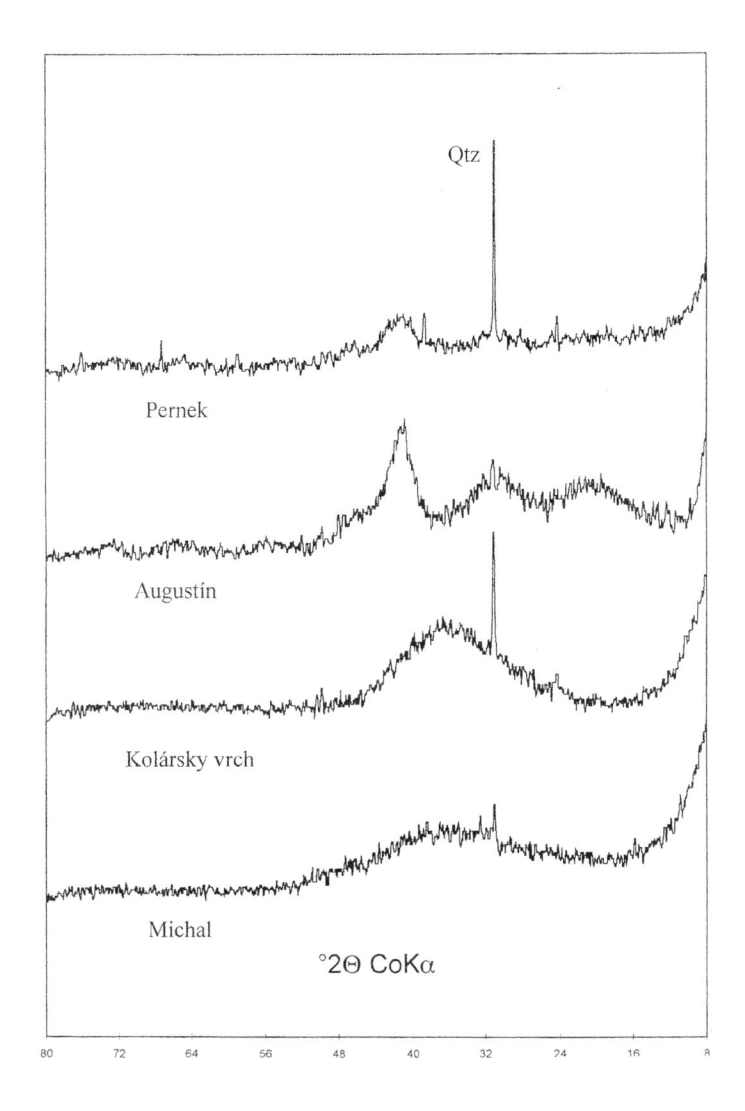

Figure 3. X-ray diffractograms of natural precipitations.

The ecological problem of the area is actually represented by the toxic metal pollution rather than by the acidification itself. Especially the disastrous rainfalls, so common in Central Europe in the past years, has increased significantly the mobility of polluted mine waters. One possible way to prevent further spreading and migration of the toxic elements is to detain the Fe precipitates soon after their formation and thereby to diminish the stocks of the potential pollutants.

Table 1. Water chemistry of the district (measure points from 1 to 7 are in Figure 2). Former data (*) are from Šucha et al. (1996) [18].

Sample	Fe2+	Fe$_{tot}$	SO4	Al	As	Sb	Cu	Zn	Mn	Ca	Na	K
							ppm					
Augustín II (point 1)*	*2.50	*3.38	*295.30	*2.11	*0.00	*0.00	*0.13	*0.41	*0.41	-	-	-
Augustín II (point 1)	-	0.75	-	-	-	-	4.15	1.90	0.75	9.02	6.35	0.80
Augustín II (point 2)	*12.66	*28.61	*791.70	*31.03	*0.00	-	*0.77	*1.27	*0.79	-	-	-
Michal (point 3)*	*6.75	*10.44	*280.20	*2.84	*0.01	*0.00	*0.06	*0.37	*0.42	-	-	-
Michal (point 3)	-	0.00	-	-	-	-	0.00	0.10	0.20	12.17	8.40	0.80
Kolársky vrch (points 4,5)	-	0.08	-	-	3.35	2.51	0.10	0.10	3.15	24.19	23.20	22.40
Pernek (points 6,7)	-	0.03	-	-	0.00	0.31	0.08	0.00	0.08	107.50	4.93	0.90

Table 2. Chemical composition of Fe ochreous precipitations, main major components, A11 and A12 – Augustín deposit, M1 – Michal deposit, K 11 and K 12 – Kolársky vrch deposit, PER1 and PER 5 – Pernek deposit, (measured points from 1 to 7 are in Figure 2)

Sample	Fe2o3	SiO2	SO3	Al2O3	As2O5
			%		
A11 (1)	50.34	1.84	1.12	0.68	0.05
A12 (1)	46.33	2.52	0.82	0.59	0.02
M1 (3)	37.18	12.88	0.23	12.47	0.02
K11 (4)	34.68	2.57	13.31	0.32	24.07
K12 (5)	36.56	6.91	12.51	1.08	21.40
PER1 (6)	51.19	19.17	3.31	8.88	0.06
PER5 (7)	55.63	10.10	2.32	3.74	0.06

Table 3. Chemical composition of Fe ochreous precipitations, minor and trace elements (measured points from 1 to 7 are in Figure 2).

Sample	Sb	Cu	Zn	Pb	Cd	P	Ti	Mn	Ca	Na	K
					ppt						
A11 (1)	239	80	400	25	13	3200	2800	0	396	600	40
A12 (1)	236	80	600	18		2700	2500	0	136	640	80
M1 (3)	191	720	840	24	12	2000	2400	80	4352	640	80
K11 (4)	527	0	606	19	472	2900	2300	1980	2832	727	323
K12 (5)	490	0	825	16	359	1600	1800	412	10837	701	124
PER1 (6)	322	523	243	37	–	1200	1100	597	1800	467	131
PER5 (7)	320	516	536	39		1000	600	1231	0	318	60

Table 4. Properties of natural Fe ochreous precipitations (*) – mole ratio, Fe$_o$ – Fe extractable in oxalate, Fe$_{tot}$ – Fe extractable in HCl. (measured points from 1 to 7 are in Figure 2).

Sample	pH	Fe$_t$	Fe$_o$	SO3	Fe$_o$/Fe$_{tot}$*	Fe$_{tot}$/S$_{tot}$*
			%			
A11(1)	2.5-3	35.20	34.16	1.12	0.97	44.95
A12(1)	2.5-3	32.40	29.28	0.82	0.90	56.50
M1(3)	6.3-6.6	24.24	24.36	13.31	1.00	2.61
K11(4)	7.5-8	25.57	24.19	12.51	0.95	2.92
K12(5)	7.5-8	26.00	24.75	0.23	0.95	161.65
PER1(6)	5.5-7	35.80	32.83	3.31	0.92	15.47
PER5(7)	5.5-7	38.90	35.40	2.32	0.91	23.98

5. Acknowledgments

This research was supported mostly by Comenius University, grant UK 1518/97. The many helpful comments by Liisa Carlson and Martin Chovan are gratefully acknowledged. We thank L. Puškelová, L. Osvald, V. Streško, and P. Andráš.

REFERENCES

1. Andraš, P. and Chovanec, V. (1985) Identification of supergenous minerals in Sb ors in Pezinok (in Slovak). *Miner. Slov.* 17(4), 335-339.
2. Bigham, J. M., Carlson, L. and Murad, E. (1994) Schwertmannite, a new iron oxyhdroxysulfate from Pyhäsalmi, Finland, and other localities. *Miner. Mag.* 58, 641-648
3. Bigham, J.M., Schwertmann, U., Carlson, L. and Murad, E. (1990) A poorly crystallized oxyhydroxysulfate of iron formed by bacterial oxidation of Fe (II) in acid mine waters. *Geochim. Cosmochim. Acta* 54, 2743-2758.
4. Bigham, J. M., Schwertmann, U. and Pfab, G. (1996a) Influence of pH on mineral speciation in bioreactor simulating acid mine drainage. *Appl. Geochim.* 11, 845-849.
5. Bigham, J. M., Schertmann, U., Traina, S. J., Winland, R. L. and Wolf, M. (1996b) Schwertmannite and the chemical modeling of iron in acid sulfate waters. *Geochim. Cosmochim. Acta* 60(12), 2111-2121.
6. Cambel, B. (1959) Hydrothermal deposits in the Malé Karpaty Mts., mineralogy and geochemistry of their ores (in Slovak). *Acta. Geol. Geogr. Univ. Comen., Geol.* Vol. 3, Bratislava, 538 pp.
7. Carlson, L. and Bigham, J. M. (1992) Retention of arsenic by precipitates from acid mine drainage. V. M. Goldshmidt Conference, Reiton, VA.
8. Carlson, L. and Schwertmann, U. (1981) Natural ferrihydrites in surface deposits from Finland and their association with silica. *Geochim. Cosmochim. Acta* 45, 421-429.
9. Chovan, M., Rojkovič, I., Andráš, P. and Hanas, P. (1992) Ore mineralization of the Malé Karpaty Mts. *Geol. Carpath.* 43, 275-286.
10. Koděra, P. ed. (1990) *Topographic mineralogy of Slovakia* (in Slovak). Vol. 2, 1st edn. SAV, Bratislava, 1590 pp.
11. Leblanc, M., Achard, B., Othman, D. B. and Luck, J. M. (1996) Accumulation of arsenic from acidic mine waters by ferruginous bacterial accretions (stromatolites). *Appl. Geochim.* 11, 541-554.
12. Letko, V., Sergejev, V. I. and Šimko, T. G. (1992) *Investigation of the pollution of environment from the tailing impoudment in Pezinok* (in Slovak). pp. 123, Manuscript IG PF UK, Bratislava.
13. Planderová, E. and Pahr, A. (1983) Biostratigraphical evaluation of weakly metamorphosed sediments of Wechsel series and their possible correlation with Harmónia Group in the Malé Karpaty Mts. *Miner. Slov.* 15(5), 358-436.
14. Polák, S. and Rak, D. (1980) Prognostic probleme of antimony metalogeny in the Malé Karpaty Mts. *In Antimony ores of Czechoslovakia*, pp. 69-87. Geol. Inst. D. Štúr, Bratislava.
15. Polák, S. (1983) Chapmanite from Pezinok (in Slovak). *Miner. Slov.* 15 (6), 565-566.
16. Schwertmann, U. (1964) differenzierung der Eisenoxide des Bodens durch Extraktion mit Ammoniumoxalat-Lösung. Zeits. *Pflanzenernähr. Düng. Bodenk.* 105, 194-202.
17. Schwertmann, U. and Murad, E. (1983) the effect of pH on the formation of goethite and hematite from ferrihydrite. *Clays & Clay minerals* 31, 227-284.
18. Šucha, V., Banásová, V., Dlapa, P., Chovan, M., Lintnerová, O., Miadoková, E., Rojkovič, I., Tríková, S., and Zlocha, M. (1996). The complex model of the environmental impact of acidification in the main Slovak mining districts (in Slovak). Manuscript. Bratislava, 118 pp.
19. Tríková, S., Chovan, M. and Kušnierová, M (1997) Oxidation of pyrite and arsenopyrite in the mining wastes (Pezinok-Malé Karpaty Mts.). *Folia Fac. Sci. Nat. Univ. Mas. Brun., Geologia* 39, 159-167.
20. Veselský, J., Forgáč, J. and Mejeed, S. Y. (1996) Contamination of the soil and stream sediments in the Malé Karpaty Mts. *Miner. Slov.* 28(3), 209-218.
21. Uher, P. (1990) Scorodite from Sb deposit Pezinok (in Slovak). *Miner. Slov.* 22(5), 451-254.

PARTITIONING OF HEAVY METALS IN SEQUENTIAL EXTRACTION FRACTIONS IN SOILS DEVELOPED OVER THE HISTORICAL "Sv. JAKOB" SILVER MINE, MOUNT MEDVEDNICA, CROATIA.

M. ČOVIĆ[1], G. DURN[1], N. TADEJ[1] and S. MIKO[2]
[1]*Faculty of Mining Geology and Petroleum Engineering*
Department for Mineralogy, Petrology and Geology of Mineral Deposit,
Pierottijeva 6, HR-10000 Zagreb, Croatia
[2]*Institute of Geology, Department for Mineral Resources, Sachsova 2 HR-10000, Zagreb, Croatia*

Abstract. Phase-selective sequential extraction techniques have been used to identify the residence sites of metals in soils developed over a historical silver-mining site. The operationally defined phases selected for extraction have been assigned to the following five fractions: adsorbed, bound to carbonates, bound to Fe-Mn oxides, bound to organic matter and residual. The following reagents had been used: NH_4-aceate, Na-acetate, hydroxylamine hydrochloride, hydrogen peroxide + nitric acid, mixed acids (HCl-HNO_3-HF), respectively. The solutions were analysed by atomic absorption flame photometry (Pb and Cd) and inductively coupled plasma atomic absorption photometry (other analysed elements). Mineralogical analysis was performed to detect Zn-Pb-bearing phases and cerussite was the only phase detected. Other mineral phases detected by XRD in the soil samples were as follows: quartz, dolomite, micas, plagioclase, K-feldspar, goethite, hydrargillite, kaolinite, chlorite, and organic matter. Assuming that mobility and biological availability are related to the solubility of the geochemical forms of the metals and the latter decreases in the order of extraction, the apparent mobility and potential metal bio-availability for these highly contaminated soils is: Cd> Pb> Zn> Cu> Ni. The distribution of Pb, Zn and Mn in mineral phases is similar in samples with both high and baseline trace-metal values. Cu, Fe, and Ni exhibit different distribution patterns in the two types of samples.

1. Introduction

Sequential extraction schemes are widely applied in both environmental and exploration geochemistry where the most liable fraction of elements, in the case of pollution studies, is attributed to anthropogenic sources, while in the case of exploration studies theses fractions are assumed to derive from a geogenic source related to mineralisation [5].

The purpose of sequential extraction in geochemical prospecting concerns the determination of metal-bearing phases in soils, the enhancement of geochemical signatures and the discrimination between lithological and environmental effects [2, 4, 10, 15]. Harrison *et al.* (1981) [7] have suggested that mobility and bioavailability of the metals decrease approximately in the order of the extraction sequence. The operationally defined extraction sequence follows the order of decreasing solubility of the geochemical forms of the metals. Hence, the exchangeable fraction may indicate which metals are the most available for plant uptake [16].

Geochemical studies that deal with distributions of various potentially toxic elements in topsoil, covering a wide area of both urban and pristine environs of Zagreb, have intensified in the last few years. One such study [12] was conducted towards the evaluation of the heavy metal content of roadside soils in the wider region. It found that a shallow soil sample (0-15 cm deep) taken in the vicinity of a silver mine that was not in operation for more than 300 years contained over 3,000 mg/kg of lead. Later studies identified anomalous concentrations of lead (from 9 to 18,000 mg/kg), zinc (from 12 to 9,000 mg/kg), copper (from 5 to 370 mg/kg), cadmium (from B.L.D. to 180 mg/kg) and mercury (from 0,1 to 1,8 mg/kg) is soils covering the area of the old mining activity [3]. The distribution of these metals in soil indicates their inherited and undisturbed geochemical relationship to the Pb-Zn sulphide mineralization. The mean baseline concentrations of these elements (aqua regia extracted) in forest soils and uncultivated soils of the region range from 5 to 546 mg/kg for Pb, Zn ranges from 12 to 1,250 mg/kg, Cd from 0,01 to 22,50 mg/kg, Cr from 9 to 161 mg/kg, Ni from 4 to 651 mg/kg, Mn from 16 to 4,533 mg/kg and Fe from 1,13 to 13,49 mg/kg [13]. The area studied is covered by typical forest brown acid soil developed on metamorphic rocks.

The use of total metal concentration cannot be a criterion to assess the potential effects of soil contamination because all forms of a given metal have not an equal impact on the environment [20]. A sequential extraction analysis was employed to delineate trace metal partitioning in soil samples with both background and high trace metal values.

The soil samples collected in the proximity of a mineral deposit are expected to have a proportion of the associated metals present as: (1) major constituents of primary ore minerals (galena, sphalerite, pyrite), and also (2) major constituents of secondary minerals (cerussite, pyromorphite, and anglesite). Those minerals are alteration products of primary ore minerals, or were formed by precipitation from metal-rich solutions. For those reasons, X-ray powder diffraction analyses of soil samples with high heavy metal concentrations were performed to identify those mineral phases.

2. Geology and mineralization

The historical "Sv. Jakob" ("Zrinski") Ag-Pb mine lies on the southeastern slopes of central Mt. Medvednica in northwest Croatia near Zagreb, capital of Croatia, as shown in Figure 1. Mt. Medvednica is a part of the Pannonian basin and according to Hass *et al.* (1995) [8] it is a SW extension of the Mid-Transdanubain terrain. Its NW part, consists of an extensive Lower Cretaceous magmatic-sedimentary rock complex that is in tectonic contact with a complex of low-grade metamorphic rocks of unknown age (Middle to Late Palaeozoic?) towards the NE and SE. These complexes are the basement of the mountain that is laterally bounded by Upper Cretaceous and Tertiary sedimentary rocks [1, 18].

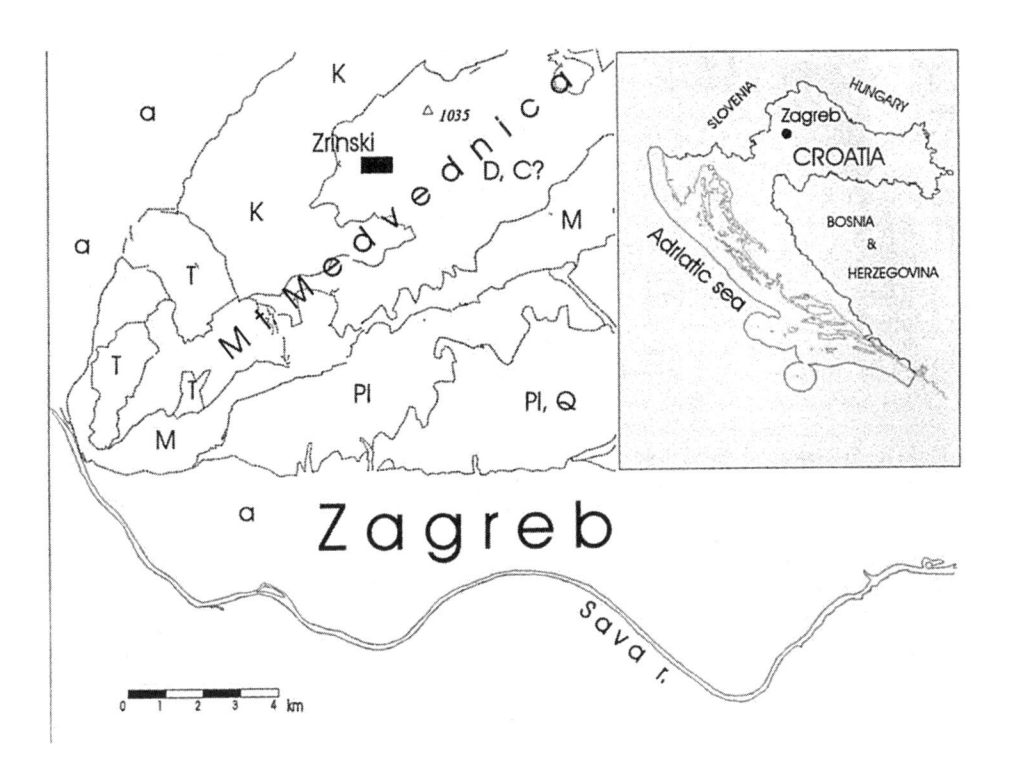

Figure 1. Geographical position of investigated area and generalized geologic map of the Mt. Medvednica. Geological legend: a – alluvium (gravel, sand and clays); Pl, Q – Pliopleistocene (gravel, sand and clays); Pl – Pliocene (marls, sand, clays and sand); M-Miocene (marly limestone, marls, limestone and sandstone); K- Cretaceous (breccia, sandstone, limestone, etc.); T – Triassic (limestone, dolomites, shale, etc.); D, C? – Devonian, Carboniferous (metamorphic volcano sedimentary complex). (After [18]).

The "Sv. Jakob" Ag-Pb deposit occurs in dolomites situated within the metamorphic rock complex at an altitude of 890 m a.s.l. The ore occurs as veinlets and lenses of galena in dolomites. It is a simple paragenesis consisting of galena, sphalerite, pyrite, dolomite, quartz and secondary minerals anglesite and cerussite [19]. The most abundant ore mineral is coarse-grained galena witch sporadically exhibits banded cleavage fractures. Sphalerite is less abundant and usually occurs on the contact between galena and dolomite or as small inclusions within galena grains. Pyrite is present in very small quantities and galena and sphalerite replace it. The galena ore gave in average 500 mg/kg of silver, and mining activities ceased in the late 17th century.

The epigenetic character of ore, the simple paragenesis, the absence of copper sulphides or their solid inclusions in bright sphalerite make this mineralization similar to Pb-Zn deposits in carbonates of Triassic age (Mississippi valley type), [19]. The galena gave a Pb-Pb model age of 293 to 331 Ma [14].

3. Methodology

3.1 FIELD METHODS (SAMPLE COLLECTION)

Ten sample sites were selected so that the results of studies performed previously [3] were used as a guideline to obtain soil samples with both low and high metal values. The distance between individual sample sites was 25 m along a profile on a north-south direction. The thickness of the soil profile in the studied area varied from 10 to 40 cm. Samples were taken from spade-dug pits below the O horizon, between 5 and 15 cm below the surface. All soil samples were air-dried, homogenized and sieved to <2.0 mm, and pulverised.

3.2 SEQUENTIAL EXTRACTION SCHEME

The sequential extraction of soil samples (1 g of pulverised sample) was performed to give the following five fractions: adsorbed (AD), bound to carbonate (CC), bound to iron and manganese oxides (FEMN), bound to organic matter (OR) and residual (RES). The procedure was slightly modified according to methods given by Tessier et al. (1979) [20] and Hall et al. (1996) [6]. The following reagents had been used: NH_4-acetate, Na-acetate, hydroxylamine hydrochloride, hydrogen peroxide + nitric acid, mixed acids (HCl-HNO_3-HF), respectively. All reagents were of analytical grade. Table 1 presents a detailed review of the procedure used in this study. It must be recognized that the results obtained are "operationally defined", i.e. selectivity is not 100% and is dependent upon such factors as the chemicals employed, the time and the nature of contact, sample to volume ratio, etc. Results will also be dependent on the grain size fraction chosen for analysis [6].

3.3 MINERALOGICAL ANALYSIS

The XRD patterns of bulk soil samples (M-1, 2, 3, 4, 5 and 8) were obtained. In order to remove dolomite and other carbonates, sample M-1 was treated by acetic acid; warm diluted HCl (pH 3-4), and warm 18% HCl. After wet sieving, the heavy fraction from sample M-1 was separated with bromoform from the particle size fraction between 0.045 and 0.1 mm.

3.4 ANALYTICAL METHODS

All solutions were analysed for Zn, Cu, Fe, Mn and Ni by a Jobin Yvon 50P simultaneous inductively coupled plasma-atomic emission spectrophotometer (ICP-AES) and Pb and Cd by a Pye Unicam SP9 flame atomic absorption spectrophotometer (AAS) using an air-acetylene flame. The analytical sensitivities for Cd, Cu, and Ni were 0.15, 0.2 and 0.2 mg/kg, respectively. Standard metal solutions were made by dilution of 1,000 stock solutions (Plasma Chem) and matrix matched. Precision and accuracy was based on the analyses of the reference geological samples GXR-5 and GXR-6. Mineral composition of selected samples was analysed with a Philips diffractometer ($CuK\alpha$ radiation) equipped with a graphite monochromator and proportional counter.

4. Results and discussion

4.1 MINERALOGICAL DATA

The results of the mineralogical analyses are shown in Table 2. The dominant minerals in all samples were quartz and micas, a fact that is a direct consequence of the weathering of a mica-schist bedrock. Dolomite was present in samples M-1 and M-2. Global sample M-2 had a diffuse diffractogram because of the presence of organic matter. Plagioclase, K-feldspar, kaolinite, goethite, and hydrargillite were present sporadically. Cerussite was the only secondary ore mineral detected in the samples, and its presence was determined in the heavy fraction of sample M-1.

4.2 PARTITIONING OF METALS IN SOILS

Table 3 shows results of the sequential extraction analysis in mg/kg. The total concentrations represent the sum of the element concentrations determined for each of the five individual extractions.

4.2.1 Lead
The results from the sequential extraction clearly show that Pb partitioning in samples with high sum of concentration and in samples with baseline content are similar, as shown in Figure 2. The CC and FEMN fractions act as predominant carriers, representing about 30-85% and 30-65% of total Pb concentration, respectively. The OR fraction is third in importance with 5-15% of total Pb. The remainder is distributed more or less equally between the AD (ca. 0.4-5% of total Pb) and the RES (ca. 0.5-5% of total Pb) fractions.

4.2.2 Zinc
The Zn partitioning is similar for samples with high concentration as for samples with baseline concentration of Zn, as shown in Figure 3. A major portion of the Zn is associated with the fraction OR (30-65% of total Zn content). Fraction FEMN is next most important with a range from 20% to 35% of the total Zn. The remainder is bound to the RES fraction (ca. 4-30% of total Zn). Less than 15% of total Zn was extracted in the first two steps.

4.2.3 Cadmium
The major part of Cd is bound to the CC and the FEMN fractions, as shown in Figure 4, with 18-48% and 23-48% of total Cd concentration, respectively. These results agree with the observations of Hickey and Kittrick (1984) [9], and are expected to come from possible solubility controls for Cd [11], and from the ease of substitution of Cd^{2+} for Ca^{2+} in calcite. The AD fraction is the next most important one with a range from 9 to 23% of total Cd content. The remainder is bound to the OR fraction (ca. 7-22% of total Cd). The Cd concentrations in the RES fractions of all samples were below the detection limit. Cd is bound to carbonates only in samples with baseline content of Cd.

476

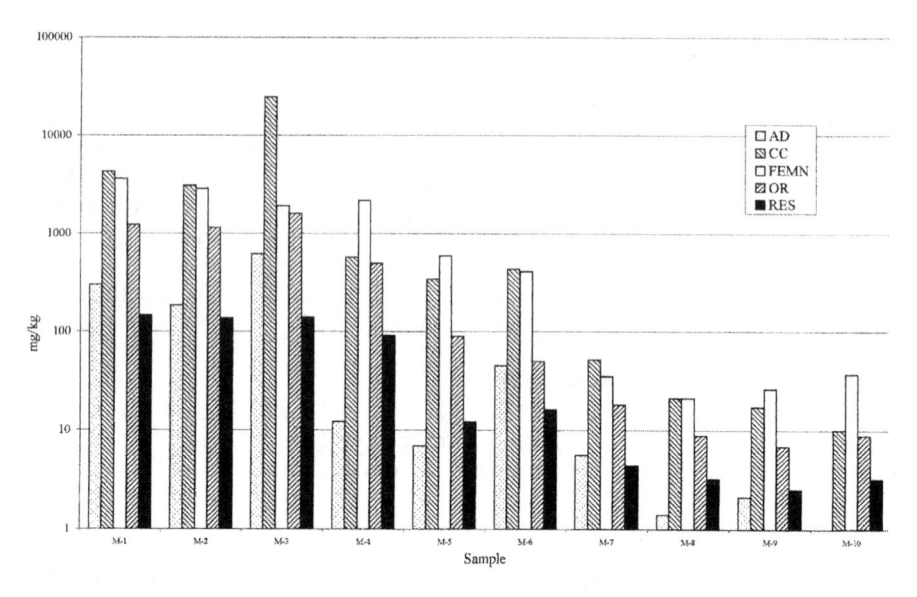

Figure 2. Concentration of lead in each of the operationally defined geochemical fractions of the experimental samples. Abbreviations are as follows: AD, adsorbed; CC, bound to carbonate; FEMN, bound to iron and manganese oxides, OR, bound to organic matter, and RES residual.

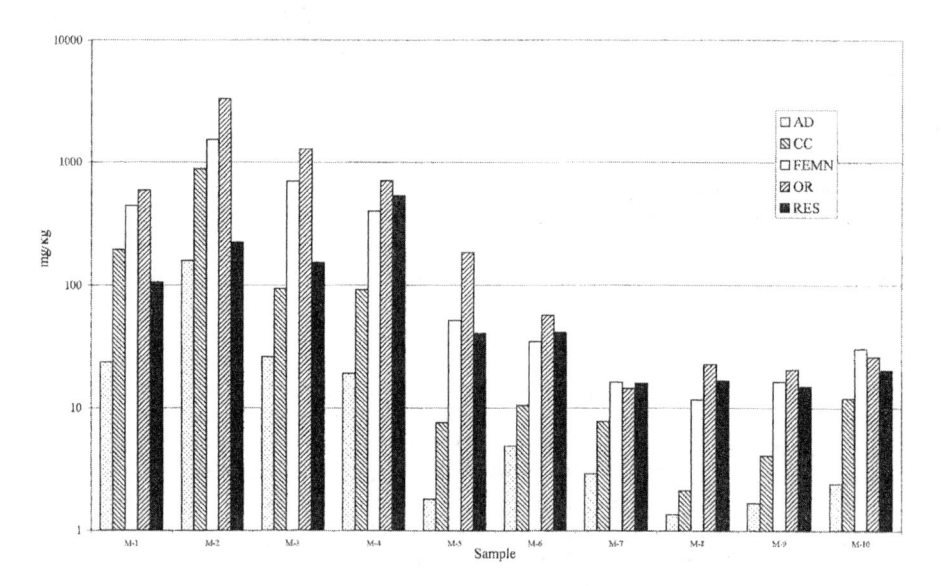

Figure 3. Concentration of zinc in each of the operationally defined geochemical fractions of the experimental samples. See Figure 2 for abbreviations.

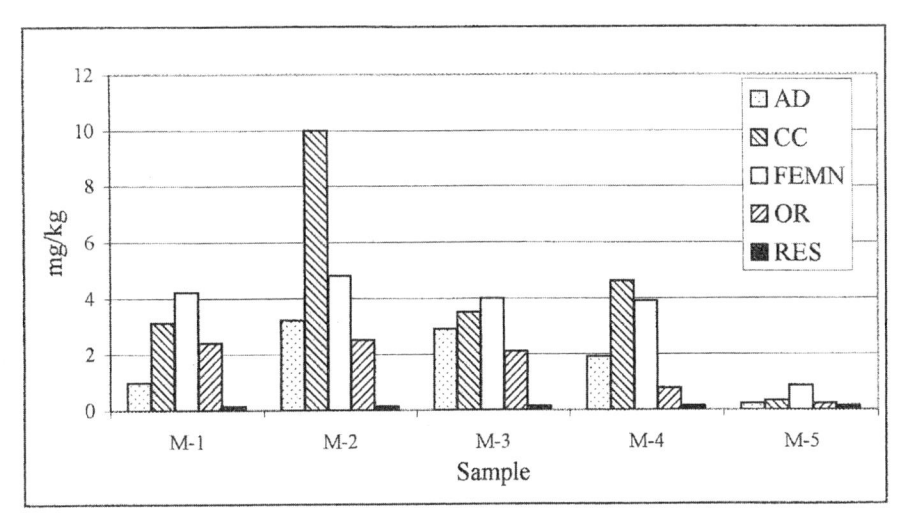

Figure 4. Concentration of cadmium in each of the operationally defined geochemicl fractions of the experimental samples. See Figure 2 for abbreviations.

4.2.4 Copper

Samples with high concentrations of Cu differ in partitioning in comparison with samples with baseline concentrations. A major portion of the Cu in samples with higher concentrations is bound to the OR fraction (ca. 22-67% of the total Cu concentration). A significant amount of Cu in organic fraction is not surprising in light of the very high values for formation constants of Cu-organic complexes [17]. The RES fraction is second in importance with 21-45% of total Cu. The next fraction is FEMN with 19-29% of total Cu content. The remainder is bound to the AD fraction (ca. 3-5.5% of total Cu). The Cu content in the OR fraction with baseline Cu concentrations is very low and in two samples below the detection limit. The most important fraction for these samples is the RES with 46-50% of total Cu concentration. The FEMN fraction is second in importance (ca. 24-33% of total Cu). The remainder is bound to the SD fraction with a range of 14-18% of total Cu content. The Cu concentrations in the CC fraction are below the detection limit.

4.2.5 Iron

Fe partitioning is different for samples with high concentration and for samples with baseline concentration of Fe. While the major portion of Fe is distributed more or less equally between the FEMN, the OR and the RES fractions in samples with baseline concentration, with respective ranges from 22 to 38%, 24-42% and 29-53% of total metal content. Samples with high concentration have major portion of Fe in the OR fraction (ca. 23-70% of total metal concentration). Fractions AD and CC contain very small quantities of Fe (max. 1.5%) in samples with both background and high trace metal values.

Table 1. Sequential extraction scheme

STEP	FRACTION	SOLUTION	VOLUME	CONDITIONS
1	exchangeable	1M NH$_4$Ac (pH7)	12ml	shake, 1 h
2	carbonates	0.1M NaOAc (pH5)	20ml	shake, 6 h
3	Fe-Mn oxides	1M NH$_2$OHxHCl in 25% HAc	15ml	boiling water bath (90°), 3h, occasionally agitation
4	organic	30% H$_2$O$_2$ + HNO$_3$	10ml+3ml	boiling water bath (86°), 2h, occasionally agitation
5	residual	aqua regia + HF	5ml+4ml	microwave digestion, 2 h.

Table 2: Mineralogical phases detected in soil sample

SAMPLE	M-1	M-2	M-3	M-4	M-5	M-8
Quartz	+++	++	+++	+++	+++	+++
Dolomite	+	++	-	-	-	-
Mica	+++	++	+++	+++	+++	+++
Plagioclase	?	+	?	+	+	+
K-feldspar	?	-	?	?	?	+
Goethite	++	+	++	++	+	+
Hydrargillite	+	-	-	-	-	?
Kaolinite	+	?	?	+	+	++
Chlorite	-	?	-	?	?	++
Cerussite	+	-	-	-	-	-
Piromorphite	?	-	-	-	-	-
Galena	-	-	?	-	-	-
Organic matter	+	+++	-	+	+	+
Pyrite	?	-	-	-	-	-

LEGEND: +++ > 10 wt%; ++ 1-10 wt %; + < 1 wt%; - mineral is not present;
? some indicates are present

4.2.6 Manganese

The major part of Mn is bound to the FEMN fraction (70-90% of total concentration). The remainder is bound to the AD, CC, OR and RES fractions with respective ranges from 0,3 to 9,8%, 2,8-22%, 4,2-13% and 0,3-6,2% of total Mn content. The distributions are similar for both types of samples.

4.2.7 Nickel

Ni partitioning is similar to the partitioning of Cu. A major portion of the Ni in samples with higher concentrations is bound to the RES, OR and FEMN fractions (ca. 36-57%, 12-30% and 25-37%, respectively). The remainder is bound to the AD fraction (ca. 7%). The Ni content in the CC fraction is below the detection limit. The RES fraction acts as predominant carrier, representing about 49-56% of the total Ni content in samples with baseline concentrations. The FEMN fraction is the next most important with a range of 26-34% of total Ni. The remainder is bound to the AD fraction (ca. 16-20%).

Table 3: Concentrations of trace metals, Mn and Fe within different fractions of soil samples

ELEMENT	SAMPLE	AD	CC	FEMN	OR	RES	SUM
Pb (mg/kg)	M-1	298	4334	3636	1227	147	9642
	M-2	183	3119	2909	1136	137	7484
	M-3	619	24330	1909	1591	140	28589
	M-4	12	275	2182	500	90	3357
	M-5	6.9	344	591	89	12	1043
	M-6	45	436	409	50	16	956
	M-7	5.6	52	35	18	4.4	115
	M-8	1.4	21	21	8.8	3.2	55
	M-9	2.1	17	26	6.8	2.5	54
	M-10	0.69	10	37	8.8	3.2	60
Zn (mg/kg)	M-1	23	194	445	591	106	1359
	M-2	157	881	1524	3298	221	6081
	M-3	26	94	697	1276	152	2245
	M-4	19	92	396	702	532	1741
	M-5	1.8	7.6	52	184	40	285
	M-6	4.9	10	35	57	41	148
	M-7	2.9	7.8	16	14	16	57
	M-8	1.4	2.1	12	23	17	56
	M-9	1.7	4.1	16	20	15	57
	M-10	2.4	12	30	26	20	90
Cd (mg/kg)	M-1	0.98	3.1	4.2	2.4	<0.15	11
	M-2	3.2	10	4.8	2.5	<0.15	21
	M-3	2.9	3.5	4	2.1	<0.15	13
	M-4	1.9	4.6	3.9	0.80	<0.15	11
	M-5	0.22	0.33	0.87	0.23	<0.15	1.8
	M-6	<0.15	0.26	0.15	<0.15	<0.15	-
	M-7	<0.15	<0.15	<0.15	<0.15	<0.15	-
	M-8	<0.15	0.17	<0.15	<0.15	<0.15	-
	M-9	<0.15	0.20	<0.15	<0.15	<0.15	-
	M-10	<0.15	0.30	<0.15	<0.15	<0.15	-
Cu (mg/kg)	M-1	2.8	<0.20	9.5	59	25	97
	M-2	2.2	<0.20	4.6	38	12	57
	M-3	2.7	<0.20	16	25	13	57
	M-4	2.8	<0.20	8	7.9	16	35
	M-5	3.2	<0.20	11	23	19	56
	M-6	3.4	<0.20	5.1	0.84	9.6	19
	M-7	3.4	<0.20	5.5	<0.20	9.6	19
	M-8	3.2	<0.20	6.1	<0.20	8.6	18
	M-9	3.5	<0.20	6.5	1.6	10	22
	M-10	3.2	<0.20	5.4	2.5	11	22
Fe (mg/kg)	M-1	40	32	5019	62710	21756	89557
	M-2	6.3	17	2500	8723	3516	14762
	M-3	4	39	20531	50044	11620	82238
	M-4	0.5	42	14761	46790	36924	98518
	M-5	4	45	7891	38319	15482	61741
	M-6	25	242	10229	8328	12012	30836
	M-7	11	322	4628	4911	10973	20845
	M-8	10	236	8620	10776	12616	32258
	M-9	4.7	212	11726	10012	9135	31090
	M-10	7.8	53	7312	11706	8449	27527.8

Table 3: Continued

ELEMENT	SAMPLE	AD	CC	FEMN	OR	RES	SUM
Mn (mg/kg)	M-1	13	244	4184	249	52	4742
	M-2	39	496	1519	179	23	2256
	M-3	73	390	10892	723	38	12116
	M-4	239	465	14849	829	162	16544
	M-5	42	250	2651	203	35	3181
	M-6	83	153	945	72	30	1283
	M-7	26	51	137	34	17	265
	M-8	29	41	392	69	15	546
	M-9	48	98	647	69	8.7	870.7
	M-10	145	366	2801	147	25	3484
Ni (mg/kg)	M-1	3.9	<0.20	22	23	25	74
	M-2	3.7	<0.20	6.5	0.38	15	26
	M-3	4.7	<0.20	20	14	23	62
	M-4	5.6	<0.20	29	9.5	34	78
	M-5	5	<0.20	17	20	24	66
	M-6	7	<0.20	8.7	<0.20	18	34
	M-7	5.5	<0.20	8	<0.20	16	30
	M-8	5.2	<0.20	8.8	<0.20	17	31
	M-9	5.5	<0.20	9	<0.20	19	34
	M-10	5.5	<0.20	12	<0.20	17	35

5. Conclusions

As the result of the historic mining activity at "Sv. Jakob", soils developed above the mine contain very high concentrations of Pb, Zn and Cd. The following chemical fractions are the major sinks for the metals in order of decreasing concentration:

Pb: CC > FEMN > OR > AD > RES
Zn: OR >> FEMN > CC > RES > AD
Cd: CC > FEMN > AD > OR
Cu: OR > RES > FEMN > AD
Fe: OR > RES > REMN >> CC > AD
Mn: FEMN >>> OR = CC > RES = AD
Ni: RES > OR > FEMN > AD

Owing to the fact that Pb, Zn and Cd in the surveyed soils are mostly associated with the CC and OR phases, the soils analysed can be considered as polluted. For the sequential extraction procedure used in this study, mobility and biological availability are assumed to decrease in the order of the metal extraction sequence. On this basis, the apparent mobility and potential bioavailability of the metals for highly contaminated soils is as follows: Cd>Pb>Zn>Cu>Ni.

Acknowledgement

Sections of this article are reprinted from Journal of Geochemical Exploration, Vol 67, Durn, G., Miko, S., Čović, M., Barudžija, U., Tadej, N., Namjesnik-Dejanović, K. & Palinkaš, L., Distribution and behaviour of selected elements in soil developed over a historical Pb-Ag mining site at Sv. Jakob, Croatia, 361-376, Copyright (1999), with permission from Elsevier Science.

References

1. Basch, O. (1981) The general geological map of SFR Yugoslavia to the scale 1:100,000, sheet Ivanić-Grad L 33-81, Institute of Geology, Zagreb, Federal Geological Survey, Belgrade.
2. Chao, T.T. and Theobald, P.K. (1976) The significance of secondary iron and manganese oxides in geochemical exploration. *Econ. Geology* 7, 1560-1569.
3. Durn, G., Palinkaš, L.A., Miko, S.F., Namjesnik-Dejanović, K and Julardžija, N. (1995) Soil contamination with heavy metals. Case study: Old mining at Sv. Jakob, Mt. Medvednica, Croatia. In *Book of Abstracts*, pp. 32. First Croatian Geological Congress, Opatija.
4. Gatehouse, S., Russell, D.W. and Van Moort, J.C. (1977) Sequential soil analysis in exploration geochemistry. *Journal of Geochemical Exploration* 8, 483-494.
5. Hall, G.E.M. (1998) Analytical perspective on trace element species of interest in exploration. *Journal of Geochemical Exploration* 61, 1-19.
6. Hall, G.E.M., Vaive, J.M., Beer, R. and Hoashi, M. (1996) Phase selective leaches for use in exploration geochemistry. In EXTECH I: *A Multidisciplinary Approach to Massive Sulphide Research in the Rusty Lake-Snow Lake Greenstone Belts, Manitoba*, (ed), G.F. Bonham-Carter A.G. Galley and G.E.M. Hall; Geological Survey of Canada, Bulletin 426, 169-200.
7. Harrison, R.M., Laxen, D.P.H. and Wilson. S.J. (1981) Chemical associations of lead, cadmium, copper, and zinc in dusts and roadside soils. *Environmental Sci. Technology* 15, 129-158.
8. Hass, J., Kovacs, S., Krystyn, L. and Lein R. (1995) Significance of Late Permian-Triassic facies zones in the Alpine-North Pannonian domain. *Tectonophysics* 242, 19-40.
9. Hickey, M.G. and Kittrick, J.A. (1984) Chemical Partitioning of Cadmium, Copper, Nickel and Zinc in Soils and Sediments Containing High levels of heavy metals. *J. Environ. Qual.*, Vol. 13, no.3, 372-376.
10. Hoffman, S.J. and Fletcher, W.K. (1979) Extraction of Cu, Zn, Mo, Fe and Mn from soils and sediments using a sequential procedure. Geochemical exploration 1978. Association of Exploration Geochemists, Rexdale, Out. 289-299.
11. Lindsay, W.L. (1979) *Chemical equilibria in soils*. John Wiley & Sons, Inc., New York.
12. Namjesnik, K. (1994) Heavy metal distributions in soils of city Zagreb. Unpublished masters thesis. Faculty of mining, geology and petroleum engineering, University of Zagreb. (in Croatian).
13. Namjesnik, K., Palinkaš, A.L., Miko, S., Durn, G., Polić, D. and Kvrgnjaš. L. (1992) Lead, zinc, nickel and mercury in soil along roadsides and of some rural and urban parts of Zagreb, Croatia. *Rudarsko metalurški zbornik*, Vol. 39, 1-2, 93-112, Ljubljana.
14. Palinkaš, L. A. (1988) Geochemical characteristic of Palaeozoic metallogenic districts Samoborska gora, Gorski kotar, Lika, Kordun and Banija. Unpublished PhD thesis. Faculty of mining, geology and petroleum engineering, University of Zagreb. (in Croatian).
15. Rose, A.W. and Suhr, N.H. (1971) Major element content as a means of allowing for background variation in stream sediment geochemical exploration. *Geochemical Exploration* Can. Ins. Min. Metall., Spec. Vol. 11, 587-593.
16. Stover, R.C., Sommers, L.E. and Silviera, D.J. (1976) Evaluation of metals in wastewater sludge. *J. Water Pollut. Con. Fed* 48, 2165-2175.
17. Stumm, W. and Morgan, J.J. (1981) Aquatic chemistry. *An introduction emphasizing chemical equilibrium in natural water*. 2nd Ed. John Wiley & Sons, Inc., New York.
18. Šikić, K., Basch, O. and Šimunić, A. (1977) The general geological map of SFR Yugoslavia to the scale 1:100,000, sheet Zagreb L 33-80, Institute of Geology, Zagreb, Federal Geological Survey, Belgrade.
19. Šinkovec, B., Palinkaš, L. and Durn, G. (1988) Ore appearance of Mt. Medvednica. *Geološki vjesnik* 41, 395-405, Zagreb. (in Croatian).
20. Tessier, A., Campbell, P.G.C. and Bisson, M. (1979) Sequential extraction procedure for the speciation of particular trace metals. *Analytical Chemistry* 51, 844-851.

PART 7. WORKING GROUP REPORTS

To facilitate greater interaction among the participants, the participants were divided into Working Groups organized around the 4 major themes. An additional 5th theme that grew out of participant interaction is the one on aggregates. The task of each Working Group was to identify talking points, discuss these, and to prepare a summary report that could be presented on the final day of the NATO ASI. The Working Group Reports that follow resulted from this effort.

REPORT OF WORKING GROUP I

Geoenvironmental Models

Compiled by DANIEL LIMPITLAW[1] and RICHARD B. WANTY[2]
[1]*University of the Witwatersrand, Johannesburg, South Africa*
[2]*United States Geological Survey, Denver, Colorado, USA*

Geoenvironmental models describe, predict, or simulate processes related to the anthropogenic flow of bulk materials to and from the earth.

Abstract. This report summarises the discussions of the Working Group on Geoenvironmental Models that were held as part of the NATO Advance Study Institute on "Deposit and Geoenvironmental Models for Resource Exploitation and Environmental Security." The Working Group was made up of experts representing 15 countries and whose expertise included the geosciences and environmental economics. The discussions focused on target audiences and uses for geoenvironmental models, the formats for presenting geoenvironmental models, and the roles of national institutes and surveys in producing geoenvironmental models. In addition, the Working Group made recommendations for future research topics that would advance the development of geoenvironmental models worldwide.

1. Introduction

The Working Group was convened to explore questions regarding the feasibility, purpose, and scope of geoenvironmental model development. Participants in the Working Group are listed in Table 1.

2. Uses and Formats of Geoenvironmental Models

The preferred use of geoenvironmental models is to communicate clearly the results of scientific research and to provide informed judgements about known or expected environmental behaviours of mineralised or altered areas, whether mined or not. The challenge is to present the results in a format appropriate to each target audience, whether citizens, interest groups, industry or public officials.

The need for geoenvironmental models is of global importance. To satisfy the global demand for raw materials, the most important first step is to develop comprehensive, reliable geoenvironmental models for characterising and predicting environmental effects of resource extraction.

Table 1. Working Group participants

Name	Organization	Country
Euro Beinat	Vrije Universiteit Amsterdam	The Netherlands
Liisa Carlson	Consultant	Finland
E. John M. Carranza	International Institute for Aerospace Survey and Earth Sciences	The Netherlands
Marta Covic	University of Zagreb	Croatia
Razia Gainutdinova	Institute of Physics NAS	Kyrgyz Republic
Nóra Gál	Hungarian Geological Institute	Hungary
Györö Jordan	Hungarian Geological Institute	Hungary
Jacek Kasi´nski	Polish Geological Institute	Poland
Kazimir Karimov	Institute of Physics NAS	Kyrgyz Republic
Daniel Limpitlaw	University of the Witwatersrand	South Africa
Árpád Lorberer	University of Budapest	Hungary
Oleg Makarynsky	Middle East Technical University	Turkey
Karol Marsina	Geological Survey of Slovak Republic	Slovakia
Charlie Moon	University of Leicester	United Kingdom
Lázsló Ódor	Hungarian Geological Institute	Hungary
Petri Peltonen	Geological Survey of Finland	Finland
Costas Ripis	Institute of Geology and Mineral Exploration	Greece
Irina Shtangeeva	St Petersburg University	Russia
Bernhard Stribrny	Federal Institute for Geosciences and Natural Resources	Germany
Andrea Szucs	Hungarian Geological Institute	Hungary
István Szücs	GEOPARD	Hungary
Stanislava Trti´ková	Faculty of Natural Sciences, Bratislava	Slovakia
Richard Wanty	U.S. Geological Survey	United States

2.1 TARGET AUDIENCES AND USES FOR GEOENVIRONMENTAL MODELS

Users of geoenvironmental models include citizens, interest groups, industry and public officials. Each group has different objectives and interests, and each has different backgrounds. As a consequence, the preparation and presentation of geoenvironmental models should be conducted at multiple levels of complexity and detail. The ideal scenario would be to provide all the information embodied in a model in a single package, but with a hierarchical structure and such that all levels of information, though accessible, were not all simultaneously displayed. It would also be useful to prepare brief summary documents that highlight key results or recommendations of a model at a specific location, thus providing an introduction to the complete body of information generated by a model.

Citizen groups may have an interest in geoenvironmental models because of the proximity of their towns and residences to mineralised areas, and perhaps to historic, current, or future mining operations. While members of society may derive some benefit from mining in the form of jobs, royalties, or taxes, they may also incur risks or expenses related to environmental impacts. The challenge of presenting geoenvironmental models to citizen groups lies in their diversity of vested interests, educational and financial backgrounds, and levels of interest or apathy. Interest groups may be considered as subsets of society that have a particular bias towards resource extraction operations, whether to preserve environments intact, thereby halting the operations, or to support the development of the operations. Geoenvironmental models must be cognisant of these competing and polarised interests, present objective scientific information, and guard against abuse of the information provided. Public officials should be able to use the information in geoenvironmental models to balance their responsibility for public welfare, economic health, and environmental preservation. For this purpose, geoenvironmental models may be only a part of the decision-making process.

A critical aspect of geoenvironmental models is their application in the field of education. Because all of the public and governments have varying degrees of interest in resource extraction and beneficiation, geoenvironmental models should educate as well as to inform. This would give model users not only a better understanding of the wide range of potential environmental effects of mining in a particular area, but also, in the context of ecoregions, appropriate environmental safeguards. At the same time, such models ensure that these safeguards were not excessive with respect to natural background levels. However, the task of public education is not solely the responsibility of the geoenvironmental modelers. Greater efforts are needed, in general, to educate the public so that informed decisions can be made. It is hoped that through greater public education, conflicts can be minimised among diverse interest groups.

2.1.1 *How can Industry Use Geoenvironmental Models?*

Geoenvironmental models can be used to predict future environmental effects associated with a mining project. In so doing, reasonable and responsible mitigation strategies can be developed. Such models can also be used to assess trans-border impacts, especially regarding fluvial and atmospheric transport of contaminants. Industry-driven projects to develop geoenvironmental models are already underway in Poland. In Hungary, hazard assessment of drinking-water wells has been conducted using common hydrogeological techniques and software.

Geoenvironmental models of the type employed by the U.S. Geological Survey (Wanty *et al.*, this volume) are considered too broad for most industrial applications as the latter commonly focus on a particular site and are not concerned with regional or cumulative impacts. Regional models are designed to place the individual environmental impacts in the context of regional geochemical variations that might be attributable to mineralised or altered rocks, mining, or other natural or human processes. If the spatial scope of the model is too narrow, regional effects are ignored. In the past, ignorance of regional effects has resulted in severe environmental degradation.

If industry is to make use of geoenvironmental models, the costs associated with their creation and presentation will have to be reduced or their benefits demonstrated by using cost-benefit analyses. Modelling must be incorporated with the life cycle of industrial activities (Figure 1). At the exploration and discovery stage of a project, companies require information about the background and baseline levels of various chemical agents, as well as information on vulnerability of environments that may be affected by any subsequent mining activity. During the production phase of a project, emission and effluent concentrations are critical, requiring knowledge of vulnerability in the context of local and regional hydrogeology and geochemistry. During mine closure and preparation for abandonment, land reuse modelling becomes important, which incorporates aspects of the baseline values (to which the local environment should return, either by design or naturally) and estimates of the longevity or persistence of environmental effects.

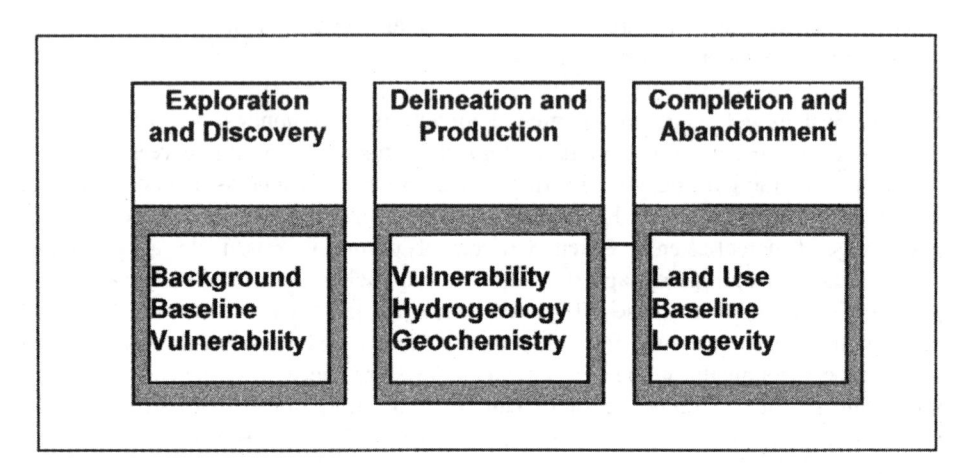

Figure 1. Components of geoenvironmental models that are relevant for the lifecycle evaluation of mineral-deposit development.

Risk and economic assessment should be an integral part of geoenvironmental models designed for industrial applications. Risk assessment of the impacts on human health in a region may be required. In many cases, industry becomes involved with models through outside professional consultants and therefore may not regard models as an ongoing management tool. Thus, the long-term effectiveness of modelling is compromised.

2.1.2 *Can Geoenvironmental Models Help Communication Between the Research Community and the Stakeholders?*

In many countries, the ability to discover new deposits is not the limiting factor, but rather the ability to obtain permission to exploit newly discovered deposits. Permission is granted only if it can be demonstrated that the environmental impacts are understood. In this regard, geoenvironmental models may be extremely useful.

Geoenvironmental models have become more important than metallogenic models in the establishment of new mines. Mineral deposit models are far more advanced than geoenvironmental models and the spatial and temporal consequences of mining environmental impacts are not fully understood. For this reason, research focussed on geoenvironmental models may foster improved communication between resource developers, interest groups and legislators, leading to greater success in achieving mineral extraction authorisation.

2.2 FORMATS FOR PRESENTING GEOENVIRONMENTAL MODELS

Geoenvironmental models must be accessible and intelligible to users and must be presented in a transparent manner. Useful formats include maps, descriptive reports, GIS presentations, animated movies, and 1-2 page summary documents. The former three formats are discussed in Wanty et al. (this volume). Members of the Working Group have had direct experience with all of these formats, with varying outcomes and degrees of success. Suggestions for a suitable format include a booklet with small maps (used successfully in the German-Polish border region), videos and cartoons and 3-D models and brief summary documents, limited to one or two pages in length. It is likely that a combination of media will be the most successful.

In general, the preferred format for geoenvironmental models depends on the target audience. At the highest levels of government, the presentations should be simple to understand yet complete in scope, so that the pertinent information can be efficiently delivered to policy makers. For this purpose, the animated movie and brief report format may be most useful. At the level of decision makers, more detailed information is required, and so the desired format may consist of maps, reports, and GIS presentations. The general public will require a wide range of formats because of the wide range of concerns and levels of understanding. To satisfy the public need for knowledge, all of the proposed formats may be necessary.

2.3 THE ROLE OF NATIONAL INSTITUTES AND SURVEYING ORGANISATIONS IN PRODUCING GEOENVIRONMENTAL MODELS

The Working Group agreed that development of geoenvironmental models was a suitable activity for national Surveys and Institutes, as well as for academe and perhaps industry as well. Moreover, involvement of national-level institutions may provide a pathway for greater international collaboration. Such collaboration would be particularly important in the European community, where many opportunities await with respect to trans-border issues.

To increase international collaboration, several recommendations are made:

- NATO and other international groups should seek out and fund multinational projects,
- International efforts should be made to standardise data bases and protocols, and make data bases more widely available; and
- International data-quality standardisation efforts should be increased.

Development and funding of more multinational projects would, of course, help in all aspects mentioned. The latter two recommendations can be achieved more readily if the first is followed. To achieve the latter, two types of activities should be encouraged; formation of international working groups to make decisions on data standards, quality assurance, and collection protocols and development of international research projects to carry out and refine the protocols and standards. We now have a 20-plus-year history throughout the world of developing environmental impact statements and environmental assessments. A retrospective look at the successes and failures of these assessments would be a starting point for developing international protocols.

The Hungarian Geological Survey is currently involved in environmental modelling of mineral deposits, primarily through the use of stream-sediment sampling programs, enhanced with sampling of other media such as surface water, agricultural crops, soils, and mine effluents. Slovakia has completed a comprehensive geochemical atlas for rocks, surface waters, and ground waters, and is embarking on an ambitious geoenvironmental modelling project for the next three years, beginning with focussed studies in mining areas and gradually increasing the spatial extent to include the surrounding countryside.

3. Where Do We Go From Here?

The recommendations of the Working Group can be divided into two categories: recommendations for future techniques and approaches to geoenvironmental modelling; and immediate recommendations for future research to facilitate the development of geoenvironmental models. The former is intended to provide guidance for the general philosophical development of geoenvironmental models and the latter is intended to provide immediate actions that should be taken to develop and implement geoenvironmental models on an international scale.

3.1 TYPES OF DATA OR FIELDS OF EXPERTISE THAT SHOULD BE ADDED TO GEOENVIRONMENTAL MODELS

The most useful disciplines to add to geoenvironmental models are those that would enhance the utility and applicability of the models. A direct consequence would be to increase the acceptance and marketability of the models. Different layers of thematic data are a key data source for geoenvironmental models. These layers require the input of a multidisciplinary team of experts. Geoscientists must restrict themselves to selecting the experts required for a comprehensive modelling exercise and not attempt to determine the methods and data required by these experts.

- Hydrometeorology–hydrologic and atmospheric flow regimes, temperature and humidity fields have a strong influence on mineral diffusion and distribution throughout the environment.
- Medicine–quantitative and qualitative impacts of mining and high natural backgrounds on human health are poorly understood. Several factors arise in this evaluation: interpretation of human health statistics is a difficult field at best, seldom resulting in unequivocal conclusions; the range of concentrations of a metal or suite of metals to which a population is exposed may be spatially and temporally variable, making dose estimates difficult and uncertain; exposure levels of the general population may be much lower than for those populations that live or work in or near mines or mineralised areas, adding uncertainty to dose-response estimates. Nevertheless, the fundamental need for medical data exists, and may in

fact drive the marketability of geoenvironmental models and environmental policy development in the future.
- Forestry–analysis of the complex interactions of floral assemblages with the inorganic environment requires specialised skills. When such assemblages are managed for economic gain, as is the case in forestry, the socio-economic impacts of mineral extraction can only be predicted in collaboration with experts from the forestry industry. Modelling results with thus have more credibility.
- Biology–different ecosystems may respond to similar perturbations in an almost limitless variety of ways. If the impacts of mining are to be reliably predicted, comprehensive understanding of the relevant portion of the biosphere is essential.
- Metallurgy–in many mining regions, the greatest environmental impact arises from the beneficiation of ore. These impacts can be greatly reduced by applying clean production technology. Geoenvironmental modelling in conjunction with expert knowledge of metallurgical processes could be used to identify the most appropriate technology for a project.
- Other data types–non-traditional geo-data, such as census or other demographic data, may provide valuable input for geoenvironmental modelling.

3.2 SUGGESTIONS FOR FUTURE ACTIVITIES

The following suggestions are made to improve the current state of geoenvironmental models:

- Publish minimum and maximum metal fluxes from various types of mines.
- Document existing case studies in order to be able to estimate the range of contaminant concentrations arising from similar, unstudied mining complexes.
- Collect data from various climate zones.
- Establish comparative models so that if, for example, one tonne of copper is required by the market, the ore source with the minimum environmental impact, porphyry, massive sulphide, or other deposit types, could be selected in a particular environment. These models should also be able to compare different types of impacts such as those arising from smelter emissions with the impacts of acid mine drainage.
- Investigate the use of geoenvironmental models to improve mining's poor image.
- Investigate the possibility of making standard materials are available for calibration of geoenvironmental analyses. Absence of these materials makes comparison of organic pollutants (etc.) difficult. If such standard materials were widely available, reliable and comparable studies would be more common.

4. Research Topics

4.1 PERSISTENCE OF ENVIRONMENTAL IMPACT OF ANCIENT MINES (L. ÓDOR, M.COVIC, N. GÁL, E.J. CARRANZA)

The objective of this research would be to study various historical mines for evaluation of the persistence of environmental impacts. Waste piles of historical mines still contain ore due to poor mining and processing techniques. The purpose of this study would be to determine, whether the different historical mines are still a source of pollution or the pollution is confined in a small area. The study would include coal, lead, gold, silver and mercury deposits under different climatic conditions.

The first step would involve a literature overview to locate historical mine sites, which have relatively good historical database on chemical analysis of stream and floodplain sediments receiving acid mine drainage. From that information, there would be a selection of mine sites, which would be studied in the second phase of the research.

The detailed study would involve sampling of spoil and tailings, water, soil, stream-, floodplain- sediments in cross-section and along the draining river. The evaluation would consider the amount of ore, type and how big area was affected.

4.2 COMPARISON OF ENVIRONMENTAL IMPACTS ARISING FROM THE EXPLOITATION OF POLYMETALLIC ORE DEPOSITS IN DIFFERENT ENVIRONMENTAL SETTINGS

The objective of this study would be to establish preliminary hazard ratings for mineral exploitation under different climatic conditions.

The study would consist of pilot plans beginning with metallic ores. The weathering of waste deposits, from similar geological settings, in different climates and their consequent impacts on the environment would be investigated.

- Pilot plan 1–Poly-metallic ore deposits in different environmental settings from neighbouring countries. Establishment of a geo-scientific network for the selection, evaluation and dissemination of geological data.
- Pilot plan 2–Poly-metallic ore deposits in coastal zones in Mediterranean countries

Such data exist in the form of general information, maps, tables, text, and in various forms and levels of information, varying from country to country. The data gathering in each country are conducted not only by the central geological surveys, but also by regional offices, which send the information to these central surveys, mainly by fax, reports, meetings and personal communication, and in rare cases via email. Thus, there is no continuous, up-to-date flow of information for evaluation and analysis.

The current trans-border exchange of geological data between institutions is conducted via announcements, exchange of scientific views, publications, conferences, workshops and seminars, as well as the exchange of scientific experts. Nevertheless, there are no ad-hoc communication facilities. Such facilities would significantly enhance the current co-operation among institutes, and provide the efficient and fast exchange of geological data, the comparison of geological data, as well as the creation of a common data-coding scheme for data analysis and evaluation. In addition to the above obstacles from the scientific point of view, important users of such data in industries on a national and international basis are not in the position to acquire up-to-date, reliably evaluated geological information on-demand.

The above problems could be addressed and resolved by the establishment of a local network, which would use the emerging telecommunication facilities and would provide a user-friendly system for the continuous processing and publication of scientific geological data.

For pilot plan 2, the network would include the countries Greece, Albania, Bulgaria and Romania. Data from the selected regions with poly-metallic ore deposits within these four countries would be collected, evaluated and disseminated. The network could take advantage of technologies and trends in the areas of network infrastructures, GIS, databases, and the Internet.

Such a network is intended to accomplish the following:

- To bring research benefits such as the introduction of new technologies and telematics infrastructures to the participating geological institutes, and especially to the participating rural regions (regional branches),
- To standardise geological data handling via the usage of a common coding scheme and data bases,
- To adapt state-of-the-art technology to the common methodologies and techniques used in the geosciences; and
- To exploit existing telecommunication/telematics solutions for the sake of research co-operation and data exchange.

In order to accomplish the above goal, the project will involve the following steps:

- Gather and assess a geological data set which is available on a national basis for Greece, Bulgaria, Romania,
- Determine a trans-national, common coding scheme for the management of the above data subset,
- Design and develop a homogeneous data processing platform for data collection, processing and evaluation based on the agreed common coding scheme,
- Establish an underlying network infrastructure, and install the data processing components,
- Design and develop a scientific information site for exchange and dissemination of the processed geological data subset; and
- Design and develop a public information site for exchange and dissemination of a publicly accessible geological data subset.

The network to be developed is shown in Figure 2.

Geological surveys in the host countries would be responsible for data collection. Analyses would be performed in a central laboratory to ensure consistency of results. The data that would be required would include:

- Geological data from geophysical and geochemical surveys
- Mining data from production records
- Environmental data from water quality, hydrologic, sedimentologic, and climatic surveys

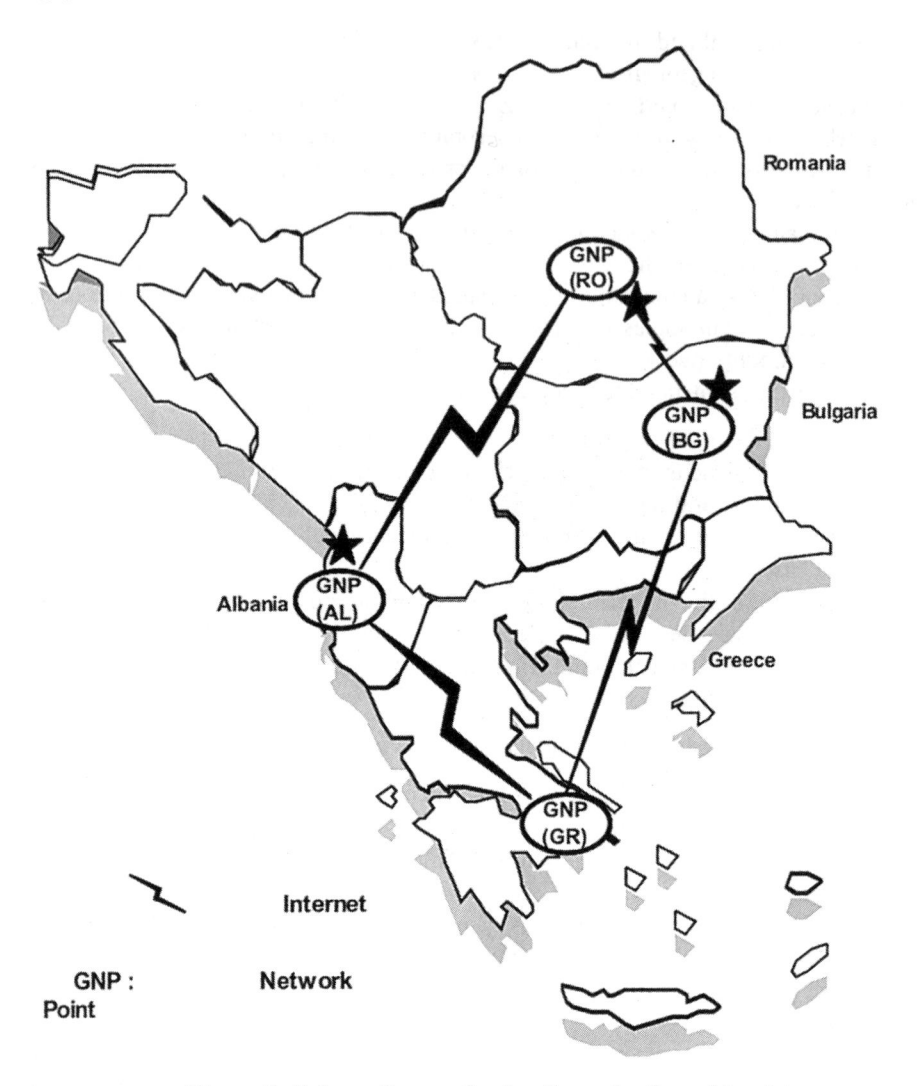

Figure 2. Schematic map for the dissemination of the data

4.3 STANDARDS FOR GEOENVIRONMENTAL MINERAL DEPOSIT MODELS (K. MARSINA)

At present, there is an urgent need for using geoenvironmental models of diverse mineral deposits for various purposes as environmental prediction and mitigation, assessment of abandoned mine-lands and mine-site remediation.

The objective of this project would be the establishment of a multinational effort (NATO countries and NATO partner countries) to compile the available national data for the purpose of developing standards for geoenvironmental models.

The project would involve the following activities:

- Determination of a general methodology for developing geoenvironmental mineral deposit models based on the work by the U.S. Geological Survey (Wanty et al., this volume) and the geoenvironmental data from other participating countries,
- Standardisation of a methodology (data selection, type and structure of a data base, media, chemical analysing procedures, outputs, and so forth) for common use in the participating countries (NATO and NATO partner countries).
- Validation of the methodology by comparing test results by applying the models to different mineral deposits that occur in the participating countries
- Final harmonisation of methodology according to the test results.
- Presentation of the results at a future NATO Advance Studies Institute.

4.4 APPLICATION OF GEOENVIRONMENTAL MODELS TO SMALL-SCALE MINES

Although the volume of waste produced by small-scale mining is much less than that of large-scale mining operations, the environmental impact is often greater on a local scale because small mining enterprises exploit the highest-grade parts of mineral deposits (high-grading). The waste produced in this way contains toxic elements and compounds in higher concentrations, which may lead to higher levels of environmental deterioration on a local scale.

Small mining companies often lack the appropriate financial capability to assess the impact of their activity on the environment. Furthermore, even if they are aware of the adverse effects of their mining activity, the lack of financial resources limits them in remediating the impacts. Small-scale mines are often left abandoned and are ignored unless serious environmental impacts arise. This is often the case in Central and Eastern European countries where appropriate legislation is lacking.

Since small-scale mines receive less publicity and they do not always have the means for proper public relations, the local population is often unaware of potential and existing dangers.

In these situations, detailed models should be constructed to address small-scale mining-induced environmental problems. The scale may limit the application of traditional GIS techniques primarily developed for regional scales. Instead, the integration of various geoenvironmental (hydro-geological, geochemical, and so forth) models must be addressed first. However, models have to place the small area studied within the context of larger-scale processes. For example, acid mine drainage generated by small mines can be transported by small streams and discharged into nearby larger rivers.

Geoenvironmental models could be used to delineate the areal extent of pollution. For example, studies in the Small Carpathian Mountains in Slovakia have shown that trace metal pollution from small mines can be transported as far as 20-50 kilometres.

Geoenvironmental models are perhaps most efficient if used in the planning phase of mining. They can be used to aid decisions regarding mining techniques, waste-handling procedures and can help guide methods for re-processing or alternative uses for wastes. Complex geoenvironmental models are essential in developing efficient methods for remediation.

Small-scale mining is a concern primarily for the local communities. Therefore, geoenvironmental modelling systems have to be capable of presenting results in a format that is accessible for the local population and the local decision-makers. This requires careful design of outputs, on one hand, and, on the other hand, flexible computer graphics techniques to produce simplified but meaningful figures and cartoons for visual communication. In this way geoenvironmental models can form the basis for information and education of the local population.

4.5 GEOENVIROMENTAL MODELLING OF FLUXES RELATED TO GLOBAL MINERAL PRODUCTION

In order to compare the fluxes of materials and human resources required to extract minerals in different parts of the world, it is proposed to identify locations around the world for which these fluxes could be minimised.
In order to accomplish this, the following is proposed:

- Acquire databases of operational data for all major producers of internationally traded mineral commodities,
- In addition to the more easily available information such as production, grade, location and cost, specify the following:

  - Land use
  - Energy use
  - Toxic by-products
  - Water use
  - Aggregate use
  - Transport
  - Infrastructure
  - People: displacement and inflow

- Evaluate the data collected in terms of the values of parameter used per tonne of mineral or metal produced; and
- Consideration of the marketing of metals with respect to the parameters involved that affect the environment.

4.6 IMPROVING COMMUNICATION WITH AND BETWEEN TARGET GROUPS (E. BEINAT)

Specialists and their models are at the interface between several target groups, namely, developers, authorities, stakeholders, citizens, and so forth. There is often a wide gap between the ideal format for communicating technical/scientific information and the best format to exchange information with different target groups. There is often limited knowledge of the information demands of different target groups.

The mismatch between information supply and information demand has several implications:

- Misunderstandings,
- Information misuse,
- Fear; and
- Social classification (friends and enemies).

These factors may favour the escalation of conflicts and become a seed for long-term opposition between groups (for example, developer and stakeholder). Effects may be an increase in development costs, the withdrawal/change of a plan, or the implementation of a plan, which is not followed by goal-conforming behaviour. The image and reputation of some players may be compromised.

These shortcomings can be related to three main factors:

- Information channels between target groups (developer, authorities, citizens, stakeholders) are not established,
- Information exchange is based on inappropriate tools/formats; and
- The information channel is unidirectional (that is, delivery versus exchange) or asymmetric (that is, mostly delivery, window dressing exchange).

The objectives of a proposed research project would be to:

- Analyse the factors which determine the success and the failure of communication between specialists and target groups,
- Organise in a systematic way the tools, procedures, and so forth, which can be used to establish which communication channels work better between which groups; and
- Develop and test these communication channels in different social, environmental and economic contexts.

To carry out such a project, the following steps would need to be taken:

- Select a number of cases, which represent successes and failures of communication and identify the target groups and their demands.
- Develop and classify the schemes for each case.
- Develop schemes to organise communication techniques and channels. Two matrices should be considered—the technique matrix, which considers which technique is most suitable for which target group (reports, video, presentation, cartoons, and so forth); and the channel matrix, which considers which channel can be most effectively established between target groups (meetings, workshops, TV programs, and so forth) as illustrated below:

TECHNIQUE MATRIX

Target groups ⇩

Communication technique ⇨

CHANNEL MATRIX

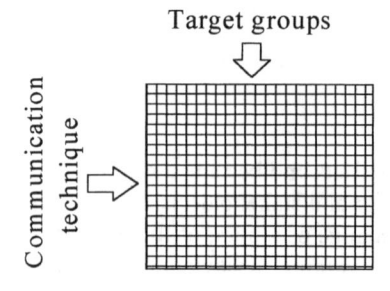

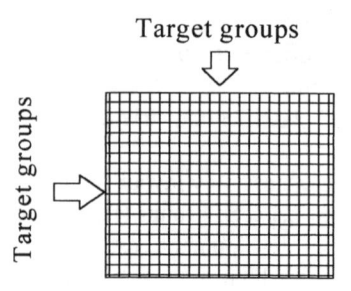

Target groups ⇩

Target groups ⇨

- Match matrices and cases. Select cases and develop the matrices on the specific cases. Design new techniques (animation, VR, games, and so forth) and channels (Internet, TV, and so forth) if necessary. Test them.
- Costs and benefits of the approach. For each case, identify:

    - Potential problems, which could occur without suitable communication channels,
    - Benefits obtained by using appropriate communication tools, and
    - Cost of their implementation (time, money, and so forth).

The following partners and expertise would be required:

- Partners—Research institutes, model developers, administration and interest groups, the private sector (for example, oil companies).
- Expertise—Technical specialised expertise (modellers, technologists/scientists, and so forth), communication experts (sociologists and psychologists, media people); possibly sector experts for the development of new instruments (for example, an Internet programmer).

The desired funding organisations for such a project include:

- EC 5th programme
- ESF and NSF
- Bilateral agreements (Embassies)
- UN
- Networks of private sector organisations (for example, NICOLE)

4.7 WAYS TO IMPROVE THE PROCESS OF BIOLOGICAL REMEDIATION (I. SHTANGEEVA)

At present, the remediation of regions polluted with heavy metals is a critical global problem. Most conventional techniques currently in use to remediate polluted soils are very expensive and usually destroy the upper horizon of the soil. It is therefore necessary to search for soft and cost-effective ways of remediation of polluted regions.

One such method is biological remediation. It is known that there are some species of plants, so called hyper-accumulators, which can accumulate high concentrations of heavy metals (cadmium, copper, zinc and some others) and thus, these plants can remove toxic element from polluted soils. This technique does not destroy the soil and at the same time it is cost-effective.

However, so far bioremediation is not widely used as a prescriptive method of soil clean up. The main factor, which prevents the application of such a method, is the long process time required. The most optimistic calculations show that it takes at least 5 years to clean up soils polluted with heavy metals. The main reason is that the majority of the toxic elements usually remain in the roots of plants and only a small fraction is transferred to the aerial portions, which are to be harvested.

In St Petersburg University, from the beginning of 1998, some pot experiments were started to determine ways to increase the rate of penetration of toxic heavy metals from roots to leaves. It was found that addition of chemicals and fertilisers and as well as some strains of micro-organisms to the soil considerably influences the distribution of heavy metals among soil, roots and leaves of plants.

The results of pot experiments will be validated in the field during the next few years. Moreover, new genetic experiments are planned which should allow the creation of new species of plants for bioremediation under various environmental conditions.

St Petersburg University is also involved in research into the use of plants enriched with chemical elements after the completion of the process of bioremediation.

Reference

Wanty, R.B., Berger, B.R., Plumlee, G.S., King, T.V.V. (this volume) *Geoenvironmental models: An Introduction.*

REPORT OF WORKING GROUP II

GIS/RS methods and techniques: A Spatial Data Laboratory Network

Compiled by Zoltán Vekerdy[1] and Chang-Jo F. Chung[2]
[1]*International Institute for Geo-Information Science and Earth Observation (ITC), Enschede, The Netherlands*
[2]*Geological Survey of Canada, Ottawa, Canada*

Abstract. This report summarizes the discussions of the Working Group on a Spatial Data Laboratory Network that were held as part of the NATO Advanced Study Institute on "Deposit and Geoenvironmental Models for Resource Exploitation and Environmental Security." The Working Group was made up of experts who represented 14 countries and had expertise in such fields as the geosciences, geoinformatics and natural resources. The discussions focused on the scope of spatial data acquisition, handling and analysis tools in the assessment and monitoring of the environmental impact of mining. After analyzing the needs of our target group, the working group suggested the set-up of a home page for facilitating the information exchange. A brief outline of the home page was also presented.

1. Introduction

The Working Group was convened to explore questions regarding spatial data acquisition, handling and analysis. Participants in the Working Group are listed in Table 1. As a consequence of the fact that this working group focused on data, in the first step, the links to the other working groups had to be analyzed. In this way the user requirements were defined.

2. Requirements of the other Working Groups towards Working Group 2

The definition of the requirements showed how the output of this Working Group would be used by the "Environmental Impact of Mining" community. This work was based on personal communication between the different working group members. The most important findings were the following:

2.1. WORKING GROUP 1

- To develop good presentations of models that could be used by the decision-makers and the community. – Visualization: graphic tools for communication (user interface).
- Data standardization, compatibility.
- Environmental effect of old and current mining.

2.2. WORKING GROUP 3

- Tools, instruments and techniques for decision support for resource assessment.
- Assessment of groundwater pollution and landscape quality (scenic visual value), soil pollution.
- Mitigation and remediation.

2.3. WORKING GROUP 4

- Training and networking tools, using several real-life examples.

Table 1. Working Group 2 participants

Name	Organization	Country
Julius BELICKAS	Geological Survey	Lithuania
Tariro P. CHARAKUPA	Environment & Remote Sensing Institute	Zimbabwe
Chang-Jo CHUNG	Geological Survey	Canada
Bernard CORNÉLIS	Département Géomatique Université de Liège	Belgium
Peng GONG	University of California	USA
Isabel GRANADO	Instituto Superior Tecnico – CVRM-IST	Portugal
Marek GRANICZNY	Geological Institute, Warszawa	Poland
Ágnes GULYÁS	Geophysical Institute (ELGI)	Hungary
Bill LANGER	US Geological Survey	USA
Greg LEE	US Geological Survey	USA
Rumen MIRONOV	Technical University of Sofia	Bulgaria
Ivars OZOLINS	Geological Survey	Latvia
Antonio PATERA	National Research Council	Italy
Juan REMONDO	University of Cantabria	Spain
Nikolay Metodier SIRAKOV	Instituto Superior Tecnico – CVRM-IST	Portugal
Irina SHTANGEEVA	Institute of Earth Crust, St. Petersburg University	Russia
Zoltán VEKERDY	International Institute for Aerospace Survey and Earth Sciences, ITC	Netherlands
Tsehaie WOLDAI	International Institute for Aerospace Survey and Earth Sciences, ITC	Netherlands

3. Contribution of Working Group 2

Based on the summary of the above mentioned requirements and the experience of our group members, Working Group 2 suggested to set-up a Spatial Data Laboratory Network with the following major tasks:

- A homepage to provide information. This would be the basis of communication and information dissemination in the network.
- Giving access to tools. It is foreseen that the network members could provide up to date information on where and how GIS/RS/spatial modeling tools can be accessed. This information would be provided via the homepage.
- Education (courses, material). Links to downloadable educational material and individuals, who develop such material, will be included in the home page.

The following issues were to be addressed:

1. Quality of digital data
2. Remediation of polluted areas
3. Structure of the spatial database
4. Creation of 3D maps
5. Link between methods and practice
6. Quantitative models and validation
7. Costs and benefits of GIS application
8. Data base management for practical problem and generalization
9. Evaluation of existing tools
10. Geological data concept
11. Classification techniques and tools for spatial multidimensional data
12. Georeferencing of remotely sensed data
13. Data exchange standards
14. Cost effective techniques for remediation
15. Simulation of environmental models
16. Data fusion and integration
17. Time aspect
18. Accessibility of data
19. Modelling of geological processes

4. The Proposed Structure of the Home Page

Working Group 2 proposed the following structure for the home page. The one below is a first draft of it. It is foreseen that modifications would be necessary during the implementation. The home page was planned for development by the Working Group 2 members from Portugal.

4.1. CENTRAL PAGE

General description of the network, main ideas, objectives, participants, etc.

- Links to the contact institutes of the countries, members of the network, contact persons, addresses, telephone, e-mail etc.
- Links to the technical information pages
- Data, data format page (see Section 5.1.1.);
- Analysis page (see Section 5.1.2.);
- GIS, RS, GPS software page (see Section 5.1.3.);
- Presentation tools page (see Section 5.1.4.).
- Link to training page (training material, links to training material providers, free software etc.)
- Link to interesting events, publications page (information about new conferences, seminars, meetings and publications.)
- Query/search page (to submit queries to the members of the network.)

4.1.1. *Data, data format page*
Data and data formats:

- Relational databases
- Standards
- Multi temporal data
- Special sensors
- Meta-data
- Data types:
- Usual GIS and RS data
- Topographical
- DEM
- Geological (lithological, surficial and structure)
- Geophysical (airborne)
- Geochemical
- SAR, TM

4.1.2. *Analysis page*
Short description of available analysis tools with links to providers, e.g.:

- o SIMULINK (package of MATLAB),
- o 3D simulation and analysis tools (GEMCOM, LINKS, VULCAN),
- o Water flow models (MODFLOW, etc.)
- o Air convection models (etc.),
- o Prediction model (Dr. Chang-Jo F. Chung's extension to PCI® software EASI-PACE),
- o Geostatistical packages.

4.1.3. *GIS, RS and GPS tools page*
GIS & RS packages:
- o Arc/Info, ArcView
- o Intergraph
- o MapInfo
- o Ilwis
- o IDRISI
- o Erdas Imagine
- o PCI
- o ER Mapper
- o ENVI

Field data capture packages:
- o GSC FieldLog
- o USGS GSMapCAD

4.1.4. *Presentation tools page*
Packages for the presentation, visualization and printing of data and analysis results:
- o Drawing packages (CorelDraw, Freehand, etc.)
- o Cartographic packages (PCI's ACE, Intergraph Map Finisher, etc.)
- o 3D packages
- o Virtual reality

REPORT OF WORKING GROUP III

Natural Resource Assessments and Resource Management

Compiled by BYRON R. BERGER[1] and ANDRÉ BOTEQUILHA[2]
[1]*United States Geological Survey, Denver, Colorado, USA*
[2]*Universidade Técnica de Lisboa, Portugal*

Abstract. This report summarizes the discussions of the Working Group on Natural Resource Assessments and Resource Management that were held as part of the NATO Advanced Study Institute on "Deposit and Geoenvironmental Models for Resource Exploitation and Environmental Security." The Working Group was made up of experts who represented 12 countries and had expertise in such fields as the geosciences and natural resources. The discussions focused on the scope of natural-resource assessments, the formats for presenting assessments, and the relation with geoenvironmental assessments. The Working Group stressed the importance of effective communication to the users of assessments and the responsibilities of the scientists who make assessments.

1. Introduction

The Working Group was convened to explore questions regarding the purpose and scope of natural-resource assessments and resource management as they relate to geoenvironmental issues. Participants in the Working Group are listed in Table 1.

2. Uses and Formats of Natural-Resource Assessments

Natural resource and geoenvironmental assessments are mechanisms for transferring scientific information and knowledge into a form that is useable in land-use decision-making, policy formulation, and public dialogue, often with international implications. Natural resource assessments usually consist of a set of procedures that result in the estimation of the quality and quantity of undiscovered deposits (i.e., mineral commodities, and energy resources) within some volume of the Earth's crust. Geoenvironmental assessments consist of a related set of procedures that result in the estimation of potential hazards owing to past resource extraction operations or anticipated in future resource extraction operations. Geoenvironmental assessments attempt to identify objectively the interactions between known or suspected deposits and the ecosystems they occupy.

Table 1. Working Group participants

Name	Organization	Country
Byron R. Berger	U.S. Geological Survey	United States
André Botequilha Leitão	Universidade Técnica de Lisboa	Portugal
Fatbardha Cara	Institute of Geological Research	Albania
Lawrence J. Drew	U.S. Geological Survey	United States
Halil Hallaci	Institute of Geological Research	Albania
Hülya Inaner	Dokuz Eylül University	Turkey
Roma Kanopiené	Geological Survey of Lithuania	Lithuania
Jyrkki Parkkinen	Geological Survey of Finland	Finland
Nyls Ponce Seoane	Institute of Geology and Paleontology	Cuba
Sankaran Rajendran	Bharathidasan University	India
Petr Rambousek	Czech Geological Survey	Czech Republic
Luis Recatalá Boix	Centro Investigaciones Sobre	Spain
Vardan Sargsyan	Yerevan State Institute of National Economy	Armenia
Janos Szanyi	Hungarian Geological Survey	Hungary
Turgay Toren	General Directorate of Turkish Coal	Turkey
Hasan Üçpirti	Sakarya University	Turkey
Mehmet Ali Yukselen	Marmara University	Turkey

To ensure a beneficial balance between natural-resource development, use, and postproduction cleanup and the protection of the environment, resource managers and policymakers must be provided with sound scientific information, advice, and counsel. For such a balance to be achieved, the scientific community needs to provide the following:

- A statement of issues and problems (scales, purpose, and so forth),
- A common set of methodologies and guidelines that provide the types and qualities of information that contributes to an effective decision-making process,
- A common set of standards and protocols by which the most appropriate assessments may be effected and accompanied by the necessary legislation in support of the assessment and the establishment of enforcement authorities,
- A sufficient understanding about deposits of natural resources and the ecosystems in which they occur that will result in meaningful assessments; and
- Efficient channel-ways through which scientific information and advice can be communicated to land-use planners and policymakers at local, regional, national, and international levels.

3. Standards and Protocols

To achieve reproducible, credible, comparable, and understandable results, it is of utmost importance to maintain consistency throughout the steps involved in assessments. That consistency is best achieved when a common set of data-gathering techniques and analytical tools are used.

In making assessments, it is also important that a common set of standards and protocols is agreed to within the scientific community. Such standards include those regarding units of measurement, quantitative analysis levels, and data analytic methods. A minimum level for quantitative assessments requires that a common set of environmental regulatory standards be met. Protocols for adopting standards include rules for counting (i.e., for example, the minimum distance necessary for considering veins in a poly-metallic vein deposit to be unrelated), for taking field measurements, and for collecting samples.

In making assessments, the existence of a comprehensive scientific understanding of the nature, origin, and mode of occurrence of the natural resources in question is implied. Similarly, to assess the health of ecosystems, there needs to be a comprehensive multidisciplinary scientific understanding of the factors that affect ecosystems, which include biological, toxicological, hydrological, climatological, geological, geochemical, pedological, and socioeconomical systems and how these systems interact locally, regionally, and globally. Under these circumstances, a holistic approach that enables scientists to answer those questions about the inevitable tradeoff inherent in environmental remediation and development mitigation strategies is best. Because mitigation is usually less costly and more efficient than remediation, it needs to be a part of the early phase of any natural-resource planning.

4. Effective Communication and Responsibilities

The effective communication of scientific information to decision-makers requires a concerted effort to express the results of assessment in language that is understandable to stakeholders who may lack formal technical training and to visually display the results in a form (i.e., GIS) that allows users to query the scientific information and relate it to other types of data such as demographics, landscape, and so forth.

Scientists must be prepared to present alternative development/environmental solutions and strategies. The remediation of sites of past natural-resource development require judgments as to the costs and environmental effects of restoration of the site to some standard ecological state (i.e., the "original" condition or rehabilitation of the site to some alternative use, such as recreation or wildlife preserve).

There are many dimensions to the decision-making process, a large proportion of which are political and cultural. To communicate effectively across national and cultural boundaries and to impact resource-management decisions, scientists must be cognizant of the differences between political constituencies and cultures. Communication is best made through well-established and meaningful multinational channels. This includes organizations within countries that are legally responsible for enforcing compliance with environmental laws and regulations, as well as those that make land-use decisions, and international organizations, such as the European Union or the United Nations under whose auspices international standards can be negotiated.

REPORT OF WORKING GROUP IV

Resource Policy and East-West Relationships

Compiled by GABOR GAÁL[1] AND SLAVKO ŠOLAR[2]
[1]*Geological Survey of Finland*
[2]*Geological Survey of Slovenia*

Abstract. This report summarizes the discussions of the Working Group on Resource Policy and East-West Relationships that were held as part of the NATO Advance Study Institute on "Deposit and Geoenvironmental Models for Resource Exploitation and Environmental Security." The Working Group was made up of experts representing 13 countries and whose expertise included the geosciences, natural resources, and economics. The discussions focused on developing a framework for resource policy, East-West relations, and proposals for future research projects. These included a proposal for developing a certification program for mines managed according to predetermined standards, and a second proposal addressing environmental hazards associated with uranium wastes. The Working Group stressed the importance of a commitment to sustainable development and the development of strategies to satisfy human needs and improve the quality of life today, while protecting those resources that will be needed in the future.

1. Introduction

The Working Group was convened to explore issues of Resource Policy and East-West Relationships. Participants in the Working Group are listed in Table 1. Through extensive discussions, the focus of group's efforts was narrowed to developing a framework for incorporating geoenvironmental models into resource policy making, and further to do this in such a way as to facilitate East-West working relationships. It was agreed that any proposed resource management strategies must be acceptable and useful to both Eastern and Western European policy makers, and further, must support the international goals of sustainable development and environmental security. The robustness of the policy framework could then be tested through field studies in Eastern and Western Europe.

Table 1. Working Group participants

Name	Organization	Country
Ilghiz Aitmatov	National Academy of Science	Kyrgyz Repub.
Djamila Aitmatova	Central Asian Econ. Comm.	Kyrgyz Repub.
Julius Belickas	Geological Survey of Lithuania	Lithuania
Gabor Gaál	Geological Survey of Finland	Finland
Vyda-Elena Gasiunienè	Geological Survey of Lithuania	Lithuania
Halil Hallaci	Institute of Geological Research	Albania
Jakup Hoxha	Institute of Geological Research	Albania
Péter Kardeván	Geological Institute of Hungary	Hungary
Kestutis Kadunas	Geological Survey of Lithuania	Lithuania
William Langer	U.S. Geological Survey	United States
André Botequilha Leitao	Universidade Técnica de Lisboa	Portugal
Eral Meral	Ege University	Turkey
Sergio Olivero	Consultant	Italy
Ivars Ozolins	Geological Survey of Latvia	Latvia
Ipo Ritsema	NITG-TNO	Netherlands
Deborah Shields	U.S. Forest Service	United States
Delia Teresa Sponza	Eylul University	Turkey
Slavko V. ● olar	Geological Survey of Slovenia	Slovenia
Janos Szanyi	Hungarian Geological Survey	Hungary
Géza Timcák	Technical University	Slovakia
Ida Torok	University of Miskolc	Hungary

2. Sustainability as a Basis for Resource Policy Formulation

Sustainability is widely acknowledged to be a desirable goal, as demonstrated by the adoption of the Forest Principles and Chapter 11 of Agenda 21 by the 1992 United Nations Conference on Environment and Development. Many countries are committing to the concepts of sustainable development and, by extension, to sustainable resource management. In response to Agenda 21, there have evolved a number of processes addressing the complex tasks of describing sustainability and identifying the related information needs, two of which are the Helsinki and Montreal Processes. The Helsinki Process started with the Second Ministerial Conference on the Protection of Forests in Europe. Pan-European criteria and indicators (C&I) for sustainable management of forests were developed within the Follow-up of the Helsinki Ministerial Conference. The Montreal Process resulted in the Santiago Declaration, which contained a consensus list of criteria and indicators for sustainable forest management agreed to by the United States and 9 other countries. Both sets of C&I represent ambitious attempts to provide a common framework within which to assess and evaluate progress towards sustainability at the national scale. Moreover, they provide the framework within which countries can engage in discussion about sustainability.

Neither the Montreal nor the Helsinki Process addressed the role of energy and mineral resources in sustainable development. Nonetheless, there is widespread agreement that nonrenewable resources deserve a place in the overall framework of sustainability. Commitment to the sustainable development paradigm necessitates integration of environmental policies and development strategies so as to satisfy human needs and improve the quality of life today, while protecting those resources that will be needed in the future. Nonrenewable resources are integral components of economic and social systems so their use is and will continue to be fundamental to satisfying human needs. Nonetheless, that very use can also negatively impact environmental systems in ways that threaten environmental security and degrade present and future quality of life. Thus it is essential that resource exploitation be undertaken in a manner consistent with an overall concept of security based on sustainable development.

Adoption and implementation of sustainable resource management by countries in the East can be facilitated by cooperation and interaction with countries in the West, particularly, members of NATO. There are, however, certain difficulties. Despite the rhetoric of sustainability, developed nations have demonstrated only limited interest in the environmental security problems of developing nations. Effective communication has been lacking. These problems have been compounded by bureaucratic inertia on both sides. At the same time, the environmental situation is deteriorating in many Eastern countries due to the economic collapse of national industries.

A partial solution to these problems is an increased cooperation between the NATO member countries and the developing countries. Successful cooperation has many facets, including linkage/partnership programs, improved information transfer, training, personal networking and communication, and perhaps most important, initiatives on the part of the NATO member countries. What is needed is a framework for cooperation on earth resource issues, specifically those related to the application of geoenvironmental models to resource policy.

3. Framework for Developing Sustainable Resource Policies

Sustainable resource management can be conceptualized as two parallel, but interrelated, systems of statements embodying (sometimes conflicting) values with regard to the environment and the needs of present and future generations (Baharuddin and Simula 1996). The higher levels of these two systems (principles and C&I) are more conceptual and theoretical, whereas lower levels (performance targets and standards) become progressively more practical and empirical.

The two paths to sustainable resource management describe the C&I and Certification approaches (Figure 1). The purpose of C&I's is to provide a framework within which to measure progress towards sustainable resource management. In contrast, Certification is awarded upon either achievement of certain expectations or a specific degree of progress toward those expectations.

The C&I path is considered first. In the case of forestry, the UNCED "Forest Principles" reside at the highest level of the system. These are broad, general guidelines with respect to forest ecosystem sustainability that have been adopted internationally. The Criteria that emerged from the subsequent Helsinki and Montreal Processes reflected that set of principles and so were quite similar. The Indicators differed due to regionally specific economic, ecological, social and cultural values and conditions, but nonetheless followed the framework laid out in the principles. At the next level of the system are national scale C&I. These are domestic refinements of the C&I that have been accepted internationally through the Helsinki and Montreal

Processes. Their purpose is to describe ecological, economic, and social system characteristics and attributes. A country's policies, and related legislation, regulations and incentives will presumably reflect the content of the C&I embraced by that nation. Performance targets and standards comprise the next level of the system. These can be thought of as quantitative or qualitative descriptions of expected ecosystem, social system and economic system performance. Management of a specific forest unit is then a combination of the externally defined targets and standards, and also internally imposed standards.

<u>C&I</u>
Intergovernmental commitments
to principles, criteria and indicators of
sustainable resource management
(UNCED. Helsinki/Montreal C&I)

↓

National criteria
and indicators
(Refinement of Helsinki/Montreal C&I)

↓

Policy Instruments:
legislation, rules and
regulations, incentives

↓

Performance targets and
standards at management
unit / compartment level

↓
↓   ↓   ↓
↓   Resource Management
↓   ↑
→ Internal standards of the resource unit ←

<u>CERTIFICATION</u>
International
certification principles
and standards

↓   ↓

National and
sub-national   →   Certifiers' own
level standards   standards

↓

Local-level assessment
standards (ecosystem or
commodity, management
unit)

↓
↓
↓

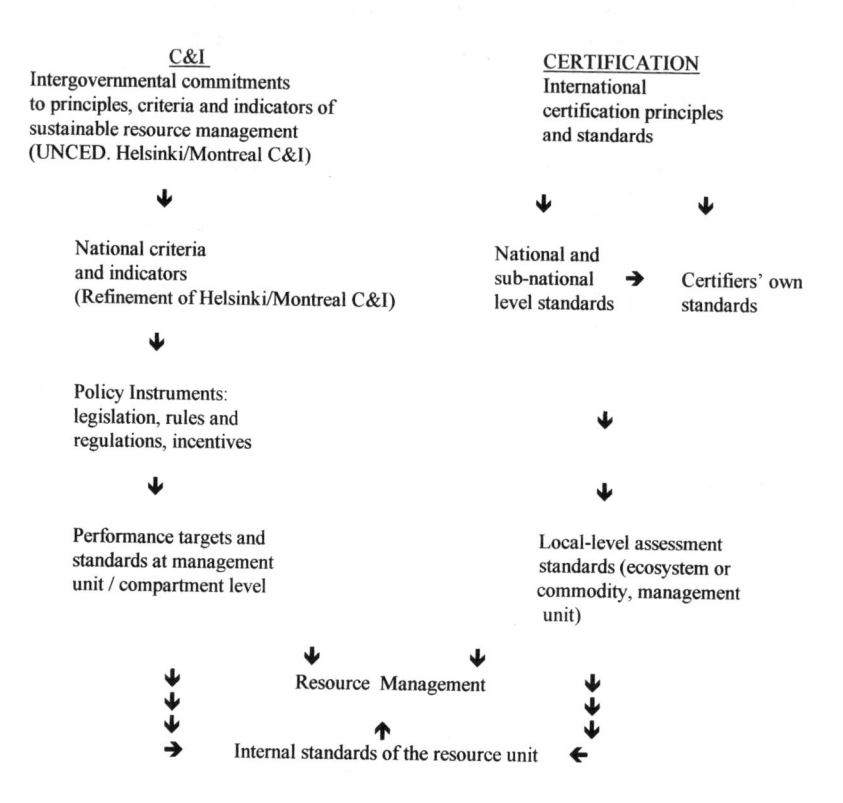

Figure 1. C&I and Certification paths to sustainable resource management.
(After: Baharuddin, H.G. and Simula, M. 1996. Timber Certification in Transition. ITTO).

The Certification perspective of sustainable resource management focuses on the process by which a third party determines when a resource, be it forest or mine, is being managed in a sustainable manner and consistent with a set of procedural and/or performance standards. A major goal of certification is to create market incentives for sustainable resource management by increasing demand for products originating from resources that are being managed in a sustainable manner. Again the process starts with internationally agreed upon principles, which in turn lead to national standards. Performance targets in the form of standards, including those developed by the

International Standardization Organization (ISO) function as incentives for compliance with sustainable management practices, while increased competitiveness acts as an external reward for private land or resource owners. These combined with the firm's internal standards and the government's management standards act in concert to ensure sustainable resource management.

A complete set of C&I and Certification guidelines for sustainable management of energy and mineral resources does not currently exist. There are no internationally agreed upon principles of sustainability for all nonrenewable resources, although a Global Energy Charter was proclaimed at the UNCED in Rio de Janeiro in 1992. Only a few countries have written national level energy and minerals C&I and begun to revise their minerals policies to be consistent with sustainability. Currently, for example, the United States, Canada, Australia and various European countries are in the process of shifting to sustainable energy and mineral resource management practices.

At present, performance targets for earth resource management are in the form of requirements for compliance with environmental regulations, but this will not necessarily ensure "best practices" behaviors or lead to outcomes consistent with sustainable development. Many countries and funding institutions also require some form of environmental analysis, an Environmental Impact Statement or Assessment (EIS, EIA), for both new and expanding mineral developments. However, these documents typically do not contain performance targets. And finally, unlike the situation within the forestry community, where sustainable management standards are widely discussed and designation as a "model forest" is of interest to industry and consumers, very little has been accomplished in the area of mine certification.

All governmental policies are political instruments that should reflect both the public's wishes and the government's domestic and international commitments. For most countries, those commitments now extend to sustainable development and sustainable resource management. The resource policies intended to implement sustainability are much more likely to be effective, to encourage the behaviors which will lead to desired outcomes, to the degree that they are science-based. This is particularly true in the case of energy and mineral resources. The purpose of the policy development framework envisioned by Working Group 4 is to facilitate the flow of relevant scientific and engineering information to decision makers charged with implementing sustainable development policies.

The general approach is shown below.

Identify general goals
⬇
Characterize the situation in question
⬇
Identify situation specific goals
⬇
Collect needed data
⬇
Conduct analysis
⬇

Recommend technical approaches that can be utilized to reach goals
↓
Define institutional and policy framework needed to implement technical approaches and reach goals
↓
Provide local capacity building

Inherent in this framework are certain assumptions about the state of the world. First, there is an international community of nations and there exist international standards with respect to resource policies and business management that are accepted by the community of nations. Second, Eastern nations desire to join the broader community of nations and understand that such participation is to some degree predicated upon adoption of an agreed set of standards. Similar logic flow can be applied to business firms, namely, that there exists an international market financed by international institutions and firms desiring to participate in this market and receive funding from the institutions realize that said participation and funding depend on the degree to which they accept and implement the business standards set forth by the international community.

3.1. RESOURCE APPLICATION OF THE POLICY DEVELOPMENT FRAMEWORK

To effectively support earth resource policy development, the general framework will need to

- Facilitate the development of resource management strategies applicable by and acceptable to both Eastern and Western European resource policy makers;
- Define and clarify the role of geoenvironmental models in policy making;
- Elucidate the economic importance of geoenvironmental models in the analyses supporting resource policy formulation and decision-making;
- Create a process that is inclusive and iterative; and
- Produce a cadre of trained individuals who can carry on with resource assessments and policy recommendations, and, in addition, transfer knowledge of the process to others in their respective countries.

The rationales for these requirements are as follows. Given that the extraction, use and disposal of energy and mineral resources have cross border effects, only policies and management strategies acceptable to, and implemented by, both Eastern and Western nations will support the long-term goal of sustainable development. Geoenvironmental models characterize and predict the environmental impacts of extraction and thus will be integral to sustainable land-management decision-making. Given that resource policy making is process not an end point, policies and management approaches in both Eastern and Western countries must be able to adapt as new information becomes available. Thus, the policy development process must be one that can be implemented iteratively so that the policies themselves can be refined and improved over time. And for iterative changes to be accepted, the process itself must be open and responsive to the input of all stakeholders. Finally, for continuous, if gradual, improvement to take place, there must be local individuals trained in applying the policy development process, so that the refinement is not dependent upon outside consultants.

With these goals in mind, the steps of the policy development process can be restated for the application to earth resources as follows.

> Identify the environmental, social, and economic (i.e., sustainability) criteria and indicators (C&I), as well as financial and management standards accepted by the community of nations;
> ↓
> Characterize the resource management problem;
> ↓
> Adapt the C&I and standards to the environment and culture at hand, with the twin goals of supporting sustainable development and meeting international standards, utilizing input from all relevant stakeholders;
> ↓
> Collect geoenvironmental and other relevant scientific data using GIS, Airborne and on-the-ground techniques;
> ↓
> Collect social and economic data using economic and social science methods such as surveys, preference elicitation, and engineering cost models;
> ↓
> Analyze the data and identify alternative resource management and policy approaches;
> ↓
> Utilize decision theoretic techniques to select recommended alternative(s);
> ↓
> Determine the institutional framework needed to implement the solution; and
> ↓
> Train partner representatives in the use of methods and tools that are developed throughout the process.

Because the solutions to earth resource policy problems may not be obvious prior to data collect and technical analysis, and due to the fact that the process of moving to full implementation in accordance with western standards can be expected to take some time, the steps outlined above should be considered as a gradual and iterative process. Relevant stakeholders would need to be kept informed and be given a voice in any policy formulation that emerged during each step in the process.

An economic component is important to bear in mind at all stages of the development of an earth resource policy. At the local level, it is important to identify costs and benefits associated with alternative remediation approaches, so that the one with the highest net present value (or the one most highly preferred) can be adopted. At the regional or national level, it is necessary to determine the willingness to pay for environmental protection in the project countries. This issue would be best addressed by using a preference or value model approach. In the case of mineral resource applications, it would involve determining the marginal cost of building geoenvironmental models and the demand for those models and the data supporting them. Initially, the demand would not be high, but information sharing (that is, advertising) could be used to remedy the situation. Regardless of the level of the application, it is essential that both the costs of resource policies and their expected benefits be determined.

4. Research Projects

The following two project proposals are illustrative of the need for East-West cooperation for addressing earth resource based environmental security issues.

4.1 ENVIRONMENTAL HAZARDS OF THE URANIUM TAILING STORAGE SITES IN CENTRAL ASIA (G. GAÁL, I. AITMATOV)

The objective of this proposed research project is to assess the risk, specifically, the geoenvironmental risks associated with abandoned uranium mines, wastes and processing facilities in Kyrgyzstan, Tajikistan and Uzbekistan, Central Asia.

Following World War II, the former Soviet Union faced an urgent problem of nuclear arms production as the result of political tensions between East and West. As a consequence, the need and the search for and subsequent production of uranium became of primary importance.

The first uranium mining operations were located in the territories of the former Soviet Union, namely, Kyrgyzstan, Uzbekistan, and Tajikistan. The operations were located in mountainous terrains near highly populated areas, including the Ferghana valley. The highly populated areas included:

- Mailuu-Suu,
- Shekaftar (Kyrgyzstan, Jalal-Abad Oblast),
- Taboshar,
- Adrasman (North Tajikistan, Khodjent Oblast); and
- Yanghiabad (Uzbekistan).

In addition, there were uranium-mining operations in other mountainous areas of the Central Asia, namely:

- Minkush,
- Karabalta; and
- Kadzhisay and others.

During this period of active uranium mining operations, the general lack of knowledge of the geological conditions, in particular, the structural and mechanical rock properties of the local bedrock in relation to the mountainous terrains and the decision to locate the radioactive waste tailings storage sites in places unfavorable for long-lasting and stable environments resulted in storage sites located on mountain slopes near the mining operations, usually, for purely economic considerations. Furthermore, the waste tailings storage sites in the territories of Mailuu-Suu and Shekaftar, as well as many other trans-border regions, were situated in river beds and floodplains, with the result that, in many cases, the materials were periodically washed downstream and, thus, causing radioactive materials to disperse downstream and into the groundwater. The waste tailings storage sites as well as the slag heaps (wastes dumps) in the Mailuu-Suu riverbed area are subject to significant debris flows. Debris flows in this area recur on an average of 1 to 5 years. The combined debris flows and flooding has resulted in the transport of significant amounts of radioactive materials throughout the Ferghana valley.

The situation is compounded further by the fact that many of the areas where the waste tailing storage sites are located are subject to landslides. Consequently, there is the additional risk of the radioactive wastes becoming incorporated in materials that would, in effect, dam the river as a result of future landslides. Were future landslides to occur, particularly in the Syrdarya and Amudarya valleys, these radioactive wastes would also be transported downstream to the more highly populated areas in Uzbekistan and Tajikistan.

In order that the environmental risks associated with the radioactive waste tailings storage sites can be more properly assessed, a research project is proposed that will have the following objectives:

- Map and collect georeferenced data and develop a GIS in the areas at risk,
- Carry out an airborne survey of sites, including radiometric, satellite, and remote – sensing; and
- Follow up with ground-based surveys using geophysics and, to a lesser extent, geochemistry.

From a human dimensions perspective, a variety of methods would need to be employed that include:

- Economic models for estimating the costs and benefits associated with alternative remediation methodologies,
- Preference analyses for determining how standards and international goals such as sustainability should be adapted and applied throughout Central Asia and for examining how alternative remediation schemes could affect stakeholders' perceptions of their own well being,
- Decision theoretic methods for supporting the policy selection process; and
- Development of a local capacity component, including but not limited to training and the implementation of the tools and methods developed as part of the project.

From an environmental perspective, other scientists who would need to be involved include:

- Biological scientists for determining the existing environmental condition (health and functioning of ecosystems), predict trends without remediation, and estimate impacts of alternative remediation approaches; and
- Ecologists for developing recommendations as to how Western standards should be modified for application in Central Asian ecosystems.

The technical data development phases of the proposed would consist of the following:

- Interpretation,
- Recommendation of alternative remediation approaches,
- Definition of possible standards and presentation of policy alternatives; and
- Dissemination of results.

A further dimension of the proposed project would be to consider the broader issues confronting the uranium industry in Central Asia. In particular, the following issues would need to be considered:

- Western standards for radioactive waste disposal need to be adopted in Central Asia taking into account the current situations facing the various former Soviet Union Republics,
- Appropriate legislation in the former Soviet Union Republics that takes into account the current environmental situation needs to be adopted; and
- Special emphasis needs to be placed on developing safe engineering practices for radioactive waste storage site remediation in the mountainous territories of the Central Asia Republics.

EU and other funding agencies are known to prefer to fund projects with cooperating partners drawn from several countries, East and West. It is expected that each partner country would receive a share of the funding. The responsibility of an overall EU coordinator would be to allocate funding to the cooperating partners; the responsibility of the regional and in-country partners would be to allocate their share of the funding to the project team members. It is to be expected there would be a staged transfer of funds; initial monies to be used during the early stages of the project; subsequently, funds would be made available upon the successful completion of each succeeding stage.

4.2 MODEL MINE CERTIFICATION: QUARRY MANAGEMENT IN KARST AREAS (D.SHIELDS, S. ŠOLAR.)

Karst is found all over the world, including Western and Eastern European countries. Mining in these areas has been part of economic activity for many centuries. Carbonate rocks (limestone) are used as aggregate or building stone and also have other industrial uses in the lime, paper, and rubber industries. Rocks and quarries on Karst areas, as well as building and natural stone, are part of a country's natural and/or cultural heritage.

Karstification, the interaction between calcium carbonate rocks and water, is a specific phenomenon that makes affected areas more environmentally sensitive and unpredictable, especially with respect to ground water and morphological features. Ground water can easily be polluted and difficult to clean up in Karst environments due to the transport system characteristics, such as caves, channels, and water-induced cracks. The ground water activity in Karst areas also leads to certain morphological features, such as sinkholes and dolinas. Thus, the risks associated with mining in active Karst areas are higher than is the case in areas where these types of geologic processes are not occurring.

Because of the widespread distribution of Karst, Eastern and Western European countries face common resource management problems and resource policy issues. Moreover, because karstification does not stop at the borders, extraction, potential pollution, and trade patterns may generate bilateral or multilateral concerns. The interrelated environmental security, economic impact, and physical safety issues associated with quarrying in Karst areas are most effectively addressed comprehensively, within a sustainable development context.

The project presented here does not address the full range of sustainable resource management issues associated with mining. Rather the focus is on development of a sustainable mining certification program and the application of that program to Karst quarries. Mine certification is envisioned as a program parallel to that for forest management certification (as differentiated from product certification.) Management certification is defined as the independent evaluation of the quality of forest-, or in this case mine-management. Both procedural standards, such as ISO, and performance standards would be part of the evaluation process. In addition to the certification program, the research project will deliver data interpretation, examples of good practice and recommendation for improvement of resource and quarry management.

This research project has general goals related to three spatial scales. At the International scale the goals are to:

- Foster East-West co-operation;
- Enhance communication; and
- Improve the environment in this sector (Karst areas).

At the national scale the goals are to:

- set standards to protect mineral resources on Karst areas;
- set up or improve surface mining monitoring in these sensitive areas; and
- reduce stakeholders conflict.

Finally, at the local scale, the goals are to:

- Increase understanding about the issues associated with quarrying and the value of these activities to the economy;

- Establish environmental standards for mining, improving the present state of art so as to minimize environmental impacts and optimize landscape design to assure effective reclamation; and
- Increase communication between producers and their stakeholders.

Therefore, in order to facilitate East-West collaboration in the development of appropriate sustainable resource management policies for quarrying in Karst areas, a research project is proposed with the following objectives:

1. Collect the relevant technical, environmental, economic and social data associated with quarrying in active Karst areas;

2. Apply sustainable development concepts to Karst resource management, focussing on the Certification perspective described in section 4 of this chapter;

    2.a. Identify appropriate Karst resource management policies, regulations, performance targets and standards based on criteria and indicators of sustainable resource management;

    2.b. Analyze the data collected in task 1, and estimate the impacts of Karst quarrying, at the local, regional, national and international scales;

    2.c. Create a sustainable resource management certification program (SRMCP) for quarries in Karst areas that includes procedural (ISO) and performance standards;

    2.d. Test the SRMCP at Karst quarries in NATO member and NATO partner countries;

    2.e. Recommend institutional framework required for application of the SRMCP;

    2.f. Train partner country representatives in the use of technical methods and scientific tools necessary for the implementation of SRMCP;

3. Formalize and extend the Policy Development Framework proposed by Working Group 4 as an approach to accomplishing objectives 1 and 2.

It is assumed that certification will require performance standards in a variety of areas. The standards will be based on resource and quarry management models for surface mining in Karst areas. These models will in turn be supported by general and site-specific geoenvironmental models. Improved technical methods for the reduction of pollution will also need to be identified and incorporated in the management models. The cultural aspects will also be addressed, including community impacts, the role of stakeholders, and their effects on resource and quarrying management. To reduce stakeholder conflicts, standards with respect to communication methods and skills will be needed. In addition, procedural standards such as ISO 9000 (quality management) and ISO 14000 (environmental management) will be incorporated in the program.

A variety of quantitative methods will be used in data collection, analysis, modeling and interpretation. GIS tools for data collecting, processing and sharing will support resource and quarry management, as well as stakeholder involvement processes, and communication. Both cost/benefit based and preference based decision models will be utilized. Finally, a training process to build up local capacity (producers, mining developers) will be established.

Methods for dissemination of results will be depend upon the targeted audience and will include: academic publishing, presentations in meetings, local capacity building, community meetings, etc. Regular project reports (progress, annual, final) will also be produced. Through information sharing, public awareness of environmental and other impacts of quarrying will be raised.

Funding will be sought from the NATO Science for Peace Program. In addition, funding will be solicited from other potential sources such as INTERREG, SME and the European Union, participating countries science and technology programs, and geological surveys, and industrial rock producers and associations dealing with Karst area mineral resources.

Reference

Baharuddin, H.G. and Simula, M. 1996. Timber Certification in Transition. International Tropical Timber Organization.

REPORT OF WORKING GROUP V

Natural Aggregate Resources – Environmental Issues and Resource Management

Compiled by WILLIAM H. LANGER[1] and SLAVKO ŠOLAR[2]
[1] *United States Geological Survey, Denver, Colorado, USA*
[2] *Geological Survey of Slovenia, Ljubljana, Slovenia*

Abstract.. This report summarizes the discussions of the Working Group on Natural Aggregate Resources (The A Group) that were held as part of the NATO Advanced Study Institute on "Deposit and Geoenvironmental Models for Resource Exploitation and Environmental Security." The working group formed spontaneously at the NATO ASI conference because many of the researchers there recognized that there are numerous international issues that relate to the identification, characterization, and development of aggregate resources; the reclamation of mined-out lands; and the environmental issues surrounding these resources. The working group was made up of 26 experts that represented 18 countries. Many of the representatives were experts in other fields and were not experts in the field of natural aggregate. However, all recognized that natural aggregate is an important resource for developing and developed countries, and shared the concern that those resources be developed in an environmentally and socially sensitive manner. The goal of the working group was to share ideas, experiences, knowledge and solutions of issues that can be used to help create a more efficient, environmentally friendly method of developing the world's most intensively produced mineral resource.

1. Introduction

The Working Group convened spontaneously to explore questions regarding the planning, exploration, development, extraction, use, recycling and disposal of natural aggregate resources, the negative environmental impacts associated with obtaining the resources, and methods to reduce those impacts. Participants in the Working Group are listed in Table 1.

2. Natural Aggregate Resources - Background

Natural aggregate is the world's number one mineral commodity (exclusive of energy resources) in terms of both volume and value. Natural aggregate consists of material composed of rock fragments that may be used in its natural state or used after mechanical processing such as crushing, washing, and sizing. Natural aggregate consists of gravel and crushed stone. Gravel generally is considered to be material whose particles are about 2.0 to 1024 millimeters in diameter[1]. Its edges tend to be rounded. Crushed stone is of the same size range, but is artificially crushed rock, boulders, or large cobbles. Most or all of the surfaces of crushed stone are produced by crushing, and the edges tend to be sharp and angular.

Table 1. Working Group participants

Name	Organization	Country
Julius Belickas	Geological Survey of Lithuania	Lithuania
André Botequilha Leitão	Universidade Técnica De Lisboa	Portugal
Liisa Carlson	Consultant	Finland
Tariro Charakupa	Environment and Remote Science Institute	Zimbabwe
Bernard Cornélis	Universite de Liege	Belgium
Marta Čović	University of Zagreb	Croatia
György Csirik	Geological Institute of Hungary	Hungary
Chang-Jo Chung	Geological Survey of Canada	Canada
Gabor Gaál	Geological Institute of Hungary	Hungary
Peng Gong	University of California, Berkeley	United States
Vyda - Elena Gasiunienė	Geological Survey of Lithuania	Lithuania
Halil Hallaci	Institute of Geological Research	Albania
Jakup Hoxha	Geological Survey of Albania	Albania
William Langer	U.S. Geological Survey	United States
Rumen Parvanov Mironov	Technical University of Sofia	Bulgaria
Sergio Olivero	CEPAS Italian National Register	Italy
Costas Ripis	Institute of Geology and Mineral Exploration	Greece
Nyls Ponce Seoane	Institute of Geology and Paleontology	Cuba
Irina Shtangeeva	St. Petersburg University	Russia
Nikolay Metodiev Sirakov	Instituto Superior Técnico de Lisboa	Portugal
Slavko Šolar	Geological Survey of Slovenia	Slovenia
János Szanyi	Hungarian Geological Survey	Hungary
Ida Torok	Miskolc University	Hungary
Stanislava Trtíková	Comenius University	Slovakia
Hasan Üçpirti	Sakarya University	Turkey
Mehmet Ali Yukselen	Marmara University	Turkey

Natural aggregate has hundreds of uses, from chicken grit to the granules on the shingles on your roof. However, most aggregate is used in cement concrete, asphalt, and for other construction purposes. The average per capita consumption of aggregate generally ranges from 5 to 15 tons per year.

Aggregate occurs where Nature put it, not necessarily where people need it. Aggregate is widely distributed throughout the world, but there are large areas where it is absent; entire countries lack suitable supplies of aggregate. In addition aggregate may not meet the quality requirements for the intended use. It must be strong and resist break-down during handling and use. It must resist the action of wetting and drying or

---

[1] Sedimentary classification: *sand* between 0.0625 mm and 2.00 mm; and *gravel* above 2.00 mm (up to 1024 mm). There are other aggregate classifications. For example, in Slovenia, aggregate producers use terms like **fine sand**, between 0.0 and 1.00 mm, **sand** for materials between 1.0 and 4.0 mm, and **gravel aggregate**, above 4.0 mm.

freezing and thawing. It must not react with cement when used as concrete or strip from bitumen when used in asphalt.

Prime aggregate resources are lost if parking lots, houses, or other buildings are constructed over the resource. Zoning and permits may restrict development of aggregate. Citizens wish to be protected from the impacts of aggregate development. Yet, for economic reasons, aggregate operations must be within reasonable distances of the market area. Some governments enact legislation to protect citizens from the impacts of aggregate extraction and to protect aggregate resource from encroachment by other land uses.

Sand and gravel commonly is mined with conventional earth moving equipment such as bulldozers, front-end loaders, tractor scrapers, and draglines. In some cases consolidated rocks heavily affected by tectonics can be extracted in the same way as sand and gravel. Most often crushed stone is quarried by drilling and blasting rock, and then is extracted using power shovels or bulldozers. After sand and gravel or stone is excavated most aggregate producers process the material by crushing it and screening it into various size categories according to specifications required by specific uses.

Aggregate is heavy and bulky. Therefore it is most economical to excavate aggregate near the point of use. Transportation can add significantly to the cost of aggregate. For example, transporting aggregate about 35 kilometers can double the price of the aggregate.

Developing aggregate resources causes environmental impacts. Most environmental impacts are not serious and can be controlled by employing careful mining practices using available technology. However, there are some geologic situations where mining aggregate may lead to serious environmental impacts, especially with regard to ground water, air, and noise pollution. Environments that are particularly prone to impacts from aggregate extraction include karst and stream channels. One of the most serious environmental problems is the dereliction of abandoned pits or quarries. The reclamation of mined-out land is an important aspect of reducing environmental impacts of aggregate extraction.

3. Natural Aggregate - Stories from Around the World

The international community shares many issues related to the planning, identification, characterization, permitting, extraction, use, recycling, disposal, and reclamation of aggregate resources. The following are stories prepared by Working Group scientists from countries around the world. By sharing ideas, experiences, knowledge, examples of good practice, and other solutions of issues related to aggregate, we can help create a more efficient, environmentally friendly method of developing the world's most intensively produced mineral resource.

3.1 MATCHING RESOURCES TO USE
    (by Stanislava Trtíková, Comenius University, Bratislava, Slovakia)

Dolomite is one of the most widespread rocks in Slovakia. In the past, Carpathian dolomitic rocks of middle to upper Triassic age were intensively exploited. For example during 1983, at 5.7 million tons, the Slovak production was the largest in the Europe. This is a large production, even in comparison with the world's leading producers, which are Canada, (9.1 million tons), and the U.S.A., (8.3 million tons).

The dolomite produced in Slovakia is of the best quality, however, it is not utilized in its most productive manner. A major part of it is used as a replacement for gravel in building and road construction and as railroad ballast. Though there is a large number of studies describing the optimal use of Carpathian dolomites, so far none of the suggested uses requiring high-quality dolomite have actually been brought into practice. According to the studies, these dolomites have a composition favorable for the heat-resistant materials used in iron casting technology, for the production of magnesium mineral fibers, and for various applications in the medicine, biology and ecology.

Currently Slovakia imports vast amounts of heat-resistant materials in spite of the fact that our finest dolomite is perfectly suited for the task.

3.2 AGGREGATE RESOURCES IN MY HOME TOWN
(by Nikolay Metodiev Sirakov, Instutito Superior Tecnico, Lisbon, Portugal)

I come from Bulgaria. Bulgaria is located on the Balkan Peninsula and is bordered to the north by Romania, to the east by the Black Sea, to the south by Turkey and Greece, and to the west by Serbia and Macedonia. The country is blessed with an abundance of aggregate. I would like to tell you about the resources near my home.

I was born in the town of Velingrad, which is located within the Rhodope Mountain Range in the southern part of Bulgaria. The land is very beautiful and is covered with forests and rivers. In the mountains near the source of the rivers there are large areas of rock. The rivers carry these rocks downstream where they are deposited alongside the streams. When my father built our house in Velingrad he used sand and stones that he collected from the river passing nearby. People still obtain aggregates from the rivers and streams for building houses and roads. Several pits and quarries for obtaining stones, marble and granite have been developed in the Rodope Mountains. However, most of them are now closed because of the economic crisis that appeared after abolishing the former communistic regime.

3.3 SUBSTITUTING WASTE DREDGE MATERIAL FOR CONSTRUCTION MATERIALS
(by Bernard Cornélis, Universite de Liege, Liege, Belgium)

Sand is being dredged from rivers in Belgium to make them safe for navigation. The dredge spoil is being used as a source of sand for construction purposes. This is one example of how to make the best use of our natural resources.

3.4 PREEMPTING RESOURCE DEVELOPMENT
(by William H. Langer, U.S. Geological Survey, Denver, Colorado, USA)

Despite the dependency of metropolitan growth on stone, sand and gravel, urban expansion often works to the detriment of the production of these basic minerals. Aggregate is a bulky, low cost commodity and because of the expense of transportation aggregate operations are located as close as possible to population centers. As a community grows, it builds out and gradually encroaches upon these established operations. The new residents located near the pit or quarry want to be isolated from the noise, dust and unsightly appearance of the pit and the hazards of increased truck traffic, and governments write regulations restricting the operation of the pit or quarry. For example, New York City is built on some of the best aggregate resources in the United States. However, the factors described above severely limit extraction. As a result aggregate is shipped by ocean freighter to New York City from as far away as Nova Scotia and Newfoundland, Canada.

3.5 USING AGGREGATE TO BUILD BEACHES ON THE ADRIATIC SEA
      (by Marta Čović, University of Zagreb, Zagreb, Croatia)

The western part of Croatia borders on the Adriatic Sea. The coastline is laced with numerous islands, indented with beautiful bays, and distinguished by gorgeous peninsulas. My mother was born in Pakoštane, which is a small coastal village between the old towns Zadar and Šibenik. The beach between Zadar and Šibenik is a beautiful resort area where people come on holiday to swim and relax on the beach.

Most people do not know, but the beach is covered with gravel that is brought there by truck. This was not always so. My mother tells me that about 50 years ago the beaches were covered with soft sand. People from islands off the coast would mine sand from the beaches and take it to the islands to make their homes. My mother also tells me that at the time, people did not like to visit the beach. Therefore, nobody missed the sand. There may be other reasons why the sand is still missing from the beaches. Nevertheless, the beaches of Zadar and Sibenik are still beautiful, and people are happy to sit on the gravel. What a pleasant use for aggregate!

3.6 GRAVEL PITS PROVIDE SAFE HAVENS FOR WILDLIFE
      (by Slavko Šolar, Geological Survey of Slovenia, Ljubljana, Slovenia)

A sand pit in the quartz sand deposit at Bizeljsko, Slovenia, produces 20,000 tons of sand per year for use as decorative silica bricks and for construction purposes. But that is not all the pit produces. It also provides protection for a rare bird called the Bee-eater. These birds nest in the walls of the sand pit. About 50 birds live here, and are the only known colony of Bee-eaters in Slovenia. The pit operators have worked with a birdwatcher society to protect the Bee-eaters, and to construct a pathway and bird watching stand so that their presence can be enjoyed.

Near Trebnje, Slovenia, an open pit called Jersovec is a source of chert, which is a hard, compact, durable type of quartz rock. It is similar to the kind of material that pre-historic people used to make arrowheads for hunting birds and animals. Most of the chert from Jersovec is used as fire-resistant material. The open pit contains a sedimentation basin to clarify the water. The basin also serves as a refuge for fish, frogs, ducks and other birds, all of which might have been hunted by prehistoric man using chert arrowheads.

3.7 TOYS OF THE GIANTS
(by János Szanyi, Geological Institute of Hungary, Budapest, Hungary)

The Mine of Nagyharsany is the biggest quarry in south Hungary. It has been in operation since the beginning of 20th Century and produces more than 100,000 tons of crushed stone each year. Much of the stone was originally used to construct railroads. The quarry is on Szarsomlyo Hill in the Villany Mountains. The geological and botanical value of the area is unique, so a part of the quarry's hill was turned into a nature reserve. When aggregate mining was finished on the east side of the hill an open-air sculptor studio was created.

Artisans quickly came to enjoy the open-air studio. Since 1967 the quarry has been the site of an International Sculptor Symposium. Many artists exhibit their statues every year. This quarry has become a very popular attraction. Nowadays you can see not only the statues, but from May to October you can watch the sculptors create works of art. The Jurassic limestone creates a picturesque setting where visitors can investigate the protected flowers, the geology, the mining operations, and works of art, all at the same place.

To me the statues in the yard are like toys of the giants.

3.8 DERELICTION OF MINED SITES AT THOUSAND LAKES
(by György Csirik, Geological Institute of Hungary, Budapest, Hungary)

A region in Hungary just south of Budapest, the capital of Hungary, has the name of Thousand Lakes. As the name implies there are numerous lakes in the area. The strange part is that most of the lakes are not natural, but were created through the mining of sand and gravel.

Most of the mining in Thousand Lakes took place during the 1960s, 1970s, and 1980s, to provide aggregate for Budapest. As gravel was mined the pits extended below the water table and quickly filled with water. When it became too difficult to continue mining, the operators abandoned the pit and moved to another place. After the political changes of the 1990s the economy slowed and demand for aggregate diminished. Many of the mining companies went out of business and many gravel pits were never reclaimed.

Today the second use of the pits is largely spontaneous. Some lakes attract swimming and jet skis. Others are used for purposes such dumping waste and washing cars, which pollute the water. All these activities are forbidden, but happen nonetheless.

There is yet another source of pollution. Water evaporates from the lakes and concentrates the natural salts. Some of the polluted water is used to irrigate farmland, which greatly reduces crop yields. It is hard to imagine that aggregate mining could affect crop yields. But this story demonstrates it can, if proper precautions are not taken. Local municipalities have plans to correct the situation, but currently lack the funds to implement the plans.

3.9 A VISIT TO THE TRAKAI CASTLE
(by Vyda-Elena Gasiunienė, Geological Survey of Lithuania, Vilnius, Lithuania)

I would like to take you on a journey. I enjoy my profession, which is like traveling to unknown places - to the entrails of the earth. You realize I am talking about geology - a profession that is at the same time science, art, and perception. I have investigated mineral deposits for many years. However, sometimes the acquisition of knowledge about the depth of the earth costs dearly. Whenever I visit the old castle and settlement of Trakai, I am always reminded of a terrible accident. During the 15th century Trakai was the residence of Lithuania's rulers, and the location of many important historical events. It is not far from our capital Vilnius. About 15,000 to 20,000 years ago glaciers from Scandinavia covered the land of my country. When the climate became warmer flowing water from melting glaciers carried sand, gravel, and boulders, and deposited them near Trakai. Our people have always needed gravel and sand to construct buildings and roads. Especially huge amounts of this material were required for rebuilding after World War II.

More than 40 years ago geologists had detected sand and gravel in the area of Trakai, and began investigating them in detail. In order to study the deposits beneath the land surface they made many excavations (up to 2 meters in diameter and 6 meters in depth) that look like wells. After the investigations were completed the holes were filled because, if left open, they could be perilous to people or animals that might by chance wander thought the area. It was resolved to take out the boards that were used for reinforcing the shaft walls. The boards in the last shaft were stuck. Spring was in the air...nobody felt the threat. One worker descended the 5-meter depth to remove the obstacle. Suddenly the mass of gravel gave way, and the man was buried up to his chest. The endeavor to liberate him from the trap of gravel was fruitless. Nothing could help. The man could not breath and quickly suffocated. Everything has rapidly finished. Since then, the gravel from these deposits has been extracted. The land has been shaped to a new relief and planted over with pine trees. The castle stands in silent testimony. Nothing remains of that tragic day except memories.

Today we are much more careful conducting our work. But this tragic story is a reminder that some types of investigations are fraught with peril, and that we must use extreme caution during our studies of the earth.

3.10 THE HAZARDS OF OPEN PITS
(by Tariro Charakupa, Environment and Remote Science Institute, Harare, Zimbabwe)

Epworth is an informal settlement in the southeast city limits of Harare the capital of Zimbabwe. This settlement has no running water or sewer system. In Epworth, as is with most informal settlements of Zimbabwe, a large percentage of the inhabitants are unemployed and turn to nature for survival. One of the major means of earning money for the men folk in Epworth is dredging, excavating, and selling sand from the river Ruwa. River sand extraction in Epworth is rampant, illegal, and uncontrolled. Apart from sand extraction, the river Ruwa also supports the lives of Epworth people with water supply, fish and recreation. The Zimbabwe Herald newspaper reported about the environmental problems brought about by all these activities being conducted at one time. I decided to go see for myself.

As it happens the river Ruwa is a hive of activity during the weekends, with young mothers doing their weekly laundry happily chatting away, kids playing just nearby, cattle drinking from the river, and people fishing, all while others are extracting sand. All this is happening within half a kilometer along the riverbank. Abandoned excavated pits are also dotted along the same distance. I was wondering if anyone could teach these people how to extract the sand in a safer way and avoid leaving abandoned pits when children screaming interrupted my thoughts. It was yet another victim of these ugly pits. One of the children fell into an abandoned pit, broke both arms and a leg, and was bleeding from a cut on the head. Almost every week an animal or a human being falls into one of these pits, sometimes resulting in fatalities. People must realize that the consequences of improper aggregate mining can be extremely severe - even fatal.

3.11 RECLAMATION – THE SOUNDS OF MINING
(by Liisa Carlson, Helsinki, Finland)

The festival stage Dalhalla at Rättvik, Sweden, was built in Draggängarna, a former limestone quarry. The quarry is 400 meters long, 175 meters wide, and 60 meters deep. The mining created a natural amphitheater that currently seats 3000 people.

In this unique setting Dalhalla is teeming with activities: operas with stage and light settings adapted to the mighty rock walls; all types of choir music; jazz and big band concerts; and symphonic and chamber music. Dalhalla demonstrates that it is possible to fill a hole in the ground - even with music!

4. Where to go from here

The Working Group committed itself to prepare a useful educational tool explaining basic issues related to natural aggregates. The goal was to "translate" scientific data to a format understandable by non-experts and present a mix of information about the culture of a number of different countries, thus reaching a broad audience. A subset of the Working Group developed the stories shown in Section 3 above. Those stories were prepared and presented on the World Wide Web as a user-friendly PowerPoint presentation. The presentation was also distributed to staff at the organizations of some of the Working Group members.

The effective communication of earth science information to decision-makers requires writing the results in scientific studies in a manner that is understandable to a non-scientific audience. Dissemination of Working Group outcomes will serve to this purpose. In addition to products, a network of earth scientists was established that, in the future, can work together on different aspects of aggregates under different organizational frameworks.